AF183575

Infectious Disease

For further volumes:
http://www.springer.com/series/7646

T. Ramamurthy · S.K. Bhattacharya
Editors

Epidemiological and Molecular Aspects on Cholera

Springer

Editors
T. Ramamurthy
National Institute of Cholera and Enteric Diseases
P-33 C.I.T. Road Scheme XM
Kolkata 700010, India
rama1murthy@yahoo.com

S.K. Bhattacharya
Indian Council of Medical Research
New Delhi 110029, India
sujitkbhattacharya@yahoo.com

ISBN 978-1-4614-2704-9 ISBN 978-1-60327-265-0 (eBook)
DOI 10.1007/978-1-60327-265-0
Springer New York Dordrecht Heidelberg London

© Springer Science+Business Media, LLC 2011
Softcover reprint of the hardcover 1st edition 2011
All rights reserved. This work may not be translated or copied in whole or in part without the written permission of the publisher (Springer Science+Business Media, LLC, 233 Spring Street, New York, NY 10013, USA), except for brief excerpts in connection with reviews or scholarly analysis. Use in connection with any form of information storage and retrieval, electronic adaptation, computer software, or by similar or dissimilar methodology now known or hereafter developed is forbidden.
The use in this publication of trade names, trademarks, service marks, and similar terms, even if they are not identified as such, is not to be taken as an expression of opinion as to whether or not they are subject to proprietary rights.
While the advice and information in this book are believed to be true and accurate at the date of going to press, neither the authors nor the editors nor the publisher can accept any legal responsibility for any errors or omissions that may be made. The publisher makes no warranty, express or implied, with respect to the material contained herein.

Printed on acid-free paper

Springer is part of Springer Science+Business Media (www.springer.com)

Preface

More than 150 years ago the cholera-causing pathogen *Vibrio cholerae* was discovered and knowledge on its characteristics, biochemistry, physiology and genetics is growing incessantly. Many of its unique features give feast for research and frustration in managing the disease. Though our understanding of the features and behaviour of *V. cholerae* are remarkable, suffering from its infection in the form of cholera is still continuing, mostly in Asian and African countries. Owing to lack of surveillance and timely reporting of cholera in endemic regions due to economic and other disincentives, the overall cholera scenario has not changed substantially. The raison d′être of its persistence in nature is mainly attributed to age old factors such as contaminated or inadequate water supply, sanitary conditions, congregation of people due to natural calamities or political unrest and by and large the poor health practises.

During the last three decades, intensive research has been made to understand the virulence properties of the organisms, genes that support the colonization and survival in the gut as well in the environments and their contribution in the epidemiology of cholera. How cholera spreads around the globe and what determines its seasonal peaks in endemic areas are not well known. These features of cholera have been hypothesized primarily due to the biotic and abiotic environmental factors and such findings are acknowledged all over the world. However, utilization of such important information still remains a gray area.

Recently, studies have also been made on the adaptive mechanisms of *V. cholerae* in the environments. The phenotypic and genetic studies showed that *V. cholerae* is relentlessly flex, resulting in instantaneous adaptation to the environment as well as in the human host. Detection of strain diversity at the molecular level and its impact of disease dynamics are the other important components in the molecular epidemiology. This field of research is growing fast facilitated by recent technological developments. However, on account of its broad spectrum of pathogenic potential, *V. cholerae* still remains as a paradigm of human pathogen.

Many cholera monographs have been published in the past describing the history of the disease and developments in the research during that time. We have kept away our sights on this erstwhile information to avoid redundancy and to focus on current developments. We hope that this book will be an abridgment for researchers in the field of cholera and other enteric diseases. We acknowledge the generous time and effort that the authors have devoted for their manuscripts and also for patience to see their great chapters in print.

Kolkata, India
New Delhi, India

T. Ramamurthy
S.K. Bhattacharya

Contents

Contributors

M. John Albert Department of Microbiology, Faculty of Medicine, Kuwait University, Safat, Kuwait, mjohnalbert@gmail.com

Martin Antonio Medical Research Council Laboratories, Banjul, The Gambia, mantonio@mrc.gm

Kalyan Banerjee Division of Biochemistry, National Institute of Cholera and Enteric Diseases, Kolkata 700010, India, banerjeekalyan2003@yahoo.co.in

Rupak K. Bhadra Infectious Diseases and Immunology Division, Indian Institute of Chemical Biology, 4 Raja S. C. Mullick Road, Jadavpur, Kolkata 700032, India, rupakbhadra@iicb.res.in

S.K. Bhattacharya Indian Council of Medical Research, V. Ramalingaswami Bhawan, Ansari Nagar, New Delhi 110029, India, sujitkbhattacharya@yahoo.com

Vincent Burrus Département de biologie, Université de Sherbrooke, 2500 boulevard de l'Université, Sherbrooke, QC, Canada J1K 2R1, vincent.burrus@usherbrooke.ca

Rita R. Colwell Maryland Pathogen Research Institute, University of Maryland, College Park, MD 20742, USA, rcolwell@umiacs.umd.edu

Alejandro Cravioto International Centre for Diarrhoeal Diseases Research, Dhaka, Bangladesh, acravioto@mail.icddrb.org

Bhabatosh Das Infectious Diseases and Immunology Division, Indian Institute of Chemical Biology, Kolkata, Jadavpur, India, dasiitkgp@yahoo.com

Gabriela Delgado Department of Public Health, Faculty of Medicine, Universidad Nacional Autónoma de México, Mexico City, Mexico, delgados@servidor.unam.mx

P. Dutta National Institute of Cholera and Enteric Diseases, Kolkata, India, drpdutta@yahoo.com

Masahiko Ehara Department of Bacteriology, Institute of Tropical Medicine, Nagasaki University, 1-12-4 Sakamoto, Nagasaki 852-8523, Japan, ehara@nagasaki-u.ac.jp

Shah M. Faruque Molecular Genetics Laboratory, International Centre for Diarrhoeal Diseases Research, Dhaka 1212, Bangladesh, faruque@icddrb.org

Nirmal Kumar Ganguly Indian Council of Medical Research, Ansarinagar, New Delhi, India, nkganguly@nii.res.in

Asoke C. Ghose National Institute of Cholera and Enteric Diseases, P-33, CIT Road, Scheme-XM, Beliaghata, Kolkata 700010, India, acghose@vsnl.net

Amit Ghosh National Institute of Cholera and Enteric Diseases, Kolkata 700010, India, amitghosh24@yahoo.com

Chris J. Grim Maryland Pathogen Research Institute, University of Maryland, College Park, MD 20742, USA, grim.christopher@gmail.com

Anwar Huq Maryland Pathogen Research Institute, University of Maryland, 3132 Bioscience Research Building #413, College Park, MD 20742, USA, huq@umd.edu

Karl E. Klose Department of Biology, University of Texas San Antonio, One UTSA Circle, San Antonio, TX 78249-1644, USA, karl.klose@utsa.edu

Inacio Mandomando Centro Investigação em Sáude Manhiça (CISM); and Instituto Nacional de Sáude, Ministério da Sáude, Maputo, Mozambique, inacio.mandomando@gmail.com

Budhaditya Mazumdar Division of Biochemistry, National Institute of Cholera and Enteric Diseases, Kolkata 700010, India, budha1979@gmail.com

John J. Mekalanos Department of Microbiology and Molecular Genetics, Harvard Medical School, 200 Longwood Avenue, Boston, MA 02115, USA, john_mekalanos@hms.harvard.edu

Jose Luis Mendez Department of Public Health, Faculty of Medicine, Universidad Nacional Autónoma de México, Mexico City, Mexico, Jlms1970@prodigy.net.mx

Elena V. Monakhova Research Institute for Plague Control, Rostov-on-Don, Russia, unicorn54@mail.ru

Rosario Morales Department of Public Health, Faculty of Medicine, Universidad Nacional Autónoma de México, Mexico City, Mexico, marosari@servidor.unam.m

Asish K. Mukhopadhyay National Institute of Cholera and Enteric Diseases, Kolkata, India, asish1967@gmail.com

G. Balakrish Nair National Institute of Cholera and Enteric Diseases, Beliaghata, Kolkata, India, gbnair_2000@yahoo.com

Ranjan K. Nandy National Institute of Cholera and Enteric Diseases, Kolkata 700010, India, nandy_rk@hotmail.com

Joe O. Oundo Center for Disease Control Kenya, Nairobi, Kenya, JOundo@ke.cdc.gov

T. Ramamurthy National Institute of Cholera and Enteric Diseases, Kolkata, India, rama1murthy@yahoo.com

Amit Sarkar Department of Microbiology, Bose Institute, Kolkata 700054, India, amitosarkar@gmail.com

Sangita Shah Infectious Diseases and Immunology Division, Indian Institute of Chemical Biology, Jadavpur, Kolkata, India, sangita_iicb@yahoo.co.in

Sumio Shinoda Faculty of Science, Okayama University of Science, 1-1 Ridai-cho, Okayama 700-0005, Japan, shinoda@dls.ous.ac.jp

Samba Sow Center for Vaccine Development, Bamako, Mali, ssow@medicine.umaryland.edu

Shanmuga Sozhamanan Genomics Department, Biological Defense Research Directorate, Naval Medical Research Center, Silver Spring, MD, USA, shanmuga.sozhamannan@med.navy.mil

D. Sur National Institute of Cholera and Enteric Diseases, Kolkata, India, dipikasur@hotmail.com

Khalid Ali Syed South Texas Center for Emerging Infectious Diseases and Department of Biology, University of Texas San Antonio, San Antonio, TX 78249, USA, syedkhalidali@hotmail.com

Yoshifumi Takeda National Institute of Cholera and Enteric Diseases, Beliaghata, Kolkata, India, takeda.yoshi@pheasant.pharm.okayama-u.ac.jp

Kamini Walia Indian Council of Medical Research, Ansarinagar, New Delhi, India, waliakamini@yahoo.co.in

Fitnat H. Yildiz Department of Environmental Toxicology, University of California, Santa Cruz, CA, USA, fyildiz@ucsc.edu

Chapter 1
General Introduction

T. Ramamurthy and S.K. Bhattacharya

Cholera is an ancient disease but continues to strike human lives in mass. Based on the published reports between 1995 and 2005, it is evident that cholera is rampant in sub-Saharan Africa (66%) than Southeast Asia (17%). Since its discovery in 1854, *Vibrio cholerae*, the causative organism of cholera, has changed its biotypes, serogroups, and strain characteristics. Thus, this organism is strongly positioned to augment problems in many aspects including epidemiology and biological complexity. Chapters 2, 3, 4, and 5 in this book describe about the cholera scenario, respectively in Asia, Africa, Commonwealth of Independent States (the then Russia), and Americas along with characteristics of *V. cholerae* strains that influenced the local epidemics.

The pathogenesis of *V. cholerae* is very complicated that involves number of virulence genes. In many studies it has been shown that attenuation of the major virulence-associated genes encoding cholera toxin (*ctx*) and hemolysin (*hlyA*) still causes diarrhea in animal models. Chapter 6 covers many of these aspects and also portrays evolution of *V. cholerae* as a human pathogen. Changes in the genetic constitution of *V. cholerae* are considered as an attribute for the recurrent cholera outbreaks and epidemics. Such changes can be monitored using several molecular tools. The significance of many such molecular epidemiological studies and molecular basis of clonal diversity are discussed in Chapter 7.

In *V. cholerae*, more than 200 serogroups have been identified and of these, only O1 and O139 are known to cause epidemics/pandemic cholera. Diversity in the somatic (O) antigen is due to the composition of monosaccharide components and biosynthesis mechanisms. Information on the genetic basis of *O*-polysaccharide (OPS)/capsule structures is developing steadily. Sequencing of specific genomic regions allowed examining genetic diversity in many *V. cholerae* strains. Chapter 8 summarizes the biochemical composition and structure of *O*-polysaccharides, genes involved in their biosynthesis, and the role of horizontal gene transfer in creating this diversity. The concept of horizontal gene transfer has been widely accepted for the evolution of new variants and acquisition of new genes in bacteria. Integrating

---

T. Ramamurthy (✉)
National Institute of Cholera and Enteric Diseases, Kolkata, India
e-mail: rama1murthy@yahoo.com

T. Ramamurthy, S.K. Bhattacharya (eds.), *Epidemiological and Molecular Aspects on Cholera*, Infectious Disease, DOI 10.1007/978-1-60327-265-0_1,

© Springer Science+Business Media, LLC 2011

conjugative elements (ICEs) are self-transmissible mobile elements that transfer between bacteria via conjugation and integrate into the chromosome. SXT-related ICEs are now present in most clinical and environmental *V. cholerae* isolates and play an important role in the spread of antibiotic resistance genes in this pathogen. The appearance of ICEs in *V. cholerae* seems to be correlated with an increased use of antibiotics in the treatment of cholera. The significance of SXT/R391 family of integrating conjugative elements in *V. cholerae* is discussed in detail in Chapter 9.

Though information on genetics and immunological aspects on *V. cholerae* has increased considerably, our knowledge on the stress-adaptive mechanisms with special reference to its survival is still fragmentary. Various intra- and extracellular small signaling molecules play critical roles in overcoming environmental onslaughts and allowing the pathogen to survive, multiply, and cause the disease. *V. cholerae* has two most important small molecules-mediated signaling pathways, which are extracellular quorum-sensing and intracellular 3′,5′-cyclic diguanylic acid (c-di-GMP) signaling systems. The intracellular (p)ppGpp molecule, called alarmone, also helps the pathogen to survive under nutritional scarcity and other stresses. In addition, universal second messenger cAMP and its receptor protein (CRP) have been implicated in virulence gene expression in *V. cholerae*. Chapter 10 describes fascinating aspects on signaling systems of *V. cholerae* during post-genome sequencing era.

Flagellar-mediated motility has been established to contribute to the pathogenesis in a number of bacteria. Information on non-motile mutants of *V. cholerae* including less fluid accumulation in the rabbit ileal loop, defects in adhesion on rabbit brush border cells, and less reactogenic in human volunteers emphasize the important role of flagellum in the pathogenesis. Chapter 11 elucidates the association between flagella-mediated motility of *V. cholerae* in its virulence, chemotaxis, and biofilm formation for environmental persistence. The filamentous phages of *V. cholerae* are well known to be associated with transmission of virulence. The aspects that are covered in Chapter 12 include classification and characteristics of filamentous phages, its role in the virulence, and use in the molecular epidemiological studies. Ever since the emergence of the novel serogroup O139, studies on *V. cholerae* non-O1, non-O139 are steadily increasing. Although the virulence mechanism of these organisms is different from the O1 and O139 serogroups, its increasing incidence as a causative agent of cholera-like diarrhea is a growing concern. Chapter 13 elaborately describes about the epidemiology, virulence mechanisms, and ecology of non-O1, non-O139 serogroups of *V. cholerae* along with other fascinating aspects on strain diversity, putative toxigenic factors, and evolutionary perspectives.

The proteases produced by pathogenic vibrios are long recognized to play pathogenic roles in the infection. Hemagglutinin/protease (HA/P) produced by *V. cholerae* is thought to play indirect pathogenic roles, which is regulated together with major pathogenic factors such as cholera toxin (CT) or toxin-co-regulated pilus (TCP) by a quorum-sensing system. Most importantly, HA/P is necessary for full expression of pathogenicity of the vibrios by supporting growth and translocation in the digestive tract. In addition to the HA/P, existence of another Zn-metalloprotease

PrtV and cysteine protease domain in RTX toxin was also demonstrated. Chapter 14 reviews the importance of *V. cholerae* protease comparing it with proteases of other vibrios. Disease manifestations in cholera are primarily attributed to the secretion of CT by *V. cholerae*. However, the discovery of additional secretory virulence factors in this organism has invoked interest in their potential role related to pathogenesis, inflammation, and immunomodulation in its disease attributes. Recent studies with *V. cholerae* strains devoid of *ctx* provide evidence that cholera also evokes inflammatory response. It is postulated that the increase in the inflammatory response to the *V. cholerae* (Δ*ctx)* infection could be due to the absence of the immunomodulatory activity of the B subunit of CT that blocks the secretion of proinflammatory cytokines by macrophages, dendritic cells, and epithelial cells. Chapter 15 on toxins of *V. cholerae* and their role in inflammation, pathogenesis, and immunomodulation reviews all the above-mentioned important aspects. *V. cholerae* cytolysin/hemolysin (VCC) is another toxin expressed by the majority of El Tor and non-O1 strains. This extracellular membrane-damaging toxin induces colloid osmotic lysis in a wide spectrum of eukaryotic cells including rabbit erythrocytes and human enterocytes. The biochemical and pathophysiological properties of this toxin are well established. However, the role of VCC in human cholera or cholera-like diarrhea is not recognized. Chapter 16 presents a summation of the recent studies on VCC that shed light on how this toxin transforms into an integral membrane protein as well as its affinity for specific protein and lipid components of the target membrane.

Till date, the oral rehydration therapy (ORT) is the main stay for the treatment of cholera patients. However, administration of antimicrobials is an adjunct to ORT as it shortens and reduces the volume and duration of diarrhea. Due to excessive use of life-saving drugs, many groups of infectious bacteria became resistance rendering these drugs ineffective. There are several mechanisms by which the bacteria become resistance to the antimicrobials and possession or acquirement of integrons is one of the well-established genetic factors for the resistance. Integrons are gene capture and expression systems by which bacteria can acquire genes for the resistance. Integrons carrying several resistance gene cassettes have been indentified in vibrios and these genes and their arrangements seem to be specific in certain regions. Chapter 17 describes about the types of integrons in *V. cholerae*, mechanisms by which the organism uptake the resistance genes, and their general distribution in different geographical locations.

Unlike the other enteric pathogens, *V. cholerae* survive and proliferate in the natural water bodies supported by tiny crustacean fauna known as zooplankton. The recent concept of cholera has included environmental aspects, as the disease is strongly associated with the change in the global climatic patterns as evidenced by El Niño events. The description of viable but non-culturable (VBNC) form in the environment and the effect of physico-chemical and biological factors on toxin gene regulation provide valuable clues to the ecology of *V. cholerae*. Chapter 18 covers many aspects of cholera including many molecular diagnoses, weather-related prediction, and simple methods for prevention. The disease management continues to be the best use of appropriate rehydration. However, many community-based studies emphasize the need for effective health services and health education. Chapter 19 describes updated management components of cholera.

To trounce the troubles mentioned in several of the chapters, the disease as well as the causative organism should be tackled with holistic strategies using field-generated data through several epidemiological investigations, informational science, and new technologies for prediction by means of computational and mathematical tools. New directions are available to overcome the emerging and reemerging infectious diseases with the strategic application of many tools and technology. Epidemiology and molecular aspects are emerging segments in the modern medicine and biology. Systems biology began to emerge with the completion of the *V. cholerae* genome projects and the resulting complete list of genes, RNAs, and proteins—thus enabling global or comprehensive measurements of the molecules in response to biological perturbations. Effective analysis and integration of this information leads to predictive models on functional aspects of *V. cholerae*. Many of the concepts about the disease as well as the evolution of virulence in *V. cholerae* have slowly been changed due to the advancement in technologies. The outcome of these developments will be fascinating in next 10–20 years and may help in making resolutions for global reduction and ultimately elimination of cholera.

Chapter 2
Asiatic Cholera: Mole Hills and Mountains

Asish K. Mukhopadhyay and T. Ramamurthy

Abstract The disease cholera has persisted in Asia since time immemorial. Almost all the pandemic phases of cholera had its origin from the Indian subcontinent. Historically, waves of cholera have wiped many million lives in this region mainly due to the general insanitary conditions and poor management of the disease. All the three cholera causing vibrios namely classical, El Tor, and the O139 have emerged from Asia at different times and one was replaced by the other by overcoming the acquired immunity. Antimicrobial resistance was not a big problem in the early 1960s as its use was very limited. With the use of third-generation drugs, *Vibrio cholerae* has acquired many resistance mechanisms over the passage of time and also due to prevailing antibiotic pressure. With its biotypes/serotypes there are considerable variations at the genetic level and many clones of *V. cholerae* have been detected. Recently, the hybrid strain of El Tor has spread in many Asian countries causing several cholera outbreaks. However, the importance of such genetic changes was not fully strengthened in epidemiological perspective. The perspectives of cholera vaccines have shown to be encouraging in many recent vaccine trials in Asia. Traditional medicine has lost its glory as it lacks the scientific evidence in curing infectious diseases. Some of the herbal formulations are now reconsidered for extensive research. The control measures for preventing cholera are yet to gain momentum in many Asian countries, as it involves coordination of government and the public with adequate funds to revamp the water supply and waste disposal systems. On the other hand, the clinical management of cholera and other diarrheal diseases are largely under the control in Asia.

2.1 Introduction

The Asian continent has been considered as the cradle of cholera for many centuries. In many publications it was shown that the high temperature, relative humidity, and

T. Ramamurthy (✉)
National Institute of Cholera and Enteric Diseases, Kolkata, 700010, India
e-mail: rama1murthy@yahoo.com

T. Ramamurthy, S.K. Bhattacharya (eds.), *Epidemiological and Molecular Aspects on Cholera*, Infectious Disease, DOI 10.1007/978-1-60327-265-0_2,
© Springer Science+Business Media, LLC 2011

intermittent rain fall form ideal climatic conditions for the incidence of cholera. Recently, this hypothesis was further strengthened with oceanographic studies, which has shown that there is a strong correlation between El Niño and the incidence of cholera. The classical, El Tor biotypes as well as the O139 serogroup had emerged from the Asian region. Cholera is still a problem in Asia, as progress toward standard of good living with all the public health facilities including uninterrupted supply of safe drinking water, environmental sanitation, implementation of efficient vaccine are somehow long-winded. The emergence and reemergence of different phenotypic and genetic variants of *Vibrio cholerae*, the causative organism of cholera, show its ability for survival in the environment and cause infection in the human host. This chapter reviews the current status of cholera in many Asian countries and many aspects on the causative organism *V. cholerae* including antimicrobial resistance and molecular epidemiology.

2.2 Cholera in the Indian Subcontinent

Cholera is an ancient disease in Indian subcontinent. Historically, it was believed that the first six classical cholera pandemics originated from the Indian subcontinent [1]. In India, the classical biotype was replaced by the El Tor from 1965 [2]. In some areas in India such as Raipur, the classical cholera prevailed till 1970 and the subsequent cholera outbreaks in 1975, 1977, 1979–1981 were caused by El Tor vibrios [3]. Continuous monitoring of cholera epidemics helped to detect frequent changes in the incidence of *V. cholerae* serogroups. Younger age group (<15 years) was the most affected population in the 1988 cholera outbreak in Delhi [4]. In Sevagram, Maharashtra, the incidence of O1 serotype Ogawa was predominant with intermittent appearance of O139 serogroup during 1992, 1997 [5]. The incidence rate of O1 and O139 serogroups in Delhi during 1992, 2000 was 81 and 14%, respectively [6]. Occurrence of O139 serogroup in Delhi was low between 1994 and 1999, but reemerged during 2000. A questionnaire-based survey conducted to estimate the water-borne infection with the use of Ganges River indicated the incidence of cholera (33 cases) among families who used the river water for many purposes including washing clothing and bathing [7]. During 2002 cholera outbreak in Chandigarh, *V. cholerae* O1 Ogawa was isolated from 18% of the hand-pump water samples [8].

The annual incidence of cholera estimated using the population census as the denominator and the age-specific number of cases as numerator showed comparatively low in Jakarta, Indonesia (0.5/1,000), than in Kolkata, India (1.6/1,000), and Beira, Mozambique (4.0/1,000) [9]. In this study, children below 5 years of age were found to be the most vulnerable group for cholera infection. *V. cholerae* O1 and O139 consecutively appeared during cholera outbreaks (2002–2003) near Karachi [10]. This study has also revealed that children less than 2 years of age were the most affected age group with O1 (49%) than O139 (21%). Remote areas such as Andaman and Nicobar Islands were free from cholera for many years. The first cholera outbreak was recorded during early 2000s due to the spread of *V. cholerae* O1 from the main land [11]. In 2002, cholera was identified due to El Tor vibrios among Nicobarese tribe in 16 villages with an attack rate of 12.8% and a case fatality ratio of 1.3% [12]. Concomitant infections by *V. cholerae* O1 and O139 serogroups were reported in 2000 from a large cholera outbreak in Ahmedabad, India [13]. In Delhi, the serotype switchover from Ogawa to Inaba has started in 2004 and 88% of the strains were identified as Inaba during 2005 [14]. Among children below 5 years of age, the incidence of cholera in Delhi was 33%. Cholera caused by the Inaba serotype was also reported from other parts of India such as Kolkata, Orissa, Andaman and Nicobar Islands [15–17].

In Bangladesh, the classical biotype was replaced by El Tor vibrios during 1964–1973 [18, 19]. During 1973–1979, cholera due to classical biotype was not detected in Matlab and Dhaka, but was predominant (79%) in southern regions of Bangladesh during 1988–1989 [20]. Intermittent appearance of classical cholera was recorded during 1979 to 1981 [21]. Classical cholera appeared in the form of large epidemic starting from Matlab, Comilla, and Dhaka during late 1982 and spread to other districts replacing the El Tor biotype [19, 21]. Phenotypically, the new classical strains were identical to the one that prevailed a decade earlier and the virulence features and seasonality resembled to that of El Tor strains prevailing at that time. It was hypothesized that the classical strains of *V. cholerae* O1 were indigenous to Bangladesh [22]. A 33-year (1966–1988) data analysis provided much information from Bangladesh [23]. Between 1966 and 1988, both classical and El Tor biotypes were prevailed and by 1988, the classical biotype disappeared. The serotype prevalence during 1988–1989 was also interesting as El Tor belonged to Inaba, whereas the classical strains to Ogawa type.

Studies conducted from 1985 to 1991 in Bangladesh indicated that the incidence of cholera was among children below 5 years (24%) and children below 2 years of age accounted for 10% of the cases [24]. The overall case fatality during epidemics was 4.0%. A mathematical Ogawa–Inaba model based on the 40 years (1966–2005) analysis explained the serotype changes in cholera case patterns in Bangladesh when the cross-immunity to one specific serotype was high [25]. It was hypothesized that intermittent appearance of Inaba serotype might be related to its long-term immunity against Ogawa.

Based on the spatial patterns and exploratory spatial data analysis, the risk factors for cholera were associated with environmental niches [26]. Environmental studies conducted during 2004 in Mathbaria and Matlab, Bangladesh, revealed that both *V. cholerae* O1 and O139 serogroups occurred predominantly as viable but non-culturable state [27]. However, culturable cells were also detected in the biofilms, which were considered as additional reservoirs of toxigenic *V. cholerae* in the aquatic environments during inter-epidemic seasons. Isolation of *V. cholerae* O1 from the aquatic environments of Bangladesh through selective enrichment using antibiotics has reemphasized the hypothesis that the humans act as reservoirs of this pathogen during inter-epidemic periods and spreading occurs through contaminated water [28].

2.3 Other Asian Countries

In Japan, cholera epidemics during 1882 and 1895 were recorded in Fukushima, northeastern part of Japan [29]. Due to cholera epidemics, the health policy has changed in Japan during Meiji period (1868–1912) [30, 31]. Changes in the cholera diffusion pattern during this period demonstrate improvements in the transportation and most importantly, growth in socio-economic systems. In Okinawa, Japan, the incidence of cholera was high in 1879 [32]. Laboratory data with *V. cholerae* O1 strains collected between 1977 and 1987 revealed low number of domestic cholera in Japan [33]. Studies conducted in various environments of Japan and imported seafood showed prevalence of non-toxigenic *V. cholerae*. In the Port of Osaka, Japan, *V. cholerae* O1 was detected during 1987–2001 and all the strains were closely related and non-toxigenic [34].

Contaminated seafood is associated with cholera in many South Asian countries. In Hong Kong, a local cholera outbreak was related to the consumption of shellfish that were kept live in contaminated seawater [35]. The cholera toxin (CT)-producing *V. cholerae* non-O1 strains were isolated from seafood in Taiwan [36]. Turtles and their breeding environment are the major reservoirs of *V. cholerae* and responsible for many outbreaks of cholera in Sichuan Province and Guangzhou area, china during 2003–2005 [37, 38]. In an investigation it was shown that turtles and other seafood harbored toxigenic *V. cholerae* O139 [39]. In Zhejiang Province, the incidence of O1 serogroup of *V. cholerae* was found to be high (9%) in turtles and cholera epidemics in this region might be associated with consumption of contaminated turtles [40].

Surveys conducted in eastern region of Saudi Arabia, mainly with expatriate workers and household contacts with *V. cholerae* infection, showed the prevalence of *V. cholerae* O1 El Tor serotype Ogawa in 113 diarrheal patients (6.0 per 100,000 population per year), 28 asymptomatic cases, and 16 of 982 household contacts of index patients [41]. Interestingly, the O1 strains isolated from asymptomatic cases were non-toxigenic as detected by CT-ELISA. Following the Iraq war, the communicable disease control program was disturbed, resulting in cholera epidemics in several districts of Basrah, Iraq, in 2003 [42]. A 6-year study (1997–2002) conducted in hospitals in Zabol city, Iran, indicated the incidence of cholera due to *V. cholerae* O1 Ogawa was 10% with maximal number of cases in 1997 (22%) [43]. Among the infected cases, there were no differences in age or social and economic strata. Almost 19% of the cases in this study were from neighboring Afghanistan. Cholera outbreak struck Kabul, Afghanistan, in 2005 and spread nationwide. The health authorities gave importance to the disease control program that included proper management and treatment supported by partner agencies that kept the mortality rate well below 0.1% [44]. Processed and raw foods imported from Asian countries have also contributed to the incidence of cholera in non-endemic areas. Three cases of cholera reported in Australia in 2006 had a link with consumption of raw whitebait imported from Indonesia [45]. In France, a total of 129 imported cases of cholera were recorded during 1973–2005 [46]. During 1980s, 94% of the patients were infected in Morocco and Algeria, and during the rest of the period, cholera was associated with patients visiting Asia and other African countries.

Since 1873, cholera was common in Sarawak, Malaysia, and the classical biotype was common prior to 1961 [47]. In Vietnam, incidence of cholera has distinct temporal and seasonal trends. Between 1991 and 2001, cholera mostly affected the central coast during May to November without having any climatic association with eight recorded outbreaks [48]. However, studies conducted from 1996 to 2002 in Bangladesh indicated that weather factors could play a role in the epidemiology of cholera with either increase or decrease during monsoon [49]. Based on a study conducted during 1993–1999, the epidemic and sporadic cholera occurs in the western regions of Indonesia during low rainfall, whereas in eastern parts, heavy rainfall has contributed to the cholera transmission [50]. A 6-year hospital-based study on the incidence of El Tor cholera in Thailand indicated that the affected age group ranged from 2 months to 15 years with an average age of 3 years [51].

2.4 Association Between *V. cholerae* and Parasites

Incidence of *V. cholerae* in association with enteric parasites has been reported in many cholera-endemic regions. In Nepal, the incidence of cholera is associated with co-infection with enteric parasites such as *Giardia lamblia* and *Ascaris lumbricoides* [52]. In a recent study, it was shown that the co-infection with geohelminth parasites reduces the magnitude of the cholera vaccine response [53]. Among Indian children, the infection due to cholera and parasites, especially with *G. lamblia*, seems common [54]. The incidence of mixed infection with *V. cholerae* O1 and *A. lumbricoides* and *G. lamblia* was more than *Trichuris trichiura* and *Entamoeba histolytica* in Kolkata [55]. This investigation also showed that the combined infection of *V. cholerae* and parasites was common in children aged between 2 and 10 years. The association between parasites and *V. cholerae* is not clear, but seems advantageous for both the pathogens. In a rat colon model, it was proved that *E. histolytica* trophozoites and cholera toxin enhanced secretion of mucin glycoproteins and stimulated colonic glycoprotein synthesis [56].

2.5 The O139 Cholera

A novel non-O1 strain of *V. cholerae* was first discovered in 1993 that has caused large epidemic in Madras, southern part of India [57]. This strain of *V. cholerae* not agglutinated with antisera from O1 to O138 was included as a new serogroup O139 [58]. This serogroup had spread quickly to Bangladesh and to other states of India within a span of about 10 months [59, 60]. Since the *V. cholerae* O139 was first discovered in the areas surrounding the Bay of Bengal (Tamil Nadu, Andhra Pradesh, West Bengal and Bangladesh), this serogroup has a synonym "Bengal." In a span of 1 year, this serogroup has been reported in many Asian countries (Fig. 2.1). The O139 infection produced severe dehydrating diarrhea, which is indistinguishable

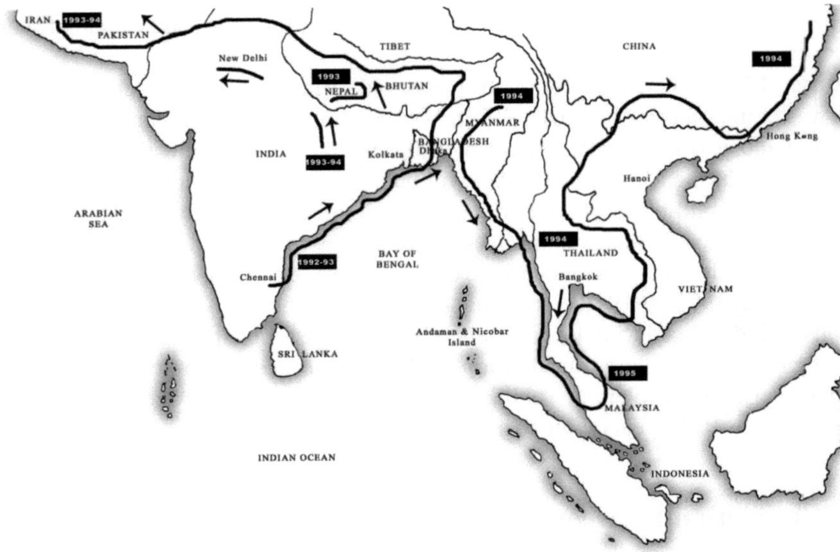

Fig. 2.1 Spread of *Vibrio cholerae* O139 in Asia during 1993–1995

from clinical cholera and does not appear to confer any cross-protection from the O1 serogroup [61]. Based on the clinical symptoms and severity of diarrhea, the disease caused by *V. cholerae* O139 is now considered as cholera [62] and the infections caused by the rest of the non-O1 and non-O139 serogroups are known as "cholera-like diarrhea."

Before its total replacement, the O139 prevailed along with the O1 serogroup in many countries. The affected age group by the O139 infection depends on the period and place of its occurrence. In Delhi the incidence of O139 and O1 cholera was frequent in children below 5 years of age [63]. In this study, the incidence of both the serogroups was detected among 1.4% of the children with cholera. A similar epidemiological observation was made in Karachi, Pakistan [64]. The incidence of O139 among children during 2003 in Pakistan was 21% and the infected patients were more likely to be febrile ($P < 0.001$) [10].

After its initial explosive epidemic during late 1992 and early 1993, occurrence of O139 serogroup had declined in many cholera-endemic regions [65, 66]. Between 1997 and 2000, incidence of cholera due to O139 serogroup decreased to 3.8% in rural Bangladesh [67]. Resurgence of O139 cholera was reported in many Asian countries including Pakistan (2000–2001), India (1997, 2001), and Bangladesh (2002), mostly affecting the older age groups [68–72]. From 1999 to 2000, most of the cholera outbreaks in India were caused by the O139 serogroup [73]. Investigations conducted in Indonesia revealed that the O139 serogroup had not invaded into this country till 1999 [50]. The first incidence of O139 was recorded in Baghdad, Iraq, in 1999, though the numbers of cases were less [74].

Imported cases of O139 cholera were reported soon after its emergence in Asia in California [75] and other parts of the USA [76], Japan [77], and Denmark [78]. The recurrent infection caused alternatively by the O1 and O139 serogroups in cholera-endemic regions emphasize the fact that the role of acquired immunity plays an important role in the emergence and dissemination of specific serogroup in a population. In addition, rapid genetic reassortment in *V. cholerae* O1 and O139 serogroups might play a role in the changing epidemiology of cholera [79]. It is still a mystery that why the so-called highly infectious O139 serogroup has not spread to the other cholera-endemic regions such as Africa.

2.6 Antimicrobial Resistance

Resistance of *V. cholerae* to antimicrobials used for the treatment of cholera is not uniform in many countries. The emergence of resistance is mainly due to the prevailing antibiotic pressure caused by overuse of antimicrobials. In several findings, it was proved that the susceptibility pattern of *V. cholerae* to the antimicrobials is constantly changing over a period of time. In India, *V. cholerae* isolated during mid-1970s were susceptible to trimethoprim–sulfamethoxazole and patients who received treatment with this drug recovered quickly [80].

In the early 1990s, *V. cholerae* O1 El Tor were susceptible to many antimicrobials including nalidixic acid, co-trimoxazole, chloramphenicol, and streptomycin in India [81]. The classical as well as El Tor strains prevailed in Bangladesh during 1988–1989. However, tetracycline resistance was detected only with the classical strains, whereas the El Tor strains remained susceptible to this drug [20].

Based on the antimicrobial resistance profiles and the resistance gene composition, *V. cholerae* O1 strains isolated during cholera outbreaks in Vietnam were described as different clones [82]. Unlike in many African countries, tetracycline-resistant strains are not common in the Asian region. However, intermittent appearance of tetracycline resistance in *V. cholerae* O1 has been reported from several Asian countries [83–86]. Cholera patients infected with tetracycline-resistant strains of *V. cholerae* purged longer with greater stool volume while receiving treatment with this drug [83].

In southern Thailand, an unusually large epidemic of cholera due to *V. cholerae* O1 was recorded during 1997–1998 [85]. All the strains were resistant to tetracycline and belong to a unique clone and this trend has not been reported in Thailand since 1993. The El Tor vibrios are resistant to polymyxin B, which is also a biotypic marker to differentiate them from classical vibrios. The El Tor strains (11%) isolated from an outbreak of cholera in Laos during 1998 was susceptible for polymyxin B and resistant to tetracycline and sulfamethoxazole–trimethoprim, which were different from those of previous strains [87]. In the following year, the incidence of polymyxin B susceptibility increased to 58%. The Kelantan (Malaysia) cholera epidemic in 1998 was identified due to tetracycline-resistant strains [84]. To control the outbreak in this region, erythromycin was substituted for tetracycline. Tetracycline resistance was uncommon in India for many years. However, 27 and

15% of Ogawa and Inaba strains from Kolkata were, respectively, resistant to this drug during 2005 [86]. A recent study conducted in Thailand showed that most of the *V. cholerae* strains isolated from cholera patients were susceptible to ceftriaxone and quinolones, which were used in the treatment of cholera [51]. However, in neighboring countries this scenario is completely different, as quinolone resistance in *V. cholerae* was common for many years.

V. cholerae O139 strains that emerged during early 1990s in several Asian countries were resistant to trimethoprim, sulfamethoxazole, and streptomycin, similar to the El Tor vibrios. However, the reemerged O139 strains were susceptible to these antimicrobials [68, 72, 88, 89]. Possible deletion of a 3.6 kb region of the SXT element in the reappeared O139 strains during 1995 in Bangladesh was thought to be responsible for susceptibility to these antimicrobials [90]. Some of the early O139 isolated in India were resistant to tetracycline, ampicillin, chloramphenicol, kanamycin, and gentamicin and a 200 kb self-transmissible plasmid carried the encoding genes for multidrug resistance [91].

V. cholerae O1 strains with reduced susceptibility to fluoroquinolones have been reported from French travelers returning from India [92]. The trend of reduced susceptibility toward ciprofloxacin seems increasing in certain states of India. About 46% of the Ogawa strains isolated in 2003 showed reduced susceptibility to ciprofloxacin [93]. In Sevagram, India, resistance to tetracycline varied from 2 to 17% [5]. Majority of the Iranian strains were resistant to streptomycin, chloramphenicol, co-trimoxazole, and tetracycline [94]. The newly emerged Inaba strains from Delhi, India, were resistant (96%) to co-trimoxazole, furazolidone, and nalidixic acid [14]. *V. cholerae* O1 appeared during 1995, 2000, and 2002 in Vietnam had different resistance profiles, as they harbored either in the class 1 integron (strains of 1992) or in SXT constin (strains of 2000) [82]. Detailed studies on integrons arecovered in Chapter 9.

2.7 Phage Typing of *V. cholerae* O1 and O139

During the pre-molecular era, phage typing of *V. cholerae* O1 was considered as one of the powerful tools in discriminating the strains of epidemiological interest. Four specific phages were used for biotyping of classical *V. cholerae* [95, 96]. Basu and Mukerjee [97] established a systematic grouping of El Tor strains. Under this scheme, a panel of five phages was used for many years. In 1971–1984 cholera outbreaks in Hyderabad, India, phage types T1, T2, and T4 were dominated during classical cholera period and types T2 and T4 dominated among El Tor strains in later years [98]. In due course of time, the classical strains disappeared in India and the El Tor strains were mostly clustered with phage types T2 and T4. Due to this limitation of Basu and Mukerjee's El Tor phages, a new phage typing scheme was introduced with a panel of five additional phages [99, 100]. In this new scheme the phage type numbers were increased to 146 when about 1000 *V. cholerae* O1 El Tor strains were tested. These phages are now being used in the current phage typing of El Tor vibrios at the National Institute of Cholera and Enteric Diseases

(NICED), Kolkata, India. Phage typing scheme for differentiating the *V. cholerae* O139 was also established in NICED with a panel of five lytic phages [101]. A comparative study of phage types in 1993–1994 and 1996–1998 showed higher percentage types T1 (40.5%) and T2 (32.1%). The phage typing scheme is undergoing rapid changes as clustering of certain types needs to be subtyped to find the epidemiological significance of strains.

2.8 Molecular Epidemiology

Molecular epidemiological studies showed frequent emergence of new epidemic clones among O1 and O139 strains isolated from the cholera patients and environmental sources, as reflected in the genomic structure analysis, location of rRNA, and CTX prophages. The detailed outlines in molecularepidemiology of cholera are described in Chapter 7.

A retrospective analysis of *V. cholerae* O1 using RFLP of enterotoxin genes revealed the presence of indigenous and exogenous strains in Hong Kong since 1978 [102]. However, the vibrios isolated during a cholera outbreak among Vietnamese refugees in Hong Kong are indistinguishable, but distinct from the strains isolated in Hong Kong prior to the outbreak [103]. *V. cholerae* O1 strains collected from domestic and imported cases of cholera that occurred between 1984 and 1997 in Aichi prefecture in Japan were subjected to PFGE and the results suggested that a new clone was introduced after 1993 from overseas and disseminated [104]. The Miri Sarawak outbreak during 1997–1998 might be due to multiple *V. cholerae* O1 clones prevailed during this period [105]. *V. cholerae* O1, O139, and rough strains were isolated from seafood samples in Malaysia during 1998–1999 and the PFGE results suggested that the O139 and rough strains would have the origin from Bengal and Thailand–Malaysia–Laos, respectively [106]. *V. cholerae* O1 strains MO1 and MO477 isolated from Madras, India, during 1992–1993 were reported as probable progenitor strains, which are neither truly classical nor El Tor in the different assay including polymyxin B susceptibility (resistant), classical and El Tor phages (resistant), *ctx* RFLP (identical to O139), copy number of *ctx* (identical to O139), and the outermembrane protein profile (identical to O139) [107].

Both toxigenic and non-toxigenic *V. cholerae* O1 were isolated in Taiwan during 1993–1995 from imported cases of cholera and food samples [108]. In this investigation, prevalence of single clone was detected by ribotyping and enterobacterial repetitive intergenic consensus (ERIC) PCR. The sequence information of *ctxB* gene is considered to be one of the useful molecular epidemiological markers based on the specific base substitutions at positions 115 and 203 [109]. In Taiwan, *V. cholerae* O1 strains isolated from imported seafood and sporadic cholera cases had *ctxB* polymorphism of genotype 1, whereas the 1962 strains isolated from cholera epidemic and soft-shelled turtles belonged to genotype 2 [110].

The O139 strains isolated from India, Bangladesh, and Thailand during 1993 were clonally similar as identified by the PFGE [111]. The Asian O139 strains isolated during 1993–1994 epidemic contained 1–4 copies of *ctx* gene with 55 genotypes as detected in the *ctx* RFLP [112]. The RFLP analysis of *ctx* was found to be superior to ribotyping method for O139 strain discrimination as the later contained only few types. The important molecular event that converted an O1 El Tor strain to O139 serogroup might be due to lateral gene transfer (LGT) event. Prevalence of 64 novel alleles among 51 sequence types from 9 sequenced loci were detected among 96 strains of O139 collected from 1992 to 2000 in Kolkata, India [113]. This study further strengthened the genetic diversity of O139 serogroup during the course of its rapid expansion and recurrent reappearance. The structure, organization, and location of CTX prophages of *V. cholerae* O139 that appeared during different years in Kolkata also revealed the fact that the genomic configuration of the pathogen is very dynamic [89]. The genomic diversity of *V. cholerae* O139 strains that appeared in India and Bangladesh between 1992 and 1998 was tested by using ribotyping and CTX genotyping methods [114]. In this study, six distinct ribotypes were identified, of which B-I–B-V types shared 11 different CTX genotypes (A–K).

The O139 serogroup had a new 35 kb genomic region replacing the O1 somatic antigenic region. Interestingly, this 35 kb region was genetically stable for many years as evidenced from RFLP analysis using many strains collected from South Asia at different periods of time [115]. In addition, the South Asian strains were different from a strain isolated in Argentina indicating the South American strain was unique and had a different origin [115]. The first O139 cholera epidemic appeared in China in 1993 after its first detection in India in 1992. Seven different ribotypes were detected with O139 strains collected in China during 1993–1999 and many strains carried two or more copies of *ctx* gene [116]. Some of the *ctx*-negative O139 strains were grouped into three separate ribotypes [116]. It is evident from this study that the O139 had multiple origins and caused epidemic and sporadic cholera in China. The remerged O139 strains during 2002 in Bangladesh belonged to a ribotype, corresponding to one of the two ribotypes identified during 1993 [72].

V. cholerae O1 El Tor strains isolated before and after the O139 epidemics in India revealed that the strains were different in their genetic profiles based on the RFLP of rRNA genes and CTX genetic element [117]. Till 1997, three different clones of *V. cholerae* O139 prevailed in India, as shown in the *Hin*dIII digestion patterns with *cxtA* probe [88]. The same molecular trend was also reported from Bangladesh [118]. The ribotype BII of *V. cholerae* O139 was responsible for a cholera outbreak in Ahmedabad in 2000, but their PFGE patterns were different from the strains isolated during 1992–1997 [13]. In the same outbreak, *V. cholerae* O1 was also isolated, but their PFGE profiles matched with clones existed in Kolkata during that time. These findings suggest that there is a continuous genetic reassortment among *V. cholerae*, whenever there is a serotype or serogroup replacement.

The epidemic O1 and O139 strains isolated between 1994 and 2002 are differentiated from non-epidemic strains by PFGE in Hong Kong [119]. Overall, 60 distinct PFGE profiles were obtained in this study, which are different from strains isolated from imported cholera cases, mainly from Indonesia, India, and Pakistan. El Tor vibrios identified from domestic and imported cases of cholera in Japan were compared by PFGE [120]. The PFGE subtypes *Not*I-AI and *Sfi*-I were found to be widely distributed in Asia.

Based on the ribotyping results, *V. cholerae* O1 Ogawa isolated in 1999–2000 and Inaba strains of 2001–2002 in Thailand exhibited different *Bgl*I profiles confirming the change in the genetic constituent [121]. The ribotyping and PFGE profiles of several outbreak strains of *V. cholerae* O1 Ogawa in metropolitan area of Kuala Lumpur, Malaysia during 1998 matched with those identified in Taiwan, Colombia, and several Asian regions (ribotype V/B21a) and Senegal (ribotype B27) [122]. However, the Inaba strains isolated during the same outbreak exhibited a new ribotype pattern (type A). The O139 strains isolated from surface waters in Malaysia uniformly harbored a 2.0 MDa non-conjugative plasmid and the *ctxA* gene, but were genetically different as determined by RAPD-PCR [123].

The O139 cholera outbreaks are related with consumption of contaminated foods in many parts of China. Outbreaks in Sichuan Province, Jiangxi Province, and Guangzhou area were due to contaminated turtle, as evidenced by PFGE analysis with strains isolated from the patients and suspected food samples [37, 38, 124]. In Guangzhou area, the genetic diversity of *V. cholerae* O1 was studied with 276 strains isolated from cholera patients, carriers, and environments during 2001–2005 [125]. As evidenced from this study, *V. cholerae* strains were categorized into three types with pathogenicity gene profiles of $ctxA^{+}$ $tcpA^{+}$ ace^{+} zot^{+} (type A), $ctxA^{-}$ $tcpA^{-}$ ace^{+} zot^{+} (type B), and $ctxA^{-}$ $tcpA^{-}$ ace^{-} zot^{-} (type C) using a multiplex PCR. The distribution of type A profile was detected in 68.5% cases with mild symptom and 22% in carriers and type C in 64% cases with mild symptom and 36% in carriers. With the environmental strains, type C profile was common (55%) than type B (36%) [125].

The incidence and pathogenic properties of *V. cholerae* non-O1, non-O139 have extensively been reported with the indication that the proportion of their incidence is constantly increasing [126–128]. Even though the non-O1, non-O139 lacks virulence properties of O1/O139, virulence machineries such as heat-stable toxin [129] and type III secretion system [128] may have some clinical significance. Some of the non-O1, non-O139 strains isolated from River Narmada at Jabalpur, India, harbored the *ctxA* and *tcpA* genes indicating the potential reservoir of the pathogen [130]. Toxigenic non-O1, non-O139 strains harboring CTX genetic element, El Tor allele of *hlyA*, and *stn* were identified in non-cholera-endemic regions such as Kerala, India [131]. However, incidences of such strains in clinical cases are not reported. The presence of viable but non-culturable (VBNC) forms of *V. cholerae* O1 in many aquatic environs was correlated well with the incidence of cholera in Vellore, southern India [132].

After its last dominance in 1989, sporadic infections caused by *V. cholerae* O1 Inaba serotype were recorded in Kolkata during 1998–1999 [133]. Ribotyping and PFGE results showed that the Inaba strains were evolved from the Ogawa serotype prevailed that time [133]. During 2004–2005, cholera caused by the Inaba serotype was recorded in 15 states of India, mostly associated in the form of outbreaks [15]. These Inaba strains had unique PFGE (pulsotype H1) and ribotype (RIV) profiles that were not recorded before. After its first appearance in July 2004, the Inaba serotype completely replaced the dominant Ogawa serotype from May 2005 in Kolkata [134]. These Kolkata Inaba strains belonged to a new clone in the ribotyping as well as PFGE, identical to the Delhi strains and had a CTX prophage with two RS elements. Similar results were obtained with Inaba strains isolated in Trivandrum, southern India, except for ribotyping, which showed that the Inaba and Ogawa strains were similar [135].

Cholera outbreaks were recorded during the late 1990s in several districts of Teheran, Iran [136]. Though the isolated strains belonged to Ogawa serotype and carried two or three copies of *ctx*, ribotyping results showed prevalence of single clone during the outbreaks. Most of the Iranian *V. cholerae* O1 strains (33–96%) harbored a large plasmid and displaced limited number of clones as evidenced from ribotyping and PFGE analysis [137]. *V. cholerae* O1 Inaba strains collected from several cholera outbreaks in Iran during 2005 were clonally identical as detected by PFGE and ribotyping [94]. Recently, frequency in the isolation of O139 serogroup is less globally. Few O139 strains isolated in India during 2003 shared similar PFGE profile with strains isolated in 2000 [93]. However, organization of the tandemly arranged prophages such as CTX (El), CTX (Cal), and truncated CTX (Cal) that had no *ctxAB* was unique in majority of the O139 strains.

Conventionally, the classical and El Tor biotypes of *V. cholerae* O1 are identified with phenotypic markers. Presently, the molecular markers were developed to differentiate the biotypes exploiting the nucleotide difference in the sequences of gene encoding the toxin-coregulated pilus (*tcpA*), *rstR*. The repeat in toxin gene *rtxC* is absent only in the classical biotype. Cholera outbreak in Mozambique during 2004–2005 was caused by hybrid strains between classical and El Tor biotypes, which were different from Bangladeshi classical strains [138]. The Mozambique hybrid strains had *rstR* allele specific for classical biotype. The cholera toxin (CT), encoded by *ctxA* and *ctxB* genes, is the main virulence factor in *V. cholerae*. Based on the amino acid residue substitution at positions 39, 46, and 68 in the B subunit of the CT (CtxB), three genotypes were identified [109]. The classical and US Gulf Coast El Tor strains were classified as genotype 1 and the Australian El Tor strains as genotype 2. Almost all the seventh pandemic El Tor strains belong to genotype 3. El Tor strains having classical CtxB has been confined to US Gulf Coast strains. In 2004, hybrid El Tor strain producing classical CT was first identified from cholera patients in Matlab, Bangladesh, and in Beira, Mozambique. In 2006, this hybrid strain replaced the prevailing El Tor biotype in Bangladesh [139]. Subsequently, the hybrid El Tor strains that produce the classical CT has been reported in many Asian and African countries [140, 141]. Genomic changes in *V. cholerae* possibly provide better survival in the environment and surpass the immuno-impediment of the host. However, the epidemiological significance of this hypothesis should be studied in detail.

2.9 Seroepidemiology

The vibriocidal antibody titer has been used as an epidemiological marker for the determination of recent cholera infection. Such markers are generally used in the cholera vaccine trials to confirm the efficacy of the vaccines. A Mexican study has demonstrated the use of seroepidemiology in studying prevalence of cholera [142]. US Gulf Coast study demonstrated that the vibriocidal titers were significantly higher in Vietnamese subjects than in non-Vietnamese subjects [143]. This indicates that the focus of cholera is confined to specific geographical regions. Compared to the serum IgG, the levels of IgA against B subunit of CT, LPS, and TcpA were high in household contacts of patients infected with *V. cholerae* O1 in Bangladesh [144]. Strong immunoglobulin IgA and IgM antibody-secreting cell (ASC) responses were reported in cholera patients against the homologous serogroup [145]. Vibriocidal assay also showed the same trend indicating that these responses did not evoke cross-protection between the infections caused by O1 and O139 serogroups.

2.10 Prospects of Cholera Vaccines in Asia

During disaster periods, it was always difficult to formulate an effective disease control measure, as this needs proper planning, extended help from local government, supply and stocking of life-saving drugs/vaccines. Under this complex situation, vaccine campaign is difficult. Surpassing all the difficulties, an oral cholera vaccine was given to people affected by tsunami in Aceh Province, Indonesia, during 2004. Almost 69% of the population received the vaccine in two doses [146]. After this successful implementation of the vaccine program, the WHO has issued recommendations in 2004 for the use of oral cholera vaccine during emergencies [146]. In cholera-endemic areas, there is an increasing demand for the oral cholera vaccine at moderate prices as protection efficacy is more than 50% and herd protection studies indicated that the unvaccinated persons were also benefitted as the overall incidence of cholera was less [147, 148]. Investigations on the safety and immunogenicity of an anti-O1 and anti-O139 killed oral whole-cell cholera vaccine (biv-WC) in Vietnam showed promising results. This vaccine gave fourfold increase in vibriocidal antibody titer in 60 and 40% of the adults against O1 and O139, respectively [149]. Similarly, the vibriocidal response was high among children 1–12 years of age with 90 and 68% seroconversion against O1 and O139 serogroups. The killed whole-cell oral cholera vaccine study conducted during 1998 in Hue, Vietnam, revealed its long-term protection (3–5 years) with high efficacy (50% range 9–63%) [150].

An analysis of cholera vaccine trials with rCTB-WC showed protection in 61–86% of people living in endemic areas for 4–6 months and up to 3 years at low levels [151]. This vaccine was proposed for workers employed in the relief or refugee camps or for those who will be traveling in epidemic areas and for those who have no medical care option. An effective vaccine against cholera has been used for public health purposes in Vietnam since the 1990s. In a cholera-endemic area in Kolkata, India, safety and immunogenicity of the reformulated bivalent killed whole-cell oral vaccine with *V. cholerae* O1 and O139 was tested in a double-blind, randomized, placebo controlled trial (healthy adults). Following immunization, 53% of adult and 80% of children vaccines showed a $\geq$4-fold rise in serum *V. cholerae* O1 vibriocidal antibody titers [152]. However, the vibriocidal antibody titers of post-immunization among vaccinees were less for *V. cholerae* O139. Remarkably, the adverse reaction for this vaccine was similar among vaccine and placebo recipients in both age groups. One of the recent studies on live cholera vaccine (VA1.3) showed that the seroconversion rate was 57% with high anti-CT response (77%) and the vaccine was highly safe in the adult volunteers [153]. A modified killed-whole-cell-oral cholera vaccine trial (with a mixture of formalin-killed and heat-killed *V. cholerae* O1 [classical and El Tor biotypes] and O139 strains) in Kolkata, India showed 68% protection efficacy among all age groups [153a]. This vaccine is licensed in India as Shancal® (Shantha Biotechnics, Hyderabad, India).

2.11 Traditional Medicine and Food Habits for Prevention of Cholera

During the fifth decade of the nineteenth century, Dr. J. Collins Browne, an army surgeon worked in British India, invented a patented medicine Chlorodyne, a compound of tincture of chloroform and morphine for the treatment of cholera patients [154]. Chlorodyne was imported to Japan during 1870 and 1873. Dr. Jyun Matsumoto supported the preparation of traditional medicine and a local preparation named "Shinyaku" that resembles Chlorodyne was marketed in Japan in 1872. In the Japanese unknown formulation, morphine hydrochloride, diluted hydrocyanic acid, and tincture of Indian hemp were replaced [154]. However, this formulation was not popular in many countries where epidemics of cholera were frequent.

Traditional herbal medicines (Kampo formulations) are being used in China and Japan for many centuries. The gallate compound from *Rhei rhizoma* inhibited cholera toxin effect in the tissue culture and fluid accumulation in the rabbit ileal loop assay [155]. The synthetic gallate has shown toxin in vivo and in vitro inhibitory effects and thus the Kampo formulation seems to be an effective adjunctive therapy with oral rehydration solution [155]. Soy sauce (Shoyu) was found to have antimicrobial activity against *V. cholerae* and other enteric bacteria along with other beneficial properties such as antihypertensive, anticarcinogenic, anticataract, antioxidant, and antiallergen [156]. Many useful herbal formulations for the treatment of cholera and other diarrheal diseases have been lost during passage of time and generations of people. Unfortunately, there was no recording system in primeval days and such formulations were kept secret for the benefit of professionals those practiced herbal medicine.

2.12 Control Measures and Health-Care Systems

For prevention of infectious diseases including cholera, the Chinese health-care model is being considered in developing countries. During 1940s, the mortality rate in China was 20 and 30–40% in children less than 1 and 5 years of age, respectively. The health care was given prioritized during later years and strengthened in 1965 after Cultural Revolution, which resulted in increasing the life span from

40 to 70 years and reducing the infant mortality very less [157]. The main health-care objectives considered in the Chinese system include the following: improving access to health care in rural areas, prevention of infection, emphasis on traditional medicine, and utilizing manpower rather than technical power in the health-care system [157]. The same system was proposed for African countries [158]. In Indonesia, several control measures were made in the anticipation of outbreaks in refugee camps. As a result of such activities, no cholera or typhoid cases were detected through routine surveillance [159].

In cholera-endemic regions, the case–control studies conducted to determine the etiological agent of outbreaks often lead to multiple sources or other sources such as river waters were periodically found to be one of the vehicles [160]. Asymptomatic cholera carriers might play an important role in the passive spreading of the disease. In Hong Kong, early investigations were made for the detection and treatment of carriers to prevent spread of sporadic cholera [161]. Food handlers were responsible for large-scale cholera outbreaks in Thailand [162]. The impact of cholera and other enteric diseases was assessed by providing deep-well tap water (DWTW) through household taps in Qidong village in China [163]. In this study, the overall incidence of enteric infection was 39% lower than the control region that had surface water supply. However, the initial cost of establishing DWTW is prohibitive due to many reasons. Cholera outbreaks in Tumpat and Kelantan, Malaysia, during 1990 were related to the use of Kelantan river water and consumption of river clams [164]. Several carriers were also identified in this outbreak investigation.

In Bangladesh, use of tube-well water played a significant role in reducing the cholera mortality in rural areas [67]. The number of cholera cases in rural areas of Bangladesh was significantly reduced when a simple filtration using old sari was tested that effectively removed the plankton and other particulate matter from the natural waters [165, 166]. Despite the rural health improvement scheme, cholera was endemic in Malaysia in 2002 as most of the rural population relied on river water supply [167]. This suggests that sanitation interventions were not as effective in reducing the water-borne diseases such as cholera. Successive O139- and O1-mediated cholera appeared in Pakistan during 2002–2003 were mainly due to person-to-person contact and through contaminated water reservoirs [10]. As the cholera management was given priority in the Vietnam Health sector, the annual mortality rate reduced from 2.0–9.65% during 1979–1983 to 1.8% in the later periods [168]. An outbreak of cholera associated with a tribal funeral in Irian Jaya, Indonesia, has been reported [169]. Generally, reports of cholera associated with attending funeral are very less in Asian regions as the burial practice of the death victims and rituals are different from African region. The case–control studies for cholera risk factors were found to be similar for O1- and O139-mediated cholera that include consumption of untreated water, uncooked pork or seafood, and food served in large gatherings [162, 170, 171]. Among children less than 6 months of age, protection associated with breast-feeding has been demonstrated for cholera and the mechanisms are related to risk of supplementary feeding that may introduce pathogens and/or immunoprotection offered by the breast milk through immunoglobulins [172].

2.13 Conclusion

Over the years, the incidence of cholera has been reduced in many Asian countries and the mortality rate has also gone down. The role of phenotypic and genetic changes in *V. cholerae* O1 and O139 serogroups and their increased incidence in few countries have not been studied in detail. Antimicrobial resistance in *V. cholerae* is emerging in Asia and this can be controlled by judicious use of drugs in the management of cholera. Several efficient methods in managing cholera outbreaks have been implemented in rural and urban areas after rigorous efficacy studies including introduction of new oral vaccines, supply of safe drinking water, sanitation, and timely administration of ORS. Success of these interventions depends on maintenance and regular practice at the community level.

References

1. Barua, D. History of cholera. 1992. In: Barua CD, Greenough III WB, editors. Cholera New York, USA: Plenum Pub. Co. pp. 1–35.
2. Datta KK, Singh J. Epidemiological profile of outbreaks of cholera in India during 1975–1989. J Commun Dis. 1990;22:151–9.
3. Darbari BS, Tiwari HL, Agarwal S. Pattern of cholera in Raipur: a twelve year appraisal. J Indian Assoc Commun Dis. 1982;5:83–7.
4. Khanna KK, Dhanvijay A, Riley LW, Sehgal S, Kumari S. Cholera outbreak in Delhi-1988. J Commun Dis. 1990;22:35–8.
5. Narang P, Mendiratta DK, Deotale VS, Narang R. Changing patterns of *Vibrio cholerae* in Sevagram between 1990 and 2005. Indian J Med Microbiol. 2008;26:40–4.
6. Das S, Gupta S. Diversity of *Vibrio cholerae* strains isolated in Delhi, India, during 1992–2000. J Health Popul Nutr. 2005;23:44–51.
7. Hamner S, Tripathi A, Mishra RK, Bouskill N, Broadaway SC, Pyle BH, Ford TE. The role of water use patterns and sewage pollution in incidence of water-borne/enteric diseases along the Ganges river in Varanasi, India. Int J Environ Health Res. 2006;16:113–32.
8. Kaistha N, Mehta M, Gautam V, Gupta V. Outbreak of cholera in and around Chandigarh during two successive years (2002, 2003). Indian J Med Res. 2005;122:404–7.
9. Deen JL, von Seidlein L, Sur D, Agtini M, Lucas ME, Lopez AL, Kim DR, Ali M, Clemens JD. The high burden of cholera in children: comparison of incidence from endemic areas in Asia and Africa. PLoS Negl Trop Dis. 2008;2:e173.
10. Siddiqui FJ, Bhutto NS, von Seidlein L, Khurram I, Rasool S, Ali M, Zafar A, Deen JL, Clemens JD, Nizami Q, Bhutta ZA. Consecutive outbreaks of *Vibrio cholerae* O139 and *V. cholerae* O1 cholera in a fishing village near Karachi, Pakistan. Trans R Soc Trop Med Hyg. 2006;100:476–82.
11. Shah WA, Shahina M, Ali N. First report of *Vibrio cholerae* infection from Andaman and Nicobar, India. J Commun Dis. 2002;34:270–5.
12. Sugunan AP, Ghosh AR, Roy S, Gupte MD, Sehgal SC. A cholera epidemic among the Nicobarese tribe of Nancowry, Andaman, and Nicobar, India. Am J Trop Med Hyg. 2004;71:822–7.
13. Chakraborty S, Deokule JS, Garg P, Bhattacharya SK, Nandy RK, Nair GB, Yamasaki S, Takeda Y, Ramamurthy T. Concomitant infection of enterotoxigenic *Escherichia coli* in an outbreak of cholera caused by *Vibrio cholerae* O1 and O139 in Ahmadabad, India. J Clin Microbiol. 2001;39:3241–6.
14. Sharma NC, Mandal PK, Dhillon R, Jain M. Changing profile of *Vibrio cholerae* O1, O139 in Delhi and its periphery (2003–2005). Indian J Med Res. 2007;125:633–40.

15. Dutta B, Ghosh R, Sharma NC, Pazhani GP, Taneja N, Raychowdhuri A, Sarkar BL, Mondal SK, Mukhopadhyay AK, Nandy RK, Bhattacharya MK, Bhattacharya SK, Ramamurthy T. Spread of cholera with newer clones of *Vibrio cholerae* O1 El Tor, serotype Inaba, in India. J Clin Microbiol. 2006;44:3391–3.
16. Pal BB, Khuntia HK, Samal SK, Das SS, Chhotray GP. Emergence of *Vibrio cholerae* O1 biotype El Tor serotype Inaba causing outbreaks of cholera in Orissa, India. Jpn J Infect Dis. 2006;59:266–9.
17. Sugunan AP, Roy S, Shahina M, Shah WA, Bharadwaj AP, Singh SS, Thanasekaran K, Sathya Prakash M, Vijayachari P. Emergence of *Vibrio cholerae* O1 Inaba in Andaman and Nicobar Islands, India. J Public Health. 2007;29:308–9.
18. Samadi AR, Huq MI, Shahid N, Khan MU, Eusof A, Rahman AS, Yunus M, Faruque AS. Classical *Vibrio cholerae* biotype displaces El tor in Bangladesh. Lancet. 1983;1:805–7.
19. Khan MU, Samadi AR, Huq MI, Yunus M, Eusof A. Simultaneous classical and El Tor cholera in Bangladesh. J Diarrhoeal Dis Res. 1984;2:13–8.
20. Siddique AK, Baqui AH, Eusof A, Haider K, Hossain MA, Bashir I, Zaman K. Survival of classic cholera in Bangladesh. Lancet. 1991;337:1125–7.
21. Samadi AR, Chowdhury MK, Huq MI, Khan MU. Seasonality of classical and El Tor cholera in Dhaka, Bangladesh: 17-year trends. Trans R Soc Trop Med Hyg. 1983;77:853–6.
22. Huq MI, Sanyal SC, Samadi AR, Monsur KA. Comparative behaviour of classical and El Tor biotypes of *Vibrio cholerae* O1 isolated in Bangladesh during 1982. J Diarrhoeal Dis Res. 1983;1:5–9.
23. Longini M Jr, Yunus M, Zaman K, Siddique AK, Sack RB, Nizam A. Epidemic and endemic cholera trends over a 33-year period in Bangladesh. J Infect Dis. 2002;186: 246–51.
24. Siddique AK, Zaman K, Baqui AH, Akram K, Mutsuddy P, Eusof A, Haider K, Islam S, Sack RB. Cholera epidemics in Bangladesh: 1985–1991. J Diarrhoeal Dis Res. 1992;10:79–86.
25. Koelle K, Pascual M, Yunus M. Serotype cycles in cholera dynamics. Proc Biol Sci. 2006;273:2879–86.
26. Ali M, Emch M, Yunus M, Sack RB. Are the environmental niches of *Vibrio cholerae* O139 different from those of *Vibrio cholerae* O1 El Tor? Int J Infect Dis. 2001;5:214–49.
27. Alam M, Sultana M, Nair GB, Sack RB, Sack DA, Siddique AK, Ali A, Huq A, Colwell RR. Toxigenic *Vibrio cholerae* in the aquatic environment of Mathbaria, Bangladesh. Appl Environ Microbiol. 2006;72:2849–55.
28. Faruque SM, Islam MJ, Ahmad QS, Biswas K, Faruque AS, Nair GB, Sack RB, Sack DA, Mekalanos JJ. An improved technique for isolation of environmental *Vibrio cholerae* with epidemic potential: monitoring the emergence of a multiple-antibiotic-resistant epidemic strain in Bangladesh. J Infect Dis. 2006;193:1029–36.
29. Kuo CL, Fukui H. Geographical structures and the cholera epidemic in modern Japan: Fukushima prefecture in 1882 and 1895. Int J Health Geogr. 2007;6:25.
30. Chemouilli P. The cholera epidemics and the development of public health in Meiji Japan. 1. Modernity, cholera, and health thought. Med Sci (Paris). 2004;20:109–14 (In French).
31. Chemouilli P. The cholera epidemics and the development of public health in Meiji Japan. 2. Strength and weakness of public health politics. Med Sci (Paris). 2004;20:236–40 (In French).
32. Fukase Y. On the prevalence of Asiatic cholera in Okinawa, 1879 and "Ryukyu-kiko". In: Tsuchiya H, editor. Nippon Ishigaku Zasshi. 1999;45:373–400 (In Japanese).
33. Minami A, Hashimoto S, Abe H, Arita M, Taniguchi T, Honda T, Miwatani T, Nishibuchi M. Cholera enterotoxin production in *Vibrio cholerae* O1 strains isolated from the environment and from humans in Japan. Appl Environ Microbiol. 1991;57:2152–7.
34. Miyagi K, Nakano T, Yagi T, Hanafusa M, Imura S, Honda T, Nakano Y, Sano K. Survey of *Vibrio cholerae* O1 and its survival over the winter in marine water of Port of Osaka. Epidemiol Infect. 2003;131:613–9.

35. Kam KM, Leung TH, Ho YY, Ho NK, Saw TA. Outbreak of *Vibrio cholerae* O1 in Hong Kong related to contaminated fish tank water. Public Health. 1995;109:389–95.
36. Wong HC, Ting SH, Shieh WR. Incidence of toxigenic vibrios in foods available in Taiwan. J Appl Bacteriol. 1992;73:197–202.
37. Liu HL, Zhang JY, Feng ZH, Li W, Cui ZG, Zhang LJ, Zhu XP, Kan B. Application of pulsed-field gel electrophoresis typing in tracing and carrying out surveillance programs on O139 cholera outbreaks. Zhonghua Liu Xing Bing Xue Za Zhi. 2006;27:102–6 (In Chinese).
38. Zhang J, Chang ZR, Zhong HJ, Wang DC, Xu J, Kan B, Ran L, Wang ZJ. Investigation on status of pollution of *Vibrio cholerae* in seafood and aquatic products in 12 provinces of China in 2005. Zhonghua Yu Fang Yi Xue Za Zhi. 2007;41:208–11(In Chinese).
39. Chang ZR, Zhang J, Wang DC, Zhong HJ, Xu J, Ran L, Wang MW, Wang ZJ, Kan B. Identification and molecular study on *Vibrio cholerae* in sea products. Zhonghua Yu Fang Yi Xue Za Zhi. 2007;41:304–6 (In Chinese).
40. Lü HK, Chen EF, Xie SY, Chai CL, Wei YD, Mo ST, Ye JL, Luo Y. Investigation on *Vibrio cholerae* carried in aquatic products of littoral areas, Zhejiang Province. Zhonghua Yu Fang Yi Xue Za Zhi. 2006;40:336–8 (In Chinese).
41. Bubshait SA, Al-Turki K, Qadri MH, Fontaine RE, Cameron D. Seasonal, non-toxigenic *Vibrio cholerae* O1 Ogawa infections in the Eastern Region of Saudi Arabia. Int J Infect Dis. 2000;4:198–202.
42. Valenciano M, Coulombier D, Lopes Cardozo B, Colombo A, Alla MJ, Samson S, Connolly MA. Challenges for communicable disease surveillance and control in southern Iraq, April–June 2003. JAMA. 2003;290:654–8.
43. Khazaei HA, Rezaei N, Bagheri GR, Moin AA. A six-year study on *Vibrio cholerae* in southeastern Iran. Jpn J Infect Dis. 2005;58:8–10.
44. Kakar F, Ahmadzai AH, Habib N, Taqdeer A, Hartman AF. A successful response to an outbreak of cholera in Afghanistan. Trop Doct. 2008;38:17–20.
45. Forssman B, Mannes T, Musto J, Gottlieb T, Robertson G, Natoli JD, Shadbolt C, Biffin B, Gupta L. *Vibrio cholerae* O1 El Tor cluster in Sydney linked to imported whitebait. Med J Aust. 2007;187:345–7.
46. Tarantola A, Ioos S, Rotureau B, Paquet C, Quilici ML, Fournier JM. Retrospective analysis of the cholera cases imported to France from 1973 to 2005. J Travel Med. 2007;14:209–14.
47. Yadav H, Chee CM. Cholera in Sarawak: a historical perspective (1873–1989). Med J Malaysia. 1990;45:194–201.
48. Kelly-Hope LA, Alonso WJ, Thiem VD, Canh do G, Anh DD, Lee H, Miller MA. Temporal trends and climatic factors associated with bacterial enteric diseases in Vietnam, 1991–2001. Environ Health Perspect. 2008;116:7–12.
49. Hashizume M, Armstrong B, Hajat S, Wagatsuma Y, Faruque AS, Hayashi T, Sack DA. The effect of rainfall on the incidence of cholera in Bangladesh. Epidemiology. 2008;19: 103–10.
50. Simanjuntak CH, Larasati W, Arjoso S, Putri M, Lesmana M, Oyofo BA, Sukri N, Nurdin D, Kusumaningrum RP, Punjabi NH, Subekti D, Djelantik S, Sukarma, Sriwati, Muzahar, Lubis A, Siregar H, Mas'ud B, Abdi M, Sumardiati A, Wibisana S, Hendarwanto, Setiawan B, Santoso W, Putra E, Sarumpaet S, Ma'ani H, Lebron C, Soeparmanto SA, Campbell JR, Corwin AL. Cholera in Indonesia in 1993–1999. Am J Trop Med Hyg. 2001;65:788–97.
51. Pancharoen C, Niwattanakanjana N, Mekmullica J, Chongsrisawat V. Hospital-based epidemiology of childhood cholera: a 6-year review in a university hospital in Bangkok, Thailand. J Med Assoc Thai. 2004;87:S59–61.
52. Tamang MD, Sharma N, Makaju RK, Sarma AN, Koju R, Nepali N, Mishra SK. An outbreak of El Tor cholera in Kavre district, Nepal. Kathmandu Univ Med J. 2005;3:138–42.
53. Cooper PJ, Chico M, Sandoval C, Espinel I, Guevara A, Levine MM, Griffin GE, Nutman TB. Human infection with *Ascaris lumbricoides* is associated with suppression of the

interleukin-2 response to recombinant cholera toxin B subunit following vaccination with the live oral cholera vaccine CVD 103-HgR. Infect Immun. 2001;69:1574–80.
54. Nagoba BS, Deshmukh SR, Dharane SM, Narute JV. Association of *Giardia lamblia* with *Vibrio* in cholera cases. Indian J Pediatr. 1992;59:386–7.
55. Saha DR, Rajendran K, Ramamurthy T, Nandy RK, Bhattacharya SK. Intestinal parasitism and *Vibrio cholerae* infection among diarrhoeal patients in Kolkata, India. Epidemiol Infect. 2008;136:661–4.
56. Chadee K, Keller K, Forstner J, Innes DJ, Ravdin JI. Mucin and nonmucin secretagogue activity of *Entamoeba histolytica* and cholera toxin in rat colon. Gastroenterol. 1991;100:986–97.
57. Ramamurthy T, Garg S, Sharma R, Bhattacharya SK, Nair GB, Shimada T, Takeda T, Karasawa T, Kurazano H, Pal A, Takeda Y. Emergence of novel strain of *Vibrio cholerae* with epidemic potential in southern and eastern India. Lancet. 1993;341:703–4.
58. Shimada T, Arakawa E, Itoh K, Okitsu T, Matsushima A, Asai Y, Yamai S, Nakazato T, Nair GB, Islam MS, Albert MJ, Takeda Y. Extended serotyping scheme for *Vibrio cholerae*. Curr Microbiol. 1994;28:175–8.
59. Albert MJ, Siddique AK, Islam MS, Faruque AS, Ansaruzzaman M, Faruque SM, Sack RB. Large outbreak of clinical cholera due to *Vibrio cholerae* non-O1 in Bangladesh. Lancet. 1993;341:704.
60. Nair GB, Ramamurthy T, Bhattacharya SK, Mukhopadhyay AK, Garg S, Bhattacharya MK, Takeda T, Shimada T, Takeda Y, Deb BC. Spread of *Vibrio cholerae* O139 Bengal in India. J Infect Dis. 1994;169:1029–34.
61. Mahalanabis D, Faruque AS, Albert MJ, Salam MA, Hoque SS. An epidemic of cholera due to *Vibrio cholerae* O139 in Dhaka, Bangladesh: clinical and epidemiological features. Epidemiol Infect. 1994;112:463–71.
62. Bhattacharya SK, Bhattacharya MK, Nair GB, Dutta D, Deb A, Ramamurthy T, Garg S, Saha PK, Dutta P, Moitra A, Mandal BK, Shimada T, Takeda Y, Deb BC. Clinical profile of acute diarrhoea cases infected with the new epidemic strain of *Vibrio cholerae* O139: designation of the disease as cholera. J Infect. 1993;27:11–5.
63. Singh J, Bora D, Sachdeva V, Sharma RS, Verghese T. *Vibrio cholerae* O1 and O139 in less than five years old children hospitalised for watery diarrhoea in Delhi, 1993. J Diarrhoeal Dis Res. 1997;15:3–6.
64. Sheikh A, Khan A, Malik T, Fisher-Hoch SP. Cholera in a developing megacity; Karachi, Pakistan. Epidemiol Infect. 1997;119:287–92.
65. Faruque AS, Fuchs GJ, Albert MJ. Changing epidemiology of cholera due to *Vibrio cholerae* O1 and O139 Bengal in Dhaka, Bangladesh. Epidemiol Infect. 1996;116:275–8.
66. Mukhopadhyay AK, Garg S, Mitra R, Basu A, Rajendran K, Dutta D, Bhattacharya SK, Shimada T, Takeda T, Takeda Y, Nair GB. Temporal shifts in traits of *Vibrio cholerae* strains isolated from hospitalized patients in Calcutta: a 3-year (1993 to 1995) analysis. J Clin Microbiol. 1996;34:2537–43.
67. Sack RB, Siddique AK, Longini IM Jr, Nizam A, Yunus M, Islam MS, Morris JG Jr, Ali A, Huq A, Nair GB, Qadri F, Faruque SM, Sack DA, Colwell RR. A 4-year study of the epidemiology of *Vibrio cholerae* in four rural areas of Bangladesh. J Infect Dis. 2003;187:96–101.
68. Jabeen K, Hasan R. Re-emergence of *Vibrio cholerae* O139 in Pakistan: report from a tertiary care hospital. J Pak Med Assoc. 2003;53:335–8.
69. Mitra R, Basu A, Dutta D, Nair GB, Takeda Y. Resurgence of *Vibrio cholerae* O139 Bengal with altered antibiogram in Calcutta, India. Lancet. 1996;348:1181.
70. Sundaram SP, Revathi J, Sarkar BL, Bhattacharya SK. Bacteriological profile of cholera in Tamil Nadu (1980–2001). Indian J Med Res. 2002;116:258–63.
71. Agrawal G, Jalgaonkar SV, Jagtap PM, Kamlakar UP, Deogade NG. Emergence and re-emergence of *Vibrio cholerae* O139: an epidemiological study during 1993–2002 at Nagpur, Central India. Indian J Med Sci. 2003;57:155–7.

72. Faruque SM, Chowdhury N, Kamruzzaman M, Ahmad QS, Faruque AS, Salam MA, Ramamurthy T, Nair GB, Weintraub A, Sack DA. Reemergence of epidemic *Vibrio cholerae* O139, Bangladesh. Emerg Infect Dis. 2003;9:1116–22.
73. Sinha S, Chakraborty R, De K, Khan A, Datta S, Ramamurthy T, Bhattacharya SK, Takeda Y, Nair GB. Escalating association of *Vibrio cholerae* O139 with cholera outbreaks in India. J Clin Microbiol. 2002;40:2635–7.
74. Al-Abbassi AM, Ahmed S, Al-Hadithi T. Cholera epidemic in Baghdad during 1999: clinical and bacteriological profile of hospitalized cases. East Mediterr Health J. 2005;11:6–13.
75. Centers for Disease Control and Prevention (CDC). Imported cholera associated with a newly described toxigenic *Vibrio cholerae* O139 strain – California, 1993. MMWR Morb Mortal Wkly Rep. 1993;42:501–3.
76. Mahon BE, Mintz ED, Greene KD, Wells JG, Tauxe RV. Reported cholera in the United States, 1992–1994: a reflection of global changes in cholera epidemiology. JAMA. 1996;276:307–12.
77. Kurazono T, Yamada F, Yamaguchi M, Ohzeki Y, Okuyama Y, Itoh K, Shimada T. The first report of traveler's diarrhea associated with a newly described toxigenic *Vibrio cholerae* O139 strain in Japan. Kansenshogaku Zasshi. 1994;68:8–12.
78. Dalsgaard A, Nielsen GL, Echeverria P, Larsen JL, Schønheyder HC. *Vibrio cholerae* O139 in Denmark. Lancet. 1995;345:1637–8.
79. Faruque SM, Sack DA, Sack RB, Colwell RR, Takeda Y, Nair GB. Emergence and evolution of *Vibrio cholerae* O139. Proc Natl Acad Sci USA. 2003;100:1304–9.
80. Dutta JK, Santhanam S, Misra BS, Ray SN. Effect of trimethoprim-sulphamethoxazole on vibrio clearance in cholera (El Tor): a comparative study. Trans R Soc Trop Med Hyg. 1978;72:40–2.
81. Saraswathi K, Deodhar LP. A study of *V. cholerae* strains isolated in Bombay. J Postgrad Med. 1990;36:128–30.
82. Ehara M, Nguyen BM, Nguyen DT, Toma C, Higa N, Iwanaga M. Drug susceptibility and its genetic basis in epidemic *Vibrio cholerae* O1 in Vietnam. Epidemiol Infect. 2004;132:595–600.
83. Glass RI, Huq I, Alim AR, Yunus M. Emergence of multiply antibiotic-resistant *Vibrio cholerae* in Bangladesh. J Infect Dis. 1980;142:939–42.
84. Ranjit K, Nurahan M. Tetracycline resistant cholera in Kelantan. Med J Malaysia. 2000;55:143–5.
85. Kondo S, Kongmuang U, Kalnauwakul S, Matsumoto C, Chen CH, Nishibuchi M. Molecular epidemiologic analysis of *Vibrio cholerae* O1 isolated during the 1997–1998 cholera epidemic in southern Thailand. Epidemiol Infect. 2001;127:7–16.
86. Roychowdhury A, Pan A, Dutta D, Mukhopadhyay AK, Ramamurthy T, Nandy RK, Bhattacharya SK, Bhattacharya MK. Emergence of tetracycline-resistant *Vibrio cholerae* O1 serotype Inaba, in Kolkata, India. Jpn J Infect Dis. 2008;61:128–9.
87. Phantouamath B, Sithivong N, Sisavath L, Munnalath K, Khampheng C, Insisiengmay S, Higa N, Kakinohana S, Iwanaga M. Transition of drug susceptibilities of *Vibrio cholerae* O1 in Lao People's Democratic Republic. Southeast Asian J Trop Med Public Health. 2001;32:95–9.
88. Mukhopadhyay AK, Basu A, Garg P, Bag PK, Ghosh A, Bhattacharya SK, Takeda Y, Nair GB. Molecular epidemiology of reemergent *Vibrio cholerae* O139 Bengal in India. J Clin Microbiol. 1998;36:2149–52.
89. Basu A, Garg P, Datta S, Chakraborty S, Bhattacharya MK, Khan A, Ramamurthy T, Bhattacharya SK, Yamasaki S, Takeda Y, Nair GB. *Vibrio cholerae* O139 in Calcutta, 1992–1998: incidence, antibiograms, and genotypes. Emerg Infect Dis. 2000;6:139–47.
90. Faruque SM, Siddique AK, Saha MN, Asadulghani, Rahman MM, Zaman K, Albert MJ, Sack DA, Sack RB. Molecular characterization of a new ribotype of *Vibrio cholerae* O139 Bengal associated with an outbreak of cholera in Bangladesh. J Clin Microbiol. 1999;37:1313–8.

91. Yamamoto T, Nair GB, Takeda Y. Emergence of tetracycline resistance due to a multiple drug resistance plasmid in *Vibrio cholerae* O139. FEMS Immunol Med Microbiol. 1995;11:131–6.
92. Tarantola A, Quilici ML; Laboratory Investigation Group. *Vibrio cholerae* O1 strains with decreased susceptibility to fluoroquinolones in travellers returning from India (Rajasthan) to France, April 2007. Euro Surveill. 2007;12:E070503.2.
93. Chatterjee S, Ghosh K, Raychoudhuri A, Pan A, Bhattacharya MK, Mukhopadhyay AK, Ramamurthy T, Bhattacharya SK, Nandy RK. Phenotypic and genotypic traits and epidemiological implication of *Vibrio cholerae* O1 and O139 strains in India during 2003. J Med Microbiol. 2007;56:824–32.
94. Pourshafie MR, Bakhshi B, Ranjbar R, Sedaghat M, Sadeghifard N, Zaemi Yazdi J, Parzadeh M, Raesi J. Dissemination of a single *Vibrio cholerae* clone in cholera outbreaks during 2005 in Iran. J Med Microbiol. 2007;56:1615–19.
95. Mukerjee S, Guha DK, Guha Roy UK, Studies on typing of cholera by bacteriophage-typing of *Vibrio cholerae* from Calcutta epidemics. Ann Biochem Exp Med. 1957;17: 161–76.
96. Mukerjee S. Bacteriophage–typing of cholera. Bull WHO. 1963;28:337–45.
97. Basu S, Mukejee S. Bacteriophage typing of *Vibrio* El Tor. Experientia. 1968;24:299–300.
98. Rathna K, Khairunnisa, Rajyalakshmi K, Naidu AS. Epidemiological patterns and incidence of bio-sero-and phage types of *Vibrio cholerae* in Hyderabad, India during 1971–1984. Acta Microbiol Hung. 1988;35:313–22.
99. Chattopadhyay DJ, Sarkar BL, Ansari MQ, Chakrabarti BK, Roy MK, Ghosh AN, Pal SC. New phage typing scheme for *Vibrio cholerae* O1 biotype El Tor strains. J Clin Microbiol. 1993;31:1579–85.
100. Sarkar BL, De SP, Saha MR, Niyogi SK, Roy MK. Validity of new phage typing scheme against *Vibrio cholerae* O1 biotype El Tor strains. Indian J Med Res. 1994;99:159–61.
101. Chakrabarti AK, Ghose AN, Nair GB, Niyogi SK, Bhattacharya SK, Sarkar BL. Development and evaluation of a phage typing scheme for *Vibrio cholerae* O139. J Clin Microbiol. 2000;38:44–9.
102. Yam WC, Lung ML, Ng KY, Ng MH. Molecular epidemiology of *Vibrio cholerae* in Hong Kong. J Clin Microbiol. 1989;27:1900–2.
103. Yam WC, Lung ML, Ng MH. Restriction fragment length polymorphism analysis of *Vibrio cholerae* strains associated with a cholera outbreak in Hong Kong. J Clin Microbiol. 1991;29:1058–9.
104. Matsumoto M, Suzuki M, Hiramatsu R, Yamazaki M, Matsui H, Sakae K, Suzuki Y, Miyazaki Y. Epidemiological investigation of a fatal case of cholera in Japan by phenotypic techniques and pulsed-field gel electrophoresis. J Med Microbiol. 2002;51:264–8.
105. Radu S, Vincent M, Apun K, Abdul-Rahim R, Benjamin PG, Yuherman, Rusul G. Molecular characterization of *Vibrio cholerae* O1 outbreak strains in Miri, Sarawak (Malaysia). Acta Trop. 2002;83:169–76.
106. Chen CH, Shimada T, Elhadi N, Radu S, Nishibuchi M. Phenotypic and genotypic characteristics and epidemiological significance of *ctx+* strains of *Vibrio cholerae* isolated from seafood in Malaysia. Appl Environ Microbiol. 2004;70:1964–72.
107. Pajni S, Sharma C, Bhasin N, Ghosh A, Ramamurthy T, Nair GB, Ramajayam S, Das B, Kar S, Roychowdhury S, Ghosh RK. Studies on the genesis of *Vibrio cholerae* O139: identification of probable progenitor strains. J Med Microbiol. 1995;42:20–5.
108. Shangkuan YH, Tsao CM, Lin HC. Comparison of *Vibrio cholerae* O1 isolates by polymerase chain reaction fingerprinting and ribotyping. J Med Microbiol. 1997;46: 941–8.
109. Olsvik O, Wahlberg J, Petterson B, Uhlén M, Popovic T, Wachsmuth IK, Fields PI. Use of automated sequencing of polymerase chain reaction-generated amplicons to identify three types of cholera toxin subunit B in *Vibrio cholerae* O1 strains. J Clin Microbiol. 1993; 31:22–5.

110. Liu DP, Chen PJ, Lin CS, Wu TN. Molecular epidemiological studies of *Vibrio cholerae* in Taiwan: genotyping by polymerase chain reaction and DNA sequencing. Zhonghua Min Guo Wei Sheng Wu Ji Mian Yi Xue Za Zhi. 1995;28:291–9.
111. Kurazono H, Okuda J, Takeda Y, Nair GB, Albert MJ, Sack RB, Chongsa-nguan M, Chaicumpa W. *Vibrio cholerae* O139 Bengal isolated from India, Bangladesh and Thailand are clonal as determined by pulsed-field gel electrophoresis. J Infect. 1994;29:109–10.
112. Dalsgaard A, Serichantalergs O, Echeverria P. Restriction fragment length polymorphisms (RFLP) of cholera toxin genes in *Vibrio cholerae* O139 recovered from patients in Thailand, India and Bangladesh. Scand J Infect Dis. 1995;27:585–8.
113. Garg P, Aydanian A, Smith D, J Glenn M Jr, Nair GB, Stine OC. Molecular epidemiology of O139 *Vibrio cholerae*: mutation, lateral gene transfer, and founder flush. Emerg Infect Dis. 2003;9:810–4.
114. Faruque SM, Saha MN, Asadulghani, Bag PK, Bhadra RK, Bhattacharya SK, Sack RB, Takeda Y, Nair GB. Genomic diversity among *Vibrio cholerae* O139 strains isolated in Bangladesh and India between 1992 and 1998. FEMS Microbiol Lett. 2000;184:279–84.
115. Falklind-Jerkérus S, Albert MJ, Weintraub A. Lack of polymorphism in a *Vibrio cholerae* O139-specific DNA region encoding the somatic antigen in strains isolated during 1993–1998. Int J Med Microbiol. 2003;292:505–11.
116. Qu M, Xu J, Ding Y, Wang R, Liu P, Kan B, Qi G, Liu Y, Gao S. Molecular epidemiology of *Vibrio cholerae* O139 in China: polymorphism of ribotypes and CTX elements. J Clin Microbiol. 2003;41:2306–10.
117. Sharma C, Nair GB, Mukhopadhyay AK, Bhattacharya SK, Ghosh RK, Ghosh A. Molecular characterization of *Vibrio cholerae* O1 biotype El Tor strains isolated between 1992 and 1995 in Calcutta, India: evidence for the emergence of a new clone of the El Tor biotype. J Infect Dis. 1997;175:1134–41.
118. Faruque SM, Ahmed KM, Abdul Alim AR, Qadri F, Siddique AK, Albert MJ. Emergence of a new clone of toxigenic *Vibrio cholerae* O1 biotype El Tor displacing *V. cholerae* O139 Bengal in Bangladesh. J Clin Microbiol. 1997;35:624–30.
119. Kam KM, Luey CK, Tsang YM, Law CP, Chu MY, Cheung TL, Chiu AW. Molecular subtyping of *Vibrio cholerae* O1 and O139 by pulsed-field gel electrophoresis in Hong Kong: correlation with epidemiological events from 1994 to 2002. J Clin Microbiol. 2003; 41:4502–11.
120. Arakawa E, Murase T, Matsushita S, Shimada T, Yamai S, Ito T, Watanabe H. Pulsed-field gel electrophoresis-based molecular comparison of *Vibrio cholerae* O1 isolates from domestic and imported cases of cholera in Japan. J Clin Microbiol. 2000;38:424–6.
121. Tapchaisri P, Na-Ubol M, Jaipaew J, Srimanote P, Chongsa-Nguan M, Yamasaki S, Hayashi H, Nair GB, Kurazono H, Chaicumpa W. Virulence genes of clinical *Vibrio cholerae* O1 isolates in Thailand and their ribotypes. J Infect. 2007;55:557–65.
122. Vadivelu J, Iyer L, Kshatriya BM, Puthucheary SD. Molecular evidence of clonality amongst *Vibrio cholerae* O1 biotype El Tor during an outbreak in Malaysia. Epidemiol Infect. 2000;124:25–30.
123. Son R, Rusul G, Samuel L, Yuherman, Senthil S, Rasip A, Nasreldin EH, Nishibuchi M. Note: characterization of *Vibrio cholerae* O139 Bengal isolated from water in Malaysia. J Appl Microbiol. 1998;85:1073–7.
124. Yang M, Diao BW, Cheng HJ, Ding S, Cui ZG, Chen FH, Xu XQ, Kan B, Yuan H. Study on the application of pulsed-field gel electrophoresis regarding infection sources identification during an outbreak of *Vibrio cholerae* in Jiangxi Province. Zhonghua Liu Xing Bing Xue Za Zhi. 2007;28:891–4 (In Chinese).
125. Wang M, Li XQ, Mo ZY, Liu YF, Deng ZA, Shen JC, Zhang XQ. Analysis of characteristics of major pathogenicity-related genes of *Vibrio cholerae* isolated in Guangzhou area from 2001 to 2005. Zhonghua Yu Fang Yi Xue Za Zhi. 2006;40:257–61 (In Chinese).
126. Ramamurthy T, Bag PK, Pal A, Bhattacharya SK, Bhattacharya MK, Shimada T, Takeda T, Karasawa T, Kurazono H, Takeda Y. Virulence patterns of *Vibrio cholerae* non-O1 strains

isolated from hospitalised patients with acute diarrhoea in Calcutta, India. J Med Microbiol. 1993;39:310–7.
127. Chandrasekhar MR, Krishna BV, Patil AB. Changing characteristics of *Vibrio cholerae*: emergence of multidrug resistance and non-O1, non-O139 serogroups. Southeast Asian J Trop Med Public Health. 2008;39:1092–7.
128. Chatterjee S, Ghosh K, Raychoudhuri A, Chowdhury G, Bhattacharya MK, Mukhopadhyay AK, Ramamurthy T, Bhattacharya SK, Klose KE, Nandy RK. Incidence, virulence factors, and clonality among clinical strains of non-O1, non-O139 *Vibrio cholerae* isolates from hospitalized diarrheal patients in Kolkata, India. J Clin Microbiol. 2009;47:1087–95.
129. Pal A, Ramamurthy T, Bhadra RK, Takeda T, Shimada T, Takeda Y, Nair GB, Pal SC, Chakrabarti S. Reassessment of the prevalence of heat-stable enterotoxin (NAG-ST) among environmental *Vibrio cholerae* non-O1 strains isolated from Calcutta, India, by using a NAG-ST DNA probe. Appl Environ Microbiol. 1992;58:2485–9.
130. Sharma A, Chaturvedi AN. Prevalence of virulence genes *(ctxA, stn, ompW* and *tcpA)* among non-O1 *Vibrio cholerae* isolated from fresh water environment. Int J Hyg Environ Health. 2006;209:521–6.
131. Kumar P, Peter WA, Thomas S. Detection of virulence genes in *Vibrio cholerae* isolated from aquatic environment in Kerala, Southern India. Appl Biochem Biotechnol. 2008;151: 256–62.
132. Jesudason MV, Balaji V, Mukundan U, Thomson CJ. Ecological study of *Vibrio cholerae* in Vellore. Epidemiol Infect. 2000;124:201–6.
133. Garg P, Nandy RK, Chaudhury P, Chowdhury NR, De K, Ramamurthy T, Yamasaki S, Bhattacharya SK, Takeda Y, Nair GB. Emergence of *Vibrio cholerae* O1 biotype El Tor serotype Inaba from the prevailing O1 Ogawa serotype strains in India. J Clin Microbiol. 2000;38:4249–53.
134. Raychoudhuri A, Chatterjee S, Pazhani GP, Nandy RK, Bhattacharya MK, Bhattacharya SK, Ramamurthy T, Mukhopadhyay AK. Molecular characterization of recent *Vibrio cholerae* O1, El Tor, Inaba strains isolated from hospitalized patients in Kolkata, India. J Infect. 2007;55:431–8.
135. Mohapatra SS, Ramachandran D, Mantri CK, Singh DV. Characterization of the genetic background of *Vibrio cholerae* O1 biotype El Tor serotype Inaba strains isolated in Trivandrum, southern India. J Med Microbiol. 2007;56:260–5.
136. Pourshafie MR, Grimont F, Saifi M, Grimont PA. Molecular epidemiological study of *Vibrio cholerae* isolates from infected patients in Teheran, Iran. J Med Microbiol. 2000;49: 1085–90.
137. Pourshafie M, Grimont F, Kohestani S, Grimont PA. A molecular and phenotypic study of *Vibrio cholerae* in Iran. J Med Microbiol. 2002;51:392–8.
138. Ansaruzzaman M, Bhuiyan NA, Safa A, Sultana M, McUamule A, Mondlane C, Wang XY, Deen JL, von Seidlein L, Clemens JD, Lucas M, Sack DA, Balakrish Nair G. Genetic diversity of El Tor strains of *Vibrio cholerae* O1 with hybrid traits isolated from Bangladesh and Mozambique. Int J Med Microbiol. 2007;297:443–9.
139. Nair GB, Qadri F, Holmgren J, Svennerholm AM, Safa A, Bhuiyan NA, Ahmad QS, Faruque SM, Faruque AS, Takeda Y, Sack DA. Cholera due to altered El Tor strains of *Vibrio cholerae* O1 in Bangladesh. J Clin Microbiol. 2006;44:4211–3.
140. Safa A, Sultana J, Dac Cam P, Mwansa JC, Kong RY. *Vibrio cholerae* O1 hybrid El Tor strains, Asia and Africa. Emerg Infect Dis. 2008;14:987–8.
141. Raychoudhuri A, Patra T, Kausik K, Ramamurthy T, Nandy RK, Takeda Y, Nair GB, Mukhopadhyay AK. Classical *ctxB* in *Vibrio cholerae* O1, Kolkata, India. Emerg Infect Dis. 2009;15:131–2.
142. González-Bonilla C, Valle-Valdez JG, Núñez-León A, Moguel-Pech L, Villanueva-Zamudio A. Seroepidemiology of cholera in Mexico. Rev Latinoam Microbiol. 1994;36:253–6 (In Spanish).

143. Hunt MD, Woodward WE, Keswick BH, Dupont HL. Seroepidemiology of cholera in Gulf coastal Texas. Appl Environ Microbiol. 1988;54:1673–7.
144. Harris JB, Larocque RC, Chowdhury F, Khan AI, Logvinenko T, Faruque AS, Ryan ET, Qadri F, Calderwood SB. Susceptibility to *Vibrio cholerae* infection in a cohort of household contacts of patients with cholera in Bangladesh. PLoS Negl Trop Dis. 2008;2:e221.
145. Qadri F, Wennerås C, Albert MJ, Hossain J, Mannoor K, Begum YA, Mohi G, Salam MA, Sack RB, Svennerholm AM. Comparison of immune responses in patients infected with *Vibrio cholerae* O139 and O1. Infect Immun. 1997;65:3571–6.
146. Chaignat CL, Monti V, Soepardi J, Petersen G, Sorensen E, Narain J, Kieny MP. Cholera in disasters: do vaccines prompt new hopes? Expert Rev Vaccines. 2008;7:431–5.
147. Islam Z, Maskery B, Nyamete A, Horowitz MS, Yunus M, Whittington D. Private demand for cholera vaccines in rural Matlab, Bangladesh. Health Policy. 2008;85: 184–95.
148. Kim D, Canh do G, Poulos C, Thoa le TK, Cook J, Hoa NT, Nyamete A, Thuy DT, Deen J, Clemens J, Thiem VD, Anh DD, Whittington D. Private demand for cholera vaccines in Hue, Vietnam. Value Health. 2008;11:119–28.
149. Trach DD, Cam PD, Ke NT, Rao MR, Dinh D, Hang PV, Hung NV, Canh DG, Thiem VD, Naficy A, Ivanoff B, Svennerholm AM, Holmgren J, Clemens JD. Investigations into the safety and immunogenicity of a killed oral cholera vaccine developed in Vietnam. Bull World Health Organ. 2002;80:2–8.
150. Thiem VD, Deen JL, von Seidlein L, Canh do G, Anh DD, Park JK, Ali M, Danovaro-Holliday MC, Son ND, Hoa NT, Holmgren J, Clemens JD. Long-term effectiveness against cholera of oral killed whole-cell vaccine produced in Vietnam. Vaccine. 2006;24: 4297–303.
151. Hill DR, Ford L, Lalloo DG. Oral cholera vaccines: use in clinical practice. Lancet Infect Dis. 2007;7:176–8.
152. Mahalanabis D, Lopez AL, Sur D, Deen J, Manna B, Kanungo S, von Seidlein L, Carbis R, Han SH, Shin SH, Attridge S, Rao R, Holmgren J, Clemens J, Bhattacharya SK. A randomized, placebo-controlled trial of the bivalent killed, whole-cell, oral cholera vaccine in adults and children in a cholera endemic area in Kolkata, India. PLoS One. 2008;3:e2323.
153. Mahalanabis D, Ramamurthy T, Nair GB, Ghosh A, Shaikh S, Sen B, Thungapathra M, Ghosh RK, Pazhani GP, Nandy RK, Jana S, Bhattacharya SK. Randomized placebo controlled human volunteer trial of a live oral cholera vaccine VA1.3 for safety and immune response. Vaccine. 2009;27:4850–6.
153a. Sur D, Lopez AL, Kanungo S, Paisely A, Manna B, Ali M, Niyogi SK, Park JK, Sarkar B, Puri MK, Kim DR, Deen JL, Holmgren J, Carbis R, Rao R, Nguyen TV, Donner A, Ganguly NK, Nair GB, Bhattacharya SK, Clemens JD. Efficacy and safety of a modified killed-whole-cell oral cholera vaccine in India: an interim analysis of a cluster-randomised, double-blind, placebo-controlled trail. Lancet. 2009;374:1694–702.
154. Ohashi K. Memoirs on Chlorodyne and Shinyaku. Yakushigaku Zasshi. 1999;34:89–96 (In Japanese).
155. Oi H, Matsuura D, Miyake M, Ueno M, Takai I, Yamamoto T, Kubo M, Moss J, Noda M. Identification in traditional herbal medications and confirmation by synthesis of factors that inhibit cholera toxin-induced fluid accumulation. Proc Natl Acad Sci USA. 2002;99: 3042–6.
156. Kataoka S. Functional effects of Japanese style fermented soy sauce (shoyu) and its components. J Biosci Bioeng. 2005;100:227–34.
157. Brown MS, Burns CE, Hellings PJ. Health care in China. Nurse Pract. 1984;39:42–4, 46.
158. Bibeau G. From China to Africa: the same impossible synthesis between traditional and western medicines. Soc Sci Med. 1985;21:937–43.
159. Winardi B, Adhyatma M. Diarrhoeal diseases among refugees in Indonesia. Southeast Asian J Trop Med Public Health. 1982;13:361–5.

160. Glass RI, Alim AR, Eusof A, Snyder JD, Jusuf B, Anwar S, Bakri Z, Helmi C, Winardi B. Cholera in Indonesia: epidemiologic studies of transmission in Aceh Province. Am J Trop Med Hyg. 1984;33:933–9.
161. Forbes GI, Lockhart JD, Robertson MJ, Allan WG. Cholera case investigation and the detection and treatment of cholera carriers in Hong Kong. Bull World Health Organ. 1968;39:381–8.
162. Swaddiwudhipong W, Akarasewi P, Chayaniyayodhin T, Kunasol P, Foy HM. Several sporadic outbreaks of El Tor cholera in Sunpathong, Chiang Mai, September–October, 1987. J Med Assoc Thai. 1989;72:583–8.
163. Wang ZS, Shepard DS, Zhu YC, Cash RA, Zhao RJ, Zhu ZX, Shen FM. Reduction of enteric infectious disease in rural China by providing deep-well tap water. Bull World Health Organ. 1989;67:171–80.
164. Isa AR, Othman WM, Ishak A. Cholera outbreak in Tumpat, Kelantan-1990. Med J Malaysia. 1990;45:187–93.
165. Huq A, Xu B, Chowdhury MA, Islam MS, Montilla R, Colwell RR. A simple filtration method to remove plankton-associated *Vibrio cholerae* in raw water supplies in developing countries. Appl Environ Microbiol. 1996;62:2508–12.
166. Colwell RR, Huq A, Islam MS, Aziz KM, Yunus M, Khan NH, Mahmud A, Sack RB, Nair GB, Chakraborty J, Sack DA, Russek-Cohen E. Reduction of cholera in Bangladeshi villages by simple filtration. Proc Natl Acad Sci USA. 2003;100:1051–5.
167. Liew KB, Lepesteur M. Performance of the rural health improvement scheme in reducing the incidence of waterborne diseases in rural Sarawak, Malaysia. Trans R Soc Trop Med Hyg. 2006;100:949–55.
168. Dalsgaard A, Forslund A, Tam NV, Vinh DX, Cam PD. Cholera in Vietnam: changes in genotypes and emergence of class I integrons containing aminoglycoside resistance gene cassettes in *Vibrio cholerae* O1 strains isolated from 1979 to 1996. J Clin Microbiol. 1999;37:734–41.
169. Korthuis PT, Jones TR, Lesmana M, Clark SM, Okoseray M, Ingkokusumo G, Wignall FS. An outbreak of El Tor cholera associated with a tribal funeral in Irian Jaya, Indonesia. Southeast Asian J Trop Med Public Health. 1998;29:550–4.
170. Swaddiwudhipong W, Akarasewi P, Chayaniyayodhin T, Kunasol P, Foy HM. A cholera outbreak associated with eating uncooked pork in Thailand. J Diarrhoeal Dis Res. 1990;8:94–6.
171. Hoge CW, Bodhidatta L, Echeverria P, Deesuwan M, Kitporka P. Epidemiologic study of *Vibrio cholerae* O1 and O139 in Thailand: at the advancing edge of the eighth pandemic. Am J Epidemiol. 1996;143:263–8.
172. Golding J, Emmett PM, Rogers IS. Gastroenteritis, diarrhoea and breast feeding. Early Hum Dev. 1997;49:S83–103.

Chapter 3
Endemic and Epidemic Cholera in Africa

Samba Sow, Martin Antonio, Joe O. Oundo, Inacio Mandomando, and T. Ramamurthy

Abstract Cholera had entered the African continent during the late 1880s mainly through trade and travel from the Asian region. The 7th cholera pandemic had started from the early 1970s and is rampant till to-date claiming many lives. Though the endemicity of cholera is common in many African and Asian countries, the morbidity and mortality are relatively high in Africa. This has been attributed mainly due to lack of health and hygiene of the population, scarcity of water, prevailing multidrug-resistant strains of *Vibrio cholerae* O1, and to a lesser extent the management of cholera in the early 1990s. Molecular epidemiological studies revealed that there is a propensity of the pathogen to change its clonality that has been attributed to the recurrent infection and waves of cholera epidemics in many African countries. One of the mysteries is why the spread of *V. cholerae* O139 serogroup is restricted only to Asian countries despite the geographical proximity and common mode of transmission. This chapter reviews cholera status in Africa viewed through many aspects of the epidemiology.

3.1 Introduction

Cholera is undulating in Africa and had spread to new countries where the disease was not reported during the last few decades. The movements of cholera epidemics are rapid due to increased travel and trade practices. Despite the best efforts, cholera is still difficult to control in many African countries. Though the causative factors remain the same, the organism *Vibrio cholerae* O1 has undergone several changes as evidenced by several genetic analyses. This chapter summarizes the recent epidemiological and molecular studies on cholera conducted in Africa, in which some of the aspects are unique for this continent.

T. Ramamurthy (✉)
National Institute of Cholera and Enteric Diseases, Kolkata 700010, India
e-mail: rama1murthy@yahoo.com

T. Ramamurthy, S.K. Bhattacharya (eds.), *Epidemiological and Molecular Aspects on Cholera*, Infectious Disease, DOI 10.1007/978-1-60327-265-0_3,
© Springer Science+Business Media, LLC 2011

3.2 Epidemics and Outbreaks

Cholera is a major health burden in many developing countries and its endemicity is mostly confined to Africa and Asia. A treatment facility-based cholera surveillance showed that annual incidence of the disease is low in Jakarta (Indonesia, 0.5/1,000 population), moderate in Kolkata (India, 1.6/1,000), and high in Beira (Africa, 4/1,000) [1]. Historical evidence showed that the 6th pandemic of cholera had reached East Africa and the mode of spread was seemingly associated with caravans and trade ships. Between 1881 and 1893, the epidemic had spread in East (Zanzibar), Central, North (Sudan), and Western parts of Africa [2–4]. The 7th El Tor pandemic had gained its entry into West Africa (Guinea) during August 1970, maybe from asymptomatic travelers from the Asian continent [4, 5]. Within a year, 25 African countries were affected by cholera with a high mortality of 16% [6]. In the following years (1972–1991), cholera was rampant in most of the African countries with case fatality ranging from 4 to 12% [6]. The African continent faced yet another big cholera epidemic in 1991 and during this year, about 20 countries reported large outbreaks amounting to more than 150,000 cases and 13,000 deaths [7]. Southern and Eastern parts (Zambia, Mozambique, Malawi, and Angola) and Western parts of Africa became the epidemic foci in 1991. After several years, Benin, Burkina Faso, Chad, Nigeria, and Togo reported high incidence of cholera during 1991 [7]. The case fatality ratio in Nigeria alone was 13%. However, in 1992, the number of reported cholera cases in Africa declined slightly with about 91,000 cases and 5,000 deaths [8].

The 7th El Tor cholera pandemic has been registered in about 175 countries and the epidemic progressed into waves of cholera with higher mortality rate. As shown in Table 3.1, epidemics of cholera are very dynamic and the list of affected countries with high morbidity and mortality of cholera is not the same during 2000–2007. Surveillance and disease burden estimates in cholera-affected countries are important components of epidemiology, which help in making timely interventions including vaccination strategies.

Based on the 632 cholera reports covering a period of about 10 years (1995–2005), it is evident that about 88% originated in sub-Saharan Africa and the rest from Southeast Asia [9]. Reintroduction of cholera in the African continent was evident from 1970. After the 1985 epidemic, cholera reemerged in Zimbabwe in 1992 following a severe drought and influx of refugees from Mozambique [10]. In West Africa, cholera epidemic occurred in Togo during 1970–1973, affecting more than 1,000 people with CFR of 4–10% [11]. Since the first outbreak during 1970 in Guinea, cholera recurred every 8 years till 1994 [12]. About the same time, cholera affected Benin [13], Burundi, Zimbabwe [14], Nigeria [15], and Senegal [16]. *V. cholerae* O1 was introduced in northern Somalia from Ethiopia during the early 1980s [17]. Recurrent cholera epidemics were reported during rainy hot seasons since 1987 in Angola [18]. Incidence of other pathogens during cholera epidemics is an important factor that should be considered. Apart from *V. cholerae* O1, the involvements of non-cholera vibrios and *V. parahaemolyticus* were identified during

Table 3.1 Incidence of cholera during 2000–2007 in the African continent

Year	No of cases (deaths)	Mostly affected country[a]	Remark	Reference
2000	118,932 (4,610)	DR Congo, Madagascar, Mozambique, Somalia, South Africa, Tanzania	Threefold decrease compared to 1999. CFR 3.9% (vs 4.2% in 1999)	[126]
2001	173,359 (2,590)	Chad, Cote d'Ivoires, DR Congo, Ghana, Madagascar, Mozambique, South Africa, Swaziland	Increase of 46% compared with 2000. CFR 1.49%	[127]
2002	137,866 (4,551)	Malawi, Mozambique, Nigeria, South Africa, Tanzania	Excluding South Africa, increase of 70% in number of cases compared with 2001. CFR 3.5%	[128]
2003	108,067 (1,884)	DR Congo, Liberia, Mozambique, Somalia, Uganda	Decrease of 21% compared with 2002. CFR 1.74%	[129]
2004	95,560 (2,331)	Cameroon, Chad, DR Congo, Mozambique, Somalia, Tanzania, Zambia	Slightly lower compared to previous years. CRF 2.4%	[130]
2005	125,082 (2,230)	DR Congo, Guinea, Guinea-Bissau, Mauritania, Senegal, Uganda	Increase of 31% compared with 2004. CFR 1.78%	[131]
2006	234,349 (6,303)	Angola, DR Congo, Ethiopia, Sudan, UR Tanzania	Increase of 87% compared with 2005 and reached the levels of 1990s. CFR 2.69%	[34]
2007	166,583 (3,994)	Angola, DR Congo, Ethiopia, Sudan, Sao Tome, and Principe	Decrease of 29% compared to 2006, but 46% increase in the mean number of cases reported during 2002–2005. CRF 2.4%	[35]

[a]Countries in alphabetical order with more than 4,000 cases during indicated years

the 1991 cholera epidemic in Nigeria [19]. *Shigella dysenteriae* type 1 was identified during the 1994 cholera epidemic in Zaire [20] and in Kenya [21].

In June 1998, cholera epidemic occurred in a Mozambican refugee population in southern Malawi [22]. The 1996 outbreak in Nigeria affected 1,384 individuals with a CFR of 5.3%, with higher number of cases during the rainy season [15]. Children were the most affected among all age groups (22% of the total cases) in Ibadan, Nigeria [23]. Based on the reported cases, cholera first appeared in Burundi,

Zaire, and Congo during 1978–1979 [24] and in South Africa in 1980 [25]. A total of 67,738 cases and 3,666 deaths (CFR 5.4%) were reported between 2000 and 2005 in eastern provinces of DR Congo [26]. Epidemic cholera in Guinea-Bissau affected more than 1,200 people during 1999 with CFR of 3.7% [27]. In this epidemic, most of the deaths were due to delayed rehydration or overhydration of the affected patients. Cholera epidemic became an annual event in Chad since the late 1990s with peak incidence between dry and rainy seasons (March–June) [28]. The infection rate and the seasonality of cholera are not constant in most of the African countries. The 7th cholera pandemic reached Madagascar in March 1999, 30 years after its appearance in East Africa [29]. In all these sites, children less than 5 years are the most affected age group. A study in Senegal conducted during 2004–2005 showed that the association of cholera and pregnancy presents high risks for the fetus and mother, as adverse events occurred due to cholera mostly leading to abortions, premature childbirths, and maternal deaths [30].

At the time of 7th pandemic during the 1960s, cholera had a strong hold in Asian countries. In the early 1970s, the pandemic reached sub-Saharan Africa and cholera remained entrenched for many decades. Thirty-one (78%) of the 40 countries, which reported cholera cases to WHO in 2005, were from sub-Saharan Africa, with the incidence rate of 166 cases/million population. This figure was 95 and 16,600 times higher than that from Asia and Latin America, respectively [31]. Recently, cholera outbreaks spread to newer African countries such as Namibia [32]. Many studies have shown that the incidence of cholera is consistently high in many African countries. The incidence and deaths due to cholera are high in Kenya [33]. Based on the Program for Monitoring Emerging Diseases (ProMED) data collected between 1995 and 2005, sub-Saharan Africa accounted for 88% of the total number of cholera outbreaks with highest multiple outbreaks in DRC Congo, Western Uganda, and South Africa, where the outbreak reports were confined to Mozambique and its borders [9]. In 2006, 234,349 cholera cases were reported from 33 African countries with 6,303 deaths (CFR 2.7%), which resembled the large epidemics of cholera during the 1990s [34]. In the following year, though the numbers of affected countries are the same, a 25% decrease in the number of cases of cholera was reported, with overall reduction of CFR from 2.7 to 2.27% in 2007 [35]. Between 2000 and 2007, the CRF remained >5% in 23 countries (Fig 3.1), and during the same period, 10 countries reported emergence and reemergence of cholera (Fig 3.2).

3.3 Risk Factors and Modes of Transmission of Cholera

It is known that transmission of cholera occurs via environments, contaminated foods, carriers of the infection, and unhygienic practices, among others. Waterborne transmission of cholera has been well documented in many studies. During 1981–1982, cholera in South Africa was associated with consumption of contaminated waters [36]. The 1978 and 1992 cholera outbreaks in Great Rift Valley in Burundi were associated with use of lake water for many purposes [37]. In Tanzania,

Fig. 3.1 High CFR (>5) reported to WHO by the African countries during 2000–2007

a prospective hospital-based matched case–control study showed that bathing in river, long distance to water source, and eating dried fish were significant risk factors for cholera [38]. The role of rainfall in the transmission and seasonality of cholera has also been established in Tanzania [39]. The role of environment and climatic factors was shown to aid spread of cholera in African countries [40, 41]. In DR Congo, lake areas have been found to be the source of cholera outbreaks [26]. Floods were also responsible for cholera outbreaks in Djibouti in 1993–1994 and Mozambique in 2000 [42, 43]. These outbreaks were favored by contaminated surface water and disturbances in the distribution of safe drinking water. In most of the epidemiological findings, lack of supply of safe drinking water and storage of water in wide-mouthed containers were significantly associated with the incidence

Fig. 3.2 Emerging and reemerging of cholera in African countries reported to WHO during 2000–2007 with number of cases and deaths in parenthesis

of cholera [33]. The nature of storage vessels plays a crucial role in the survival and dissemination of *V. cholerae*. In a study it was shown that the vibrios survived long (about 22 days) in rusted iron drums than the clay pots (up to 5 days) [44].

Case–control studies showed that consumption of leftover foods is one of the major factors for cholera infection in many epidemics, e.g., leftover crabs in Guinea-Bissau [45], millet gruel-mediated infection in drought-affected areas in Mali [46], and peanut sauce contamination in Guinea [47]. Domestic cholera has been reported in Germany due to consumption of fresh fish imported from Nigeria [48] and fresh sorrel from West Africa in France [49]. Street-vended water and not washing hands with soap before eating food are possible reasons for the 1995–1996 cholera outbreaks in Kano State, Nigeria [50]. The drinking water sold by water vendors was associated with an increased risk of contacting cholera in Lebowa [51]. Early outbreaks of cholera in Katsina, Nigeria, were associated with fecal contamination of well waters and water from sellers [52]. Wells and piped water seemed the principal cause of water-borne cholera outbreaks ($p<0.02$) compared to borehole wells during 1995–2001 [53]. In a study, consumption of raw vegetables was reported as the causative agent of cholera ($p=0.003$) in Lusaka, Zambia, during 2003–2004 [54]. The Senegal cholera outbreak during 2004 might have been due to consumption of contaminated food or water [55]. Large-scale farming and mining activities led to eutrophication and chemical pollution in Bindura, Zimbabwe. Studies conducted in this area during 2000–2003 with groundwater supply through borehole indicated higher fecal coliforms and aerobic bacterial counts, and these factors correlated well with the incidences of cholera in the same region [56]. Matched-pair case–control study revealed that the transmission of cholera in Lebowa was due to consumption of water from the Gumpies River [51].

Cholera outbreaks are directly related with sanitation, personal hygiene, and good supply of potable water. Due to political unrest, large population movements have taken place in many countries in Africa since the 1970s. Outbreaks of cholera have occurred in many refugees camps due to overcrowding, poor sanitation, water supply, and malnutrition in the affected population. Many deaths were reported due to inappropriate medical care due to either dehydration or overhydration [57]. These conditions are always deficient in refugee camps and slums [58]. In Malawi, cholera cases were reported with high attack rate (2.6%) with CFR of 3.3% during 1988 [59]. This outbreak was related to consumption of contaminated shallow waters. In 1990, an epidemic of cholera continued among Mozambican refugees in Malawi causing 1,931 cases and 68 deaths (CFR 3.5%). Several factors were identified in this outbreak including use of river water and eating stale food [60]. The most important African cholera is the 1994 epidemic in Rwanda. More than 1 million refugees fled to the North Kivu region in Zaire. The total cholera cases in this epidemic was 36,471 and 60% of the deaths were related to cholera alone. Of the many factors, scarcity of water was the main one and the average available water was evaluated as 200 ml per day per person. Cholera struck again in Rwanda in 1994 and 1996 in many refuges camps. About 553,000 refugees repatriated from Zaire in November 1996 and in this outbreak the mortality was low (3,586 deaths), perhaps due to acquired immunity from the 1994 outbreak [61]. The other factors responsible for cholera deaths among Rwandan refugees include emergence of multidrug-resistant strains of *V. cholerae* O1, inadequate use of oral rehydration therapy, inappropriate intravenous fluids, and most importantly poor experience of health workers in management of cholera [20]. In addition, *Shigella dysenteriae* type 1 was also involved in this outbreak in Zaire [62] and in Kenya [21].

Overcrowded and unhygienic conditions lead to cholera outbreaks in hospital settings [63, 64]. In hospital settings, spread of cholera occurs through carriers [65]. Generally, person-to-person transmission of cholera is not considered to cause epidemics, as the carrier rate should be enormously high. In Nigeria, the fecal excretion of *V. cholerae* during convalescence of cholera patients was 2–28 weeks with mean excretion of about 10^3 vibrios per gram of the stool [66]. Many epidemiological studies showed that family contacts with index cases play a crucial role in the transmission of cholera during outbreaks [67]. In Tanzanian and Kenya, traditional ritual and religious practices, especially the procedure of burying the victims, played a key role in transmission of the cholera [68, 69]. Handling the bodies of cholera victims was shown as one of the factors in spreading the infection. Epidemiological investigations carried out during the 1994 cholera epidemic in Guinea-Bissau indicated that cholera was strongly associated with eating at a funeral with non-disinfected corpse or transporting and washing the dead bodies [70]. Cholera outbreak (111 cases and 11 deaths) occurred in a village of Guinea-Bissau that was associated with eating a rice-based meal made by the persons who had prepared a cholera victim's body for burial [45]. Cholera and other infectious diseases were introduced in Mauritius due to movement of French soldiers and laborers from the Indian subcontinent [71]. In Somaliland, introduction of cholera and other infectious diseases was due to Imperial army movement and displacement of native people [72]. Contribution of travel to cholera endemic areas is a minor factor in transmission of the disease to non-cholera areas. Studies based on the imported cases in Europe and North America during 1975 and 1981 showed that the incidence per journey for foreign travelers visiting Africa or Asia was about 1 in 500,000 [73, 74].

The relevance of various climatic conditions as supportive factors for cholera epidemics is not fully established. Data collected from remotely sensed weather parameters and cholera epidemics in many regions have shown to be useful in the development of early-warning systems and epidemic preparedness. El Niño/Southern Oscillation (ENSO) events trigger cholera and other tropical diseases in many coastal areas around the world. As proposed in many findings, timely detection of global climate anomalies has significant implications for public health. The El Niño phenomenon was considered for major cholera epidemics in Djibouti, Kenya, Mozambique, Somalia, Uganda, and Tanzania in 1997 [75]. El Niño phenomenon causing increased rainfall and flooding has been linked to outbreaks of cholera in Uganda during 2002–2003 [76]. The Lake Victoria Basin study (East Africa) supported with several hydrological features indicated the association of cholera epidemics and El Niño effect [77]. This study showed that the low economic status of the population living around the lake and poor healthcare system and abnormal high temperature in two consecutive years followed by slight cooling triggered epidemics of cholera. The coastal cholera transmission hypothesis was further supported with studies conducted in KwaZulu-Natal, South Africa, where the sea surface temperature and rainfall showed strong association with the cholera outbreaks [40]. In Calabar, Nigeria, the incidence of cholera and other diarrheal diseases mostly occurred during the dry season followed by subsidence at the onset of rainy season [78]. In hilly regions (Mwanza) of Tanzania, the incidence of cholera was associated with heavy rains coupled with poor sanitary conditions and sewage disposal system [79]. The South African cholera epidemics occurred during summer rainfalls of 1980–1987 [80]. A 10-year study (1981–1990) showed that the cholera peak was during hot and rainy season in Dakar, Senegal [81]. A natural disaster such as volcanic eruption was also reported to be associated with cholera outbreak in Congo during 2002 [43]. In the African setting, the prime factors identified using ProMED data (1995–2005) were heavy rainfall and flooding followed by refugees or internally displaced persons setting [9].

Multivariate analysis showed that the 1997 cholera epidemic in Western Kenya was due to consumption of water from Lake Victoria, sharing food with persons with diarrhea, and attending funeral feasts of cholera victims [82]. In a hypothetical model it was shown that spread of cholera outbreaks due to *V. cholerae* O1 in Africa (1970–1971, 2005–2006) and *V. cholerae* O139 in India (1992–93) was through wind in specific directions [83]. This study emphasized the role of flying insects (aeroplakton) as one of the factors in the dissemination of cholera, as they were carried along with the wind directions.

The association of parasite infection and cholera was not studied in detail. Some of the Tanzanian children who had not undergone ascariasis treatment were coinfected with cholera and vomited or passed *Ascaris lumbricoides* [79]. Immune suppression caused by HIV might be a predisposal factor for cholera and other infectious diseases. A case–control study conducted in Mozambique has shown the association between HIV infection and increased risk for cholera [84].

3.4 The Organism

Based on the phenotypic markers, *V. cholerae* O1 strains are divided into classical and El Tor biotypes [85]. The 5th and 6th cholera pandemics were caused by the classical biotype while the El Tor has become dominant since its first appearance in 1961, replacing the classical biotype. In addition to the phenotypic traits, several molecular markers have recently been used for the identification of biotypes, which include toxin co-regulated pilus (TCP) gene, *rtxC* gene that encodes part of the repeat in toxin (RTX). Using these molecular markers hybrid El Tor strain having classical cholera toxin was identified first in Bangladesh and Mozambique [86–88]. The El Tor hybrid strain was also identified in India [89] and other Asian and African countries [88]. The clinical features of the disease caused by El Tor hybrid expressing classical cholera toxin appear more severe and associated with large cholera outbreaks in Asian and African countries. It is known that survival capacity of El Tor strains is comparatively high and that the classical cholera toxin is more severe. These two best combinations might have supported the El Tor hybrids to survive and spread the disease with high morbidity and mortality that were never seen in recent years.

V. cholerae O1 Inaba was associated with African cholera for many years [33]. Early cholera outbreaks (1971–1975) in Algeria were caused by the serotype Ogawa [90]. Cholera outbreak in Mozambican refugee camp, Malawi, was caused by *V. cholerae* O1 Inaba during 1988 [59]. In Kano, Nigeria, the Inaba serotype prevailed for a long time (1995–1999) [53]. In Zaria, Nigeria, Hikojima serotype that reacts with both Ogawa and Inaba antisera was prevalent from 1976 to 1978, but Ogawa became dominant from 1984 to 1986 [91]. The classical biotype of *V. cholerae* O1 was reported in Calabar, Nigeria, during 1988 [78].

3.5 Seroepidemiology

Vibriocidal antibody titer in cholera patients, convalescent cases, and healthy carriers is a useful marker in seroepidemiological studies as it reflects acquirement of immunity against *V. cholerae.* The seropositivity of healthy contacts remained high during and after cholera epidemics in Algeria (43.3%), Constantine (53%), and Mali (46%), indicating persistence of vibriocidal antibodies in cholera endemic areas for many years [92]. Prevalence of vibriocidal antibodies (38%) was not significant between cases, carriers, and contacts in Algeria [92]. In addition, the healthy contacts and children born during non-cholera epidemic period showed that asymptomatic infection is frequent and that the pathogen may be circulating in populations between the epidemics [92, 93].

3.6 Use of Antimicrobials

Antimicrobial resistance is an integral mechanism developed by the pathogens, usually due to excessive use of drug(s) for the treatment of diseases. Since the emergence of antimicrobial resistance may significantly affect the strategies for controlling cholera, continuous monitoring of antimicrobial resistance is important. The toxigenic *V. cholerae* is generally susceptible to many antimicrobials including tetracycline/doxycycline, amoxicillin, and fluoroquinolones. During the early stages of cholera epidemic (1977–1978) in Tanzania, *V. cholerae* O1 strains were susceptible to tetracycline. Due to therapeutic and prophylactic use of this drug, resistance developed in a large proportion (75%) and there were treatment failures when the drug was used [94]. The Ministry of Health used 1,788 kg of tetracycline during this period for treatment and prophylaxis [94]. *V. cholerae* O1 isolated in Kenya during 1982–1985 were resistant to tetracycline, ampicillin, and trimethoprim/sulfamethoxazole [95]. The strains isolated from Nyanza Province, Kenya, in 1983 were mostly sensitive to chloramphenicol and nalidixic acid but 75% of the strains were resistant to tetracycline, streptomycin, and ampicillin [96]. The O1 strains isolated during the 1983 cholera epidemics in Somali were resistant to many drugs, especially ampicillin, tetracycline, and kanamycin, which were replaced by susceptible strains to these drugs during 1986 [17]. Most of the O1 strains isolated between 1987 and 1990 in Angola were resistant to ampicillin, streptomycin, and trimethoprim/sulfamethoxazole but susceptible to chloramphenicol and tetracycline [97]. In Senegal O1 strains isolated from 1981 to 1990 were also susceptible to sulfamids and tetracycline [81]. *V. cholerae* O1 Ogawa isolated during the 1991 cholera epidemic in Benin are widely susceptible to tetracycline, sulfamid, and ampicillin [98]. In Guinea-Bissau and Senegal, the newly emerged clones of *V. cholerae* O1 in 1997 were susceptible for tetracycline but remained resistant to trimethoprim/sulfamethoxazole and chloramphenicol [99]. The epidemic *V. cholerae* O1 strains isolated from Rwandan refugees in Zaire were resistant to several antimicrobials used at that time including tetracycline, aminopenicillins, and trimethoprim/sulfamethoxazole. Only fluoroquinolones remained active on these strains, but the use of this drug in outbreak-affected areas is prohibitive [62].

When the *V. cholerae* O1 strains isolated between 1994 and 1996 from six countries from Eastern Africa region (Kenya, Sudan, Somalia, Tanzania, Rwanda, and Mogadishu) were tested for antimicrobial susceptibility, there was no uniformity in the pattern of resistance [100]. The main observation in this finding was that the strains from Kenya and Sudan were susceptible to tetracycline, whereas the strains from Tanzania and Rwanda were resistant. The Kenyan and Somalian strains gained resistance for chloramphenicol and trimethoprim/sulfamethoxazole during 1996 but remained susceptible to nalidixic acid and erythromycin [100].

In *V. cholerae* O1, low level of resistance to tetracycline was recorded (2–3%) during 1990–1991 in Zambia and in the following years, the resistance increased to this drug (95%) as well as to other important drugs such as chloramphenicol (78%), doxycycline (70%), and trimethoprim/sulfamethoxazole (97%) [101]. Following the replacement of erythromycin for tetracycline in treating the cholera cases, there was a significant drop in the tetracycline resistance [101]. The Kenyan strains associated with large outbreaks in 1998–1999 were susceptible to tetracycline, but were resistant to chloramphenicol, streptomycin, and trimethoprim/sulfamethoxazole [102]. Most of the *V. cholerae* O1 strains from cholera outbreaks are resistant to chloramphenicol, streptomycin, and trimethoprim/sulfamethoxazole [55]. *V. cholerae* O1 isolated from cholera outbreaks from Dakar, Senegal, during 1995–1996 [103] and Dar Es Salaam, Tanzania, during 1997–1999 [104] were resistant to chloramphenicol, streptomycin, and trimethoprim/sulfamethoxazole. Since the *V. cholerae* O1 Ogawa strains are mostly resistant to tetracycline, trimethoprim/sulfamethoxazole (97% each), and chloramphenicol (58%), quinolones and third-generation cephalosporins are being considered as supportive therapy for cholera in Mozambique [105].

Antibiotic prophylaxis is usually not part of cholera interventions in many countries because it does not prevent contamination and is limited by contraindications, costs, mode of administration, and spread of resistant organisms. In Douala, Cameroon, systematic use of chemoprophylaxis since 1983 has led to the selection of *V. cholerae* O1 strains resistant to sulfamide and tetracycline [106]. In the same area, antimicrobials were used to treat the patients and contacts during 2004. The data generated from a study have shown that the rate of cholera in contact group fell from 30 to 0.2% [107]. It is recommended that the use of antimicrobials should be restricted to close contacts, as this group is most likely to get the infection. In a comprehensive study with cases and contacts, it was shown that development of resistance to the antimicrobials seems unlikely especially when they were under controlled use [108, 109]. However, short-term studies might not indicate the acquisition of resistance genes or mutations in the encoding genes.

In most rural populations in many African countries, traditional medicine (mostly from herbal products) is still being practiced for the treatment of cholera and other diarrheal diseases. Ethnobotanical studies conducted on 21 medicinal plants showed at least two of them (*Punica granatum* and *Indigofera daleoides*) with strong antimicrobial properties [110]. Of several medicinal plants tested, *Terminalia avicennioides* from Nigeria showed higher vibriocidal activity than the others [111]. Use of traditional medicine should be approached scientifically, since disease such as cholera can kill a patient if not treated properly.

3.7 Molecular Findings

Molecular epidemiological studies on *V. cholerae* are being carried out mostly using randomly amplified polymorphic DNA (RAPD) patterns, ribotyping, and pulsed-field gel electrophoresis (PFGE). The amplified fragment length polymorphisms (AFLP) identified two major clusters with *V. cholerae* O1 El Tor strains isolated during the 1960s and 1970s and the other strains from 1980s to 1990s, indicating the change in the clonality of El Tor vibrios over a period of 30 years [112]. The AFLP also established to check independent introduction of a strain that caused large outbreaks during the 1970s and 1990s in Africa. The outbreak strains from Southern and Eastern Africa are markedly different from those from Western Africa [112]. In most of the African and Asian O1 strains, the ribotype B5a was found to be common; however, a new ribotype B27 with a toxinogenotype TB31 was identified in Senegal during the 1996 cholera outbreak [113]. The Guinea-Bissau cholera outbreak preceded the one from Senegal with ribotype B5a, supporting the view that infection has spread from Guinea-Bissau [99]. Ribotyping analysis showed that the cholera epidemic in Guinea-Bissau during 1994–1995 was due to the introduction of a new *V. cholerae* O1 strain that was distinct from the 1987 strains [114].

RAPD results showed that the strains associated with the largest cholera epidemics in Kenya (Eastern Africa) during 1998–1999 were clonally related with early strains identified from several outbreaks in West Africa during 1994 [102]. Using PFGE, it was shown that a single clone of *V. cholerae* caused several cholera outbreaks in Kenya during 2005 [33]. The serotype switchover from O1 Ogawa to Inaba and vice versa is common in many cholera endemic areas. Studies conducted in South Africa with epidemic strains of Ogawa and Inaba serotypes showed existence of different clones of *V. cholerae* O1 during cholera epidemics in 2001–2002 and from 1980 to 1987, though the 2001–2002 strains were related irrespective of serotypes [115]. *V. cholerae* O1 Inaba strains isolated from Namibia outbreaks displayed identical PFGE profile [32].

During the mid-2000s, new hybrid strains of *V. cholerae* O1 that do not belong to either classical or El Tor biotypes have been identified in Bangladesh and Mozambique [86, 87]. The hybrid strains gave *rstR* gene allele of classical type. Interestingly, the hybrid strains isolated in Bangladesh and Mozambique are clonally different from each other and also from the classical and El Tor biotypes as detected in the PFGE [86].

3.8 Strategies to Curtail Cholera Outbreaks

In slum settings, implementation of proper hygienic practices might decrease the risk of *V. cholerae* infection. Studies conducted in Guinea-Bissau and Guinea showed the importance of hand washing in prevention of cholera at the family level

[45, 47]. In addition to clean drinking water supply and not using hands for eating, using limes in the main meal and washing hand with soap showed protective effect in the initial phase of cholera epidemic in Guinea-Bissau [116, 117]. In Nigeria, simple strategies such as accurate diagnosis at the laboratory, registration of suspected cases, case management, and public health measures targeting personal hygiene and water treatment showed a reduction in case fatality rates to 15% in 1995–1996, 5% in 1997, and 2% in 1999 [53]. Considering the prevalence of HIV in many cholera endemic African regions, the usefulness of cholera vaccine was under dispute for many years. A recent mass cholera vaccine trial conducted in Beira, Mozambique, was found to be protective (78%, $p=0.004$) against cholera in all age groups during the 2004 outbreak [118]. Simple methods such as solar heating of drinking water by exposing the stored water in a clear plastic bottle under sunlight can be adopted easily by the households. Adopting this strategy, a controlled study has shown drastic reduction in the risk of cholera in Kenyan children [119]. As evidenced from a study conducted in Guinea-Bissau, lactating mothers suffering from cholera should be encouraged to breastfeed their children, as this might reduce the risk of illness in children [120]. Household contamination is one of the major factors for cholera and other enteric diseases as the stored water is severely exposed to contamination during use. Use of special containers that prevent direct contact has been shown to reduce risk of infection by cholera [121]. In-home chlorination seems very effective as this preventive method reduced the burden of cholera for 3 years (2000–2003) in Zambia [122].

Despite the improved safety and effectiveness of oral cholera vaccine, its use in preventing epidemics, especially in refugee settings, is still controversial. Managing cholera at the refugee camps with preemptive therapy such as mass vaccination is cost-effective [123]. In situations where no rehydration therapy is suitable for severe cholera, the vaccination appears to cost less than the reactive therapy [123]. However, mass vaccination might disrupt other priority interventions [124]. The recent case–control oral cholera vaccine (WC/rbs) studies conducted in Mozambique (Beira) demonstrated that the protective efficacy was 78% [118].

3.9 Conclusions

While preparing this manuscript, Zimbabwe has declared an emergency and appealed for international help to battle a cholera outbreak that has killed 2928 people, (CFR 4.88%), with 60,055 reported cases of the disease since August 2008 [125]. The disease has spread to neighboring South Africa, Mozambique, Zambia, and Botswana. Without timely implementation of proper management strategies, combating cholera will be a big problem in the future. Cholera caused by the newly emerged serogroup O139 has spread like a forest fire in many Asian countries in 1992–1993. However, infection caused by this serogroup has not been reported in Africa, despite the geographical proximity and common mode of transmission. Apart from the geographical barrier, other mechanisms such as immunological protection and absence of carriers of O139 in the population are considered some of the important points for its absence in the African continent.

References

1. Deen JL, von Seidlein L, Sur D, Agtini M, Lucas ME, Lopez AL, Kim DR, Ali M, Clemens JD. The high burden of cholera in children: comparison of incidence from endemic areas in Asia and Africa. PLoS Negl Trop Dis. 2008;2:173.
2. Pollitzer RC. Monograph No. 43. Geneva: World Health Organization; 1959.
3. Stock RF. Cholera in Africa: diffusion of the disease 1970–1975 with particular emphasis on West Africa. Special report 3. London: International African Institute; 1976.
4. Swerdlow DL, Isaäcson M. The epidemiology of cholera in Africa. In: *Vibrio cholerae* and cholera: molecular to global perspective. Washington, USA: ASM Press; 1994. pp 297–307.
5. Isaäcson M, Clarke KR, Ellacombe GH, Smit WA, Smit P, Koornhof HJ, Smith LS, Kriel LJ. The recent cholera outbreak in the South African gold mining industry: a preliminary report. S Aft Med J. 1974;48:2557–60.
6. World Health Organization. Cholera in Africa. Wkly Epidemiol Rec. 1991;66:305–11.
7. World Health Organization. Cholera in Africa. Wkly Epidemiol Rec. 1992;67:253–60.
8. World Health Organization. Cholera in Africa. Wkly Epidemiol Rec. 1993;68:149–55.
9. Griffith DC, Kelly-Hope LA, Miller MA. Review of reported cholera outbreaks worldwide, 1995–2005. Am J Trop Med Hyg. 2006;75:973–7.
10. Bradley M, Shakespeare R, Ruwende A, Woolhouse ME, Mason E, Munatsi A. Epidemiological features of epidemic cholera (El Tor) in Zimbabwe. Trans R Soc Trop Med Hyg. 1996;90:378–82.
11. Bockemühl J, Schröter G. The El Tor cholera epidemic in Togo (West Africa) 1970–1972. Tropenmed Parasitol. 1975;26:312–22.
12. Boiro MY, Lama N, Barry M, Diallo R, Morillon M. Cholera in Guinea: the 1994–1995 epidemic. Med Trop (Mars). 1999;59:303–6.
13. Foundohou J, Josse R, Anagonou S, Comlanvi E, Mevel A, Catraye J, Izurietta H, Gbaguidi E, Diallo S, Dodin A. Cholera in Benin (epidemic of 1991). Med Trop (Mars). 1993;53: 341–9.
14. Centers for Disease Control and Prevention (CDC). Epidemic cholera—Burundi and Zimbabwe, 1992–1993. Morb Mortal Wkly Rep. 1993;42:407–9,415–6.
15. Lawoyin TO, Ogunbodede NA, Olumide EA, Onadeko MO. Outbreak of cholera in Ibadan, Nigeria. Eur J Epidemiol. 1999;15:367–70.
16. Roquet D, Diallo A, Kodio B, Daff BM, Fenech C, Etard JF. Cholera epidemic in Senegal in 1995–1996: an example of geographic approach to health. Sante. 1998;8:421–8.
17. Coppo A, Colombo M, Pazzani C, Bruni R, Mohamud KA, Omar KH, Mastrandrea S, Salvia AM, Rotigliano G, Maimone F. *Vibrio cholerae* in the horn of Africa: epidemiology, plasmids, tetracycline resistance gene amplification, and comparison between O1 and non-O1 strains. Am J Trop Med Hyg. 1995;53:351–9.
18. Colombo MM, Francisco M, Ferreira BD, Rubino S, Cappuccinelli P. The early stage of the recurrent cholera epidemic in Luanda, Angola. Eur J Epidemiol. 1993;9:563–5.
19. Eko FO, Udo SM, Antia-Obong OE. Epidemiology and spectrum of *Vibrio* diarrheas in the lower cross river basin of Nigeria. Cent Eur J Public Health. 1994;2:37–41.
20. Siddique AK, Salam A, Islam MS, Akram K, Majumdar RN, Zaman K, Fronczak N, Laston S. Why treatment centres failed to prevent cholera deaths among Rwandan refugees in Goma, Zaire. Lancet. 1995;345:359–61.
21. Iijima Y, Oundo JO, Taga K, Saidi SM, Honda T. Simultaneous outbreak due to *Vibrio cholerae* and *Shigella dysenteriae* in Kenya. Lancet. 1995 Jan 7;34:69–70.
22. Hatch DL, Waldman RJ, Lungu GW, Piri C. Epidemic cholera during refugee resettlement in Malawi. Int J Epidemiol. 1994;23:1292–9.

23. Falade AG, Lawoyin T. Features of the 1996 cholera epidemic among Nigerian children in Ibadan, Nigeria. J Trop Pediatr. 1999;45:59–62.
24. Yala F, Dodin A, Diana Y. Role of interhuman infection during the cholera epidemic in the People's Republic of the Congo (1978–1979). Bull Soc Pathol Exot Filiales. 1982;75: 345–51.
25. Küstner HG, Gibson IH, Carmichael TR, Van Zyl L, Chouler CA, Hyde JP, du Plessis JN. The spread of cholera in South Africa. S Afr Med J. 1981;60:87–90.
26. Bompangue D, Giraudoux P, Handschumacher P, Piarroux M, Sudre B, Ekwanzala M, Kebela I, Piarroux R. Lakes as source of cholera outbreaks, Democratic Republic of Congo. Emerg Infect Dis. 2008;14:798–800.
27. Gunnlaugsson G, Angulo FJ, Einarsdóttir J, Passa A, Tauxe RV. Epidemic cholera in Guinea-Bissau: the challenge of preventing deaths in rural West Africa. Int J Infect Dis. 2000;4:8–13.
28. Richard V, Tosi C, Arzel B, Kana N. Endemic cholera in Chad: a real public health problem. Med Trop (Mars). 1999;59:169–172.
29. Champetier de Ribes G, Rakotonjanabelo LA, Migliani R, Pfister P, Rakotondramarina LD, Ranjalahy J, Ratsimbazafimahefa Rahantalalao H. One-year assessment of the cholera epidemic in Madagascar, from March 1999 to March 2000. Sante. 2000;10:277–86.
30. Diop SA, Manga NM, Dia NM, Gaye S, Ndour CT, Seydi M, Soumare M, Diop BM, Sow PS. Cholera and pregnancy: epidemiological, clinical, and evolutionary aspects. Med Mal Infect. 2007;37:816–20.
31. Gaffga NH, Tauxe RV, Mintz ED. Cholera: a new homeland in Africa? Am J Trop Med Hyg. 2007;77:705–13.
32. Smith AM, Keddy KH, De Wee L. Characterization of cholera outbreak isolates from Namibia, December 2006 to February 2007. Epidemiol Infect. 2007;29:1–3.
33. Mugoya I, Kariuki S, Galgalo T, Njuguna C, Omollo J, Njoroge J, Kalani R, Nzioka C, Tetteh C, Bedno S, Breiman RF, Feikin DR. Rapid spread of *Vibrio cholerae* O1 throughout Kenya, 2005. Am J Trop Med Hyg. 2008;78:527–33.
34. World Health Organization. Cholera, 2006. Wkly Epidemiol Rec. 2007;82:273–84.
35. World Health Organization. Cholera, 2007. Wkly Epidemiol Rec. 2008;83:269–84.
36. Sitas F. Some observations on a cholera outbreak at the Umvpti Mission Reserve. Natl S Afr Med J. 1986;70:215–8.
37. Birmingham ME, Lee LA, Ndayimirije N, Nkurikiye S, Hersh BS, Wells JG, Deming MS. Epidemic cholera in Burundi: patterns of transmission in the Great Rift Valley Lake region. Lancet. 1997;349:981–5.
38. Acosta CJ, Galindo CM, Kimario J, Senkoro K, Urassa H, Casals C, Corachán M, Eseko N, Tanner M, Mshinda H, Lwilla F, Vila J, Alonso PL. Cholera outbreak in southern Tanzania: risk factors and patterns of transmission. Emerg Infect Dis. 2001;7:583–7.
39. Mhalu FS, Mntenga WM, Mtango FD. Seasonality of cholera in Tanzania: possible role of rainfall in disease transmission. East Afr Med J. 1987;64:378–87.
40. Mendelsohn J, Dawson T. Climate and cholera in KwaZulu-Natal, South Africa: the role of environmental factors and implications for epidemic preparedness. Int J Hyg Environ Health. 2008;211:156–62.
41. Constantin de Magny G, Guégan JF, Petit M, Cazelles B. Regional-scale climate-variability synchrony of cholera epidemics in West Africa. BMC Infect Dis. 2007, 19;7:20.
42. Morillon M, De Pina JJ, Husser JA, Baudet JM, Bertherat E, Martet G. Djibouti, history of 2 epidemics of cholera: 1993–1994. Bull Soc Pathol Exot. 1998;91:407–11.
43. Naidoo A, Patric K. Cholera: a continuous epidemic in Africa. J R Soc Health. 2002;122: 89–94.
44. Patel M, Isaäcson M. Survival of *Vibrio cholerae* in African domestic water storage containers. S Afr Med J. 1989;76:365–7.
45. Shaffer N, Mendes P, Costa CM, Larson E, Anderson PA, Bean N, Tauxe RV, Blake PA. Epidemic cholera in Guiena-Bissau: importance of foodborne transmission. Abst 1459. Prog Abst 28th Intersci Conf Antimicrob Agents Chemother 1982.

46. Tauxe RV, Holmberg SD, Dodin A, Wells JV, Blake PA. Epidemic cholera in Mali: high mortality and multiple routes of transmission in a famine area. Epidemiol Infect. 1988;100:279–89.
47. St Louis ME, Porter JD, Helal A, Drame K, Hargrett-Bean N, Wells JG, Tauxe RV. Epidemic cholera in West Africa: the role of food handling and high-risk foods. Am J Epidemiol. 1990;131:719–27.
48. Schürmann D, Ebert N, Kampf D, Baumann B, Frei U, Suttorp N. Domestic cholera in Germany associated with fresh fish imported from Nigeria. Eur J Clin Microbiol Infect Dis. 2002;21:827–8.
49. Infuso A, de Valk H, Bigot Marie C, Fournier JM, Cadiot G, Andremont A, Desenclos JC. Cholera in a Paris resident with no history of travel. A case report. Presse Med. 1998;27: 202–4.
50. Hutin Y, Luby S, Paquet C. A large cholera outbreak in Kano City, Nigeria: the importance of hand washing with soap and the danger of street-vended water. J Water Health. 2003;1: 45–52.
51. Sinclair GS, Mphahlele M, Duvenhage H, Nichol R, Whitehorn A, Küstner HG. Determination of the mode of transmission of cholera in Lebowa. An epidemiological investigation. S Afr Med J. 1982;62:753–5.
52. Umoh JU, Adesiyun AA, Adekeye JO, Nadarajah M. Epidemiological features of an outbreak of gastroenteritis/cholera in Katsina, Northern Nigeria. J Hyg (Lond). 1983;91: 101–11.
53. Usman A, Sarkinfada F, Mufunda J, Nyarango P, Mansur K, Daiyabu TM. Recurrent cholera epidemics in Kano—northern Nigeria. Cent Afr J Med. 2005;51:34–8.
54. DuBois AE, Sinkala M, Kalluri P, Makasa-Chikoya M, Quick RE. Epidemic cholera in urban Zambia: hand soap and dried fish as protective factors. Epidemiol Infect. 2006;134: 1226–30.
55. Ndour CT, Manga NM, Ká R, Dia-Badiane NM, Fortez L, Seydi M, Soumaré M, Sow AI, Diop BM, Sow PS. Cholera epidemic of 2004 in Dakar, Senegal: epidemiologic, clinical and therapeutic aspects. Med Trop (Mars). 2006;66:33–8.
56. Nyati H. Evaluation of the microbial quality of water supplies to municipal, mining and squatter communities in the Bindura urban area of Zimbabwe. Water Sci Technol. 2004;50:99–103.
57. Swerdlow DL, Malanga, Begokyian G, Jonkman M, Toole M, Tauxe RV, Waldman R, Puhr N, Blake PA. Epidemics of antimicrobial resistant *Vibrio cholerae* O1 infection in a refugee camp-Malawi, abstract 529. Program Abstr. 31st Intersci Conf Antimicrob Agents Chemother.1991.
58. Djeddah C, Miozzo A, Di Gennaro M, Rosmini F, Martino P, Pasquini P. An outbreak of cholera in a refugee camp in Africa. Eur J Epidemiol. 1988;4:227–30.
59. Moren A, Stefanaggi S, Antona D, Bitar D, Etchegorry MG, Tchatchioka M, Lungu G. Practical field epidemiology to investigate a cholera outbreak in a Mozambican refugee camp in Malawi, 1988. J Trop Med Hyg. 1991;94:1–7.
60. Swerdlow DL, Malenga G, Begkoyian G, Nyangulu D, Toole M, Waldman RJ, Puhr DN, Tauxe RV. Epidemic cholera among refugees in Malawi, Africa: treatment and transmission. Epidemiol Infect. 1997;118:207–14.
61. Banatvala N, Roger AJ, Denny A, Howarth JP. Mortality and morbidity among Rwandan refugees repatriated from Zaire, November, 1996. Prehosp Disaster Med. 1998;13:17–21.
62. Cavallo JD, Niel L, Talarmin A, Dubrous P. Antibiotic sensitivity to epidemic strains of *Vibrio cholerae* and *Shigella dysenteriae* 1 isolated in Rwandan refugee camps in Zaire. Med Trop (Mars). 1995;55:351–3.
63. Mhalu FS, Mtango FD, Msengi AE. Hospital outbreaks of cholera transmitted through close person-to-person contact. Lancet. 1984;2:82–4.
64. Cliff JL, Zinkin P, Martelli A. A hospital outbreak of cholera in Maputo, Mozambique. Trans R Soc Trop Med Hyg. 1986;80:473–6.

65. Chapman JA, Collocott LP. Cholera in children at Eshowe Hospital. S Afr Med J. 1985;68:249–53.
66. Utsalo SJ, Eko FO, Umoh F, Asindi AA. Faecal excretion of *Vibrio cholerae* during convalescence of cholera patients in Calabar, Nigeria. Eur J Epidemiol. 1999;15:379–81.
67. Cans C. An epidemic of cholera in the city of Pemba (Mozambique) in 1983. Rev Epidemiol Sante Publique. 1986;34:419–26.
68. Mandara MP, Mhalu FS. Cholera control in an inaccessible district in Tanzania: importance of temporary rural centres. Med J Zambia. 1980–1981;15:10–3.
69. Paugam H. Cholera++ epidemic in Kenya. Can Nurse. 1999;95:42–6.
70. Gunnlaugsson G, Einarsdóttir J, Angulo FJ, Mentambanar SA, Passa A, Tauxe RV. Funerals during the 1994 cholera epidemic in Guinea-Bissau, West Africa: the need for disinfection of bodies of persons dying of cholera. Epidemiol Infect. 1998;120:7–15.
71. Parahoo KA. Early colonial health developments in Mauritius. Int J Health Serv. 1986;16:409–23.
72. Mohamed J. Epidemics and public health in early colonial Somaliland. Soc Sci Med. 1999;48:507–21.
73. Synder JD, Blake PA. Is cholera a problem for US travelers? JAMA. 1982;247:2268–9.
74. Morger H, Steffen R, Schär M. Epidemiology of cholera in travellers, and conclusions for vaccination recommendations. Br Med J (Clin Res Ed). 1983;286:184–6.
75. World Health Organization. Cholera in 1997. Wkly Epidemiol Rec1998;73:201–8.
76. Alagio SO, Nakayama J, Erume J. Cholera in epidemic districts in Uganda during El Niño rains: 2002–2003. Afr Health Sci. 2006;93–7.
77. Olago D, Marshall M, Wandiga SO, Opondo M, Yanda PZ, Kanalawe R, Githeko AK, Downs T, Opere A, Kavumvuli R, Kirumira E, Ogallo L, Mugambi P, Apindi E, Githui F, Kathuri J, Olaka L, Sigalla R, Nanyunja R, Baguma T, Achola P. Climatic, socio-economic, and health factors affecting human vulnerability to cholera in the Lake Victoria basin, East Africa. Ambio. 2007;36:350–8.
78. Utsalo SJ, Eko FO, Antia-Obong EO. Features of cholera and *Vibrio parahaemolyticus* diarrhoea endemicity in Calabar, Nigeria. Eur J Epidemiol. 1992;8:856–60.
79. Matiko CM. Cholera reveals helminthiasis problem [letter]. Afr Health. 1998;20:2.
80. Küstner HG, Du Plessis G. The cholera epidemic in South Africa, 1980–1987. Epidemiological features. S Afr Med J. 1991;79:539–44.
81. Sow AI, Cisse MF, Sanou I, Diallo-Gaye A, Diop PD, Boye CS, Mboup S, Samb A. Evaluation of *Vibrio cholerae* isolation over 10 years in CHU Fann. Dakar Med. 1992;37:113–6.
82. Shapiro RL, Otieno MR, Adcock PM, Phillips-Howard PA, Hawley WA, Kumar L, Waiyaki P, Nahlen BL, Slutsker L. Transmission of epidemic *Vibrio cholerae* O1 in rural western Kenya associated with drinking water from Lake Victoria: an environmental reservoir for cholera? Am J Trop Med Hyg. 1999;60:271–6.
83. Paz S, Broza M. Wind direction and its linkage with *Vibrio cholerae* dissemination. Environ Health Perspect. 2007;115:195–200.
84. von Seidlein L, Wang XY, Macuamule A, Mondlane C, Puri M, Hendriksen I, Deen JL, Chaignat CL, Clemens JD, Ansaruzzaman M, Barreto A, Songane FF, Lucas M. Is HIV infection associated with an increased risk for cholera? Findings from a case-control study in Mozambique. Trop Med Int Health. 2008;13:683–8.
85. Kaper JB, Morris JG, Jr, Levine MM. Cholera. Clin Microbiol Rev. 1995;8:48–86.
86. Ansaruzzaman M, Bhuiyan NA, Safa A, Sultana M, McUamule A, Mondlane C, Wang XY, Deen JL, von Seidlein L, Clemens JD, Lucas M, Sack DA, Balakrish Nair G. Genetic diversity of El Tor strains of *Vibrio cholerae* O1 with hybrid traits isolated from Bangladesh and Mozambique. Int J Med Microbiol. 2007;297:443–9.
87. Nair GB, Qadri F, Holmgren J, Svennerholm AM, Safa A, Bhuiyan NA, Ahmad QS, Faruque SM, Faruque AS, Takeda Y, Sack DA. Cholera due to altered El Tor strains of *Vibrio cholerae* O1 in Bangladesh. J Clin Microbiol. 2006;44:4211–3.

88. Safa A, Bhuyian NA, Nusrin S, Ansaruzzaman M, Alam M, Hamabata T, Takeda Y, Sack DA, Nair GB. Genetic characteristics of Matlab variants of *Vibrio cholerae* O1 that are hybrids between classical and El Tor biotypes. J Med Microbiol. 2006;55:1563–9.
89. Raychoudhuri A, Patra T, Ghosh K, Ramamurthy T, Nandy RK, Takeda Y, Nair GB, Mukhopadhyay AK. The prototype seventh pandemic El Tor biotype of *Vibrio cholerae* O1 is completely replaced from 1995 by El Tor strains that have classical type CtxB in Kolkata, India. Emerg Infect Dis. 2009;15:131–2.
90. Guechi Z, Mered B. Some aspects of the evolution of cholera in Algeria from 1971 to 1975. Arch Inst Pasteur Alger. 1978–1979;53:241–6.
91. Onyemelukwe GC, Lawande RV. Serotype variation in *Vibrio cholerae* el tor diarrhoea in northern Nigeria. Cent Afr J Med. 1991;37:186–9.
92. Bendib A, Bougoudogo F, Messast M, Rezkallah D, Hamchi L, Bellouche Y, Aidoud NS, Keddouri A, Fournier JM, Dodin A. Value of vibriocidal antibody research in endemic areas of *Vibrio cholerae* 0:1. Bull Soc Pathol Exot. 1994;87:33–7.
93. Bendib A, Nora M, Mahrouf H, Saib M, el Abassi K, Dodin A. Seroepidemiological survey of El Tor cholera in an endemic region of Algeria. Bull Soc Pathol Exot. 1994;87:143–7.
94. Mhalu FS, Mmari PW, Ijumba J. Rapid emergence of El Tor *Vibrio cholerae* resistant to antimicrobial agents during first six months of fourth cholera epidemic in Tanzania. Lancet. 1979;1:345–7.
95. Finch MJ, Morris JG Jr, Kaviti J, Kagwanja W, Levine MM. Epidemiology of antimicrobial resistant cholera in Kenya and East Africa. Am J Trop Med Hyg. 1988;39:484–90.
96. Ichinose Y, Ehara M, Watanabe S, Shimodori S, Waiyaki PG, Kibue AM, Sang FC, Ngugi J, Kaviti JN. The characterization of *Vibrio cholerae* isolated in Kenya in 1983. J Trop Med Hyg. 1986;89:269–76.
97. Ferreira E, Costa M, Vaz Pato MV. Resistance to antibiotics of *Vibrio cholerae* strains isolated in Angola. Pathol Biol (Paris). 1992;40:561–5.
98. Anagonou SY, Josse R, Foundohou J, Helynck B, Massougbodji A, Sadeler BC. Bacteriological aspects of cholera in Benin (epidemic of 1991). Bull Soc Pathol Exot. 1993;86:333–5.
99. Aidara A, Koblavi S, Boye CS, Raphenon G, Gassama A, Grimont F, Grimont PA. Phenotypic and genotypic characterization of *Vibrio cholerae* isolates from a recent cholera outbreak in Senegal: comparison with isolates from Guinea-Bissau. Am J Trop Med Hyg. 1998;58:163–7.
100. Materu SF, Lema OE, Mukunza HM, Adhiambo CG, Carter JY. Antibiotic resistance pattern of *Vibrio cholerae* and *Shigella* causing diarrhea outbreaks in the eastern Africa region: 1994–1996. East Afr Med J. 1997;74:193–7.
101. Mwansa JC, Mwaba J, Lukwesa C, Bhuiyan NA, Ansaruzzaman M, Ramamurthy T, Alam M, Balakrish Nair G. Multiply antibiotic-resistant *Vibrio cholerae* O1 biotype El Tor strains emerge during cholera outbreaks in Zambia. Epidemiol Infect. 2007;135:847–53.
102. Scrascia M, Maimone F, Mohamud KA, Materu SF, Grimont F, Grimont PA, Pazzani C. Clonal relationship among *Vibrio cholerae* O1 El Tor strains causing the largest cholera epidemic in Kenya in the late 1990s. J Clin Microbiol. 2006;44:3401–4.
103. Sow AI, Cissé MF, Gaye M, Kébé A, Sy OK, Dia NM, Komo A. Bacterial diversity during the cholera epidemic in Dakar, Senegal (1995–1996). Bull Soc Pathol Exot. 1997;90:160–1.
104. Urassa WK, Mhando YB, Mhalu FS, Mjonga SJ. Antimicrobial susceptibility pattern of *Vibrio cholerae* O1 strains during two cholera outbreaks in Dares Salaam, Tanzania. East Afr Med J. 2000;77:350–3.
105. Mandomando I, Espasa M, Vallès X, Sacarlal J, Sigaúque B, Ruiz J, Alonso P. Antimicrobial resistance of *Vibrio cholerae* O1 serotype Ogawa isolated in Manhiça District Hospital, southern Mozambique. J Antimicrob Chemother. 2007;60:662–4.
106. Garrigue GP, Ndayo M, Sicard JM, Fonkoua MC, Lemao G, Durand JP, Dodin A. Antibiotic resistance of strains of *Vibrio cholerae* eltor isolated in Douala (Cameroon). Bull Soc Pathol Exot Filiales. 1986;79:305–12.

107. Guévart E, Noeske J, Sollé J, Mouangue A, Bikoti JM. Large-scale selective antibiotic prophylaxis during the 2004 cholera outbreak in Douala (Cameroon). Sante. 2007;17:63–8.
108. Guévart E, Solle J, Mouangue A, Noeske J, Bita A, Fonkoua MC, Ndayo Wouafo M. Antibiotic susceptibility of *Vibrio cholerae* O1: evolution after prolonged curative and preventive use during the 2004 cholera epidemics in Douala. Med Mal Infect. 2006;36:329–34.
109. Noeske J, Guévart E, Kuaban C, Solle J, Fonkoua MC, Mouangue A, Fouda AB. Routine use of antimicrobial drugs during the 2004 cholera epidemic in Douala, Cameroon. East Afr Med J. 2006;83:596–601.
110. Mathabe MC, Nikolova RV, Lall N, Nyazema NZ. Antibacterial activities of medicinal plants used for the treatment of diarrhoea in Limpopo Province, South Africa. J Ethnopharmacol. 2006;105:286–93.
111. Akinsinde KA, Olukoya DK. Vibriocidal activities of some local herbs. J Diarrhoeal Dis Res. 1995;13:127–9.
112. Lan R, Reeves PR. Pandemic spread of cholera: genetic diversity and relationships within the seventh pandemic clone of *Vibrio cholerae* determined by amplified fragment length polymorphism. J Clin Microbiol. 2002;40:172–81.
113. Aidara-Kane A, Boye CS, Koblavi S, Grimont F, Grimont P. Isolation of a new variant of *Vibrio cholerae* O1: *V. cholerae* O1 ribotype B27 toxinogenotype TB31 during the last cholera epidemic in Senegal. Jpn J Med Sci Biol. 1997;50:227–32.
114. Dalsgaard A, Mortensen HF, Mølbak K, Dias F, Serichantalergs O, Echeverria P. Molecular characterization of *Vibrio cholerae* O1 strains isolated during cholera outbreaks in Guinea-Bissau. J Clin Microbiol. 1996;34:1189–92.
115. Keddy KH, Nadan S, Govind C, Sturm AW; Group for Enteric, Respiratory and Meningeal Disease Surveillance in South Africa. Evidence for a clonally different origin of the two cholera epidemics of 2001–2002 and 1980–1987 in South Africa. J Med Microbiol. 2007;56:1644–50.
116. Rodrigues A, Brun H, Sandstrom A. Risk factors for cholera infection in the initial phase of an epidemic in Guinea-Bissau: protection by lime juice. Am J Trop Med Hyg. 1997;57: 601–4.
117. Rodrigues A, Sandström A, Cá T, Steinsland H, Jensen H, Aaby P. Protection from cholera by adding lime juice to food—results from community and laboratory studies in Guinea-Bissau, West Africa. Trop Med Int Health. 2000;5:418–22.
118. Lucas ME, Deen JL, von Seidlein L, Wang XY, Ampuero J, Puri M, Ali M, Ansaruzzaman M, Amos J, Macuamule A, Cavailler P, Guerin PJ, Mahoudeau C, Kahozi-Sangwa P, Chaignat CL, Barreto A, Songane FF, Clemens JD. Effectiveness of mass oral cholera vaccination in Beira, Mozambique. N Engl J Med. 2005;352:757–67.
119. Conroy RM, Meegan ME, Joyce T, McGuigan K, Barnes J. Solar disinfection of drinking water protects against cholera in children under 6 years of age. Arch Dis Child. 2001;85:293–5.
120. Qureshi K, Mølbak K, Sandström A, Kofoed PE, Rodrigues A, Dias F, Aaby P, Svennerholm AM. Breast milk reduces the risk of illness in children of mothers with cholera: observations from an epidemic of cholera in Guinea-Bissau. Pediatr Infect Dis J. 2006;25:1163–6.
121. Roberts L, Chartier Y, Chartier O, Malenga G, Toole M, Rodka H. Keeping clean water clean in a Malawi refugee camp: a randomized intervention trial. Bull World Health Organ. 2001;79:280–7.
122. Centers for Disease Control and Prevention (CDC). Cholera epidemic associated with raw vegetables—Lusaka, Zambia, 2003–2004. MMWR Morb Mortal Wkly Rep. 2004;53:783–6.
123. Naficy A, Rao MR, Paquet C, Antona D, Sorkin A, Clemens JD. Treatment and vaccination strategies to control cholera in sub-Saharan refugee settings: a cost-effectiveness analysis. JAMA. 1998;279:521–5.
124. Legros D, Paquet C, Perea W, Marty I, Mugisha NK, Royer H, Neira M, Ivanoff B. Mass vaccination with a two-dose oral cholera vaccine in a refugee camp. Bull World Health Organ. 1999;77:837–42.

125. World Health Organization. Cholera, 2008. Wkly Epidemiol Rec. 2009;84:309–24.
126. World Health Organization. Cholera, 2000. Wkly Epidemiol Rec. 2001;76:233–40.
127. World Health Organization. Cholera, 2001. Wkly Epidemiol Rec. 2002;77:257–68.
128. World Health Organization. Cholera, 2002. Wkly Epidemiol Rec. 2003;78:269–76.
129. World Health Organization. Cholera, 2003. Wkly Epidemiol Rec. 2004;79:281–8.
130. World Health Organization. Cholera, 2004. Wkly Epidemiol Rec. 2005;80:261–8.
131. World Health Organization. Cholera, 2005. Wkly Epidemiol Rec. 2006;81:297–308.

Chapter 4
Phenotypic and Molecular Characteristics of Epidemic and Non-epidemic *Vibrio cholerae* Strains Isolated in Russia and Certain Countries of Commonwealth of Independent States (CIS)

Elena V. Monakhova

Abstract This chapter reviews *Vibrio cholerae* strains of different serogroups circulating in Russia and CIS countries with special reference to their pathogenicity and molecular characteristics. Certain epidemics, outbreaks, and sporadic cases of cholera, which took place during the 6th and 7th pandemics, are described. The data of epidemiological analysis as well as variable microbiological and molecular features of toxigenic and non-toxigenic strains isolated in different territories authenticated the imported nature of cholera manifestations. In the 21st century the spread of diseases caused by *V. cholerae* in Russia tends to decrease, but due to the high morbidity taking place in a number of Asian and African countries and migration of people the danger of cholera importations and spread still exists. The possibility of long-term preservation of the cholera agent in the environment and formation of stable or temporary endemic foci in climatic conditions of Russia and other CIS countries are discussed.

4.1 Introduction

The history of cholera epidemics in Russia is associated with the importation of the infection, which is stipulated by the intensity of people's migration (pilgrimage, tourism, shop-tours, trading business, etc.). During the 7th pandemic, sporadic cases of cholera in Russia and other USSR republics (today's CIS countries) were mainly caused by imported strains of *Vibrio cholerae* O1 El Tor Ogawa and Inaba and O139 serogroup carrying main virulence genes *ctxAB* and *tcpA*. Strains that lack these genes are usually considered to have no epidemic potential. At the same time, the isolation of *ctxAB*$^{-}$ strains from cholera patients was reported on repeated occasions including mixed infections [1, 2]. Numerous cases of diarrhea invoked by non-toxigenic strains of *V. cholerae* were described in many countries [3–5], indicating their worldwide distribution. Considerable interest has been shown in the analysis of the structure and function of their genomes [6, 7], genetic relations of toxigenic strains [4], and epidemic and etiological potential [5, 8]. The *ctxAB*$^{-}$

E.V. Monakhova (✉)
Research Institute for Plague Control, Rostov-on-Don, Russia
e-mail: unicorn54@mail.ru

T. Ramamurthy, S.K. Bhattacharya (eds.), *Epidemiological and Molecular Aspects on Cholera*, Infectious Disease, DOI 10.1007/978-1-60327-265-0_4,
© Springer Science+Business Media, LLC 2011

strains though unable to generate massive epidemics still may cause sporadic cases and even focal outbreaks of cholera-like diseases obviously due to a number of accessory toxins and additional virulence factors. The main properties and biological activity of *V. cholerae* toxins described so far are observed in many works [9–22]. The major genes and their products are listed in Table 4.1.

Table 4.1 Genetic determinants of certain *V. cholerae* proteins referred to in this chapter

Gene	Product or function	Reference
ctxAB[a]	A and B subunits of the cholera toxin (CT), the main virulence factors of epidemic *V. cholerae*	[9, 10]
cep[a]	Core encoded pilin, adhesin, and major CTXφ capsid protein	[11, 12]
orfU[a] ($gIII^{CTX}$)	$pIII^{CTX}$, minor CTXφ coat protein	[13]
ace[a]	Accessory cholera enterotoxin, pore-forming toxin, and a minor CTXφ capsid protein	[9–11, 14]
zot[a]	Zonula occludens toxin, modulator of intestinal tight junctions, and inner membrane protein required for the assembly of CTXφ virions	[9–11, 14, 15]
rstR[b,c]	RstR, repressor of CTXφ	[11, 16]
rstA[b,c]	RstA, required for CTXφ replication	[11, 16]
rstB[b,c]	RstB, required for CTXφ integration	[11, 16]
rstC[c]	Gene, specific for RS1φ	[11, 16]
*att*RS	Site of specific CTXφ integration	[11, 16]
rtxA[d]	RtxA, cytotoxic factor	[14, 17]
rtxC[d]	RtxC, activator of RtxA	[14, 17]
ACD-*rtxA*[d]	Actin cross-linking domain of RtxA	[17]
ACD-*vgrG*	Actin cross-linking domain of VgrG, a "twin" and putative ancestor of ACD-rtxA with the same activity	[17]
tcpA[e]	Toxin coregulated pili, colonization factor, and CTXφ receptor	[10, 12]
toxT[e]	ToxT, regulatory protein	[10, 12]
int (VC1758)[f]	Integrase, 5′-end sequence of VPI-2	[18]
nanH[f]	Neuraminidase	[18]
vce[f]	3′-end sequence of VPI-2	[18]
toxR	ToxR global regulator	[9, 10]
hapA	Hemagglutinin/protease (HA/P)	[14]
cef	CHO cell-elongating factor, cytotonic toxin	[14]
mshA	Mannose-sensitive pilin, adhesin	[10]
tolQRA	TolQ, TolR, TolA, proteins required for CTXφ infection and maintenance of bacterial outer membrane integrity	[20]
stn/sto	Heat-stable toxin, guanylate cyclase activator	[14, 15]
slt1A	Shiga-like toxin	[9, 21]
tdh	Thermostable direct hemolysin (TDH) of *V. parahaemolyticus*	[9, 22]
trh	TDH-related hemolysin of *V. parahaemolyticus*	[9, 22]

The marked genes represent constituents of the following gene clusters:
[a]CTXφ core region
[b]RS2 element, a component of CTXφ
[c]RS1φ
[d]RTX cluster
[e]VPI, *Vibrio* pathogenicity island
[f]VPI-2, the second *Vibrio* pathogenicity island

This chapter reviews the strains circulating in Russia and CIS countries with special reference to their pathogenicity and molecular characteristics.

4.2 Toxigenic (CTXφ⁺) *V. cholerae* Strains

4.2.1 O1 Serogroup

4.2.1.1 Classical Strains

The data on the properties of the classical biotype vibrios circulating in Russia during six pandemics are scanty. From 1823 to 1896 totally 4,837,236 patients had cholera in different Russian territories, of whom 1,984,049 had died [1]. These facts suggest that the pathogen possessed a powerful epidemic potential due to expression of a wide set of virulence factors, presumably the cholera toxin (CT). The large-scale epidemics occurred both in numerous rural settlements and in big cities. In the beginning of the 19th century rampant cholera was believed to be indigenous in Europe but this scenario was not reported in Russia. The imported origin of the infection was well documented underlining the main routs of its spread. Cholera penetrated into the Asian territories of Russia from India via Iraq, Syria, Iran, Afghanistan, China, Korea, and other countries of Central Asia with subsequent spread to European territories as well by "loop roads" from Western Europe brought by pilgrims, seamen, trade people, and other migrating groups of population [1, 23]. The only "indigenous" cholera outbreak was from Vladivostok in 1938 with 57 cases [24].

The years of the Second World War are characterized by numerous cholera outbreaks especially among the evacuated civil people and prisoners in a number of USSR republics [1]. The last Asiatic cholera outbreak in Russia took place in 1942 in the lower Volga region and was characterized by specific peculiarities such as atypical clinical picture of the disease and negligible spread of infection. Most cases lacked significant dehydration or algid and the morbidity rate was very low. Due to this reason, the outbreak remained unidentified for a long time. Retrospective studies on classical biotype of *V. cholerae* showed its ability of high-level production of CT and toxin-coregulated pilus (TCP). In spite of these factors, majority of the strains showed less virulence in the infant rabbit model. All the strains studied, in contrast to the typical cholera pathogens, were auxotrophs and needed purine and/or amino acids for growth in minimal medium. Moreover, these strains containing the structural gene *hapA* produced no soluble hemagglutinin protease (HAP), which enables the vibrios to disseminate in the environment. The described peculiarities of the *V. cholerae* strains were likely to be responsible for the unusual infectious and epidemic processes observed during the Volga cholera outbreak. The propagation of vibrios in the host might be inhibited by the low concentration of nutrients in the small intestine due to alimentary dystrophy during the war years of famine. The other reason might be stipulated by the inability of production of HAP which is responsible for CT activation. Such mutant strains might appear as a result of extensive application of cholera phages for treatment and prevention of cholera at that time [25, 26].

During recent times, no cases of cholera caused by classical vibrios were registered in Russia. Nonetheless, two toxigenic strains harboring the classical biotype genomic features, namely CTXφ^+RS1φ^- *rtxA*$^-$*rtxC*$^-$ genotype and presence of *tcpA* of classical type, were isolated from the Black Sea in Sochi in 1996 and from the river water in Rostov-on-Don in 1999 (Monakhova EV, Smolikova LM, unpublished data). Their origin remains unknown. *V. cholerae* O1 strains having similar genetic features were reported in Bangladesh (since 2001) and Mozambique (2004) [27–30]. The strains isolated in Russia did not spread probably owing to the fact that the environment of the country is usually not favorable for long-term survival of toxigenic *V. cholerae*. However, such strains demand special attention as they may represent the agent of a new more severe form of cholera [31].

4.2.1.2 El Tor Strains

The 7th pandemic outbreaks and sporadic cases in Russia were mainly due to the import of cholera from India, Syria, Iran, Pakistan, Sri Lanka, and China with subsequent formation of active disease areas (Fig. 4.1). Maximal importations were registered in Central and Volga Federal districts [32]. Statistical data on cholera in USSR are available from the WHO Bulletin OMS [33]. Briefly, between 1965 and 1989 a total of 10,723 cholera cases and carriers were reported from 11 Soviet republics. Thereafter, numerous communications were presented in the press but here we describe the most significant events.

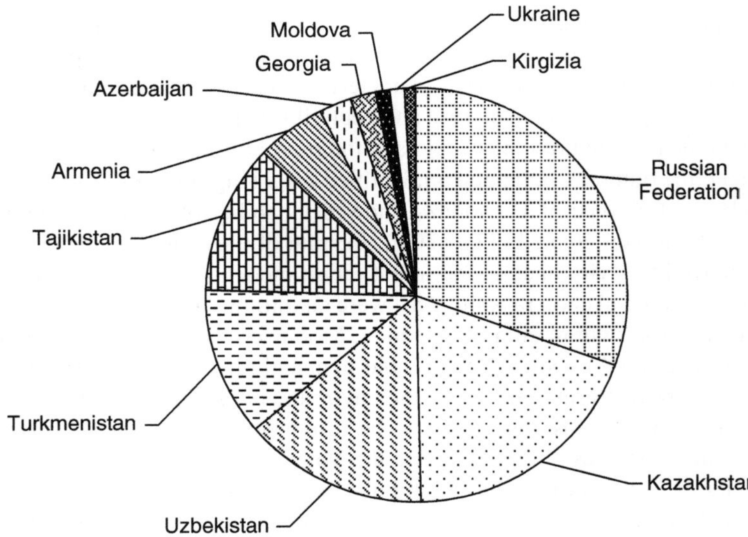

Fig. 4.1 Imported cases of cholera in CIS countries

Large-scale (1973–1975) and local outbreaks (1990) as well as repeated sporadic cases (1967–2001) of cholera occurred in the Rostov-on-Don region. The variable number of tandem repeat (VNTR) analysis of 82 clinical strains revealed 22 genotypes. The predominant cluster (F) was presented not only by human isolates but also included cholera vibrios circulating in surface water reservoirs within the same period [34]. It was assumed that these strains might have emerged in 1973 as a causative agent of cholera epidemic persisting up to 1992, indicating its long-term survival in the natural aquatic environment due to the ecosystem specificity.

In 1990, a unique waterborne cholera outbreak took place in the Stavropol territory being brought by Syrian hired house builders staying in a tourist camp. As a result of the sewerage system faultiness, cholera vibrios penetrated into a spring, which became the source of infection. All 49 patients and most of the carriers used the water from this spring. Despite the crucial anti-epidemic measures, some infected tourists carried the agent to several districts of Russia but due to continuous cholera surveillance and timely treatment, further spread of cholera was prevented [35, 36]. *V. cholerae* isolates shared 16S ribotypes identical to those of the strains isolated from the water of Danube River during the same year suggesting their common origin. Interestingly, the toxigenic strains isolated during the Azov (Rostov region) outbreak in 1990 showed similar but different 16S ribotypes indistinguishable from that of the Columbian human isolate of 1991 [37].

One of the most problematic territories of Russia concerning cholera is the Dagestan Republic, where cases of the disease were registered during all pandemics with the exception of the fourth one [38]. For the first time, import of this infection by pilgrims, returning from their Hajj by motor transport via the countries of southwest Asia, was registered in 1994. In the epidemic process, 184 settlements of 27 regions, 8 towns, and 1 housing estate were involved with 2,327 registered cholera cases and *Vibrio* carriers. High infection rates were detected in regions situated in different geographical zones. Frequent contacts and the alimentary route were the main pathways of transmission during this epidemic, while the role of the water route was considerably less significant [39]. Unexpected prolongation of the epidemic process of cholera in Dagestan was suggested to be due to climate-geographical, social-demographical, and sanitary-hygienic peculiarities [40]. Among the strains isolated from 19 regions of Dagestan, a high proportion was found to have resistance to tetracycline (65%) and chloramphenicol (28.6%) and some were found resistant to furagin and erythromycin. Out of 242 strains, 163 were found to have multiple resistance. Gentamicin, ciprofloxacin, and doxycycline were shown to have high in vitro activity [41]. Strains isolated during 1994–1998 from different districts of Dagestan showed identical or similar genotypes in mono-primer PCR, indicating that they belonged to a single clone. All shared multiple resistance to tetracycline, chloramphenicol, streptomycin, trimethoprim/sulfamethoxazole and susceptibility to doxycycline, rifampicin, gentamicin, amikacin, kanamycin, and quinolones [42, 43]. In addition, all toxigenic strains formed a single cluster (A) in the VNTR analysis. In spite of variability in their genotypes, the members of the epidemic cluster were thought to be the descendants of a single clone [44].

In 1994, cholera vibrios penetrated from Dagestan to Azerbaijan and caused a local outbreak. However, cholera frequently occurred in Azerbaijan from 1970 to 1995 [45]. In Turkmenistan, two outbreaks of cholera caused by toxigenic *V. cholerae* with all the virulence-associated genes were registered in 1969 and 1972. Strains with the same genotype were also isolated from carriers and the environment [46].

The epidemic manifestations of cholera in Ukraine and Moldova were reported during 1970–1991 [47, 48]. In Ukraine, the high morbidity due to cholera was recorded in 1970, which was brought into the ports of Odessa and Kerch and followed by subsequent spread to other territories. The greatest number of cholera cases was registered in Vilkovo (Odessa region). The next outbreak started in 1991 and covered Odessa (Vilkovo, Kiliya), Nikolaev, and Chersonese regions. Cholera in these regions was transmitted through water, mainly by sea. All human isolates shared identical properties and were highly toxigenic in the infant rabbit model [47]. However, 16S ribotyping with *Bgl*I revealed differences in hybridization patterns of strains collected from the Vilkovo, Kiliya, Nikolaev, and Chersonese regions. These results confirmed the epidemiological analysis, which suggested the common origin of cholera in Vilkovo and Kiliya foci and its subsequent spread from Kiliya to Nikolaev, but the Chersonese outbreak had a different origin. The strains isolated in 1989 from the Danube River had different ribotypes as compared to the strains responsible for the large-scale outbreak in Stavropol in 1990. Hence, the hypothesis of their persistence in the environment and relation to the mentioned Ukraine outbreaks was denied [37]. Nonetheless, the outbreak is believed to be Danube waterborne and might be imported from Romania (where the cases of cholera were reported) considering the time of the outbreak and geographical proximity [49].

In Moldova, local outbreaks of water origin were registered during 1970–1972 with subsequent prolonged (more than 16 years) circulation of non-toxigenic El Tor vibrios [48]. As compared to this period (57 cases), a large outbreak of cholera occurred in 1995 (240 cases), while from 1980 to 2000 only sporadic cases were reported [50]. In CIS countries, the number of cholera cases gradually decreased from 1997 to 2007 (Fig. 4.2). Outbreaks and sporadic cases were registered in Kazakhstan (1997, 2005), Turkmenistan and Uzbekistan (1997), Azerbaijan and Armenia (1998), and Ukraine (1999–2001, 2003) [32]. During 1994–2004, more than 80 imported cases of cholera were registered in different parts of Russia. Toxigenic *V. cholerae* were isolated from the environment, mainly in the districts where the outbreaks prevailed but sometimes during non-epidemic periods [51]. Import of cholera to Primorye might be due to infected drivers, who arrived from China in 1999. In 2000, a cholera outbreak with 32 patients and 22 carriers was recorded in this region mainly brought by Russians with mild diarrhea, returning from China [52].

V. cholerae El Tor strains were isolated in Kazan during 2001 at different periods of the outbreak. A version of the emergence of the cholera focus was confirmed, namely water route of transmission as a result of bathing in a water reservoir contaminated by sewage. In this outbreak, 52 patients with acute diarrhea and

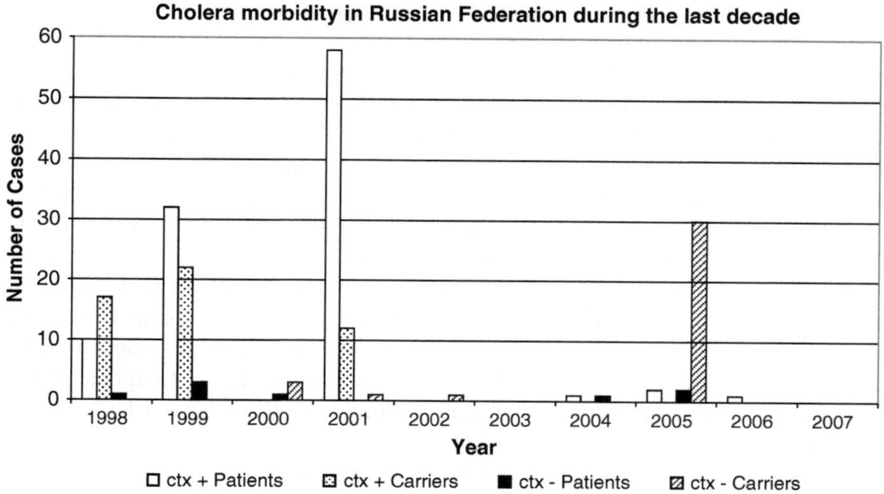

Fig. 4.2 Cholera morbidity in Russian Federation during 1998–2007

18 carriers were identified [53]. Strains isolated from patients, carriers, and environments were phenotypically identical, susceptible to ciprofloxacin, tetracycline, rifampicin, β-lactams and resistant to trimethoprim–sulfamethoxazole, streptomycin, furazolidone, and nalidixic acid (which may be regarded as phenotypic markers). Retrospective PCR analysis revealed the single genotype, i.e., with the profile of CTXφ$^+$ RS1φ$^+$ *att*RS$^+$ VPI$^+$ VPI-2$^+$ *rtxA*$^+$ *rtx*C$^+$ ACD-*rtxA*$^+$ ACD-*vgrG*$^+$ *cef*$^+$ *hapA*$^+$ *mshA*$^+$ *toxR*$^+$ *tolQRA*$^+$ *stn/sto*$^-$ *slt1*$^-$ [8, 43, 54]. These strains also showed identical patterns of polymorphic DNA fragments in the mono-primer PCR. VNTR analysis of their DNAs made it possible to classify all these strains as members of one cluster (A), which also included Dagestan strains [44]. However, strains from Dagestan and Kazan formed separate sets of genotypes in the dendrogram, indicating their divergence. The epidemiological relation between the causative agents of cholera in these two regions remains unclear but the data obtained suggest the existence of a hypothetical precursor for both clones and confirm the unique clone involved in the Kazan outbreak [55].

In 2005, *V. cholerae* was imported from Tajikistan to Tver region and Moscow without subsequent spread of the infection [56]. Strains isolated from patients belonged to Inaba serotype and carried all the common virulence-associated genes. The clinical strain from Bashkortostan (imported from India in 2004) differed only by the deletion of the distal part of VPI-2 [8]. The strain isolated in Tver was highly resistant to nalidixic acid, streptomycin, ampicillin, and trimethoprim/sulfamethoxazole and showed cross-resistance to fluoroquinolones (ciprofloxacin, ofloxacin, pefloxacin, and norfloxacin) and moderate resistance to ceftriaxone and cefotaxime. The only potent antibiotics for this strain were tetracycline and aminoglycosides other than streptomycin [43].

In 2001, four focal points of cholera were detected in Kazakhstan with 23 patients and 20 carriers. The clinical manifestations varied from severe hypovolemic shock to mild subclinical forms [57]. Importation of cholera from India was also registered in Bashkortostan in 2004 and Murmansk in 2006 and from Tajikistan to Tver region and Moscow in 2005 [8]. Certain epidemic strains isolated in Dagestan (1994, 1996, and 1998), Primorye Territory and Sakhalin (1999), Kazan (2001) as well as strains imported from India to Bashkortostan (2004) and from Tajikistan to Tver region (2005) were analyzed for distribution of the VNTR in VcA, VcB, VcC, VcD, and VcG loci. A single VNTR cluster was identified, which differed from non-toxigenic O1 strains. Overall, 41 VNTR genotypes were revealed within the cluster. These results confirm the genetic relatedness of the toxigenic *V. cholerae* O1 strains and their clonal diversity [8]. Thus, the data of epidemiological analysis as well as variable microbiological and molecular features of toxigenic strains isolated in different territories authenticate the imported character of El Tor cholera manifestations occurring in most of the CIS countries.

Almost all toxigenic *V. cholerae* O1 El Tor strains, which were tested in the infant rabbit model, caused a typical choleragenic effect and frequent death of infected animals with the clinical picture indistinguishable from the one caused by strains from India and other Asian countries [58], i.e., abundant accumulation of clear colorless serous liquid (sometimes with floccular formations) in the intestine. The vibrios propagated in the intestine but did not penetrate beyond its bounds. Ultrastructural alterations in the gut were typical for CT effect. In semi-thin sections of the small intestine of infant rabbits infected with an El Tor *Vibrio* isolated from a clinical source in Taganrog (Rostov region) in 1992 selective lesions of villi were observed as well as ballooning degeneration of epitheliocytes, i.e., the cells looked like giant vacuoles filled with liquid. In ultrathin sections, nuclear pyknosis, lysis, and degradation of organelles were revealed. Intestinal villi capillaries were characterized by extremely thin endothelial cells with increased cytoplasmic osmiophilia and numerous pinocytic vesicles [59, 60]. The tested strain isolated during the cholera epidemic belonged to the Inaba serotype, was non-hemolytic, produced CT in vitro, and preserved its virulence for many years. Its CTX core and RS blot hybridization patterns were likely to indicate the presence of three tandem copies of the prophage [61].

4.2.2 O139 Strains

Soon after the first communication on the emergence of a novel epidemic *V. cholerae* O139 (Bengal) in Southern India [62], few cases of severe cholera (including one algid patient) caused by strains belonging to this serogroup were registered in Russia and Kirgizia. For the three identical strains isolated from patients in Azov (Rostov region, Russia) in 1993, the spread from Southern India was reliably established while the fourth was obtained at the same time in another district of Rostov region from a patient who had no contacts with the three mentioned above. Two strains isolated in Kirgizia in 1994 are also likely to be of Indian origin. These six strains were highly choleragenic on the infant rabbit model and showed dissemination of vibrios out of the gut to other organs (liver, spleen, lungs) and blood [59, 63], a phenomenon frequently observed for non-O1 *V. cholerae* [64]. Ultrastructural changes in the small intestine of infected animals included dramatic vacuolization of epitheliocytes, significant cytoplasmic hydration along with vascular hematological shifts such as white and red blood cell injuries, degranulation of platelets, hemolysis of erythrocytes, enhancement of the transepithelial pinocytosis, thinning of the endothelium, labilization of vascular wall, diapedesis of erythrocytes, and plasmohemorrhagia. The microorganisms were sometimes revealed among intracellular organelles, which confirm the higher invasive ability of O139 as compared to O1 vibrios [59].

The PCR genotypes of all six strains were identical to that of Indian MO45 strain used in comparative studies, namely *ctxAB*$^+$ *cep*$^+$ *orfU*$^+$ *ace*$^+$ *zot*$^+$ *attRS*$^+$ *rstR*$^+$ *rstA*$^+$ *rstB*$^+$ *rstC*$^+$ *tcpA*$^+$ *toxT*$^+$ *rtxA*$^+$ ACD-*rtxA*$^+$ *rtxC*$^+$ ACD-*vgrG*$^+$ *nanH*$^-$ *mshA*$^+$ *toxR*$^+$ *cef*$^+$ *hapA*$^+$ *tolQRA*$^+$ *stl1*$^-$*stn/sto*$^-$ *tdh*$^-$ *trh*$^-$ [65; Monakhova EV, Mikhas NK, unpublished]. Additionally, genes of two pathogenicity islands VSPI and VSPII as well as *tlc, mshQ,* and some of the O-antigen cluster genes were detected in one of the Azov strains, which had the same genotype as those of the Indian and Bangladeshi O139 strains isolated during 1992–1993 [66]. Thus, the human isolates besides CTXφ and VPI contained a rather broad set of genes for accessory virulence factors with the exception of neuraminidase (NANase), which was shown as a characteristic feature of the O139 strains isolated after 1992 [18]. It was postulated that the deletion of the *nan-nag* region from the VPI-2 of Bengal strains might provide one more explanation of the displacement of O139 serogroup by El Tor biotype as the *nan-nag* region was considered to contribute to the survival of the bacterium in different ecological niches. Nevertheless, the Bengal strains were shown to produce other neuraminidases having no genetic or immunological relatedness to NANase and possessing distinct physicochemical properties and substrate specificity [67]. The neuraminidase purified from a toxigenic human isolate appeared to be a high molecular weight glycoprotein unable to hydrolyze the synthetic substrate 2-(4′-methyl-umbelliferyl)-*N*-acetyl-α-D-neuraminic acid (MUA) unlike NANase of *V. cholerae* O1 and non-O1, non-O139 but highly active toward Tween 20. On the other hand, the neuraminidase purified from a non-toxigenic environmental O139 strain though immunologically related to that of the toxigenic strain was able to cleave MUA but not Tween 20 [68, 69]. Genetic determinants of these neuraminidases were not yet identified and their relevance to disease as well as to survival of vibrios in the environment remains to be determined.

A comparative analysis of hybridization patterns of *Hind*III and *Bgl*II DNA restriction digestions with molecular CTXφ core and RS probes showed significant diversity. Strains isolated in two districts of the Rostov region in 1993 as well as the reference strain MO45 contained two CTXφ prophages arranged in tandem, but hybridization patterns and the numbers of RS elements present in their genomes were different. The three Azov strains representing a single clone contained two tandem copies of CTX core genes flanked by RS elements and the third RS was located between the core genes [70]. The presence of three RS elements in these strains was also shown in Southern blots of the *Pst*I–*Bgl*II restriction digestions [71]. The strain isolated in another district of Rostov region had dissimilar hybridization patterns and was likely to contain additional RS copy(*ies*) located in different sites of the chromosome [70, 71] similar to the Indian El Tor strains isolated before the spread of O139 vibrios [72]. Both clones isolated in Russia had two tandem CTX copies like the reference MO45 strain [73] but the latter had another restriction site pattern outside the prophage genome.

The O139 strains isolated in Kirghizia in 1994 seem to harbor a single prophage flanked by RS at both ends. In 1994 the epidemic Bengal vibrios carrying two tandem prophage copies were replaced by the novel El Tor clone in the majority of Asian countries but remained in Bangladesh [73]. El Tor strains with a single copy were isolated in India in 1998, which indicated the continuous change in the structure and organization of the CTXφ prophage [74]. Nonetheless, it was suspected that the Kirgizian clone was rather imported from Southern Asia than being spread from Russia and had genome shifts during its survival in the environment. Strains isolated in Rostov region and one of the two obtained from Kirgizia shared a common 23S ribotype, identical to that of an Indian isolate SG24 and significantly different from the classical 569B. On the other hand, 16S ribotypes of the Rostov region and one Kirgizian isolate were identical to each other but differed from that of SG24 [75]. This ribotype designated R-I was considered as a new type compared to the ribotypes of Indian strains isolated during 1992–2000 [76], while SG24 had the B-I ribotype like the majority of the Bengal strains. The 16S ribotype of the second Kirgizian strain differed from both mentioned above [75]. All strains isolated in Russia and Kirgizia formed a unique VNTR cluster (B) together with those obtained from India and had no genetic relatedness to non-toxigenic environmental O139 strains belonging to another variable cluster (C) that also did not harbor *cxtAB* and *tcpA* [77]. The "Russian" O139 strains were susceptible to ampicillin but resistant to trimethoprim/sulfamethoxazole and streptomycin, i.e., had an STX element [43, 71] which was confirmed later by PCR [66]. Similar results were also obtained for strains isolated in Southern Asia [43, 62, 74].

In spite of the large incidence of O139 cholera in Asian and other countries, there were no more importations of toxigenic Bengal strains to most of the Russian territories. Only few imported cases of the infection in Kalmykiya, Uzbekistan, and Kazakhstan were reported [32, 78], but the properties of the O139 isolates were not presented. Numerous CTX$^-$ O139 strains constantly circulating in the surface waters in various parts of Russia lacked epidemic potential and were characterized by drastic differences in their genome organization [75, 77].

4.2.3 CTXφ$^+$ Non-O1, Non-O139 Strains

V. cholerae non-O1, non-O139 containing CTX element represent the minority of toxigenic strains and in spite of CT production, they do not tend to epidemic spread [79]. However, reports on the isolation of such strains from sporadic cases of cholera-like diarrhea in different countries are available [79–82].

The non-O1, non-O139 *V. cholerae* human strains carrying CTXφ prophage and *tcpA* gene of either El Tor or classical type were obtained only from Uzbekistan during 1987–1990 but not from the other CIS territories [83]. All the 22 strains were isolated from patients with manifestations of acute diarrhea of different severity. Retrospective studies showed that the majority of these strains were hemolytic. Twelve strains belonged to O9, O15, O28, and O74 serogroups while the rest did not react with the available set of O2–O84 antisera prepared in the Rostov-on-Don Institute for Plague Control [83]. All the strains caused a significant choleragenic effect and dissemination in the infant rabbit model. The supernatants of the gut contents of infected animals caused hemolysis of sheep erythrocytes. Unlike the O1 strains, toxigenic non-O1, non-O139 *V. cholerae* simultaneously produced CT and soluble hemolysin both in vitro and in vivo [84].

Electron microscopy of the small intestine of infant rabbits challenged with a toxigenic strain isolated in Uzbekistan revealed ultrastructural changes similar to those caused by O139 strains. Epithelial cells and capillaries appeared to be especially sensitive to the infection. Epitheliocytes decreased in size and reduction of microvilli followed by clasmatosis of underlying cytoplasm took place. Usually, subapical plasma membrane contained large osmiophobic zones lacking any organelles or inclusions. Various lesions were observed in mitochondria, which sometimes transformed into myelin figures detecting irreversible changes. Characteristic features of damage were large cavities containing membrane structures or plasma-like substance. The liquid was delivered into intestinal lumen by means of vacuoles and cavities as well as via enlarged intercellular spaces. Corresponding alterations synchronously occurred in all parts of the microcirculatory bed. As a result, plasma and blood cells penetrated into stroma. Polymorphonuclear leukocytes and macrophages prevailed among cellular elements [59]. Thus, the virulence of *ctxAB*$^{+}$ non-O1, non-O139 *V. cholerae* strains is similar to that of toxigenic O1 mainly due to the production of CT, which upsets the water–salt balance giving rise to the secretion of liquid into the gut's lumen and abundant diarrhea. However, the complex pathological manifestations revealed by electron microscopy are thought to be caused by CT and possibly other factors, as the tested strain was likely to contain only one copy of CTX prophage [61].

As detected by colony hybridization and PCR, all strains carried CTXφ, RS1φ, VPI, intact or truncated VPI-2 and RTX cluster (including ACD-*rtxA), mshA, mshQ, cef, hapA,* and *toxR* genes. The presence of *tcpA* gene was detected in colony hybridization, but not in PCR with the primers for *tcpA* of classical or El Tor types [61, 85], probably due to the genetic diversity of *tcpA* common among different *V. cholerae*, especially the non-O1, non-O139 strains [86]. Some of the toxigenic non-O1, non-O139 strains, however, were shown to carry *tcpA* of classical type [66]. The other genes of VPI such as *toxT, aldA,* and *tcpP* were present in all strains. The genes *rig, gmh*D, and *orf2* flanking the O-antigen gene cluster were also detected as well as the insertion sequence IS*1385* located inside this cluster. Sequencing and comparative analysis of *ctxA, ace, zot, rstC, toxT, tlc, rig,* and *orf2* fragments showed slight differences between strains isolated in 1986–1987 and 1990, but all of them were significantly distinct from the sequences of *V. cholerae* O1 N16961 and the sequence of virulent *V. cholerae* O37 1322-69 available in GenBank [85]. Based on these data, it was assumed that the strains isolated in Uzbekistan in 1990 represent a single clone having no close relation to the *V. cholerae* O1. However, molecular typing of several strains carried out by the Pasteur Institute (France) revealed significant variability of CTX types and 16S and 23S ribotypes (Koblavi S, 1992, personal communication), which suggests their non-clonal origin. The VNTR analysis also revealed genetic diversity, which was comprised by 11 individual genotypes and showed multiplicity of expression of the locus VcF located between *zot* and *ctxA* genes. The latter finding was established as a possible additional parameter useful in defining the epidemiological value of the toxigenic strains [87].

Uzbekistan with its hot climate remains as the CIS country where occurrence of *V. cholerae* non-O1, non-O139 is frequent but the isolation of CTXφ^+ strains was not reported after 1990.

4.3 Non-choleragenic Strains

4.3.1 Pre-CTXφ^+ O1 and Non-O1, Non-O139 Strains

Among the non-choleragenic (*ctxAB*$^-$) *V. cholerae* strains, of special interest are those carrying the derivatives of the CTXφ precursor, pre-CTXφ. First revealed by Boyd EF et al. [88] in the genomes of two clinical and one environmental *V. cholerae* O1 and non-O1, non-139 strains and later by Rui et al. [89] in one O139 strain (GenBank access number AF416590), pre-CTXφ remains a rather rare finding.

Out of several thousand *V. cholerae* strains investigated in Rostov-on-Don Research Institute for Plague Control, only 18 O1 and 8 non-O1, non-O139 were shown to contain one to two copies of the pre-CTXφ prophage [61, 90–92]. This group of strains included nine O1 isolates, which caused a local outbreak of acute diarrhea in Uzbekistan (Katta-Kurgan, Samarkand region) in 1990 presumably among children, as well as one O1 and six non-O1, non-O139 (O13 and O untypable) from sporadic cases from different regions in 1987–1988. One O8 strain was obtained from an acute diarrhea patient in Rostov-on-Don, Russia, in 1981. The rest eight O1 and one non-O1, non-O139 strains were isolated from sewage and river water in Moscow (1992), Rostov-on-Don (1982, 1987), Crimea, Ukraine (1991), and Chechen Republic of Russia (1995). In the infant rabbit model, both clinical and environmental strains caused a significant enteropathogenic effect sometimes resembling the choleragenic one, and in a number of animals, dissemination of the tested bacteria into various organs and tissues was detected. The sterile cultural supernatants of the majority of the strains stipulated cell rounding in CHO cell assay (one caused 100% death) and disconnection of cells in McCoy and L-929 monolayers as well as an increase of skin permeability in Craig's test [93].

Besides the genes of pre-CTXφ, all the strains contained *hapA, cef, rtxA,* and *rtxC* genes. The *rtxA* and *vgrG* ACD domains, genes of pathogenicity islands VPI and VPI-2, *mshA* were present in diverse combinations. *tcpA* was detected in all strains by colony hybridization but only partly by PCR (as described above for toxigenic non-O1, non-O139 strains). None carried *slt1, tdh,* or *trh* genes. Thus, the diarrheic syndrome of different severity registered in patients from whom these strains were isolated and signs of virulence revealed on the laboratory models are likely to be due to expression of the genes of the accessory pathogenicity factors including Cep, Ace, and Zot [93].

A comparative analysis of DNA restriction patterns with molecular probes recognizing pre-CTXφ core genes and flanking RS sequences showed heterogeneity in the number of truncated prophage copies and their localization as well as position of *Hind*III and *Bgl*II restriction sites within them. Among 17 clinical strains, 5 etiologically significant clones were recognized, each associated with a definite period and region of isolation. Among these clones, one was a causative agent of an acute diarrhea outbreak in Uzbekistan. These strains were shown to carry a derivative of the CTXφ precursor containing a complete set of genes required for phage production and the RS2 element. RS1φ prophage was present only in two strains, one of which constantly formed its replicative form as well as the pre-CTXφ RF [91]. Variability of the integrated prophages depends mainly on the type of RS elements. At least half of the investigated strains possessed a potential for generation of infectious virions and horizontal transfer of virulence genes through "conventional" (pre-CTXφ infection) and "unconventional" (common transduction) pathways [61, 91]. Further investigations confirmed their ability to produce RFs and phage particles as a result of mitomycin C induction (Monakhova EV, unpublished data).

The origin of four environmental pre-CTXφ-carrying O1 strains with identical PCR genotypes isolated during July–August 2007 from sewage and river waters near Rostov-on-Don [92] remains unclear. No acute diarrheal diseases caused by strains of similar characteristics were registered this time. It seems unlikely that the organism survived in the environment for 20–25 years, after the last isolation of such strains in the same region. One clinical and three environmental *ctxAB*$^-$ *ace*$^+$ *zot*$^+$ *tcpA*$^+$ *toxT*$^+$ *attRS*$^+$ strains were identified in Turkmenistan during the 1970s–1980s [46], but there is no sufficient evidence that they are not deletion mutants but real carriers of pre-CTXφ.

4.3.2 *CTXφ$^-$/Pre-CTXφ$^-$ O1 Strains*

Long-term studies conducted in Russia showed that *ctxAB*$^-$ *tcpA*$^-$ strains are usually isolated from sporadic diarrheal cases [58]. Such isolates may contain or lack *tcpA* and other genes of VPI. Few *ctxAB*$^-$ *tcpA*$^+$ strains were isolated from sporadic cases of acute diarrhea and healthy carriers in Rostov-on-Don (1974), Volgograd region (1999), and Astrakhan (2001) [32]. An outbreak caused by *V. cholerae* O1 *ctxAB*$^-$ *tcpA*$^+$ strains was recorded in Kamensk district of Rostov region in 2005 and it was associated with contamination at the source of water supply. *V. cholerae* O1 was isolated from 2 patients and 30 carriers. In addition, *V. cholerae* isolated from the surface water reservoir within the area of water intake showed the strain identity. As indicated by the results of genomic and VNTR analysis, the responsible strains differed from *ctxAB*$^-$ *tcpA*$^-$ and *ctxAB*$^-$ *tcpA*$^+$ strains, which were isolated earlier from cholera patients, carriers, and the environment both in the presence and in the absence of epidemiological complications and constituted a unique group of strains with a specific genotype. These findings strongly supported the view that these strains were evolved through importation. Indeed, three carriers were revealed among workers who arrived from Tajikistan to this region not long before the beginning of the outbreak.

Both clinical and environmental strains had the same gene profile, i.e., *ctxAB*$^-$ *cep*$^-$ orfU$^-$ *ace*$^-$ *zot*$^-$ *attRS*$^+$ *rstR*$^-$ *rstA*$^-$ *rstB*$^-$ *rstC*$^-$ *tcpA*$^+$ *toxT*$^+$ *rtxA*$^+$ ACD-*rtxA*$^+$ *rtxC*$^+$ ACD-*vgrG*$^+$ *int*$^+$ *nanH*$^+$ *vce*$^-$ *mshA*$^+$ *toxR*$^+$ *cef*$^+$ *hapA*$^+$ *tolQRA*$^+$ *sltI*$^-$ *stn/sto*$^-$, but different from strains previously isolated in several Russian territories and formed a unique cluster of seven similar VNTR genotypes. These genotypes suggested the imported origin not only of *ctxAB*$^+$ *tcpA*$^+$ but also of *ctxAB*$^-$ *tcpA*$^+$ and *ctxAB*$^-$ *tcpA*$^-$ strains. In spite of the apparent clonal nature of the outbreak-associated strains, some variations were detected in antibiotic resistance. Strains isolated from patients and from a carrier were resistant to chloramphenicol and majority (26 of 32) of them was resistant to sulfamethoxazole/trimethoprim [8]. Production of TCP was thought to be the main reason of *Vibrio* carrier status. However, many outbreaks were caused by *tcpA*$^-$ strains as well.

Spread of non-epidemic *V. cholerae* O1 in Donetsk (Ukraine) from June to September 1971 was associated with consumption of infected milk produced by one of the city's factories. During this period, 14 patients and 278 carriers were identified. Almost all the strains isolated from humans, milk, and water shared several phenotypic properties such as hemolytic, hemagglutinating, proteolytic, and lecithinase activity and caused moderate enteropathogenic effect in infant rabbits and rabbit ileal loop (RIL) models [6]. The cultural supernatants caused cell rounding and destruction in L-929 cell cultures (Alekseeva LP, personal communication). Furthermore, these strains shared the same genotype: CTXφ^-RS1φ^- *tcpA*$^-$ *toxT*$^-$ *rtxA*$^+$ *rtxC*$^+$ ACD-*rtxA*$^+$ ACD-*vgrG*$^-$ *int*$^+$ *nanH*$^+$ *vce*$^-$ *mshA*$^+$ *toxR*$^+$ *cef*$^+$ *hapA*$^+$ *tolQRA*$^+$ *sltI*$^-$ *stn/sto*$^-$ *trh*$^-$. The sequences of *tdh* were revealed in blot hybridization but not in PCR (Monakhova EV, Smolikova LM, unpublished). Thus, the strains obviously belonged to a single clone, with the exception of two clinical isolates, which appeared to be non-hemolytic, caused elongation of L-929 cells, and contained ACD-*vgrG*. The outbreak was brought under control after stopping milk production and by closing the factory. Six strains of identical genotype were isolated from the environment the next year, which indicates long-term persistence of the suspected strain, but there were no more patients or carriers in the subsequent years [6].

Several *ctxAB*$^-$ *tcpA*$^-$ strains isolated from cholera patients in Ukraine (1991), Moldova (1991), Azerbaijan (1989), and Uzbekistan (1989) causing a typical choleragenic effect in most infected infant rabbits were highly adhesive and invasive. Electron microscopy of the small intestine of animals infected with a representative strain of this group revealed dramatic affect on gut epithelium accompanied by severe circulatory disturbances comparable with those caused by choleragenic strains. Cholera vibrios, however, were seen not only in intestinal lumen, but also in villus stroma. Ultrastructural investigation revealed destruction of microvilli forming epitheliocyte brush border, which resulted in extensive uncovering of apical plasma membrane. The labilization of the latter promoted clasmatosis of cytoplasmic parts and individual epithelial cell desquamation. Inside the intestinal lumen, the microorganisms were observed together with cellular fragments and separated organelles. It is notable that in spite of invasion, vibrios were found rarely in the epithelial cytoplasm. Nevertheless, such an interaction resulted in irreversible dystrophic injuries and necrosis in epithelium. The most interesting fact is that the vibrios penetrated into the submucous layer in great numbers and even localized between smooth muscle cells. In stroma they arrange close to the capillaries and smooth muscle cells of the intestine or were phagocytized by macrophages. The cases of incomplete phagocytosis were also observed. When the diarrheal syndrome was completely developed, remarkable changes in microcirculation and endothelium took place. In the capillary lumen, mixed platelet–erythrocyte thrombi were observed, as well as frequent adhesion and aggregation of degranulated platelets combined with fibrin precipitation. In endothelial cells, micropinocytosis intensification, cytoplasm vacuolization, formation of numerous microvilli followed by their clasmatosis, and local foci of plasma membrane destruction usually accompanied by platelet adhesion occurred. The lesions in the intracellular contact zone were terminated by disjunction of endotheliocytes and increasing vascular permeability of the hemato-enterocytic barrier. Extravasal damages were characterized most frequently by release of liquid part of blood, fibrin, and blood cells into the interstitial space [59, 60, 94, 95].

The above-tested strain was isolated from a cholera patient during the epidemic of 1991 in Ukraine caused by a toxigenic clone. However, it lacked all the genes of CTX and RS elements and VPI as well as *stn/sto, slt1, trh,* and *tdh*, but contained *att*RS, intact RTX cluster, and VPI-2, *mshA*, *cef*, *hapA*, *toxR*, ACD-*vgrG*, *tolQRA*. Probably, in the absence of CT, some of these genes showed high level of expression and the observed effects appeared as a result of the combined action of several factors. At least drastic cytoplasm vacuolization similar to that described above was observed in the intestine of suckling mice challenged with *V. cholerae* hemagglutinin/protease [96], and the cultural supernatants of this and some other non-toxigenic strains caused notable elongation of CHO cells [97] comparable to that elicited by Cef [98, 99]. The observed effects might also be induced by other toxins with non-identified genetic determinants such as WO7 [100] or dermonecrotic factor [101] as well as by still unknown factors.

Cases of mild diarrhea caused by non-toxigenic hemolytic strains were regularly registered in Azerbaijan on the background of circulation of such strains in surface water reservoirs [45]. In the beginning of the 21st century, non-toxigenic strains were isolated from individual patients and carriers during the cholera outbreaks in Turkmenistan [46], sporadic diseases in Krasnodar Territory, Astrakhan, Volgograd, Kirov, Sakhalin regions, Chelyabinsk, Kalmykia, and others. The "sporadic" isolates differed from each other by the absence of certain genes from the tested set (*rtxA*, ACD-*rtxA*, ACD-*vgrG*, *int*, *nanH*, *vce*, and *mshA*) but all contained distal part of *rtxA, rtxC, toxR, cef, hapA, tolQRA*, lacked *slt1* and *stn/sto*, and showed diverse VNTR genotypes [8].

The severity of sporadic cases of cholera caused by *ctxAB*$^-$ *tcpA*$^-$ strains varies from mild diarrhea to significant dehydration. The latter mainly concerns immunodeficient humans weakened by concomitant or chronic illnesses. For example, a non-toxigenic strain was isolated in Chelyabinsk during 2000 from a cholera patient with hormone-dependent bronchial asthma and chronic gastritis. The signs of cholera were fierce including abundant watery diarrhea, vomiting, and muscle cramps. However, the agent lacked not only CTX element and VPI but also a major part of VPI-2, *mshA* (in the presence of *mshQ*), ACD-*vgrG, stn/sto,* and *slt1*. From a set of tested genes the agent contained only RTX cluster (including ACD-*rtxA*) and *hapA* [7, 8]. In the infant rabbit model, the strain caused moderate enteropathogenic effect in response to 10^9 colony forming units (CFU) and in vitro showed proteolytic, hemolytic, and phospholipase activity. So the severe diarrhea might be completely or partly stipulated by these factors and RtxA in highly sensitive humans [7].

Another strain isolated in Kalmykia during 2002 from a case of mixed cholera–salmonellosis infection had similar phenotype and PCR genotype but caused a significant enteropathogenic effect in infant rabbits challenged with 10^7 CFU. Nonetheless, its contribution to diarrhea observed in the patient is still unclear. Molecular typing of both strains by means of mono- and multiprimer PCR has demonstrated that they differed from each other as well as from *V. cholerae* human and environmental strains isolated in other parts of Russia [7].

It is still not clear whether non-toxigenic human pathogens are the deletion mutants of epidemic strains somehow preserved after certain outbreaks or endemic clones imported from other countries.

4.3.3 CTXφ$^-$/Pre-CTXφ$^-$ Non-O1, Non-O139 Strains

As well as reported in many countries [102–109], in Russia and CIS countries the role of *ctxAB*$^-$ *V. cholerae* non-O1, non-O139 in human pathology is also restricted by sporadic cases and local outbreaks of acute enteric diseases [110] and, rarely, generalized fatal infections [111], which, however, would not be ignored. Numerous non-O1, non-O139 cases were registered presumably in most of the CIS countries for more than a century.

In the former USSR, non-O1, non-O139 infections were registered virtually on all territories excluding the most northern regions. A number of focal outbreaks occurred in the 1960s–1970s in Rostov region, Moldova, Kazakhstan, and Uzbekistan; the other cases were rather sporadic though larger in numbers with several fatal cases [110]. Multiple gastroenteritis diseases caused by *V. cholerae* O47 were registered in Rostov region in 1970 [112] and in Minsk region (Byelorussia) in 1979 [113]. Local outbreaks took place in Minsk among the participants of weddings with predominance of asymptomatic carriers. Additionally, 11 sporadic cases including 4 mild and 7 of median severity were revealed. The isolated strains were susceptible to tetracycline, chloramphenicol, streptomycin, monomycin, kanamycin, erythromycin, penicillin, ampicillin, and furazolidon. Only one strain was resistant to chloramphenicol, erythromycin, and furazolidon. All were lethal and enteropathogenic in the infant rabbit model.

Interestingly, the non-O1, non-O139 *V. cholerae* was susceptible to most of the antibacterial drugs. For instance, all the 983 strains isolated from humans in different parts of Ukraine from 1999 to 2003 showed variable levels of susceptibility to most of the 16 tested antibiotics but only few were resistant to low concentrations of streptomycin, chloramphenicol, polymyxin, and benzylpenicillin [114].

Unfortunately, most publications reporting about isolation of non-O1, non-O139 vibrios from patients and carriers in Russia and other CIS countries lack any molecular characteristics of the agents, minimizing the information on their pathogenic potential. Only during the last two to three decades, certain retrospective molecular studies were conducted by few groups of investigators [83, 84, 115, 116].

Cases of gastroenteritis caused by non-toxigenic non-O1, non-O139 strains were found in Astrakhan almost every year. PCR analysis of strains isolated from 1973 to 2003 revealed the absence of not only CTX, RS1, and VPI but also VSPI and VSPII, *nanH*, and *mshA* in 26 out of 27 tested isolates [115]. Only seven strains carried *mshQ* and *att*RS while *rtxA, hapA,* and *toxR* were detected in all strains. One strain isolated in 1976 had a different genotype containing three genes of VPI (*aldA*, *tcpP*, and *tcpH* in the absence of *tcpA* and *toxT*), *nanH,* and *mshA*. RAPD-PCR patterns of strains isolated in 2000 and 2002 were variable with the exception of two strains isolated in 2000 showing no clonal relatedness between them.

As mentioned earlier, the majority of strains from patients with acute diarrhea are obtained from Uzbekistan. However, 249 of the recent (2000–2001) human isolates tested did not contain genetic determinants of any particular virulence factors such as *stn/sto* or *slt1*. *ctxAB* genes were detected in none of the strains and only five of them were positive for *tcpA*. The etiology of acute enteric disease was confined to serogroups O18, O62, O82, and O37 but the total number of identified serogroups was 32 among 84% of strains [83, 117]. In fact, no distinct correlation was seen between the affiliation to certain serogroup and etiological outcome of the agent. In Kalmykia and Astrakhan, serogroups O2, O5, O8, O18 and O83, O13, O65 were frequently detected among non-cholera *Vibrio*-mediated infections. Human isolates from India were also included in this investigation and showed prevalence of O5, O11, and O47 among 15 common serogroups [83].

Strains isolated in CIS countries caused variable manifestations of the enteropathogenic effect in the infant rabbit model [84]. Further investigations showed that non-toxigenic *V. cholerae* non-O1, non-O139 usually carry limited sets of virulence genes as compared to O1 serogroup. Nevertheless, they are able to cause human diseases of different severity but the main responsible factors remain undetermined. As a rule, they contain RTX cluster, *cef* and *hapA* genes. We failed to find *stn/sto, tdh,* or *trh* genes among the CIS isolates while their presence in non-O1, non-O139 genomes is reported elsewhere [9, 118, 119]. In the absence of VSPI, VSPII was identified in one O50 strain from Rostov-on-Don isolated in 1974 [66].

Recently a type III secretion system (TTSS) that is related to the TTSS2 gene cluster found in a pandemic clone of *Vibrio parahaemolyticus* was identified in the genomes of certain non-O1, non-O139 strains including isolates from patients with severe watery diarrhea. The genes for this *V. cholerae* TTSS system appear to be present in many non-O1, non-O139 strains, both clinical and environmental [119, 120]. The TTSS genes were localized in a 48 kb gene *vcs* cluster on chromosome I [119] and were predicted to encode the structural components of the TTSS membrane-associated pore-forming complex (*vcs*CJRTQVU) and an ATPase that provides energy to drive secretion by means of the apparatus (VcsN) [120].

PCR screening of non-toxigenic *V. cholerae* O9, O17, O41, O50, O85, O87 and other serogroups isolated in Astrakhan, Rostov, Saratov and Kuibyshev (Samara) regions, Kalmykia, Bashkortostan, Turkmenia, and Uzbekistan in 1968–2002 (as well as several strains from Southeastern Asia) mainly from patients for the presence of TTSS genes showed that nearly 30% carried all four tested genes, *vcsC2, vcsN2, vcsV2,* and *vspD*, suggesting the presence of the intact *vcs* cluster, which might contribute to the pathogenic properties of these strains. In contrast, toxigenic O1, O139, and non-O1, non-O139 strains investigated so far including 22 isolates from Uzbekistan contained none of these genes [116]. Thus, among $CTX^{+}TCP^{+}$ *V. cholerae* up to now only representatives of O141 were found to be positive for TTSS [120]. The reasons for enteropathogenicity of other strains lacking *vcs* genes remain to be elucidated.

4.4 Environmental Strains: Harmless Refugees or Ambushing Bandits?

Due to the existing danger of cholera import and spread, rigorous monitoring is in practice in all the territories conducted by Sanitary and Epidemiological Services of Russia and CIS countries. Tens and hundreds of O1, O139, and non-O1, non-O139 strains are being isolated from surface water reservoirs every year. The large-scale PCR detection of *ctxAB* and *tcpA* genes is aimed at the revelation of potentially epidemic agents. However, environmental toxigenic strains usually accompany cholera outbreaks and are rarely detected in the disease-free periods [58] while the non-toxigenic strains are so numerous that they are considered as representatives of indigenous microflora. On the other hand, some authors presumed subsistence of long-existing temporary natural foci of infection in some cholera-prone areas supported by favorable ecosystem [34, 40] and even a view was given considering cholera as an environment-related sapronose infection [121].

During cholera outbreaks, the causative agents contaminating water represent real danger even in CIS countries with temperate climate. Their long persistence may be related to the favorable environmental conditions. As seen in many investigations, *V. cholerae* reside in the intestine of carriers. Indeed, cases of prolonged carriers of *V. cholerae* were observed in Russia and Ukraine [122]. However, in experimental models it was shown that the formation of endemic niche of cholera is impossible in climatic conditions of European Russia. It was concluded that the organism's ability to temporarily accumulate in water ecosystems as a result of recurrent fecal contaminations has no connection with real nidification [78]. Such a possibility seems to be probable for southern CIS countries with hot climate (Uzbekistan, Kazakhstan, Tajikistan, Kirgizia) [46] but the epidemiological analysis usually reveals the imported nature of cholera from the neighboring countries where endemicity for the disease prevails with constant circulation of cholera vibrios belonging to different serogroups, carrying diverse combinations of virulence genes. Nevertheless, only few strains were shown to have epidemic potential and even they have to undergo an enrichment phase in the intestinal environment of a mammalian host before their spread [123].

In Russia, *ctxAB*$^-$ *tcpA*$^+$ strains are also considered to possess certain epidemic potential as possible recipients of CTXφ [124]. This possibility cannot be excluded but CTXφ infects recipient *V. cholerae* El Tor strains much more effectively in vivo than in vitro [10, 11]. Recently, it was shown that the wild-type *ctxAB*$^-$ *tcpA*$^-$ strains may uptake CTXφ due to the products of *tolQRA* genes [20], which are present in all wild-type *V. cholerae* strains investigated so far [125].

Attempts are being made to explain the possibility of long-term persistence of *V. cholerae* by its transfer to non-culturable forms with subsequent reversion with all its virulence features. The non-culturable forms of toxigenic *V. cholerae* were detected by means of PCR in the natural water reservoirs in Crimea, Kharkov region (Ukraine), and Azerbaijan [126]. Though the ability of their reversion was shown in experimental models [127–129] the real danger of such reverted forms of *V. cholerae* was not confirmed. The first results of a recent study on biofilm formation by *V. cholerae* strains isolated in the Russian Federation showed their potential ability to generate cell monolayers as an intermediate state of biofilms [130].

In any case, most of the CIS environmental strains once being expatriated from humans seem to become "poor outcasts" forced to adapt somehow to the new surroundings, frequently irreparably losing their virulence genes while the minority of them residing in favorable conditions of asymptomatic carriers may represent "ambushing war criminals" awaiting for an occasion to take revenge.

4.5 Conclusions

Epidemic situation and cholera morbidity in Russia and other CIS countries show that from the very end of the 20th century, the spread of diseases caused by epidemic *V. cholerae* strains tends to decrease, but infrequent imported sporadic cases continue. However, the existence of stable and temporary endemic foci was identified in a number of Asian and African countries and migration of people increase the threat of import of infection and consequent prediction of cholera in Russia and other CIS countries.

Both toxigenic and non-toxigenic *V. cholerae* strains are epidemiologically very important. Though usually unable to cause a large-scale epidemic, the non-toxigenic vibrios must not be ignored as the symptoms of diarrhea are frequently severe.

The exact reasons of elevated virulence of certain *ctxAB*$^-$ strains are not yet clear but obviously are associated with the production of a number of accessory toxins and additional virulence factors, which are likely to be "interchangeable" in the absence of CT that usually "masks" their activity in toxigenic strains. In animal models, the non-toxigenic strains with different sets of virulence genes often show the same or very similar effects, maybe due to diverse expression levels of variable genes by individual strains and/or to the competition of the products for appropriate secretion systems. Cholera vibrios of distinct serogroups with diverse sets of virulence genes are considered by some authors as possible natural sources of genetic determinants for the emergence of new epidemic clones. Taking into account the extremely high plasticity of *V. cholerae* genome as well as emergence of O139 and O141 as epidemic strains the probability of further "surprises" is anticipated. More investigations are needed to ascertain whether the historically "endemic areas" of Russia and other CIS countries may become "cradles" for the newly emerged and most virulent strains of *V. cholerae*.

References

1. Onischenko GG, Lomov YuM, Moskvitina EA, Podosinnikova LS. Cholera. In: Pokrovsky VI, Onischenko GG, Cherkassky BL, editors. Evolution of infectious diseases in Russia in the XX century. Moscow: "Meditsina"; 2003.pp. 103–36 [in Russian].
2. Onischenko GG, Lomov YuM, Moskvitina EA, Fedorov YuM, Podosinnikova LS, Gorobets AV. Cholera at the beginning of the XXI century. Prognosis. Zh Mikrobiol Epidemiol Immunobiol. 2005;3:44–8 [in Russian].
3. Bachelor RA, Wignall FS. Nontoxigenic O1 *Vibrio cholerae* in Peru: a report of two cases associated with diarrhea. Diagn Microbiol Infect Dis. 1988;10:135–8.
4. Pal A, Saha PK, Nair GB, Yamasaki S, Takeda T, Takeda Y, Bhattacharya SK, Ramamurthy T. Clonal analysis of non-toxigenic *Vibrio cholerae* O1 associated with the outbreak of cholera. Indian J Med Res. 1999;109:208–11.
5. Pichel M, Rivas M, Chinen I, Martin F, Ibarra C, Binsztein N. Genetic diversity of *Vibrio cholerae* O1 in Argentina and emergence of a new variant. J Clin Microbiol. 2003;41: 124–34.
6. Puzanova NG, Smolikova LM, Monakhova EV, Podosinnikova LS, Bozhko NV, Arutyunov YuI, Golenischeva EN, Kudryakova TA, Sanamyants EM, Mikhas NK. Retrospective characteristics of atoxigenic cholera vibrios O1 isolated from patients with cholera manifestations. In: Cholera and *Vibrio* Human Pathogens, Materials of the Cholera Problem Committee, Rostov-on-Don, vol 18; 2005. pp. 57–9 (in Russian).
7. Smirnova NI, Kostromitina EA, Toporkov AV, Kutyrev VV. Atypical *Vibrio cholerae* El Tor strains isolated from human beings. Epidemiol Infect Dis. 2005;5:15–20 (in Russian).
8. Onishchenko GG, Lomov YuM, Moskvitina EA, Podosinnikova LS, Monakhova EV, Vodopyanov AS, Vodyanitzkaya SYu, Prometnoy VI, Telesmanich NR, Dudina NA. Epidemic manifestations of cholera associated with *ctxAB*$^-$ *tcpA*$^+$ *Vibrio cholerae* O1. Zh Mikrobiol Epidemiol Immunobiol. 2007;1:23–9 (in Russian).
9. Sears CL, Kaper JD. Enteric bacterial toxins: mechanisms of action and linkage to intestinal secretion. Microbiol Revs. 1996;60:167–215.
10. Faruque SM, Albert MJ, Mekalanos JJ. Epidemiology, genetics and ecology of toxigenic *Vibrio cholerae*. Microbiol Mol Biol Revs. 1998;62:1301–14.
11. Waldor MK, Mekalanos JJ. Lysogenic conversion by a filamentous phage encoding cholera toxin. Science. 1996;272:1910–4.

12. Karaolis DKR, Kaper JB. Pathogenicity islands and other mobile virulence elements of *Vibrio cholerae.* In: Kaper JB, Hacker J, editors. Pathogenicity islands and other mobile virulence elements, vol 9. Washington DC: Amer Soc for Microbiol; 1999. pp. 167–87.
13. Heilpern AJ, Waldor MK. $pIII^{CTX}$, a predicted CTXφ minor coat protein, can expand the host range of coliphage fd to include *Vibrio cholerae.* J Bacteriol. 2003;183:1037–44.
14. Monakhova EV, Pisanov RV. Toxins of *Vibrio cholerae.* Mol Gen Microbiol Virusol. 2005;1:7–18. (in Russian).
15. Uzzau S, Fasano A. Cross-talk between enteric pathogens and the intestine. Cell Microbiol. 2000;2:83–9.
16. Faruque S.M., Asadulghani, Kamruzzaman M, Ghosh AN, Nair GB, Mekalanos JJ, David A. Sack DA. RS1 element of *Vibrio cholerae* can propagate horizontally as a filamentous phage exploiting the morphogenesis genes of CTXφ. Infect Immun. 2002;70:163–70.
17. Sheahan KL, Cordero CL, Satchell KJ. Identification of a domain within the multifunctional *Vibrio cholerae* RTX toxin that covalently cross-links actin. Proc Natl Acad Sci USA. 2004;101:9798–9803.
18. Jermin WS, Boyd EF. Characterization of a novel *Vibrio cholerae* pathogenicity island (VPI-2) encoding neuraminidase (*nanH*) among toxigenic *Vibrio cholerae* isolates. Microbiology. 2002;148: 3681–93.
19. Monakhova EV, Pisanov RV Comparative analysis of literature and computer data on the properties of cytotoxic factor NMDCY and hemagglutinin/protease of *Vibrio cholerae.* Problemy OOI, Saratov. 2006;91:15–20 (in Russian).
20. Heilpern AJ, Waldor MK. CTXφ infection of *Vibrio cholerae* requires *tolQRA* gene products. J Bacteriol.2000;182:1739–47.
21. O'Brien AD, Holmes RK. Shiga and shiga-like toxins. Microbiol Revs. 1987;51:206–20.
22. Honda T, Iida T. The pathogenicity of Vibrio parahaemolyticus and the role of the thermostable direct haemolysin and related haemolysins. Rev Med Microbiol. 1993;4:106–13.
23. Burgasov PN. Cholera El Tor (Mannual for doctors), Moscow: "Meditsina". 1976 (in Russian).
24. Murnachev GP. Regularity of cholera spread in the Far East of Russia. In: Cholera and *Vibrio* Human Pathogens, Materials of the Cholera Problem Committee, Rostov-on-Don, vol 20; 2007. pp. 32–3 (in Russian).
25. Smirnova NI, Cheldyshova NB, Zadnova SP, Kutyrev VV. Molecular-genetic features of *Vibrio cholerae* classical strains that caused the Asiatic cholera epidemics in Russia in 1942. Mol Gen Mikrobiol Virusol. 2001;4:12–8 (in Russian).
26. Smirnova NI, Cheldyshova NB, Zadnova SP, Kutyrev VV. Molecular-genetic peculiarities of classical biotype *Vibrio cholerae,* the etiological agent of the last outbreak Asiatic cholera in Russia. Microb Pathog. 2004;36:131–9.
27. Nair GB, Faruque SM, Bhuiyan NA, Kamruzzaman M, Siddique AK, Sack DA. New variants of *Vibrio cholerae* O1 biotype El Tor with attributes of the classical biotype from hospitalized patients with acute diarrhea in Bangladesh. J Clin Microbiol. 2002;40: 3296–9.
28. Nair GB, Qadri F, Holmgren J, Svennerholm A-M, Safa A, Bhuiyan NA, Ahmad QS, Faruque SM, Faruque ASG, Takeda Y, Sack DA. Cholera due to altered El Tor strains of *Vibrio cholerae* O1 in Bangladesh. J Clin Microbiol. 2006;44:4211–3.
29. Lee JH, Kyung Ho Han KH, Choi SY, Lucas MES, Mondlane C, Ansaruzzaman M, Nair GB, Sack DA, von Seidlein L, Clemens JD, Song M, Chun J, The Mozambique Cholera Vaccine Demonstration Project Coordination Group, Kim DW. Multilocus sequence typing (MLST) analysis of *Vibrio cholerae* O1 El Tor isolates from Mozambique that harbor the classical CTX prophage, J Med Microbiol. 2006;55:165–70.
30. Faruque SM, Tam VC, Chowdhury N, Diraphat P, Dziejman M, Heidelberg JF, Clemens JD, Mekalanos JJ, Nair GB. Genomic analysis of the Mozambique strain of *Vibrio cholerae* O1 reveals the origin of El Tor strains carrying classical CTX prophage. Proc Natl Acad Sci USA. 2007;104:5151–6.

31. Nair GB, Bhuiyan NA, Nusrin S, Alam N, Safa A, Qadri F, Faruque SM, Siddique AK, Takeda Y, Sack DA. Molecular basis of the emergence of a new more severe form of cholera. *In: Vibrio* 2007, Abstract book of the second conference on the biology of vibrios, Paris 2008;S4-1.
32. Moskvitina EA. Current trends in the evolution of the seventh cholera pandemics. Problemy OOI. 2008;1:22–6 (in Russian).
33. Narkevich MI, Onischenko GG, Lomov YuM, Moskvitina EA, Podosinnikova LS, Medynsky GM. The seventh pandemic of cholera in the USSR, 1961–1989. WHO Bulletin OMS. 1993;71:189–96.
34. Mishan'kin BN, Vodopianov AS, Lomov YuM, Romanova LV, Vodopianov SO, Suchkov IYu, Mishan'kin MB, Cherepakhina IYa, Duvanova OV, Shishiyanu MV. Retrospective VNTR analysis of genotypes of *Vibrio cholerae* O1 strains isolated during the 7th cholera pandemic in Rostov region. Mol Gen Mikrobiol Virusol. 2004;4:28–33 (in Russian).
35. Bogdanov IK, Grizhebovsky GM, Isaev VP, Yundina GP. Past and present of cholera in the Stavropol territory. *In:* Cholera, Materials of Russian scientific conference, Rostov-on-Don, 1995. pp. 29–31 (in Russian).
36. Efremenko VI, Belyaev EN, Nikiforov VN, Gnutov IN, Bogdanov IK, Postovoy PP. Cholera outbreak in the city of Stavropol in 1990. In: Cholera, materials of Russian scientific conference, Rostov-on-Don; 1995. pp. 40–3 (in Russian).
37. Monakhova EV., Vlasov VP, Moskvitina EA, Alekseenko VV. Prospects for use of genome fingerprinting in *Vibrio cholerae* studies. In: Cholera, Materials of Russian scientific conference, Rostov-on-Don; 1995. pp. 100–3 (in Russian).
38. Onischenko GG, Belyaev EN, Moskvitina EA, Rezaikin VI, Lomov YuM, Medinsky GM. Cholera in dagestan. In: Medinsky GM, editors. Rostov-on-Don: Polygraph; 1995 [in Russian].
39. Onishchenko GG, Lomov YuM, Moskvitina EA. Cholera in the Republic of Dagestan. Zh Mikrobiol Epidemiol Immunobiol. 1995;2:3–8 [in Russian).
40. Onishchenko GG, Lomov YuM, Moskvitina EA, Chiburaev VI, Liakhover ML, Rokhoev MR, Terent'ev AN, Asvarov BM, Chernyshev SN, Gadzhieva SZ. The epidemiological characteristics of cholera in the Republic of Dagestan. An assessment of the epidemic control measures. Zh Mikrobiol Epidemiol Immunobiol. 1995;2:9–22 (in Russian).
41. Podosinnikova LS, Mazrukho BL, Lomov YuM, Khanum'ian TA, Voronezhskaya LG, Kudryakova TA, Makedonova LD, Mironova AV. The characteristics of *Vibrio cholerae* isolated in Dagestan in 1994. Zh Mikrobiol Epidemiol Immunobiol. 1995;2:74–8 (in Russian).
42. Mishan'kin BN, Romanova LV, Vodopianov SO, Ryzhko IV, Kudryakova TA, Cherepakhina IYa. Comparative analysis of *Vibrio cholerae* strains isolated in Dagestan from humans in 1994 and 1998. In: Cholera and *Vibrio* Human Pathogens, Materials of the Cholera Problem Committee, Rostov-on-Don, vol 12; 1999. pp. 36–9. (in Russian).
43. Ryzhko IV, Dudina NA, Lomov YuM, Shutko AG, Tsuraeva RI, Anisimov BI. Activity of 22 antibacterials against O1 and O139 serogroup *Vibrio cholerae* strains isolated from humans within 1927–2005 in various regions of the world. Antibiotics and Chemotherapy. 2005;50:38–42 (in Russian).
44. Vodopianov SO, Vodopianov AS, Oleinikov IP, Goncharov EK, Duvanova OV, Suchkov IYu, Cherepakhina IYa, Mishan'kin BN. Multilocus VNTR-typing of *Vibrio cholerae* strains isolated during epidemic outbreaks in Russian territory. In: Cholera, Materials of the VIII Russian scientific conference on cholera problem, Rostov-on-Don, 2003. pp. 114–9 (in Russian).
45. Kurbanov ShKh. Epidemic manifestation of cholera in Azerbaijan Republic in the last decade. In: Cholera and *Vibrio* Human Pathogens, Materials of the Cholera Problem Committee, Rostov-on-Don, vol 14; 2001. pp. 14–5. (in Russian).
46. Smirnova NI, Kostromitina EA, Cheldyshova NB, Kutyrev VV. Differences in virulence genes in *Vibrio cholerae* El tor strains isolated from different sources in Turkmenistan territory. Mol Gen Mikrobiol Virusol. 2002;4:12–8. (in Russian).

47. Lomov YuM, Characterization of *Vibrio cholerae* strains isolated in Ukraine in 1991. *In:* Cholera, Materials of Russian scientific conference, Rostov-on-Don; 1992. pp. 71–3 (in Russian).
48. Onishchenko GG, Lomov YuM, Moskvitina EA, Podosinnikova LS, Moiseeva AV, Alekseenko VV, Chebotar' IF, Shestialtynova IS, Vlasov VP, Kudriakova TA. Cholera in Ukraine and Moldova during the 7th pandemic. Zh Mikrobiol Epidemiol Immunobiol. 1993;2:52–7 (in Russian).
49. Moskvitina EA, Lomov YuM, Golubev BP, Pasyukov VV, Kravchenko AN, Fetsailova OP, Karbyshev GL. Cholera outbreak in Vilkovo, Odessa region, in 1991. In: Cholera, Materials of Russian scientific conference, Rostov-on-Don; 1992. pp. 44–7 (in Russian).
50. Evtodienko VG, Slyusar' VN, Gutsu AV, Kozokaru RI, Benesh OA, Sandul GN. On the problem of organization of epidemic cholera control in Moldova republic. *In:* Cholera and *Vibrio* Human Pathogens, Materials of the Cholera Problem Committee, Rostov-on-Don, vol 20; 2007. pp. 37–8 (in Russian).
51. Moskvitina EA, Lomov YuM, Gorobets AV, Podosinnikova LS, Ivanova SM. Epidemic situation on cholera in the world, CIS countries and Russia (2001–2005). In: Cholera and *Vibrio* Human Pathogens, Materials of the Cholera Problem Committee, Rostov-on-Don, 2005. pp. 15–8 (in Russian).
52. Moskvitina EA, Lomov YuM, Prometnoy VI. Importations of cholera to Russia (1998–2007). In: Cholera and *Vibrio* Human pathogens, materials of the cholera problem committee, Rostov-on-Don, vol 21; 2008. pp. 16–8 (in Russian).
53. Onishchenko GG, Lomov YuM, Mishan'kin BN, Mazrukho BL, Podosinnikova LS, Kudryakova TA, Moskvitina EA, Vodopianov SO, Ryzhko IV, Kazakova ES, Sharova IN, Plotnikova EA, Davydova NA, Abramova EG, Korolev IuS, Shestialtynova IS, Sharifulina DM, Kuriaeva NIu, Iumangulova EF, Chernyshova AV, Bugorkova TV, Rusakova TG, Maslenikova AL, Milova MV, Zakharova LI, Bilalova TG, Shut'ko AG, Kachkina GV. Characterization of *Vibrio cholerae* El Tor in the city of Kazan in 2001. Zh Mikrobiol Epidemiol Immunobiol. 2002;2:3–6 (in Russian).
54. Dudina NA, Ryzhko IV, Lomov YuM, Tsuraeva RI, Shut'ko AG. Activity of antibacterial preparates *in vitro* and *in vivo* against *Vibrio cholerae* strains isolated in Kazan city in 2001. Uspekhy Sovremennogo Estestvoznaniya. 2003;6:48–9 (in Russian).
55. Mishan'kin BN, Vodopianov AS, Lomov YuM, Vodopianov SO, Romanova LV, Cherepakhina IYa, Suchkov IYu, Duvanova OV, Shishiyanu MV. Multilocus VNTR-genotyping of the *Vibrio cholerae* isolated in Kazan during the outbreak of cholera in summer of 2001. Zh Mikrobiol Epidemiol Immunobiol. 2003;6:11–5 (in Russian).
56. Moskvitina EA, Lomov YuM, Gavrineva AV, Podosinnikova LS, Andrusenko IT, Vodyanitskaya SYu, Smolikova LM, Ivanova SM, Shestialtynova IS, Monakhova EV. Cholera in the world, CIS countries and Russia: morbidity rate and prognosis. In: Cholera and *Vibrio* Human Pathogens, Materials of the Cholera Problem Committee, Rostov-on-Don, vol 19; 2006. pp. 10–2 (in Russian).
57. Sagymbek UA, Askarov AM, Suleimenov BM, Musagalieva RS, Ospanov KS, Ivanova RS, Kosovtsev VYa, Vystrepov VN, Zhdanov IG, Mustapaev EA, Afanasiev LI, Muzarbekov AM, Neizer VE, Beisen AK, Korkembaeva G, Khamzin GKh. An outbreak of cholera in Aktau and Zganaozen cities of Stausk region of Kazakhstan republic in 2001. *In:* Cholera and *Vibrio* Human Pathogens, Materials of the Cholera Problem Committee, Rostov-on-Don, vol 20; 2007. pp. 39–42 (in Russian).
58. Podosinnikova LS, Lomov YuM, Vlasov VP, Monakhova EV, Moskvitina EA, Lobanova LI, Bardykh ID, Shestialtynova IS, Glyan'ko EV. To the virulence of *Vibrio cholerae* isolated in different epidemic situations. Zh Mikrobiol Epidemiol Immunobiol. 1993;3:48–51 (in Russian).
59. Vlasov VP, Bardakhchian EA, Kharlanova NG, Monakhova EV, Vinokur NI. Comparative studies of vct^{+} and vct^{-} *Vibrio cholerae* virulence. In: Cholera, materials of Russian scientific conference, Rostov-on-Don, 1995. pp. 92–8 (in Russian).

60. Vlasov VP, Kharlanova NG, Monakhova E.V, Vinokur NI. Bardakhchian EA. Ultrastructural changes in the small intestine caused by choleragenic and non-choleragenic *Vibrio cholerae* strains. Doklady Akademii Nauk. 1997;352:278–80. (in Russian).
61. Monakhova EV, Pisanov RV, Mikhas' NK. Investigation of genome polymorphism of *ctxAB⁻Vibrio cholerae* strains carrying proximal part of the CTX element. Problemy OOI, Saratov. 2001;2:90–6 (in Russian).
62. Ramamurthy T, Garg S, Sharma R, Bhattacharya SK, Nair GB, Shimada T, Takeda T, Karasawa T, Kurazano H, Pal A, Takeda Y. Emergence of a novel strain of *Vibrio cholerae* with epidemic potential in southern and eastern India. Lancet. 1993;341:703–4.
63. Mironova AV, Men'shikova EA. Comparative studies of the invasive properties of O1, O139 and non-O1 vibrios. In: Cholera and *Vibrio* human pathogens, materials of the cholera problem committee, Rostov-on-Don, vol. 14; 2001. pp. 57–8 (in Russian).
64. Voronezhskaya LG, Somova AG, Podosinnikova LS, Kozyreva LA. Experimental generalized NAG infection. Acute Intestinal Infections. Leningrad. 1982;6:84–8 (in Russian).
65. Monakhova EV, Mikhas NK, Mazrukho BL. Diversity of the genome integrated CTX element in some toxigenic *Vibrio cholerae* Bengal strains. Problemy OOI, Saratov. 2003;85:90–6 (in Russian).
66. Eroshenko GA, Kukleva LM, Shavina NIu, Kutyrev VV. Molecular-epidemiological characteristic and possible origin of *Vibrio cholerae* non-O1, non-O139 with complete and limited set of virulence genes. Zh Mikrobiol Epidemiol Immunobiol. 2007;5:24–8 (in Russian).
67. Shimanyuk NYa, Duvanova OV, Suchkov IYu, Mishan'kin BN. Neuraminidase of *Vibrio cholerae* O139 "Bengal": discovery, purification and certain properties. Biotechnology. 1999;3:56–62 (in Russian).
68. Duvanova OV, Shimanyuk NYa, Atarova GT, Mishan'kin BN. Neuraminidase and tweenase activity of *Vibrio cholerae* O139 Bengal serogroup of different origin. *In:* Problems of sanitary and epidemic protection of territories of the CIS countries, Abstracts of the International Scientific Conference, Saratov 1998;160–1 (in Russian).
69. Duvanova OV, Shimanyuk NYa, Mishan'kin BN. The ability of tween 20 cleavage as a differential feature of O139 serogroup vibrios of different origin. Clin Lab Diagnostics. 2000;5:48–9 (in Russian).
70. Monakhova EV., Mikhas' NK., Mazrukho B.L Diversity of the genome integrated CTX element in some toxigenic *Vibrio cholerae* Bengal strains. Problemy OOI, Saratov. 2003;85:90–6 (in Russian).
71. Eroshenko GA, Osin AV, Cheldyshova NB, Smirnova NI. Genetic peculiarities of avirulent and virulent *Vibrio cholera* serogroup O139 strains isolated on the territory of Russia. Problemy OOI, Saratov. 2001;1:69–76 (in Russian).
72. Sharma C, Nair GB, Mukhopadhyay AK, Bhattacharya SK, Ghosh RK, Ghosh A. Molecular characterization of *Vibrio cholerae* O1 biotype El Tor strains isolated between 1992 and 1995 in Calcutta, India: evidence for the emergence of a new clone of the El Tor biotype. J Infect Dis. 1997;175:1134–41.
73. Faruque SM, Ahmed KM, Siddique AK, Zaman K, Alim AR, Albert MJ. Molecular analysis of toxigenic *Vibrio cholerae* O139 Bengal strains isolated in Bangladesh between 1993 and 1996: evidence for emergence of a new clone of the Bengal vibrios. J Clin Microbiol. 1997;35:2299–306.
74. Basu A, Garg P, Datta S, Chakraborty S, Bhattacharya T, Khan A, Ramamurthy T, Bhattacharya SK, Yamasaki S, Takeda Y, Nair GB. *Vibrio cholerae* O139 in Calcutta, 1992–1998: Incidence, antibiograms, and genotypes. Emerging Infect Dis. 2000;6: 139–7.
75. Eroshenko GA, Osin AV, Shchelkanova EYu, Smirnova NI. A comparative analysis of genomes of virulent and avirulent *Vibrio cholerae* O139. Mol Gen Microbiol Virusol. 2004;2:11–6 (in Russian).
76. Faruque SM, Siddique AK, Saha MN, Asadulghani, Rahman MM, Zaman K, Albert MJ, Sack DA, Sack RB. Molecular characterization of a new ribotype of *Vibrio cholerae*

O139 Bengal associated with an outbreak of cholera in Bangladesh. J Clin Microbiol. 1999;37:1313–8.

77. Mishan'kin BN, Romanova LV, Lomov YuM, Shimaniuk NIa, Vodop'ianov SO, Cherepakhina IIa, Suchkov IYu, Duvanova OV. *Vibrio cholerae* O139 isolated from humans and the water from open reservoirs: a comparative genotyping. Zh Mikrobiol Epidemiol Immunobiol. 2000;3:3–7 (in Russian).
78. Ochirov VKh, Sikhov VV, Savchenko ST, Lobanov AN, Yashkulov KB. To the endemicity of cholera in the moderate climatic conditions of the European part of Russia on the example of Volgograd region and republic Kalmykia. In: Cholera and *Vibrio* human pathogens, materials of the cholera problem committee, Rostov-on-Don, vol. 18; 2005. pp. 34–6 (in Russian).
79. Morris JG. Non-O group 1 *Vibrio cholerae*: a look at the epidemiology of an occasional pathogen. Epidemiol Rev. 1990;12:179–91.
80. Mukopadhyay AK, Garg S, Mitra R, Basu A, Rajendran K, Dutta D, Bhattacharia SK, Shimada T, Takeda T, Takeda Y, Nair GB. Temporal shifts in traits of *Vibrio cholerae* strains isolated from hospitalized patients in Calcutta: a 3-year (1993 to 1995) analysis. J Clin Microbiol. 1996;34:2537–43.
81. Bhattacharya MK, Dutta D, Bhattacharya SK, Deb A, Mukopadhyay AK, Nair GB, Shimada T, Takeda Y, Chowdhury A, Mahalanabis D. Association of a disease approximating cholera caused by *Vibrio cholerae* of serogroups other than O1 and O139. Epidemiol Infect. 1998;120:1–5.
82. Dalsgaard A., Serichantalergs O, Forslund A., Lin W, Mekalanos JJ, Mintz E, Shimada T., Wells JG. Clinical and environmental isolates of *Vibrio cholerae* serogroup O141 carry the CTX phage and the genes encoding the toxin-coregulated pili. J Clin Microbiol. 2001;39:4086–92.
83. Avdeeva EP, Mazrukho BL, Voronezhskaya LG, Tsedova EG, Monakhova EV, Mikhas' NK, Nyematov AS, Inogamova IA, Obrotkina NF, Malkova GI, Davydova VK, Agapov BL. The specific features of originally different *Vibrio cholerae* non-O1, non-O139 serogroups. Epidemiol Infect Dis. 2006;2:19–22 (in Russian).
84. Vlasov VP, Voronezhskaya LG, Mikhas' NK, Ivanova LV, Monakhova EV. Isolation and characterization of toxigenic (*vct+*) strains of enteropathogenic *Vibrio cholerae* non-O1 from different territories. In: Cholera, Materials of Russian scientific conference, Rostov-on-Don, 1992. pp. 144–7 (in Russian).
85. Eroshenko GA, Shavina NYu, Krasnov YaM, Guseva NP, Fadeeva AV, Valova TV. Kutyrev VV. Genetic peculiarities of virulent cholera vibrios of non-O1, non-O139 serogroups isolated in Uzbekistan. In: Cholera and *Vibrio* Human Pathogens, Materials of the Cholera Problem Committee, Rostov-on-Don, vol 21; 2008. pp. 63–7 (in Russian).
86. Vodopianov SO, Vodopianov AS, Suchkov IYu, Goncharov EK, Mishan'kin MB, Oleinikov IP, Mishan'kin BN. Genetic diversity of *tcpA* gene in representatives of *Vibrio* genus. In: Kutyrev VV, Saratov, editors. Current technologies in the diagnostics of particularly dangerous infectious diseases, Materials of the 4th Inter-State Scientific and Practical conference of CIS member States, 2003. pp. 38–42 (in Russian).
87. Onishchenko GG, Mishan'kin BN, Lomov YuM, Vodopianov SO, Voronezhskaya LG, Suchkov IYu, Cherepakhina IYa, Duvanova OV, Shishiyanu MV. Serogroups non-O1 *Vibrio cholerae* isolated in Uzbekistan in 1987–1990: a retrospective VNTR analysis. Epidemiol Infect Dis. 2003;6:25–9 (in Russian).
88. Boyd EF, Heilpern AJ, Waldor MK. Molecular analysis of a putative CTXφ precursor and evidence for independent acquisition of distinct CTXφs by toxigenic *Vibrio cholerae*. J Bacteriol. 2000;182:5530–8.
89. Rui Y., Kan B, Gao s, Liu Y, Qi.G. Core region of non-CTXphi genome in *Vibrio cholerae* O139 (JS 9803). Original Source Database: Genbank, Division: PHG, direct submission 2001: AF416590.
90. Monakhova EV, Mikhas' NK, Smolikova LM, Mishan'kin BN. Investigation of genetic determinants of the virulence cassette toxins and neuraminidase of cholera vibrios by means of molecular probes. Zh Microbiol Epidemiol Immunobiol. 2001;2:25–9 (in Russian).

91. Monakhova EV, Lomov YuM, Mikhas' NK, Pisanov RV. Structure and variability of the truncated CTX-element of the cholera vibrios. Problemy OOI, Saratov. 2007;1:58–61 (in Russian).
92. Kruglikov VD, Monakhova EV, Avdeeva EP, Tsedova EG, Mazrukho BL, Shestialtynova IS, Mazrukho AB, Nepomnyaschaya NB, Shalu OA. Genotypic characteristics of *Vibrio cholerae* O1 strains revealed in the environmental water reservoirs of the Rostov-on-Don territories in 2007. In: Cholera and *Vibrio* Human Pathogens, Materials of the Cholera Problem Committee, Rostov-on-Don, vol 21. 2008. pp. 58–60 (in Russian).
93. Monakhova EV, Mironova AV, Alekseeva LP, Mazrukho AB. The study of pre-CTXφ carrying *Vibrio cholerae* virulence: genotypic and phenotypic characteristics. Zh Microbiol Epidemiol Immunobiol. 2008;4:27–32 (in Russian).
94. Bardakhchian EA, Kharlanova NG, Monakhova EV. Ultrastructural changes in the small intestine caused by *Vibrio cholerae* strains producing cholerogen and alternative toxin Russian J Gastroenterol Hepatol Coloproctol. 1997;VII(5 app 4):75 (in Russian).
95. Bardakhchian EA, Lomov YuM, Kharlanova NG, Monakhova EV, Vinokur NI. Ultrastructural alterations in infant rabbit small intestine exposed to *vct*$^{-}$ strain of cholera *Vibrio*. Archives of Pathology (Arkhivy patologii). 2000;1:24–9 (in Russian).
96. Bardakhchian EA, Monakhova EV, Kharlanova NG, Sayamov SR, Pisanov RV. Ultrastructural changes in the small intestine of suckling mice caused by hemagglutinin/protease of *Vibrio cholerae.* Bull Experim Biol Med. 2008;145:475–80 (in Russian).
97. Monakhova EV, Mikhas' NK, Pisanov RV, Salnikova OI, Mazrukho AB, Ovsova LM, Mazrukho BL. Characterization of accessory diarrheagenic and cytotoxic factors produced by *Vibrio cholerae.* Problemy OOI, Saratov. 2004;1:51–4 (in Russian).
98. McCardell BA, Kothary MH, Hall RN, Sathyamoorthy V. Identification of a CHO-elongating factor produced by *Vibrio cholerae* O1. Microb Pathogen. 2000;29:1–8.
99. Monakhova EV, Lomov YuM, Pisanov RV, Alekseeva LP, Markina OV, Mironova AV, Verkina LM, Mikhas' NK, Aseeva LE, Kagramanov VS. Cloning of the *Vibrio cholerae* CHO-cell elongating factor gene in *Escherichia coli* and its expression under the control of P_{BAD}-promoter. Biotechnology. 2005;5:12–8 (in Russian).
100. Walia K, Ghosh S, Singh H, Nair GB, Ghosh A, Sanhi G, Vohra H, Ganguly NK. Purification and characterization of novel toxin produced by *Vibrio cholerae* O1. Infect Immun. 1999;67:5215–22.
101. Atarova GT, Pomukhina OI, Uraleva VS, Cherepakhina IYa. Isolation and certain properties of *Vibrio cholerae* dermonecrotic factor. In: Cholera, Materials of Russian scientific conference, Rostov-on-Don, 1992. pp. 82–3 (in Russian).
102. Finch MJ, Valdespino JL, Wells JG, Perez-Perez G, Arjona F, Sepulveda A, Bessudo D, Blake PA. Non-O1 *Vibrio cholerae* infections in Cancun, Mexico. Am J Med Hyg. 1987;36:393–7.
103. Dalsgaard A, Albert MJ, Taylor DN, Shimada T., Meza R, Serichantalergs O, Echeverria P. Characterization of *Vibrio cholerae* non-O1 serogroups obtained from an outbreak of diarrhea in Lima, Peru. J Clin Microbiol. 1995;33:2715–22.
104. Sharma C, Thungapathra M, Ghosh A, Mukhopadhyay AK, Basu A, Mitra R, Basu I, Bhattachaya SK, Shimada T, Ramamurthy T, Takeda T, Yamasaki S, Takeda Y, Nair GB. Molecular analysis of non-O1, non-O139 *Vibrio cholerae* associated with an unusual upsurge in the incidence of cholera-like disease in Calcutta, India. J Clin Microbiol. 1998;36:756–63.
105. Namdary H., Klaipset CR, Hoghes J.L. A cytotoxin-producing strain of *Vibrio cholerae* non-O1, non-O139 as a cause of cholera and bacteriemia after consumption of raw clams. J Clin Microbiol. 2000;38:3518–9.
106. Singh DV, Matte MH, Matte GR, Jiang S, Sabeena F, Shukla BN, Sanyal SC, Huq A, Colwell RR. Molecular analysis of *Vibrio cholerae* O1, O139, non-O1 and non-O139 strains: clonal relationship between clinical and environmental isolated. Appl Environ Microbiol. 2001;67:910–21.

107. Bidinost C, Saka HA, Aliendro O, Panzetta-Duttari G, Carranza P, Echenique J, Patrito E, Bocco JL. Virulence factors of non-O1, non-O139 *Vibrio cholerae* isolated in Cordoba, Argentina. Rev Argent Microbiol. 2004;36:158–63.
108. Lukinmaa S, Mattila K, Lehtinen V, Hakkinen M, Koskela M, Sitonen A. Territorial waters of Baltic sea as a source of infections caused by *Vibrio cholerae* non-O1, non-O139: report 3 hospitalized cases. Diagn Microbiol Infect Dis. 2006;54:1–6.
109. Restrepo D, Huprikar SS, VanHorn K, Bottone EJ. O1 and non-O1 *Vibrio cholerae* bacteriemia produced by hemolytic strains. Diagn Microbiol Infect Dis. 2006;54: 145–8.
110. Onischenko GG, Ganin VS, Golubinsky EP. Vibrios of non-O1 serological group and their significance in human pathology. Moscow: GOU VUNMC RF. 2001 (in Russian).
111. Milyutin VN, Somova AG, Medinsky GM, Podosinnikova LS, Marchuk LM, Smolikova LM, Podsedlovsky TS, Emdina IA, Arkhipova TV, Rugene KL, Badzilauskene AP, Verekone SA, Kelminskene SA, Kirdeev VK, Uoginatas GP. NAG Vibrio, serovar 2, causing generalized infection. Zh Mikrobiol Epidemiol Immunobiol. 1982;3:24–5. (in Russian).
112. Medinsky GM, Voronezhskaya LG, Somova AG, Filatov HG. Diarrheal diseases caused by *Vibrio cholerae* nonO1 serogroup. Topical problems of microbiology, diagnostics and prophylaxis of cholera, abstracts of the all-union scientific conference, Rostov-on-Don, 1988. pp. 276–9 (in Russian).
113. Zaidenov AM, Grachev YuA, Medinsky GM, Sudzhaev GA, Ryzhikov MI, Smolikova LM, Voronezhskaya LG, Puchkina LA, Somova AG, Gritsenko AN, Podosinnikova LS, Senchuk TT, Kolyadko PI, Guschina IM, Bogdshich FF, Zhoglo NN, Emdina IA, Lobanova LN. Acute gastroenteritis diseases caused by NAG-vibrios of O47 serotype. In: Votyakov VI, Minsk, editors. Voprosy sanitarno-epidemiologicheskogo nadzora, 1979: 20–5 (in Russian).
114. Davidenko SN, Khaitovich AB, Ilyichev YuA. Antibiotic sensitivity of *Vibrio cholerae* serogroup non-O1 isolated in Ukraine from humans during the period from 1999 to 2003. *In*: Cholera and *Vibrio* Human Pathogens, Materials of the Cholera Problem Committee, Rostov-on-Don, vol 18; 2005. pp. 54–6 (in Russian).
115. Eroshenko GA, Koukleva LM, Shavina NYu, Kutyrev VV. Genetic peculiarities of *Vibrio cholerae* strains non-O1, non-O139 serogroups isolated from patients in Astrakhan in 1976–2003. Problemy OOI, Saratov. 2006;2:41–4. (in Russian)
116. Eroshenko GA, Kutyrev VV, Fadeeva AV, Shavina NYu, Stepanov AV. Distribution of type III secretion system genes in cholera vibrios of different serogroups. Zh Mikrobiol Epidemiol Immunobiol. 2008;4:23–7. (in Russian).
117. Avdeeva EP, Mazrukho BL, Voronezhskaya LG, Monakhova EV, Kudryakova TA, Tsedova EG, Neimatov AS, Inogamova IA, Yakubova IS. Characterization of enteropathogenic *Vibrio cholerae* nonO1/O139 isolated in Uzbekistan. Zh Mikrobiol Epidemiol Immunobiol. 2005;3:80–2 (in Russian).
118. Pal A, Ramamurthy T, Bhadra RK, Takeda T, Shimada T, Takeda Y, Nair GB, Pal SC, Chakrabarti S. Reassessment of the prevalence of heat-stable enterotoxin (NAG-ST) among environmental *Vibrio cholerae* non-O1 strains isolated from Calcutta, India, by using a NAG-ST DNA probe. Appl Environ Microbiol. 1992;58:2485–9.
119. Chen Y, Johnson JA, Pusch GD, Morris JG, Stine OC. The genome of non-O1 *Vibrio cholerae* NRT36S demonstrates the presence of pathogenic mechanisms that are distinct from O1 Vibrio cholerae. Infect Immun. 2007;75:2645–7.
120. Dziejman M, Serutto D, Tam VS, Sturtevant D, Diraphat P, Faruque SM, Rahman MH, Heidelberg JF, Decker J, Li Li, Montgomery KT, Grills G, Kucherlapati R, Mekalanos JJ. Genomic characterization of non-O1, non-O139 *Vibrio cholerae* reveals genes for a type III secretion system. Proc Natl Acad Sci USA. 2005;102:3465–70.
121. Litvin VYu. Cholera as natural nidal sapronose infection. *In*: Cholera, Materials of Russian scientific conference on cholera problem, Rostov-on-Don, 2005. pp. 24–7 (in Russian).
122. Lomov YuM, Medinsky GM. Preservation of the cholera agent in the inter-epidemic season. *In*: Cholera, Materials of Russian scientific conference on cholera problem, Rostov-on-Don, 2005. pp. 17–24 (in Russian).

123. Faruque SM, Chowdhury N, Kamruzzaman M, Dziejman M, Rahman MH., Sack DA, Nair GB, Mekalanos JJ. Genetic diversity and virulence potential of environmental *Vibrio cholerae* population in a cholera-endemic area. Proc Natl Acad Sci USA. 2004;101: 2123–8.
124. Lomov YuM. Cholera pandemics and evolution of the agent. *In*: Cholera, Materials of the VII Russian scientific conference on cholera problem, Rostov-on-Don 2003. pp. 13–23 (in Russian).
125. Monakhova EV, Mikhas'NK. Tol-cluster of *Vibrio cholerae* and its role in horizontal transmission of CTX element. *In*: Cholera, Materials of the VII Russian scientific conference on cholera problem, Rostov-on-Don; 2003.pp. 134–7 (in Russian).
126. Chetina EV, Grizhebovsky GM, Bryukhanov AF, Khaitovich AB, Kurbanov ShKh, Gintzburg AL. On the possible mechanism of endemic nature of modern cholera (role of noncultivated forms of *Vibrio cholerae* O1). Mol Gen Mikrobiol Virusol. 1993;6:18–22 (in Russian).
127. Nikoleishvili LR, Chepkova EA, Olekhova OYu, Sokolenko AV, Titova SV. Investigation of conditions inducing *Vibrio cholerae* transfer to non-culturable forms and reversion to the initial state. In: Cholera and *Vibrio* Human Pathogens, Materials of the Cholera Problem Committee, Rostov-on-Don, vol 14; 2001. pp. 44–5 (in Russian).
128. Urbanovich LYa, Maramovich AS, Mironova LV, Golubinsky EP, Sappo SG, Ganin VS, Kulikalova ES, Beloborodov YuV. Characterization of biological features of *Vibrio cholerae* non-culturable forms and their ability to reversion in experimental conditions. In: Cholera and *Vibrio* Human Pathogens, Materials of the Cholera Problem Committee, Rostov-on-Don, vol 17; 2004. pp. 45–8 (in Russian).
129. Osina NA, Bugorkova TV, Kulichenko AN. On the problem of reversion of non-culturable *Vibrio cholerae* forms. Problemy OOI. 2006;2:44–8 (in Russian).
130. Chemisova OS, Alekseeva LP, Mazrukho BL, Shestialtynova IS, Tatarenko OA. Studies of biofilm formation by *Vibrio cholerae* El Tor strains. In: Cholera and *Vibrio* Human Pathogens, Materials of the Cholera Problem Committee, Rostov-on-Don' vol 21; 2008. pp. 72–74 (in Russian).

Chapter 5
The Re-emergence of Cholera in the Americas

Gabriela Delgado, Rosario Morales, Jose Luis Mendez, and Alejandro Cravioto

Abstract In the history of cholera, seven cholera pandemics occurred during the nineteenth and twentieth centuries and five of these affected the American continent. Ships transporting European immigrants in the nineteenth century probably brought the disease to the American continent during the second pandemic. Between 1973 and 1991, infrequent indigenous cholera cases were reported in different parts of the USA including Louisiana, Florida, Georgia, Maryland, which have been linked to the Gulf Coast reservoir. In 1991, cholera resurfaced in Latin America accounting for 66% (396,536 cases) of all cholera cases reported worldwide that year. Unexpectedly, cholera re-emerged in Peru during early 1991 and quickly turned into an epidemic that extended to Bolivia, Brazil, Chile, Colombia, Ecuador, El Salvador, Guatemala, Honduras, Panama, Venezuela, and Mexico. During this time, several important environmental factors including El Niño were attributed to the increased incidence of cholera. During and after the cholera outbreaks in South America, intensive measures were taken that include epidemiological surveillance, setting up of a laboratory network, proper patient care, improvements in basic sanitation, and clean water supply systems. Such timely measures helped to control cholera and other diarrheal diseases. Several molecular techniques have allowed the study of clonal variations in *Vibrio cholerae* for the first time. Molecular tools helped in identifying the epidemiological links with import of cholera by travelers as well as of the environmental origin. Several serotypes of *V. cholerae* exist in costal environments and seroconversion may play an important role in the epidemiology of cholera.

5.1 Introduction

Vibrio cholerae O1 is the etiological agent of cholera, which is an acute and severe gastrointestinal infection characterized by the abrupt appearance of severe watery diarrhea, vomiting, and rapid dehydration that can lead to death within a few hours

---

A. Cravioto (✉)
International Centre for Diarrhoeal Diseases Research, Dhaka 1212, Bangladesh
e-mail: acravioto@mail.icddrb.org

T. Ramamurthy, S.K. Bhattacharya (eds.), *Epidemiological and Molecular Aspects on Cholera*, Infectious Disease, DOI 10.1007/978-1-60327-265-0_5,
© Springer Science+Business Media, LLC 2011

[1]. Cholera is one of the epidemic diseases that have been documented in social and political history for more than 2000 years, with descriptions being found in ancient Greek, Sanskrit, and Chinese [2]. The epidemiology of cholera centers on the biological and ecological properties of *V. cholerae* and the complex patterns of human behavior in environments that, for climatic reasons or poverty, allow the transmission of the pathogen from one human being to another. The epidemiology of cholera in Americas is different from Asiatic and African cholera. This chapter reviews the salient aspects of cholera in Americas.

5.2 Epidemiology of Cholera in Americas

Seven cholera pandemics occurred during the nineteenth and twentieth centuries, with five of these affecting the American continent. Ships transporting European immigrants in the nineteenth century probably brought the disease to the American continent during the second pandemic, with the first cases being detected on 8 June 1832 in Québec and on 10 June in Montreal. In the same month, the disease was first reported in the United States of America (USA) in New York, with more cases appearing in Philadelphia in July. The disease probably crossed the border into Mexico via the American colonialists who were settling in Texas or via ships that came from New Orleans to Mexican ports in the Gulf of Mexico, the Yucatan peninsula or the port of Havana. Cholera first appeared in Mexico around July and August of 1833 in the city of Saltillo, the then capital of the Province of Coahuila–Texas. In August of 1833, cholera cases were detected in Mexico City with over a thousand people dying of the disease in that month alone. The second pandemic also affected Peru, Chile, Cuba, Nicaragua and Guatemala [3]. Of the seven pandemics for which records exist, the third was one of the most catastrophic with 140,000 deaths in France, 20,000 in England and more than 200,000 in Mexico. In the USA, the disease accounted for an increased number of human deaths with New Orleans mourning the loss of 5,000 citizens in August 1853 [3].

Mexico was once again affected by cholera during the fourth and fifth pandemics at a time when the country was fighting both internally and against the invasion of the French army. In a bizarre way, the disease helped to fight the invaders during the fourth pandemic when cholera appeared among the French troops forcing them to seek treatment in large numbers [3]. The fifth pandemic in 1882 affected mostly the south-east of the country. Cases were detected in Chiapas, Tabasco, and the Isthmus de Tehuantepec. On 12 February 1883, the Epidemiology Commission of the Superior Health Council reported an increase of 75% in the case-fatality rate of cholera cases. This high mortality rate and the severe measures used by the government to control the disease ended the epidemic in the same year [3]. Other Latin American countries such as Brazil, Chile, Argentina, and Uruguay continued to battle cholera until 1895 [3]. Cholera then disappeared from the continent for almost 100 years.

Between 1973 and 1978, autochthonous cases of cholera were reported in different parts of the USA [4]. In the latter part of 1978, 11 people were infected by

V. cholerae El Tor in Louisiana after eating cooked crabs from different sites along the coastal marshes, and in the same year, two isolated cases were reported in Florida [5] and Georgia [6] associated with eating contaminated raw oysters. In 1981, there was a small outbreak of severe diarrhea associated with *V. cholerae* El Tor isolated from 16 people who had eaten raw shellfish from the Gulf of Mexico [7]. In 1983, an American tourist returning from Cancun, Mexico, became ill with a cholera-like syndrome and a toxigenic *V. cholerae* O1 Inaba strain was isolated from the feces. A subsequent epidemiological study in the same area revealed that *V. cholerae* O1 could be found in 16% of the environmental samples but none of these strains were able to produce cholera toxin (CT) and thereby were unable to cause disease in humans [8]. In 1984, another isolated case was reported in Maryland following the consumption of infected crabs caught along the Texas coast [9]. In 1986, 18 people living in 12 different places in Louisiana had a cholera-like syndrome after eating crabs and shrimps collected from different locations along the Louisiana coast [6]. In the autumn of 1991, a single cholera case was identified on an oilrig barge in Texas, followed by 13 secondary cases of cholera and one asymptomatic infection associated with the same organism [10]. Between 1973 and 1991, a total of 65 cholera cases have been linked to the Gulf Coast reservoir of the USA. Before 1991, there were only two reports of *V. cholerae* O1 isolation in other parts of the American continent, one toxigenic strain isolated from sewage water but not humans in Rio de Janeiro, Brazil, and a non-toxigenic *V. cholerae* O1 isolated from two American tourists visiting Peru.

In 1991, cholera resurfaced in Latin America accounting for 66% (396,536 cases) of all cholera cases reported worldwide that year [11]. For some unknown reason, cholera re-emerged in Peru on 23 January 1991 and quickly turned into an epidemic that extended to Bolivia, Brazil, Chile, Colombia, Ecuador, El Salvador, Guatemala, Honduras, Panama, Venezuela, and Mexico [12–14]. The infection rate for the continent was 521 cases per million people with a mortality rate per case of 1%. Peru was the most seriously affected country with almost 300,000 cases in the first year and 45,000 new cases per week in the first few weeks of the epidemic, which severely affected the economy of the country [15, 16]. Fifteen years later, the origin of the epidemic remains unknown, although there are various hypotheses as to how cholera arrived and spread throughout Peru: (1) through the discharge of sewage water from the ships that came into port carrying passengers or crew who were asymptomatic and who had arrived from cholera-infected areas. This discharge contaminated the sea and food sources such as fish, molluscs, and crustaceans infecting them with *V. cholerae*; (2) through the emptying of non-potable water containing *V. cholerae* from different compartments of ships into the sea; (3) by non-O1 *Vibrio* strains that acquired virulence genes through infection with phages; (4) from plankton in coastal areas infected with toxigenic *V. cholerae* O1 in a viable but non-culturable (VBNC) form or in a culturable form but in very low concentrations that did not allow its recovery in the laboratory [17]. Environmental factors, such as El Niño or changes in the salinity and/or concentrations of nutrients in the sea, could have favored the dissemination of the infecting strains in these areas, along with poor sanitation, malnutrition, and poverty in the local population [18].

On 12 June 1991, the first case of cholera in Mexico linked to the same epidemic in Peru was identified in a rural community in the mountains of the State of Mexico [19, 20]. At the beginning of the outbreak, a total of 8,622 cases in 16 locations in Mexico were identified with the largest number of patients appearing in the States of Hidalgo, Puebla, Chiapas, Guerrero, Veracruz, Yucatan, and Mexico. The average mortality rate was 1.15%. The most frequent form of transmission was by food handlers who were asymptomatic carriers moving along commercial routes and through the consumption of non-chlorinated water [19].

In 1991, the cholera epidemic entered the USA with travellers returning from Latin America. Food has been cited as the mode of transmission of cholera from Latin America to the United States in several instances. An index case of cholera was identified in New York in 1991. This case was followed by three secondary cases of symptomatic infection with *V. cholerae* O1 [21]. The index case had traveled to Ecuador and had returned with some boiled crabs, which were contaminated with the *V. cholerae* O1 strain prevailing in Latin America at the time [10]. The molecular characteristics of this strain of *V. cholerae* O1 were completely different from those isolated from the coastal areas of Texas and Louisiana. In another instance in 1991, 76 airline passengers returning from Argentina to Los Angeles were infected with *V. cholerae* O1 after eating contaminated shrimp and fish salads prepared in Lima, Peru [22]. Although no secondary spread occurred, 37 passengers became ill, and of those one died. A total of 41 imported cases of cholera have been documented in the United States between 1961 and 1990. Most of these cases were associated with travel to Latin American countries or with eating seafood brought from there.

In 1992, a total of 358,174 cases of cholera and 2,602 resulting deaths were reported in the Americas with the largest countries affected being Bolivia, Brazil, Colombia, Ecuador, Guatemala, and Peru. In the same year, Argentina, Belize, Costa Rica, and Guyana were added to the list of countries reporting cholera cases, while Surinam reported only 12 cases within its territory [11]. In 1994, around 127,180 cases and 1,321 deaths were recorded, with Brazil, Guatemala, and Peru continuing to be the most affected countries [11]. At the beginning of 1995, there was a noticeable drop in the number of cholera cases in most countries. Paradoxically, Mexico reported its greatest number of cases in 1995 with 16,430 cases and 137 deaths [11].

By 1997, 15 countries in the Americas reported cholera cases of which 7 (Costa Rica, Belize, Chile, Ecuador, Honduras, and the USA) reported sporadic cases only. During that year, a total of 17,760 cholera cases and 225 related deaths were reported officially to the World Health Organization (WHO) indicating a reduction of 25% in the number of cases and 36% in the number of related deaths compared with the previous year. Brazil and Peru continued to present the majority of cases, followed by Mexico and Venezuela, although as stated, all countries showed a marked reduction compared with 1996. The global case-fatality rate in 1997 was 1.3%, which was very similar to that in 1996 of 1.4%, with rates ranging from 0% in some countries to 2.8% in Nicaragua and 4.6% in Ecuador [23].

In 1998, the incidence of cholera increased considerably, probably due to the natural disasters that occurred, which were linked to climatic phenomena such as El Niño and Hurricane Mitch in Central America. There was resurgence of the cholera outbreaks in Peru in the first three months of 1998, which spread once again to neighboring countries. A total of 57,106 cases and 558 related deaths were reported to the WHO accounting for a 300% global increase in the number of cases and a 100% increase in the number of related deaths. Peru reported 71% (41,717) of the total cases in the Americas with Guatemala reporting 5,970, Ecuador 3,724, Brazil 2,571, and Nicaragua 1,437 cases, while the USA and Canada reporting sporadic cases only [24].

In 1999, a considerable reduction in the incidence of cholera was once again reported by the WHO with total cases falling to 8,126 and related deaths to 103 worldwide. The most dramatic reductions were seen in Colombia, Ecuador, and Peru, although there were increases seen year on year in Brazil and El Salvador. However, the number of countries reporting cases fell to 12 with Argentina, Bolivia, Canada, and Chile reporting no cases and the USA reporting 6 cases only, all of which were imported [25].This trend in the reduction of cases and related deaths due to cholera continued during the year 2000 with only 3,101 cases worldwide and 40 deaths being reported by the WHO. The greatest reductions during this year were seen in Brazil, Ecuador, Guatemala and Nicaragua but once again, El Salvador reported an increase in cases (631) and two related deaths. As the number of cases decreased, so did the list of countries reporting cases [26].

By 2001, the total number of cases reported by WHO was down to 535 with no related deaths. Peru showed a marked reduction of 47%, reporting only 494 cases in the same 12 months. The other cases that year were found in Brazil, Ecuador, and Guatemala, with 7, 9, and 13 cases, respectively [27]. The following three years have shown a continuous reduction in the number of reported cases: in 2002, Canada and the USA reported 4 and 2 imported cases only, Guatemala reported 1 case, and Peru 16 cases [28]; in 2003, Peru did not report any cases of cholera for the first time since 1991, while Canada, Ecuador, Guatemala and the USA continued to report a small number of cases [29]; and in 2004, the countries reporting cholera cases were Brazil [21], Canada (three imported), Colombia (two), Ecuador (five), and the USA (five imported) [30]. According to figures from the WHO between 1991 and 2004 the Americas reported a total of 1,277,833 cases of cholera with 12,742 related deaths, with Peru accounting for the largest number of both [11]. Although epidemic levels have decreased considerably since cholera first reappeared in the Americas, the Regional Surveillance, Prevention, and Control Committee, established to monitor the spread of the disease, has continued to function.

5.3 The Environmental Aspects of Cholera

During the last cholera pandemic, contaminated seafood was identified as being the main source of infection in several outbreaks with raw shrimp, crabs, oysters, clams, shellfish, and mussels being the most widely cited sources [9, 31, 32]. Reports of cholera outbreaks from the US Gulf Coast during the last two decades provide epidemiological evidence of a definitive role of fish and shellfish in the transmission of cholera.

The epidemic and seasonal character of cholera in the continent could be related to the survival of *V. cholerae* O1 as a VBNC state or in localized ecological niches in aquatic environments during the periods between epidemics [33–37]. In order to understand the ecology of *V. cholerae*, it is necessary to know something about the aquatic ecosystems that give shelter to this microorganism and help maintain the endemic presence of cholera in Latin America. In addition, some experts believe that it is necessary to avoid human exposure to the natural reservoirs of toxigenic *V. cholerae* O1 in order to maintain a suitable level of control [38].

Toxigenic *V. cholerae* O1 naturally seeks out the optimum conditions for growth and survival in its aquatic environments. These conditions, warmth, a richness in nutrients, moderate salinity, and a neutral or slightly alkaline pH, are typically found in the estuaries and coastal marshes in equatorial, tropical, and sub-tropical regions where phytoplankton, zooplankton, fish, molluscs, and crustaceans can be found in abundance [39, 40]. Actually, the environmental niche for *V. cholerae* O1 is found in estuaries where the microorganism can survive even in the absence of human fecal contamination. In the USA, toxigenic *V. cholerae* O1 biotype El Tor has been isolated from shrimps and crabs in the State of Louisiana [41] and from oysters and fish guts caught in Mobile Bay, Alabama [42]. In Florida, non-toxigenic strains of the O1 serogroup and El Tor biotype have been isolated from oysters (*Crassostrea virginica*), which were found in the estuarial waters [43]. In 1991, *V. cholerae* O1 was detected in fish and molluscs living in the coastal waters of Peru [44]. The chitinous surface of crustaceans can provide an adequate substrate for the multiplication of this microorganism [45].

On these same lines, a study to determine the frequency of isolation of *Vibrio* spp. was conducted during the months of November 1997 (rainy period) and May 1998 (dry season) in Mexico [46, and unpublished observations]. The study involved the collection of water, oyster, and sediment samples from six different points in the Mecoacán lagoon, which is located in the State of Tabasco in the south-eastern part of Mexico. The results indicated a higher frequency of isolation of different *Vibrio* species during the rainy period when compared with the dry season, with *Vibrio parahaemolyticus* being the most frequently isolated strain, followed by *Vibrio alginolyticus* and non-O1 *V. cholerae* strains [46]. When the O antigen of these *V. cholerae* strains was determined, 39% did not agglutinate with any of the specific antisera, 30% were identified as belonging to serogroup O130, 8% to serogroup O37 and 4% to serogroup O150. There was a statistically significant correlation between the presence of *V. parahaemolyticus* and non-O1 *V. cholerae* with an increase in the water temperature and the amount of dissolved oxygen in the sedimentary samples [46]. Non-O1 *V. cholerae* strains have been previously associated with gastroenteritis in humans [47], as well as being responsible for an outbreak of cholera in the Sudan [48]. In another study conducted in Mexico [49], non-O1 *V. cholerae* strains belonging to serogroups O5, O37, and O112 were also identified in samples of water and sediment from different sites in the State of Campeche, also in the south-eastern part of Mexico. It is interesting to note that both Tabasco and Campeche border the Gulf of Mexico in its most southern part, as Louisiana and Texas do in the North.

In Argentina, a VBNC form of *V. cholerae* O1 was identified in water samples from the estuary of the Rio del Plata and from the adjacent Atlantic Ocean reef [50]. In addition to being found in water samples, these VBNC forms were also detected in phytoplankton and zooplankton, mainly copepods. These findings seem to indicate that under favorable conditions, the VBNC forms could revert to a pathogenic and transmissible state that could then be able to cause disease in humans exposed to these organisms.

Given the epidemic nature of cholera in some Latin American countries, it is clear that certain microbiological and epidemiological studies of different aquatic ecosystems are needed to locate the biotic reservoirs that favor the persistence of toxigenic *V. cholerae* O1 during periods between epidemics. It would also be important to determine the relationship that might exist between the seasonal variations of the abiotic parameters (temperature, nutrients, salinity, oxygen saturation, etc.) and the start and end of these outbreaks. Furthermore, studies are needed to determine if communities that use river water that contains VBNC organisms, or that consume seafood infected with these forms, are at risk of developing symptomatic or asymptomatic infections.

5.4 Management of Cholera and Other Related Diarrheal Infections

More than a decade since the last cholera epidemic in Latin America, several lessons have been learned. In the case of Mexico, the Mexican Government considered that cholera and epidemic outbreaks represented a national security risk and implemented a number of initiatives within the health sector, such as epidemiological surveillance, the setting up of a laboratory network and changes in patient care, along with campaigns for public awareness and improvements in basic sanitation and clean water supply systems, that have helped with the control of cholera and other diarrheal diseases. These control strategies have had an overall impact in the general incidence of mortality due to diarrhea among children less than 5 years of age. These measures have also had an economic impact through an increase in the number of tourists wishing to visit the country [51].

5.5 Molecular Characterization of *V. cholerae*

For a long time, epidemiological surveillance of cholera was severely limited due to the poor sensitivity of the typing methods in existence. In most laboratories, *V. cholerae* O1 strains have been identified by bacteriological methods and classified into two main serotypes, Inaba and Ogawa, and two biotypes, classical and El Tor, which are based on the phenotypic characteristics of the bacteria [52, 53]. A third serotype has been described, Hikojima, which is really a combination of the other two. These basic identification systems can be complemented with other more sensitive phenotypic methods, such as biotyping, serotyping with monoclonal antibodies, sensitivity testing to specific bacteriophages, fimbriae characterization, resistance to antimicrobials or the analysis of patterns of protein extracts, outer membrane proteins, carbohydrates or lipids obtained by electrophoresis in different substrates. Important as they might be to identify the bacteria, these techniques are unable to determine the allelic variations in specific genes in these organisms and for this reason, these methods are useless for analyzing the genetic structure of the genus [52, 53].

In 1992, Stroeher et al. [54] reported that the *V. cholerae* O1 Inaba causing the epidemic in Latin America could change its serotype from Inaba to Ogawa and *vice versa*. Moreover, they determined that this change in phenotype was due to a deletion or an insertion of a single nucleotide in the *rfbT* gene that acted as a stop codon and led a truncated product. This finding indicated that even sophisticated phenotypic methods were not sufficient to understand the mechanism by which *V. cholerae* behaved during the course of an epidemic. In addition, they showed a need for genetic techniques to study the more intrinsic mechanisms of adaptation of the bacteria to different environments and to develop more useful methods for its control.

During the last 20 years, techniques for molecular sub-typing such as multilocus enzyme electrophoresis (MLEE), ribotyping, pulsed-field gel electrophoresis (PFGE), and the use of PCR and specific molecular probes of DNA have allowed the study of clonal variations in these bacteria [55, 56]. These techniques have also allowed natural populations of this microorganism to be studied on a large scale to determine the genetic diversity of *V. cholerae* strains [53]. Although MLEE has been shown to be the least discriminatory method compared with other molecular subtyping techniques, such as ribotyping or PFGE, the data produced is reproducible and suitable for establishing the evolutionary significance of the genetic relationship that exists between *V. cholerae* strains. In addition, MLEE creates an efficacious platform for interpreting the results of other sub-typing techniques applied to this species.

The unexpected appearance of cholera in the Americas between 1978 and 1993 was perfect timing for the application of these genetic techniques to the epidemiological surveillance of the disease. In 1993, Wachsmuth et al. [57] used various molecular techniques such as ribotyping, restriction fragment length polymorphism (RFLP) of the *ctx*A gene, sequencing of the B subunit of the cholera toxin (*ctx*B), and MLEE to characterize 197 strains of *V. cholerae* O1 isolates from different parts of the world. The results of this study showed that there were at least four clones of *V. cholerae* O1 biotype El Tor worldwide: an Australian clone (ET 1); a clone from the Gulf of Mexico in the USA (ET 2); the clone identified as the agent involved in the seventh epidemic (ET 3); and a Latin American clone isolated during the reappearance of cholera in the 1990s (ET 4). The ET 4 clone was probably a variant of the ET 3 clone prevalent in the Western hemisphere, while the ET 2 clone had apparently evolved in a separate form and was unrelated genetically to the previous two.

Popovic et al. [58] conducted a second study in which they applied ribotyping to identify the origin of the *V. cholerae* O1 strains and to follow their geographical movement as mentioned above. Their results supported that some ET3 strains and all ET4 strains belonged to the same ribotype 5, indicating that they are all genetically related. What was interesting was that the ET3 strains had been isolated some 15 years prior to the last pandemic in Latin America from different countries around the world, none of which was part of that continent.

Prior to the aforementioned studies, Chen et al. [59] carried out a study in 1991 to determine the genetic relationships among toxigenic and non-toxigenic isolates of *V. cholerae* O1 strains isolated in the Western hemisphere. Using the MLEE technique they concluded that while all the toxigenic and some non-toxigenic O1 strains belonged to a single clone, the majority of non-toxigenic isolates were more genetically diverse belonging to a wider variety of clones. In 1994, Cravioto et al. used MLEE to study the genetic diversity and clonal relationship between 133 *V. cholerae* strains isolated from different parts of Mexico after the start of the 1991 epidemic [60]. The results showed that 45 of the 133 strains were identified as *V. cholerae* O1 biotype El Tor of which 35 belonged to the Ogawa serotype and 10 to the Inaba serotype. Both groups had a clonal behavior. The remaining 88 strains belonged to different non-O1 *V. cholerae* serogroups that had been isolated from sporadic cases of diarrhea or from environmental sources [60]. The data showed that the *V. cholerae* O1 Ogawa and Inaba El Tor strains isolated in Mexico were closely related genetically since of the 16 loci studied by MLEE; the two groups differed only in the mobility of leucine aminopeptidase (Fig. 5.1).

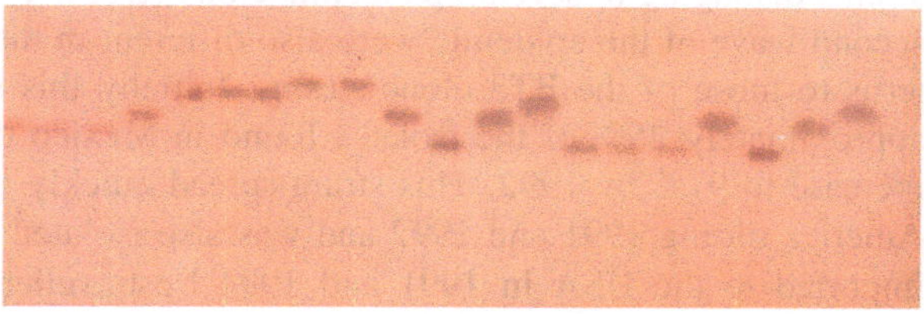

Fig. 5.1 Electrophoresis of protein extracts from *V. cholerae* strains in a starch gel stained to determine the mobility of leucine aminopeptidase (LAP)

Cravioto et al. [60] also found that the *V. cholerae* O1 Ogawa strain causing the epidemic in the American continent was genetically identical by MLEE to the *V. cholerae* O139 Bengal strain that had recently appeared in south-east Asia (Fig. 5.2). Genotypic techniques, such as PFGE, did allow the differentiation of the major ETs of *V. cholerae* O1 strains isolated during the Latin American epidemic [61]. It was a faster method, which was more accessible to laboratories than was MLEE and therefore could be recommended for use in epidemiological surveillance studies. With the use of PFGE, Cameron et al. [61] were able to determine the presence of a second *V. cholerae* O1 clone isolated during the latter part of the Latin American epidemic that was not related to the ET3 clone. This indicated a second source of cholera of unknown origin causing disease simultaneously in the same populations.

Evins et al. [62] also looked at the genetic relationship between toxigenic and non-toxigenic *V. cholerae* O1 strains isolated in different countries in the Americas following the outbreak of the 1991 epidemic. Their results supported the findings of Cameron et al. indicating that the second epidemic wave that took hold in Latin America was caused by a *V. cholerae* O1 Ogawa El Tor belonging to the ET3, but with an uncommon antimicrobial resistance pattern different to the one found in

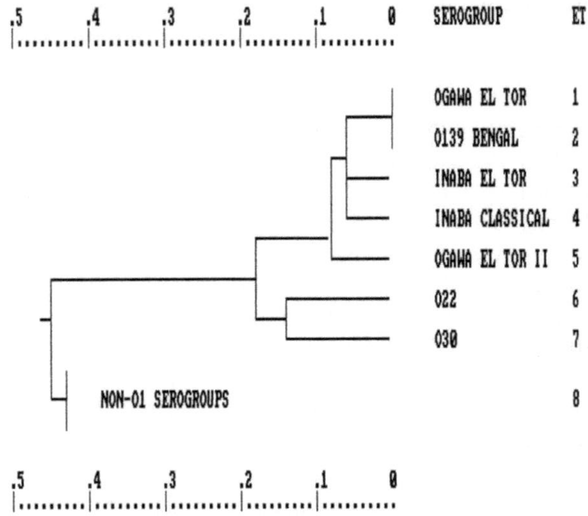

Fig. 5.2 Genetic relationship between O1 and non-O1 *V. cholerae* strains isolated in Mexico: 1, El Tor Ogawa; 2, O139 Bengal; 3, El Tor Inaba; 4–7, classic Inaba, El Tor Ogawa II, O22, and O30, respectively; 8, 86 non-O1 strains in total with 55 belonging to O5-O, O12, O14, O24, O26, O29, O31, O34, O35, O37, O39, O41, O43, O44, O48, O53, O62, O64, O68, O76, O79, and O80 serotypes, and 31 not belonging to any of the 83 non-O1 serotypes for which prototype strains were available

strains belonging to seventh pandemic clone (ET3). The strains isolated during the second wave of the epidemic were also different in their ribotypes and PFGE patterns to those of the ET3 clone strains. Initially this second strain accounted for approximately 25% of the isolates found in Mexico during 1991 [1, 63] with an increase to 97% by 1992. This strain spread quickly through Mexico and Central America during 1991 and 1992 and was also isolated in Romania in 1991. It was imported to the USA in 1991 and 1994 by travellers returning from India and south-east Asia respectively. Evins et al. [62] also concluded that the majority of the non-toxigenic strains analysed in their study were genetically unrelated to the epidemic toxigenic strains causing disease in Latin America.

Despite all the information generated during the cholera epidemic in Latin America, the capacity of *V. cholerae* to spread throughout the environment and infect large groups of the population remains largely unexplained. Some of these issues were addressed by Beltrán et al. [64] in 1999 when they analyzed using MLEE the genetic relationships between 397 *V. cholerae* strains isolated between 1932 and 1995 from human and environmental sources in different parts of the world. The collection included 143 of the initial 193 strains that had been used as prototype strains in the serological reference system for *V. cholerae* established by Shimada et al. and Yamai et al. [65, 66] together with 254 *V. cholerae* strains isolated from people with diarrhea in different countries of the world. The authors found that the 397 strains studied belonged to 279 different ETs, indicating that *V. cholerae* strains are genetically very diverse (Fig. 5.3). The difference between these findings and those reported by other authors [67] was probably related to the diversity of non-O1 serotypes, sources of isolation and geographical regions analyzed by Beltran and her co-workers. The study showed that 37 ETs formed a tight conglomerate (A), which included all the strains belonging to serogroups O1 and O139 that are able

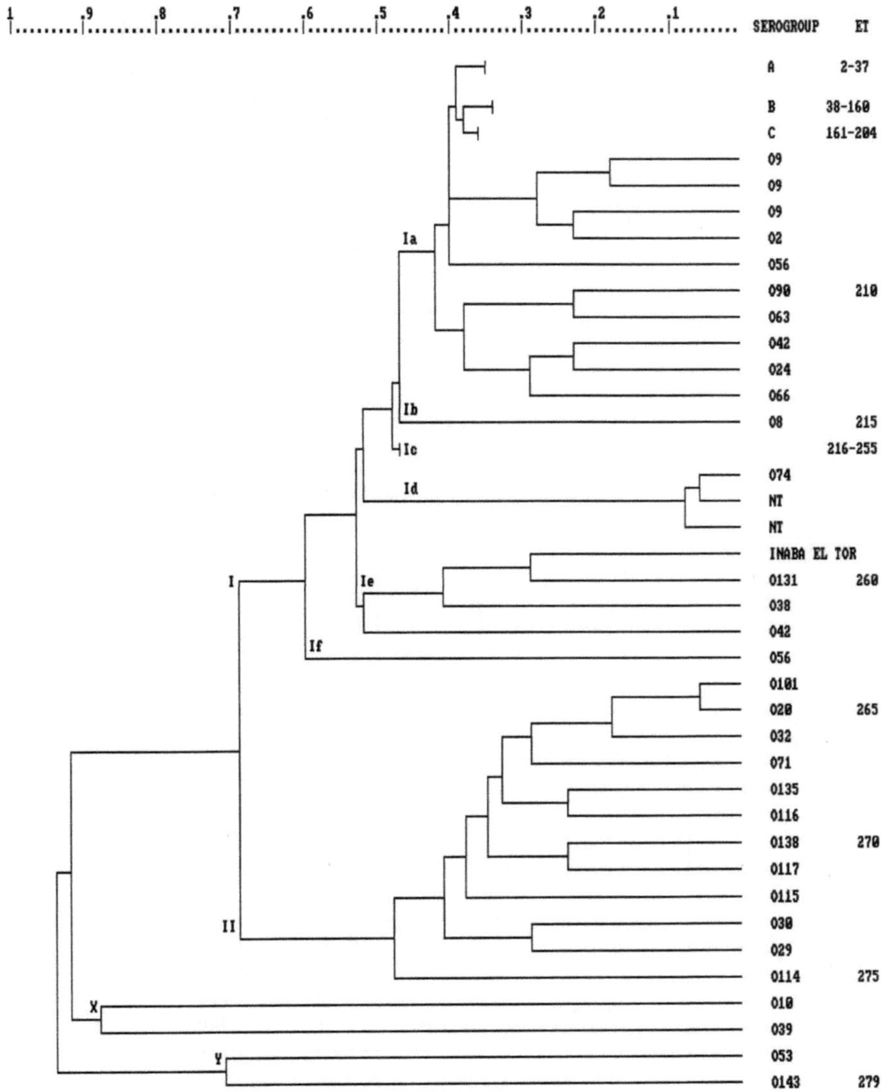

Fig. 5.3 Dendrogram showing genetic relationships of the ETs of *V. cholerae*, based on MLEE analysis (17 loci). The dendrogram was constructed from a matrix of pairwise genetic distances using UPGMA method. The lineages of subdivision Ic and conglomerates A, B, and C are truncated

to cause cholera epidemics (Fig. 5.4). The findings also indicated that, at least in the strains included in conglomerate A, the rate of recombination and horizontal transfer of genes related to basic metabolic functions in these bacteria were sufficiently high to avoid the development and maintenance of distinctive alleles for

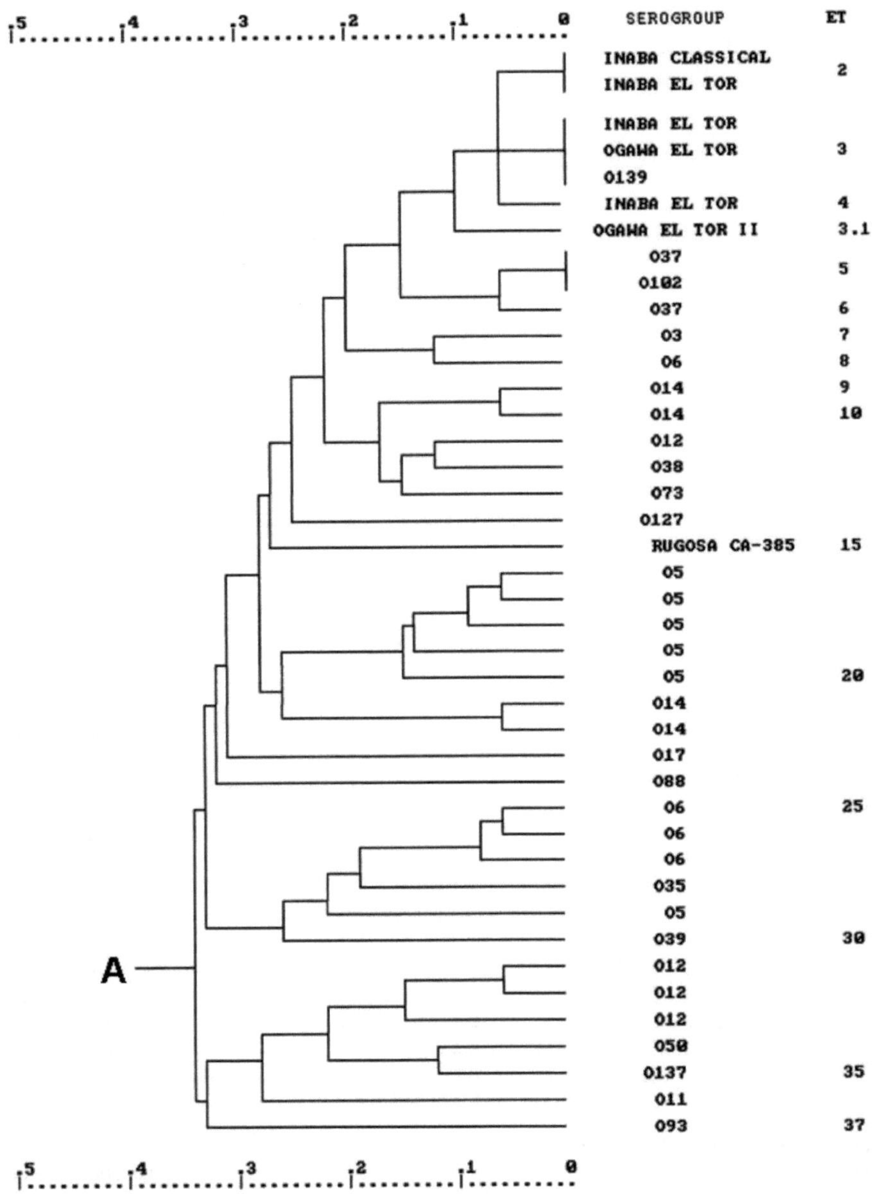

Fig. 5.4 Dendrogram showing genetic relationships of the 37 ETs of *V. cholerae* in conglomerate A, based on MLEE analysis (17 loci). The dendrogram was constructed from a matrix of pairwise genetic distances using UPGMA

extended periods of time. Although the data also suggested that the frequency of intragenic recombination and the distribution of genes related to basic metabolism in *V. cholerae* were higher than in other enteric organisms, the results indicated the existence of defined clonal lineages that have persisted over long periods of time.

The most obvious examples of the latter results would be the strains of *V. cholerae* of serotypes O1 and O139 associated with cholera pandemics.

A further finding reported by Beltrán et al. [64] was that *V. cholerae* strains belonging to the same serotype could often be found in divergent lineages with little genetic relationship between them or that strains belonging to different serotypes could be found in the same ET. Recombination or spontaneous mutations in the *rfb* genes that code for the somatic antigen in these strains were offered as a plausible explanation for these findings. Results from previous studies support this premise showing that the *rfb* genes of *V. cholerae* can suffer mutations due to horizontal transfer with relatively high frequency [68, 69].

5.6 *Vibrio cholerae* Non-O1 Associated with Cholera-Related Diarrhea

Until 1992, *V. cholerae* non-O1 strains had been isolated only from sporadic cases and small outbreaks of diarrhea but never during epidemics. The appearance and spread in India and Bangladesh of clinical cases similar to cholera, associated with the isolation of a strain of *V. cholerae* belonging to serotype O139, marked a change in the currently accepted paradigm of the disease [70].

Using MLEE, Cravioto et al. [60] demonstrated that the genotype of this O139 strain was identical to that of *V. cholerae* O1 Ogawa biotype El Tor strains, which were causal agents of the cholera epidemic in Latin America during the early 1990s (Fig. 5.2). Other comparative analysis showed similar findings, indicating the existence of a common ancestral parent for both strains that acquired external genes through horizontal transfer and/or suffered gene deletions or chromosomal rearrangements that gave rise to this new epidemic clone [71–74]. Faruque et al. [75] showed that the O139 *vibrios* came from different lineages and that this serogroup included both epidemic and non-epidemic strains deriving from different progeny, namely *V. cholerae* O1 and *V. cholerae* non-O1.

Another example of this difference in lineages was seen in a *V. cholerae* strain of serotype O37, which was the causal agent of a cholera outbreak in the Sudan in 1968 [76]. In 1995, Bik et al. [69] demonstrated that this strain was genetically related to the *V. cholerae* O1 strains with a classic biotype that had been previously implicated in large cholera outbreaks. Further studies also indicated that another O37 *V. cholerae* strain, isolated in 1969 in India, was similar to the one found in Sudan and to the classic biotype O1 strains isolated during one of the cholera pandemics. As with the O139 strains, the O37 isolates seem to have acquired the genes that code for a new somatic antigen through horizontal transfer and rearrangement of genes in the *rfb* region.

The close evolutionary relationship between the O1, O139, and O37 clones indicates that new epidemic or more virulent clones can emerge through lineage modification. It seems evident that the O139 strains evolved from an O1 El Tor clone that acquired a transposon containing the genes, which code for a new lipopolysaccharide (LPS) identified in the serological system as O139, following the loss of most of the genes that regulate the synthesis of the O1 LPS [48].

Colwell et al. [77] have proposed that under special conditions, the non-O1 *V. cholerae* strains can be converted into the O1 serogroup and *vice versa* and that this mechanism could be involved in the survival of these strains in the environment. The proposed mechanism by which these changes would occur is spontaneous mutation of the genes that code for the somatic antigen and/or induction or repression of these genes as a response to physical, chemical or biological changes in the environment. This seroconversion could play an important role in the epidemiology of cholera since it could be clearly related to the ability of humans to develop and maintain a protective immune mechanism against the disease for long periods of time. The implications of these findings for the development of vaccines for use in public health are obvious.

For the time being, the only way to keep abreast of future cholera outbreaks in places of the world like Latin America is through well-conducted epidemiological surveillance systems with well-established laboratories that can monitor the presence and virulence characteristics of *vibrios* in both human and environmental samples.

Acknowledgments The research was supported by grants from the Consejo Nacional de Ciencia y Tecnología de México (project 2397 PB) and the UNAM Dirección General de Apoyo al Personal Académico (project IN211496).

References

1. Giono S, Gutierrez Cogno L, Rodriguez Angeles G, del Rio Zolezzi A, Valdespino Gonzalez JL, Sepulveda Amor J. *Vibrio cholerae* O1 fenotipos y genotipos en México. Gac Med Méx. 1995;131:28–35.
2. Shears P. Cholerae. Ann Trop Med Para. 1994;88:109–22.
3. Rodríguez SE, Ruiz RM, Oineda MR, Murguía MR, Valtierra RA. Historia. En Manual para la Vigilancia del Colera. In: Luna Rafael, editor. Dirección General de Epidemiología. Secretaria de Salud, 1999. pp. 11–8.
4. Weissman JB, DeWitt WE, Thompson J, Muchnick CN, Portnoy BL, Feeley JC, Gangarosa EJ. A case of cholera in Texas 1973. Am J Epidemiol. 1974;100:487–98.
5. Klontz KC, Tauxe RV, Cook WL, Riley WH, Wachsmuth IK. Cholera after the consumption of raw oysters. Ann Intern Med. 1987;107:846–8.
6. Lowry PW, Pavia AT, McFarland LM, Peltier BH, Barrett TJ, Bradford HB, Quan JM, Lynch J, Mathison JB, Gunn RA, Blake PA. Cholera in Louisiana. Widening spectrum of seafood vehicles. Arch Intern Med. 1989;149:2079–84.
7. Pavia AT, Campbell JF, Blake PA, Smith JD, McKinley TW, Martin Dl. Cholera from raw oysters shipped interstate. J Am Med Assoc. 1987;285:2374.
8. CDC. Epidemiologic notes and reports cholera in a tourist returning from Cancún, Mexico—New Jersey. MMWR. 1983;32:357.
9. Lin F-Y, Morris GJ Jr., Kaper JB, Gross T, Michalski J, Morrison C, Libonati JO, Israel E. Persistence of cholera in the United States: isolation of *Vibrio cholerae* O1 from a patient with diarrhea in Maryland. J Clin Microbiol. 1986;23:624–6.
10. Johnston JM, Martin DL, Perdue J, McFarland LM, Caraway CT, Lippy EC, Blake PA. Cholera on a Gulf Coast oil ring. N Engl J Med. 1983;309:523–6.
11. Number of Cholera Cases in the Americas (1990–2004). Ministries of Health. http://www.paho.org/Spanish/AD/DPC/CD/cholera-1990-2004pdf
12. CDC. Cholera—Peru, 1991. MMWR. 1991;40:108–10.

13. CDC. Update: cholera outbreak—Peru, Ecuador, and Colombia. MMWR. 1991;40:225–7.
14. CDC. Update: cholera—Western hemisphere, 1991. MMWR. 1991;40:860.
15. Seas C, Gotuzzo E. Cholera overview of epidemiologic, therapeutic, and preventive issues learned from recent epidemic. Int J Inf Dis. 1996;1:38–46.
16. Anonymous. The economic impact of cholera epidemic, Peru 1991. Epidemiol Bull. 1992;13:9–11.
17. Seas C. Miranda J, Gil AI, Leon-Barua R, Patz J, Huq A, Colwell RR, Sack RB. New insights on the emergence of cholera in Latin America during 1991: the Peruvian experience. Am J Trop Med Hyg. 2000;62:513–7.
18. Cowell RR. Global climate and infectious disease: the cholera paradigm. Science. 1996;274:2025–31.
19. Situación epidemiológica de cólera. Republica Mexicana. Secretaria de Salud. Febrero 1992.
20. Kumate J, Sepúlveda J, Gutiérrez G. El cólera: epidemias, endemias y pandemias. McGraw-Hill, México: Interamericana, 1993.
21. CDC. Cholera—New York. 1991. MMWR. 1991;40:516–8.
22. CDC. 1992. Cholera associated with an international airline flight. MMWR. 1992;41:134–5.
23. Weekly Epidemiological Record. Cholera 1997. WHO, Geneva. 1998;73:201.
24. Weekly Epidemiological Record. Cholera 1998. WHO, Geneva. 1999;74:257–64.
25. Weekly Epidemiological Record. Cholera 1999. WHO, Geneva. 2000;75:249–56.
26. Weekly Epidemiological Record. Cholera 2000. WHO, Geneva. 2001;76:233–40.
27. Weekly Epidemiological Record. Cholera 2001. WHO, Geneva. 2002;77:257–68.
28. Weekly Epidemiological Record. Cholera 2002. WHO, Geneva. 2003;78:269–76.
29. Weekly Epidemiological Record. Cholera 2003. WHO, Geneva. 2004;79:281–8.
30. Weekly Epidemiological Record. Cholera 2004. WHO, Geneva. 2005;80:261–8.
31. CDC. Toxigenic *Vibrio cholerae* O1 infections—Louisiana and Florida. MMWR. 1986;35:606–7.
32. Gergatz SJ, McFarland LM. Cholera on the Louisiana Gulf Coast: historical notes and case report. J La State Med Soc. 1989;141:29–34.
33. Huq A, Colwell RR, Rahman R, Ali A, Chowdhury MA, Parveen S, Sack DA, Russek-Cohen E. Detection of *Vibrio cholerae* O1 in the aquatic environment by fluorescent-monoclonal antibody and culture methods. Appl Environ Microbiol. 1990;56:2370–3.
34. Colwell R, Huq A. Vibrios in the environment: viable but non-culturable *Vibrio cholerae*. In: Kaye Wachsmuth I, Blake P, Olsvik O, editors. Vibrio cholerae *and cholera: Molecular to global perspectives*. Washington, DC: American Society for Microbiology; 1994. pp. 117–33.
35. Islam M, Drasar B, Bradley Sack R. The aquatic environment as a reservoir of *Vibrio cholerae*: a review. J Diarrhoeal Dis Res. 1993;11:197–206.
36. Miller C, Feachem R, Drasar B. Cholera epidemiology in developed and developing countries: new thoughts on transmission, seasonality, and control. Lancet. 1985;1:261–2.
37. Miller CJ, Drasar BS, Feachem RG, Hayes RJ. The impact of physico-chemical stress on the toxigenicity of *Vibrio cholerae*. J Hyg (Cambridge). 1986;96:49–57.
38. Glass R, Black R. The epidemiology of cholera. In: Barua D, Greenough III W, editors. Cholera. New York: Plenum Medical Company; 1992. pp. 129–54.
39. Nebel B, Wright R. Environmental science: the way the world works. 4a. ed. New Jersey: Prentice Hall; 1993.
40. Tyler Miller Jr. G. Environmental science: sustaining the earth. 3a ed. San Francisco: Wadsworth Publishing Company; 1991.
41. Blake P, Allegra D, Snyder. Cholera: a possible focus in the United Stated. N Engl J Med.1980;302:305–9.
42. DePaola A, Capers G, Motes ML, Olsvik O, Fields PI, Wells J, Wachsmuth IK, Cebula TA, Koch WH, Khambaty F, et al. Isolation of Latin American epidemic strain of *Vibrio cholerae* O1 from US Gulf Coast. Lancet. 1992;339:624.
43. Hood M, Ness G, Rodrick G. Isolation of *Vibrio cholerae* serotype O1 from the eastern oyster, *Crassostrea virginica*. Appl Environ Microbiol. 1981;41:555–8.

44. Tamplin M, Carrillo C. Environmental spread of *Vibrio cholerae* in Peru. Lancet. 1991;338:1216–7.
45. Baine WB, Mazzotti M, Greco D, Izzo E, Zampieri A, Angioni G, Di Gioia M, Gangarosa EJ, Pocchiari F. Epidemiology of cholera in Italy, 1973. Lancet. 1974;2:1370–4.
46. Ramírez PM. Frecuencia de Aislamiento de *Vibrio* spp. En muestras de agua, ostión y sedimento en la laguna de Mecoacán, Tabasco, México. Tésis de Maestría, Universidad Nacional Autónoma de México, 2003.
47. Wilson R, Lieb S, Roberts A, Stryke S, Janowski H, Gunn R, Davis B, Riddler CF, Barrett T, Morris JG, Blake PA. Non-O group 1 *Vibrio cholerae* gastroenteritis associated with eating raw oyster. Am J Epidemiol. 1981;114:293–8.
48. Bik Em, Gouw RD, Mooi FR. DNA fingerprinting of *Vibrio cholerae* strains with a novel insertion sequence element: a tool to identify epidemic strain. J Clin Microbiol. 1996;34:1453–61.
49. Isaac-Marquez AP, Ledezma-Davila CM, Eslava CC, Navarro-Ocaña A, Cravioto-Quintana A. Serotypes of *Vibrio cholerae* non-O1 isolated from water supplies for human consumption in Campeche, Mexico and their antibiotic susceptibility patter. Mem Inst Oswaldo Cruz. 1998;93:17–22.
50. Binsztein N, Costagliola MC, Pichel M, Jurquiza V, Ramirez FC, Akselman R, Vacchino M, Huq A, Colwell R. Viable but nonculturable *Vibrio cholerae* O1 in the aquatic environment of Argentina. Appl Environ Microbiol. 2004;70:7481–6.
51. Sepulveda J, Valdespino JL, Garcia-Garcia L. Cholera in Mexico: the paradoxical benefits of the last pandemic. Intern J Infec Dis. 2006;10:4–13.
52. Wachsmuth K, Olsvik O, Evins G, Popovic T. Molecular epidemiology of cholera. In: Wachsmuth K, Blake P, Olsvik O, editors. *Vibrio cholerae* and cholera. Molecular to global perspectives. Washington, DC: American Society for Microbiology; 1994.
53. Selander RK, Caugant DA, Ochman H, Musser JM, Gilmour MN, Whittam T. Methods of multilocus enzyme electrophoresis for bacterial population genetics and systemics. Appl Environ Microbiol. 1986;51:873–84.
54. Stroeher UH, Karageorgos LE, Morona R, Manning PA. Serotype conversion in *Vibrio cholerae* O1. Proc Natl Acad Sci USA. 1992;89:2566–70.
55. García-Martínez J, Martínez-Murcia AJ, Rodríguez Varela F, Zorraquino A. Molecular evidence supporting the existence of two major groups in uropathogenic *Escherichia coli*. FEMS Immunol Med Microbiol. 1996;14:231–44.
56. Lan R, Reeves P. Recombination between rRNA operons created most of the ribotype variation observed in the seventh pandemic clone of *Vibrio cholerae*. Microbiology 1998;144:1213–21.
57. Wachsmuth K, Evins G, Fields P, Olsvik O, Popovic T, Bopp C, Wells J, Carrillo C, Blake P. The molecular epidemiology of cholera in Latin America. J Infect Dis. 1993;167:621–6.
58. Popovic T, Boop Ch, Olsvic O, Wachsmuth K. Epidemiologic application of standarized ribotype scheme for *Vibrio cholerae* O1. J Clin Microbiol. 1993;31:2474–82.
59. Chen F, Evins GM, Cook WL, Almeida R, Hargrett-Bean N, Wachsmuth K. Genetic diversity among toxigenic and non toxigenic *Vibrio cholerae* O1 isolated from the Western hemisphere. Epidemiol Infect. 1991;107:225–33.
60. Cravioto A, Beltrán P, Delgado G, Navarro A, Eslava C, Leon S, González A. Non-O1 *Vibrio cholerae* O139 "Bengal" is genetically related to *Vibrio cholerae* O1 El Tor Ogawa isolated in Mexico. J Infect Dis. 1994;169:1412–3.
61. Cameron DN, Khambaty FM, Wachsmuth K, Tauxe RV, Barret TJ. Molecular characterization of *Vibrio cholerae* O1 strains by pulsed field gel electrophoresis. J Clin Microbiol. 1994;32:1685–90.
62. Evins G, Cameron D, Wells J, Green K, Popovic T, Giono S, Wachsmuth K, Tauxe R. The emerging diversity of electrophoretic types of *Vibrio cholerae* in the Western hemisphere. J Infect Dis. 1995;172:173–9.

63. Giono S, Rodriguez MG, Gutierrez L, Valdespino JL. Phenotypic and genotypic characterization of *Vibrio cholerae* O1. Rev Latinoam Microbiol. 1994;36:243–51.
64. Beltrán P, Delgado Gabriela, Navarro A, Trujillo F, Selander R, Cravioto A. Genetic diversity and population structure of *Vibrio cholerae*. J Clin Microbiol. 1999;37:581–90.
65. Shimada T, Arakawa E, Itoh K, Okitsu T, Matsushima A, Asai Y, Yamai S, Nakazato T, Nair B, Albert MJ, Takeda Y. Extended serotyping scheme for *Vibrio cholerae*. Curr Microbiol. 1994;28:175–8.
66. Yamai S, Okitsu T, Shimada T, Katsube Y. Distribution of serogroups of *Vibrio cholerae* non-O1 non-O139 with specific reference to their ability to produce cholera toxin, and addition of novel serogroups. Kansenshogaku Zasshi. 1997;71:1037–45.
67. Salle CA, Momen H. Identification of *Vibrio cholerae* by enzyme electrophoresis. Trans R Soc Trop Med Hyg. 1991;85:544–7.
68. Karaolis DKR, Lan R, Reeves PR. The sixth and seventh cholera pandemics are due to independent clones separately derived from environmental non toxigenic, non-O1 *Vibrio cholerae*. J Bacteriol. 1996;177:3191–8.
69. Bik EM, Bunschoten AE, Gouw RD, Mooi FR. Genesis of the novel epidemic *Vibrio cholerae* O139 strain: evidence for horizontal transfer of genes involved in polysaccharide synthesis. EMBO J. 1995;14:209–16.
70. Delgado G. Análisis molecular de cepas de *Vibrio cholerae*. Tésis de Maestría. Universidad Nacional Autónoma de México, 2005.
71. Berche P, Poyart C, Abachin E, Lelievre H, Vandepitte J, Dodin A, Fournier JM. The novel epidemic strain O139 is closely related to the pandemic strain O1 of *Vibrio cholerae*. J Infect Dis. 1994;170:701–4.
72. Popovic T, Fields P, Olsvik O, Wells JG, Evins GM, Cameron DN, Farmer III JJ, Bopp CA, Wachsmuth K, Sack RB, Albert MJ, Nair B, Shimada T, Feeley JC. Molecular subtyping of toxigenic *Vibrio cholerae* O139 causing epidemic cholera in India and Bangladesh, 1992–1993. J Infect Dis. 1995;171:122–7.
73. Calia K, Murtagh, Ferraro M, Calderwood S. Comparison of *Vibrio cholerae* O139 with *Vibrio cholerae* O1 Classical and El Tor biotypes. Infect Immun. 1994;62:1504–6.
74. Stroeher UH, Jedani KE, Dredge BK, Morona R, Brown MH, Karageorgos LE, Albert MJ, Manning PA. Genetic rearrangements in the *rfb* regions of *Vibrio cholerae* O1 and O139. Proc Natl Acad Sci USA. 1995;92:10374–8.
75. Faruque S. Manujendra S. Asadulghani, Sack D. Sack B, Takeda Y, Nair B. The O139 serogroup of *Vibrio cholerae* comprises diverse clones of epidemic and non epidemic strains derived from multiple *V. cholerae* O1 or non-O1 progenitors. J Infect Dis. 2000;182:1161–8.
76. Karaolis DKR, Johnson JA, Bailey CC, Boedeker EC, Kaper JB, Reeves PR. A *Vibrio cholerae* pathogenicity island associated with epidemic and pandemic strains. Proc Natl Acad Sci USA. 1998;95:3134–9.
77. Colwell RR, Huq A, Chowdhury MA, Brayton PR, Xu B. Serogroup conversion of *Vibrio cholerae*. Can J Microbiol. 1995;41:946–50.

Chapter 6
The Evolution of *Vibrio cholerae* as a Pathogen

John Joseph Mekalanos

Abstract *Vibrio cholerae* is the causative agent of cholera, a severe gastrointestinal infection that is a major public health problem in developing countries where water quality and sanitation are poor. Not all strains of *V. cholerae* are highly pathogenic for humans and genetic analysis has shown that the difference between pathogen and nonpathogenic strains resides in their virulence gene content. In this chapter, I will review the field's knowledge of the major virulence factors of *V. cholerae*, the genetic elements that encode them, and the evolutionary process that leads to the emergence of toxigenic *V. cholerae*. Ecological factors (e.g., lytic phage sensitivity, chitin utilization, and quiescent environmental states) will not be discussed in this chapter given that these aspects of *V. cholerae* biology will be discussed in detail in other chapters. Instead, this chapter will focus on aspects of the organism that relate to its evolution as a pathogen with emphasis on its known and putative virulence factors, the genetic elements that encode them, and how this knowledge has shaped cholera vaccine development efforts.

6.1 Introduction

Vibrio cholerae, the causative agent of cholera, is a gram-negative bacterium that is the paradigm for enterotoxic gastrointestinal pathogens [1]. An enterotoxic pathogen like *V. cholerae* gains its entry through oral injection of food or water contaminated with the fecal material of a cholera victim. Thus, epidemics caused by *V. cholerae* are a major public health problem in developing countries where water quality and sanitation are poor. Not all *V. cholerae* are highly pathogenic for humans and aquatic environments can carry significant numbers of avirulent *V. cholerae* that

J.J. Mekalanos (✉)
Department of Microbiology and Molecular Genetics, Harvard Medical School, 200 Longwood Avenue, Boston, MA 02115, USA
e-mail: john_mekalanos@hms.harvard.edu

T. Ramamurthy, S.K. Bhattacharya (eds.), *Epidemiological and Molecular Aspects on Cholera*, Infectious Disease, DOI 10.1007/978-1-60327-265-0_6,
© Springer Science+Business Media, LLC 2011

seldom if ever cause human disease. A sudden outbreak of cholera can occur in a region by introduction of pathogenic strains for the first time or by the emergence of a new pathogenic strain perhaps after de novo genetic recombination events. Thus, pathogenic strains emerge most efficiently in regions of the world where transmission between humans occurs and most often this emergence is associated with the introduction of a fully virulent strain into a susceptible population (e.g., the Latin American epidemic of 1991). Occasionally, strains of low pathogenicity acquire virulent genes and thus emerge as a unique new pathogenic clone (e.g., the emergence of El Tor O1 disease in Indonesia in 1961). Over the last few decades, dramatic progress has been made in understanding the molecular aspects that account for virulence of this organism and how virulent strains emerge in nature. Many aspects on the ecology of *V. cholerae* have also been revealed and some of these clearly play a role in the seasonal pattern of outbreaks of the disease in endemic areas. However, the ecology and the epidemiology of cholera are topics that will be covered extensively in other chapters of this book. This chapter will focus on aspects of *V. cholerae* that relate to its evolution as a pathogen with particular focus on its known and putative virulence factors as well the genetic elements that encode them.

6.2 Pathogenesis

In its simplest form, the pathogenesis of cholera involves two steps: (i) colonization of the upper intestine by toxigenic *V. cholerae* and (ii) secretion of cholera toxin (CT), a potent enterotoxin capable of causing the profuse watery diarrhea that is characteristic of the disease [2]. In the four decades since the purification of CT [3], much has been learned about the virulence factors that *V. cholerae* produce (Fig. 6.1). In brief, the toxin causes a secretory diarrhea by altering ion fluxes across the intestinal mucosa. This is accomplished by elevating cAMP levels in intestinal cells through the activation of a G protein (Gsα) that controls host cell adenylate cyclase activity. The toxin accomplished this by first binding GM1 ganglioside receptors on host cells through its B subunit, then translocating its active A subunit to the cytosol of the target cell. There, the A subunit catalyzes the ADP ribosylation of the host G protein Gsα, which in turn activates host cell adenylate cyclase. The cAMP that accumulates in target cells activates protein kinases, which in turn phosphorylate membrane proteins one of which is the CFTR chlorine and bicarbonate conductance channel. Active Cl^- and $HCO_3{}^-$ transport into the lumen of the intestine produces an osmotic movement of water out of the tissues and thus the profuse secretory diarrhea characteristic of this disease.

The genes encoding the toxin (*ctxAB*) were initially shown to reside on an accessory 7–9-kilobase chromosomal DNA segment termed the CTX genetic element which structurally resembles a compound transposon [4] and is absent in avirulent environmental strains of *V. cholerae*. This element was eventually found to be the composite genomes of helper and satellite filamentous phages termed CTXΦ and RS1Φ, respectively [5].

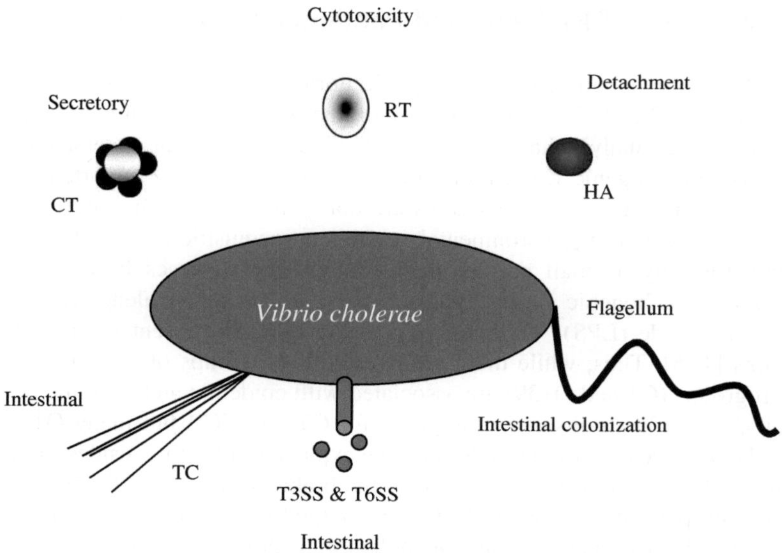

Fig. 6.1 Major known and hypothetical virulence factors of *V. cholerae*. Molecules are denoted as follows: CT, cholera toxin; RTX, the RTX toxin; HAP, HA protease; TCP, toxin coregulated pilus; T3SS, type-III secretion gene cluster; T6SS, type-VI secretion system

Filamentous phages use pili as receptors and it was found that CTXΦ efficiently infects only *V. cholerae* strains that express a specific pilus. Remarkably, this pilus is also a virulence factor of the organism called toxin-coregulated pilus or TCP, an essential intestinal colonization factor of *V. cholerae* [6]. Expression of TCP is coordinately expressed with CT in a regulatory cascade that seems to become more complex year to year [7]. Under in vitro conditions, CT is expressed when TCP is expressed. But in humans, the temporal expression of TCP is likely to occur early in the infection cycle [8].

The "one two punch" of TCP and CT explains much of cholera pathogenesis and the evolution of *V. cholerae* as a pathogenic organism. Numerous other mutations affect the virulence of the organism in animal models but these two factors play the most important role in human disease based on the collective experience gained from human volunteer models for cholera. However, despite the completion of the *V. cholerae* genome sequence in 2000 [9], the story of cholera as a disease and *V. cholerae* as a pathogen is far from complete. The *V. cholerae* genome sequence has allowed new methods to be applied that have revealed changes that the organism has undergone that have been linked to its emergence as a pathogen [10, 11]. Some of these genetic changes probably define important virulence-altering traits. A discussion of all the proposed virulence traits of *V. cholerae* is beyond the scope of this short chapter and therefore focused on those that shaped the field's thoughts about cholera vaccine development and the evolution of this pathogen.

6.3 Emergence of Endemic and Pandemic *Vibrio cholerae*

Vibrio cholerae is an autochthonous aquatic flora in estuarine and brackish water environments [12]. With the completion of the *V. cholerae* genomic sequence, microarray-based analysis has provided a reliable framework for understanding the emergence of pathogenic *V. cholerae* strains (Fig. 6.2). The most important conclusions one can reach from these studies are that pathogenic strains of *V. cholerae* are quite different from environmental strains and acquisition of virulence genes accounts for only a small part of the variation one observes between strains. For example, pathogenic strains typically carry genes for virulence factors and lipopolysaccharide (LPS) "O" antigens that are typically absent in environmental strains [1, 5]. Thus, while there are over 200 serogroups of *V. cholerae*, only two serogroups (O1 and O139) are associated with epidemic and endemic cholera. Virtually, all O139 strains carry the genes for CT and TCP, but many O1 strains (particularly in the environment) do not. The explanation for this resides in the fact that the O139 serogroup is a derivative of the highly successful CT^+ TCP^+ O1 El Tor "seventh pandemic clone" [13]. O139 emerged in the Bay of Bengal in 1992 [14], an area that has been the home of cholera for centuries. At that time, cholera in

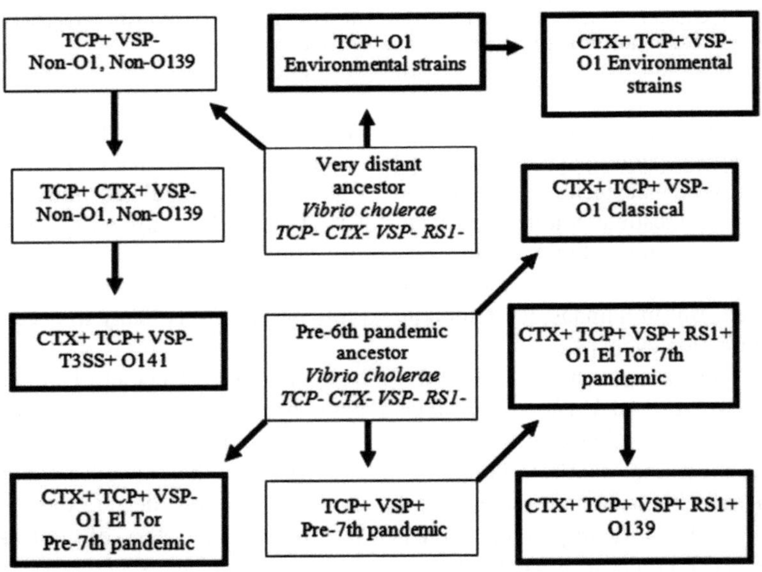

Fig. 6.2 Proposed evolution of major classes of *V. cholerae*. This scheme is based on a combination of microarray-based comparative genomic data and gene allelic variations. Phenotypes in bold boxes are those of strains, serogroups, and biotypes of *V. cholerae* that are known to exist today, while phenotypes in light boxes represent those of the most likely hypothetical precursor strains. Arrows show the simplest path for the emergence of strains. Genotypes are as follows: CT, cholera toxin and other CTX phage sequences; TCP, toxin coregulated pilus and other TCP island genes; VSP, seventh pandemic islands I and II; RS1, RS1 satellite phage; T3SS, type-III secretion gene cluster

South Asia was largely caused by a group of closely related pathogenic strains now referred to as the seventh pandemic El Tor O1 clone [5]. Seven distinct pandemics of cholera have occurred since 1817 and the seventh pandemic clone emerged in Indonesia in 1961. It spread across the Indian subcontinent and within two decades displaced resident "classical" biotype *V. cholerae* O1 strains that existed there for perhaps centuries. Both classical and El Tor biotypes are CTX^+ and TCP^+, so the explanation for the evolutionary success of the seventh pandemic clone remains largely a mystery. The emergence of the O139 was assumed to mark the beginning of the eighth pandemic but so far this organism has co-existed with its parental O1 El Tor strains only in South Asia and yet to spread across the globe. Thus, it seems likely that O139 and O1 El Tor strains are here to stay and comparable in their virulence and evolutionary fitness. Both clones are likely more adapted to cholera "ecology" than are earlier classical strains and the explanation for this may reside in their virulence gene makeup and other innate properties (e.g., phage resistance) as discussed below. TCP^- CT^- O1 strains that exist in the environmental waters of Bangladesh are not closely related to the pathogenic clones causing endemic disease [15]. Thus, the emergence of pathogenic classical and El Tor strains occurred long ago coincident with their capture of the genes for TCP and CTX (Fig. 6.2). This is not to state that new clones cannot emerge from environmental *V. cholerae*. Indeed, a CT^+, TCP^+ clone of *V. cholerae* that is endemic to the Gulf Coast of the United States, appears to be an emergent clone that is closely related to a CT^-, TCP^+ strain that co-exists within this same aquatic environment [16]. Despite being pathogenic and in the aquatic environment for decades, this Gulf Coast clone did not cause the epidemic in Peru in 1991; the seventh pandemic clone that did cause this epidemic was introduced from Asia and is clearly more ecologically adapted to causing epidemic and endemic disease. Emergence of new toxigenic strains "de novo" just does not seem to happen very often despite the fact that some O1 and non-O1, non-O139 strains carry genes encoding TCP [17].

In sum, emergence of pathogenic strains of *V. cholerae* with endemic and pandemic potential has been exclusively linked to the acquisition of the *TCP* and *CTX* genes by only a limited spectrum of O1 and O139 strains. This fact alone suggests that the precursors of these pandemic clones probably display traits that are lacking in typical environmentally adapted *V. cholerae* regardless of their serogroup. Analysis of the genome content of environmental and pathogenic O1, non-O1, and non-O139 supports this conclusion [10, 18]. Unfortunately, the strains that represent these true ancestors of the pandemic *V. cholerae* classical and seventh pandemic O1 clones are missing from the microbiological record at this time (Fig. 6.2).

6.4 Horizontal Transfer of *Vibrio cholerae* Virulence Genes

As noted earlier, the *ctxAB* genes encoding CT reside in the genome of a lysogenic filamentous phage called CTXΦ [16]. The receptor for CTXΦ is the major colonization factor, TCP. The CTXΦ efficiently recognizes only cells expressing TCP and consequently shows high frequencies of infection of TCP^+ recipient cells in the host intestinal environment where TCP is optimally expressed [6, 17]. TCP is itself encoded by a pathogenicity island that was once proposed to correspond to the genome of another filamentous phage [19]. However, current thought in the field is that this conclusion was incorrect [5, 20]. There is evidence that the island can minimally excise and therefore presumably integrate using a phage-like integrase and attachment site [21]. How TCP island DNA moves between *V. cholerae* strains in nature remains a mystery although both transduction [22] and transformation [23] remain possible mechanisms.

The CTXΦ genome is composed of several open reading frames (ORFs), including *ace, zot, ctxAB, cep*, and *orfU*. These ORFs are located on a 4.5-kb "core region" of the CTX element, which is essential for the morphogenesis of CTXΦ particles [6, 24]. However, originally both *zot* and *ace* were thought to be functional toxin genes (see below). Adjacent to the core is the RS2 region encoding ORFs *rstR, rstA2*, and *rstB2*. These genes encode products required for the replication, and regulatory functions of CTXΦ [25]. In some strains of *V. cholerae* (as well as all seventh pandemic O1 and O139 strains), one or more copies of an element called RS1 appear integrated next to the CTXΦ genome. This element is related to RS2 but encodes an additional ORF called *rstC* [26]. RS1 has been found to correspond to a satellite phage that can package its genome into particles constructed by the CTXΦ [5, 26]. Remarkably, RS1 phage can cause CTXΦ phage to replicate in lysogens through the action of RstC, which acts as an anti-repressor of RstR, which in turn negatively controls CTXΦ lysogeny [26].

Recent dramatic progress has been made in understanding the CTX phage acquisition process from the point of view of phage recombination [85–88]. In brief, it has been recently recognized that the chromosomal attRS1 insertion site of CTX phage actually corresponds to the *dif* recombination sequence which functions in chromosome dimer resolution using XerC and XerD recombinases [85–87]. We found that acquisition and chromosomal integration of another satellite filamentous phage called TLC can introduce a functional *dif* site into *V. cholerae* strains that lack this site. In doing so, these strains become permissive for integration of CTX phage. Thus, five filamentous phages appear to cooperate in the emergence of toxigenic *V. cholerae* [88].

There have been numerous reports of *CTXΦ* genes found in the genomes found in nonpathogenic *V. cholerae* environmental strains and even other non-*Vibrio* species. In many cases, the genes identified are *zot* and *ace* because investigators have employed probes for these two putative toxin genes [27]. However, the *zot* gene of CTXΦ corresponds to the most well-conserved gene ortholog of the filamentous phages (gene I) and *ace* corresponds to another less well-conserved filamentous phage gene ortholog (gene VI) [6, 28]. Indeed both *zot* and *ace* are absolutely required for production of CTXΦ particles [6, 29] and thus like their homologs in other isolates are likely involved in filamentous phage morphogenesis; indeed many filamentous phages of other *Vibrio* species carry homologs of these *CTXΦ* genes. Thus, reports that *ace* and *zot* genes exist in various bacterial isolates have led to much confusion about whether such isolates carry *V. cholerae* virulence genes or whether these genes are simply indicative of the presence of a filamentous phage genome. Accordingly, a brief history of the discovery of *zot* and *ace* may provide the reader on some perspective on this confusing topic.

6.5 Reactogenicity of Nontoxigenic Vaccine Prototypes in Humans

Observations made initially with an undefined *V. cholerae* mutant expressing only the B subunit of CT [30] and then confirmed with a variety of defined deletion mutants (Levine et al. [31]) indicate that simple loss of production of active CT does not fully attenuate the virulence of *V. cholerae* in volunteer subjects. Two explanations were put forward early on to explain residual adverse symptoms or "reactogenicity" caused by *Δctx* vaccine strains in humans: (i) *V. cholerae* must produce another enterotoxin besides CT and (ii) colonization of the small intestine by *V. cholerae* (which includes the multiple steps of adherence to mucus and different epithelial cell types, as well as survival and multiplication within small intestinal micro-environments) was sufficient to induce a secretory response without direct involvement of any novel enterotoxin per se.

The first hypothesis stimulated the search for new toxins. One of the first promising leads came in the form of a report that some *V. cholerae* strains appeared to produce toxic activity that was immunologically related to *Escherichia coli Shiga*-like toxin [32]. However, the completion of the *V. cholerae* genome [9] revealed no genes with similarity to *slt*. Similarly, the proposal that El Tor hemolysin [33] was the enterotoxin responsible for reactogenicity of *Δctx* vaccine strains is not supported by the high reactogenicity displayed by various vaccine strains specifically deleted in the *hly* and *ctx* coding sequences [34, 35].

6.6 The Discovery of Zot and Ace

In 1988, investigators at the University of Maryland Center for Vaccine Development (CVD) established that a series of vaccine candidates derived from seventh pandemic El Tor strains and classical strains deleted in all or part of the *ctxAB* operon were highly reactogenic in volunteers and caused a dose-dependent, moderate-to-severe diarrheal disease syndrome [31, 34, 35]. In 1983, our laboratory reported the construction of recombinant cholera vaccine candidate O395-N1, which carries defined internal deletions in its two copies of the *ctxA* gene [36], but was not tested in volunteers at CVD until 1988 [37]. O395-N1 surprisingly caused only one mild (280 mls) episode of diarrhea in 1 of 13 individuals that ingested this vaccine while clearly colonizing all volunteers [37]. The reduced ability of strain O395-N1 to induce diarrhea in human subjects was in striking contrast to CVD101 [31], a *ΔctxA* derivative of the same parental strain (Ogawa 395) used to construct O395-N1. Strain CVD101 induced moderate-to-severe diarrhea in volunteers at doses that were actually lower than those used for O395-N1 [31, 38]. The CVD group reasoned that perhaps the low diarrheal response seen with O395-N1 might be explained by the existence of an enterotoxin produced by CVD101 that for unknown reasons was not elaborated by O395-N1. This hypothesis prompted [39] to perform Ussing chamber experiments that led to the demonstration of an activity present in supernatant fluids of CVD101 that was absent in similar culture fluids of O395-N1. The toxin responsible for this activity was designated zonula occulens toxin (Zot), because its effect on mucosal slices could be traced morphologically to the disruption of the tight junctions between rabbit enterocytes [40]. No biochemical evidence was provided in the initial description of Zot that gave any clues about the nature of Zot activity (e.g., its molecular size, sensitivity to proteases, neutralization by antibody).

Because Zot activity was elaborated by their isogenic CVD101 strain but not by O395-N1, Kaper and colleagues further reasoned that the structural gene for Zot might have been mutated during the construction of O395-N1. This logic directed their focus to DNA sequences located near the *ctxA* gene (where our genetic manipulations of O395-N1 had been done), and in 1993 their group reported that Zot activity could be detected in supernatant fluids of *E. coli* cultures carrying plasmids with inserts corresponding to DNA located immediately upstream of *ctxA* [40]. Sequencing of this DNA insert revealed an open reading frame that was designated the *zot* structural gene. Genetic evidence suggested that mutations in *zot* eliminated Zot biological activity detectable in *E. coli* supernatants in Ussing chambers. Although *E. coli* is notorious for its inability to secrete heterologous polypeptides, no data were presented on the localization of a Zot-related polypeptide in *E. coli* or *V. cholerae* supernatant fluids. Furthermore, an examination of the *zot* gene product suggests the absence of an N-terminal signal sequence and more likely an inner membrane localization [28]. Critically, deletion of the *zot* gene in CVD101 (the strain originally designated as a producer of Zot activity) followed by Ussing chamber analysis of supernatants was not performed. Since then, all studies on Zot have been performed with materials produced in *E. coli*. These have led to the discovery of putative eukaryotic homologs of Zot, the identification of a "Zot receptor," and the use of Zot-derived peptides in therapeutic applications [41–43]. This is quite remarkable given that no confirming data from other groups have provided any evidence that Zot plays a role in *V. cholerae* pathogenesis.

Further DNA sequencing upstream of the *zot* gene revealed two additional open reading frames. The first of these (~98 amino acid residues) was again associated with the production of activity in *E. coli* supernatant fluids that was detectable in Ussing chambers but was different from the biological activity previously designated Zot toxin. Accordingly, the toxin responsible for biological activity was designated accessory cholera enterotoxin (Ace) and was determined to act through induction of chloride ion secretion by enterocytes [44]. Again, no biochemical evidence was initially presented to show that the *ace* gene product was directly associated with toxic activity being elaborated by *V. cholerae* (e.g., antibody neutralization, protein purification). Eventually, recombinant Ace was produced in yeast cells and shown to have biological activity [45]. The question however remains whether *V. cholerae* produces biologically active Ace.

The insights noted above came together into two distinct strategies for construction of nonreactogenic, live attenuated cholera vaccine derivatives of El Tor strains. Deletion of the entire CTX element in several vaccine candidates [13, 46, 47] was envisioned to be the best way to prevent recombinant *V. cholerae* vaccine strains from reacquiring the CTX element. In parallel, the CVD group constructed a Δ*zot*, Δ*ace* derivative (CVD110) that lacked only the core of the CTX element [35] but retained RS elements and therefore two attRS1 sites where CTXΦ could in theory re-insert. In a series of definitive clinical studies, the CVD group found that their strain CVD110 (deleted for *zot*, *ace*, *cep*, *orfU*, and *ctxA*) was not significantly lower in reactogenicity than previous were El Tor vaccine strains deleted in *ctxA* and *hlyA* alone [35]. Thus, volunteer studies have provided no evidence that the *zot* or *ace* gene products are enterotoxins. Indeed, the nonreactogenic live cholera vaccine CVD103-HgR retains two copies of the coding sequences for both *zot* and *ace*. No evidence has since been reported for the existence of any point mutations in the *zot* or *ace* genes that might account for the low reactogenicity of either CVD103-HgR [38] or the original "Zot-deficient" strain O395-N1 [39]. Thus, while the original reason for deleting the entire CTX element (i.e., the CTXΦ) in vaccine strains remains sound in terms of blocking its reacquisition [46], on the basis of these volunteer studies, one must conclude that deletion of *zot* and *ace* does nothing to improve the measurable reactogenicity of live attenuated cholera vaccines.

How do we explain the biochemical and genetic data suggesting that the *zot* and *ace* gene products are new cholera enterotoxins while at the same time reconciling their known role as phage morphogenesis proteins? First, Zot and Ace biological activity in Ussing chambers may actually not be associated with these gene products per se but is due to contaminating substances that may be released as a result of expression of these membrane proteins in a heterologous bacterial host. For example, a substance as simple as adenosine is a highly active segretogue and proteases of many specificities can disrupt tight junctions (see below). Furthermore, a variety of bacterial products including LPS, lipoproteins, bacterial DNA, and peptidoglycan fragments are known to have potent biological activities for different cell types (e.g., macrophages and epithelial cells) by activating Toll-like receptors (TLRs) [49]. TLR activation can cause induction of proinflammatory responses, which are known to mediate dramatic changes in the integrity of tight junctions between epithelial cells. Therefore, *zot* and *ace* gene products may simply change the distribution or the production of such non-specific, *E. coli*-derived or host cell-derived inflammatory mediators. More recently, chemically pure Ace- and Zot-derived polypeptides have been purported to have biological activity [42, 45]. Even if chemically pure *zot*- and *ace*-encoded polypeptides or derivatives can be shown to have activity in a given assay, the biological significance in regard to *V. cholerae* may still be questioned. Are these proteins produced by any strain of *V. cholerae* and does genetics or biochemistry support a role for them in virulence factors or in an enterotoxic response in vivo? Finally, more recent work strongly supports the conclusion that *V. cholerae* produces other toxic molecules including a protease and a toxin that clearly cause disruption of tight junctions in cell monolayer systems [50–52]. This author proposes that until more direct evidence of a role of Zot or Ace in *V. cholerae* pathogenesis is reported, that the field re-consider attributing toxic activity to these gene products in the future.

6.7 Other Toxins and Virulence Loci in the Evolution of *Vibrio cholerae*

Hemolysin. The hemolysin produced by some strains of *V. cholerae* is called HlyA and is clearly established as a toxin active on mammalian cells that has enterotoxic activity in the rabbit ileal loop assay [53]. It is a pore-forming toxin and found to be produced by numerous *Vibrio* species and has been implicated as a diarrheal disease-associated factor, particularly by non-O1 and non-O139 *V. cholerae* [54]. Logically, it was deleted from the genome of several early live attenuated *V. cholerae* vaccine candidates. However, deletion of *hlyA* did not significantly attenuate JBK70 or CVD110 [31, 48]. It remains possible that HlyA is a virulence factor for some *Vibrio* species but seems to be of dubious importance to toxigenic *V. cholerae*. Indeed classical biotype strains naturally carry deletion mutations in the locus [55] and many El Tor strains produce no detectable hemolysin on blood agar plates for reasons that have not been determined at the molecular level.

Hap protease. The hemagglutinin protease (HA protease or Hap) is a zinc metalloprotease produced by some virulent and many avirulent *V. cholerae*. Mutations in *hap* (the structural gene for the protease) do not attenuate *V. cholerae* but do render the organism less prone to detachment from the mucosal surface (Fig. 6.2). Regulation of Hap expression is through quorum sensing which occurs at high bacterial cell density and thus might logically occur late in the infection cycle [56, 57]. A role for Hap in cholera pathogenesis emerged in studies directed at understanding more about Zot and Ace. In an effort to detect Zot or Ace activity in culture supernatants of defined mutants of *V. cholerae*, we studied transcellular epithelial resistance in polarized T84 polarized epithelial cells [51]. In these studies, the evidence supported that all "Zot-like" activity produced by repeat in toxin (RTX) strains of *V. cholerae* could be accounted for by production of Hap; independent biochemical studies reached similar conclusions about wild-type *V. cholerae* strains [50]. These data suggested that Hap might be associated with the reactogenicity of live attenuated CTX^{-} Hap^{+} vaccine strains. Recent work has explored this possibility in animal studies [52] but obtained little evidence for this. Furthermore, at least one vaccine strain (Peru-15) is known to be Hap^{+} but yet in hundreds of volunteers appears to be totally nonreactogenic [58–60]. In a study it was shown that the purified HAP from a non-O1, non-O139 strain was biologically active in rabbit ileal loop, Ussing chamber, and tissue culture assays [61]. Hap is clearly an interesting potential *V. cholerae* virulence factor but given that many naturally occurring pathogenic strains are HapR negative (the essential positive regulator of Hap expression) and fail to produce Hap in vitro [56], one has to conclude, for now, that this factor probably plays a minor role at best in the virulence of TCP^{+} CTX^{+} *V. cholerae*.

RTX toxin. The first example of a putative virulence factor to be revealed by the *V. cholerae* genome sequencing project came in 1999 with the recognition that the strain being sequenced (N16961) carries a cluster of genes that apparently encode a toxin active on mammalian tissue culture cells [62]. This toxin was called *V. cholerae* RTX toxin or RtxA because it was related to the RTX family of toxins produced by numerous other gram-negative organisms. Curiously, the cluster of genes responsible for RtxA production is immediately adjacent to the chromosomal insertion site (*attRS1*) of the CTXΦ and therefore has been deleted in a whole series of *V. cholerae* vaccine derivatives carrying *attRS1* deletions [62]. Given that these derivatives seem marginally lower in their reactogenicity [47] compared to other previously studied *ctxA* deletion strains such as JBK70 [31], it would appear that RtxA may indeed be a new *V. cholerae* virulence factor active in human disease. In vitro studies clearly show that this toxin is produced by multiple strains of *V. cholerae* [52, 62]. Studies conducted with T84 cell monolayers confirmed that production of RtxA was correlated with deficit of the integrity of the paracellular junction [52], an activity that has been attributed to Zot. The biochemical mode of action of RtxA was reported by Fullner and Mekalanos [63] to involve covalent cross-linking of actin and this conclusion was further confirmed by subcloning of the actin cross-linking domain of the toxin [64]. Classical strains of *V. cholerae* carry a deletion mutation in genes encoding part of the RTX toxin region but seventh pandemic El Tor strains and most other *V. cholerae* strains have this locus intact [62]. Given that many CT^-, non-O1, non-O139 strains make RTX, the toxin may be a virulence factor that is particularly important in CT^- strains. In this regard, no phenotype for RtxA mutants of *V. cholerae* has so far been reported in an intestinal model of cholera but this may be in part because the action of CT overwhelms one's ability to detect an effect due to other virulence factors. Animal models have been developed that clearly establish RtxA as an important in vivo virulence factor involved in induction of inflammation [65]. In the mouse pulmonary challenge model, RtxA is more important than Hap and HlyA in inducing a profound and ultimately lethal inflammatory response in the lung [65]. It is interesting to speculate that RtxA might be involved in causing inflammatory diarrhea by non-O1 and non-O139, CT^- strains, but this idea has not been fully investigated.

Mannose-sensitive hemagglutinin (MSHA). MSHA corresponds to a pilus adhesin that is widely distributed among El Tor biotype *V. cholerae* strains [66] and its expression by these strains suggested that MSHA is a virulence factor. However, strains carrying mutations in genes that encode MSHA are fully virulent and produce no intestinal colonization defect in animals [67] and in human volunteer studies [68]. Nonetheless, the locus does seem to be important as a receptor for some filamentous phages and is involved in biofilm formation in vitro and perhaps in aquatic environments.

Motility. *Vibrio cholerae* is a highly motile organism that uses chemotaxis to move from the lumen toward the intestinal mucosa. Motility and chemotaxis are inversely regulated with other virulence phenotypes such as production of TCP and CT [69]. *Vibrio cholerae* has nearly 50 chemotaxis receptors and some of these are coordinately regulated with CT and TCP [70, 71]. Microarray-based analysis of *V. cholerae* genome transcriptional patterns suggests that motility and chemotaxis genes are highly expressed in both animals and humans [8, 72]. A number of studies have found that motility plays a role in virulence but its contribution is complex to say the least. Downregulation of chemotaxis genes has been proposed as a possible mechanism linked to hyperinfectivity of vibrios present in cholera stools

[73]. Nonmotile *V. cholerae* mutants clearly colonize animals and human volunteers [58–60, 69, 74]. Flagellin is a known TLR agonist and it seems possible that this molecule may be proinflammatory in the intestine [49]. This may explain why nonflagellated strains like Peru-15 and Bengal-15 are nonreactogenic in human volunteer studies compared to their flagellated parental strains Peru-3 and Bengal-3 [47, 58–60, 74].

Most recently, evidence that this hypothesis is correct has been obtained using a new animal model for reactogenic diarrhea. Rui et al. [89] have shown that deletion of all five flagellin genes from *V. cholerae* results in a nonreactogenic strain in a suckling rabbit model for residual diarrheal symptoms caused by CTX-negative derivatives of *V. cholerae* [90]. Thus, flagellin produced by live attenuated vaccine strains may be thought of as a toxic factor given its ability to activate the mucosal innate immune and induce inflammatory diarrheal symptoms.

6.8 Conclusion

Vibrio cholerae comparative genomics has begun to provide insights into the evolution of this organism in the past and perhaps also hints of changes that may become more common in the future. For example, the seventh pandemic El Tor O1 clone carries two chromosomal islands that are lacking in strains of the classical biotype as well as most non-O1, non-O139 isolates [10, 18]. What do these islands encode? Time will tell since these islands are finding their way into other *Vibrio* species and thus may influence their virulence and epidemiology as well [75]. Recently, it has been recognized that some non-O1, non-O139 strains carry an island of DNA that encodes a type III secretion system (T3SS) [18]. This T3SS system is similar to one that has been described in a pandemic clone of *Vibrio parahaemolyticus* and known to be involved in virulence [76]. We have shown that *V. cholerae* T3SS is required for intestinal colonization [77] but its role in pathogenesis remains to be determined. This T3SS is present in all strains of the O141 serogroup and this is quite ominous because these strains are clonal and distributed throughout the world as a cause of sporadic diarrheal disease [18]. More recently, another secretion system termed "type VI secretion" (T6SS) has been discovered in a non-O1, non-O139 strain that caused epidemic cholera in Sudan in 1963 [78]. This system is responsible for macrophage cytotoxicity and recently we have shown that T6S mediates the translocation of the actin cross-linking effector domain into phagocytic target cells [91] and that this activity correlates with induction of a profound inflammatory response in the intestinal mucosal [92]. However, the genes for T6SS are widely distributed in pathogenic gram-negative species [79] and most recently have been shown to be involved in the virulence of some of these bacteria [80]. It seems likely that these T3SS and T6SS systems will be integrated with other virulence mechanisms and result in the emergence of new pathogenic clones of *V. cholerae*. Thus, surveillance for new strains of *V. cholerae* seems a prudent undertaking and will likely help us stay one step ahead of evolutionary adaptations that may change the pattern of cholera epidemiology in the future. The recent recognition that phage predation may play a role in controlling cholera epidemics [81, 82] also suggests that changes in this organism may be driven by factors outside of the human host. The emergence of new serogroups (e.g., O139) and the displacement of old biotypes (e.g., the classical by the El Tor) may be driven by environmental forces that include phages, propensity to form biofilms on aquatic matter, and even interaction with marine animals [12, 83, 84]. It seems inevitable that the natural history of cholera will provide inspiration for research questions for years to come.

Acknowledgments I offer sincere thanks to my many students, fellows, and collaborators that have shared a passion for understanding *Vibrio cholerae*. Your hard work, creativity, and devotion to the challenge have made working on this problem an enjoyable adventure. I also want to thank the Harvard University, National Institutes of Health, The Ellison Medical Foundation, The International Vaccine Institute, The Melinda and Bill Gates Foundation, Avant Immunotherapeutics, and all others that have supported research in my laboratory and those of my colleagues throughout the world.

References

1. Faruque SM, Albert MJ, Mekalanos JJ. Epidemiology, genetics, and ecology of toxigenic *Vibrio cholerae*. Microbiol Mol Biol Rev. 1998;62:1301–14.
2. De Haan L, Hirst TR. Cholera toxin: a paradigm for multi-functional engagement of cellular mechanisms. Mol Membr Biol. 2004;21:77–92.
3. Finkelstein RA, LoSpalluto JJ. Pathogenesis of experimental cholera. Preparation and isolation of choleragen and choleragenoid. J Exp Med. 1969;130:185–202.
4. Mekalanos JJ. Duplication and amplification of toxin genes in *Vibrio cholerae*. Cell. 1983;35:253–63.
5. Faruque SM, Mekalanos JJ. Pathogenicity islands and phages in *Vibrio cholerae* evolution. Trends Microbiol. 2003;11:505–10.
6. Waldor MK, Mekalanos JJ. Lysogenic conversion by a filamentous phage encoding cholera toxin. Science. 1996;272:1910–4.
7. Klose KE. Regulation of virulence in *Vibrio cholerae*. Int J Med Microbiol. 2001;291:81–8.
8. Larocque RC, Harris JB, Dziejman M, Le X, Khan AI, Faruque AS, Faruque SM, Nair GB, Ryan ET, Qadri F, Mekalanos JJ, Calderwood SB. Transcriptional profiling of *Vibrio cholerae* recovered directly from patient specimens during early and late stages of human infection. Infect Immun. 2005;73:4488–93.
9. Heidelberg JF, Eisen JA, Nelson WC Clayton RA, Gwinn ML, Dodson RJ, Haft DH, Hickey EK, Peterson JD, Umayam L, Gill SR, Nelson KE, Read TD, Tettelin H, Richardson D, Ermolaeva MD, Vamathevan J, Bass S, Qin H, Dragoi I, Sellers P, McDonald L, Utterback T, Fleishmann RD, Nierman WC, White O. DNA sequence of both chromosomes of the cholera pathogen *Vibrio cholerae*. Nature. 2000;406:477–83.
10. Dziejman M, Balon E, Boyd D, Fraser CM, Heidelberg JF, Mekalanos JJ. Comparative genomic analysis of *Vibrio cholerae*: genes that correlate with cholera endemic and pandemic disease. Proc Natl Acad Sci USA. 2002;99:1556–61.
11. Raskin DM, Seshadri R, Pukatzki SU, Mekalanos JJ. Bacterial genomics and pathogen evolution. Cell. 2006;124:703–14.
12. Colwell RR, Spira WM. The ecology of *Vibrio cholerae*. In: Barua D, Greenough III WB, editors. Cholera. New York: Plenum Publishing Corporation; 1992. pp. 107–27
13. Waldor MK, Mekalanos JJ. Emergence of a new cholera pandemic: molecular analysis of virulence determinants in *Vibrio cholerae* O139 and development of a live vaccine prototype. J Infect Dis. 1994;170:278–83.

14. A Cholera Working Group, International Centre for Diarrhoeal Diseases Research, Bangladesh. Large epidemic of cholera-like disease in Bangladesh caused by *Vibrio cholerae* O139 synonym Bengal. Lancet. 1993;342:387–90.
15. Faruque SM, Chowdhury N, Kamruzzaman M, Dziejman M, Rahman MH, Sack DA, Nair GB, Mekalanos JJ. Genetic diversity and virulence potential of environmental *Vibrio cholerae* population in a cholera-endemic area. Proc Natl Acad Sci USA. 2004;101:2123–8.
16. Popovic T, Bopp C, Olsvik O, Wachsmuth K. Epidemiologic application of a standardized ribotype scheme for *Vibrio cholerae* O1. J Clin Microbiol. 1993;31:2474–82.
17. Faruque SM, Asadulghani M, Saha MN, Alim AR, Albert MJ, Islam KM, Mekalanos JJ. Analysis of clinical and environmental strains of nontoxigenic *Vibrio cholerae* for susceptibility to CTXPhi: molecular basis for origination of new strains with epidemic potential. Infect Immun. 1998;66:5819–25.
18. Dziejman M, Serruto D, Tam VC, Sturtevant D, Diraphat P, Faruque SM, Rahman MH, Heidelberg JF, Decker J, Li L, Montgomery KT, Grills G, Kucherlapati R, Mekalanos JJ. Genomic characterization of non-O1, non-O139 *Vibrio cholerae* reveals genes for a type III secretion system. Proc Natl Acad Sci USA. 2005;102:3465–70.
19. Karaolis DK, Somara S, Maneval DR, Jr., Johnson JA, Kaper JB. A bacteriophage encoding a pathogenicity island, a type-IV pilus and a phage receptor in cholera bacteria. Nature. 1999;399:375–9.
20. Faruque SM, Zhu J, Asadulghani M, Kamruzzaman M, Mekalanos JJ. Examination of diverse toxin-coregulated pilus-positive *Vibrio cholerae* strains fails to demonstrate evidence for *Vibrio* pathogenicity island phage. Infect Immun. 2003;71:2993–9.
21. Rajanna C, Wang J, Zhang D, Xu Z, Ali A, Hou YM, Karaolis DK. The *Vibrio* pathogenicity island of epidemic *Vibrio cholerae* forms precise extrachromosomal circular excision products. J Bacteriol. 2003;185:6893–901.
22. O'Shea YA, Boyd EF. Mobilization of the *Vibrio* pathogenicity island between *Vibrio cholerae* isolates mediated by CP-T1 generalized transduction. FEMS Microbiol Lett. 2002;214: 153–7.
23. Meibom KL, Blokesch M, Dolganov NA, Wu CY, Schoolnik GK. Chitin induces natural competence in *Vibrio cholerae*. Science. 2005;310:1824–7.
24. Davis BM, Waldor MK. Filamentous phages linked to virulence of *Vibrio cholerae*. Curr Opin Microbiol. 2003;6:35–42.
25. Waldor MK, Rubin EJ, Pearson GD, Kimsey H, Mekalanos JJ. Regulation, replication, and integration functions of the *Vibrio cholerae* CTXphi are encoded by region RS2. Mol Microbiol. 1997;24:917–26.
26. Davis BM, Kimsey HH, Kane AV, Waldor MK. A satellite phage-encoded antirepressor induces repressor aggregation and cholera toxin gene transfer. EMBO J. 2002;21: 4240–9.
27. Singh DV, Matte MH, Matte GR, Jiang S, Sabeena F, Shukla BN, Sanyal SC, Huq A, Colwell RR. Molecular analysis of *Vibrio cholerae* O1, O139, non-O1, and non-O139 strains: clonal relationships between clinical and environmental isolates. Appl Environ Microbiol. 2001;67:910–21.
28. Koonin EV. The second cholera toxin, Zot, and its plasmid-encoded and phage-encoded homologues constitute a group of putative ATPases with an altered purine NTP-binding motif. FEBS Lett. 1992;312:3–6.
29. McLeod SM, Kimsey HH, Davis BM, Waldor MK. CTXphi and *Vibrio cholerae*: exploring a newly recognized type of phage–host cell relationship. Mol Microbiol. 2005;57:347–56.
30. Levine MM, Black RE, Clements ML, Lanata C, Sears S, Honda T, Young CR, Finkelstein RA. Evaluation in humans of attenuated *Vibrio cholerae* El Tor Ogawa strain Texas Star-SR as a live oral vaccine. Infect Immun. 1984;43:515–22.
31. Levine MM, Kaper JB, Herrington D, Losonsky G, Morris JG, Clements M, Black RE, Tall B, Hall R. Volunteer studies of deletion mutants of *Vibrio cholerae* O1 prepared by recombinant techniques. Infect Immun. 1988b;56:161–7.

32. O'Brien AD, Chen ME, Holmes RK, Kaper J, Levine MM. Environmental and human isolates of *Vibrio cholerae* and *Vibrio parahaemolyticus* produce a *Shigella dysenteriae* 1 (Shiga)-like cytotoxin. Lancet. 1984;1:77–8.
33. Manning PA, Brown MH, Heuzenroeder MW. Cloning of the structural gene (*hly*) for the haemolysin of *Vibrio cholerae* El Tor strain 017. Gene. 1984;31:225–31.
34. Levine MM, Kaper JB. Live oral vaccines against cholera: an update. Vaccine. 1993;11: 207–12.
35. Tacket CO, Losonsky G, Nataro JP, Cryz SJ, Edelman R, Fasano A, Michalski J, Kaper JB, Levine MM. Safety and immunogenicity of live oral cholera vaccine candidate CVD 110, a *ΔctxA Δzot Δace* derivative of El Tor Ogawa *Vibrio cholerae*. J Infect Dis. 1993;168: 1536–40.
36. Mekalanos JJ, Swartz DJ, Pearson GD, Harford N, Groyne F, de Wilde M. Cholera toxin genes: nucleotide sequence, deletion analysis and vaccine development. Nature. 1983;306:551–7.
37. Herrington DA, Hall RH, Losonsky G, Mekalanos JJ, Taylor RK, Levine MM. Toxin, toxin-coregulated pili, and the *toxR* regulon are essential for *Vibrio cholerae* pathogenesis in humans. J Exp Med. 1988;168:1487–92.
38. Levine MM, Kaper JB, Herrington D, Ketley J, Losonsky G, Tacket CO, Tall B, Cryz R. Safety, immunogenicity, and efficacy of recombinant live oral cholera vaccines, CVD 103 and CVD 103-HgR. Lancet. 1988a;2:467–70.
39. Fasano A, Baudry B, Pumplin DW, Wasserman SS, Tall BD, Ketley JM, Kaper JB. *Vibrio cholerae* produces a second enterotoxin, which affects intestinal tight junctions. Proc Natl Acad Sci USA. 1991;88:5242–6.
40. Johnson JA, Morris JG, Jr., Kaper JB. Gene encoding zonula occludens toxin (*zot*) does not occur independently from cholera enterotoxin genes (*ctx*) in *Vibrio cholerae*. J Clin Microbiol. 1993;31:732–3.
41. Fasano A. Regulation of intercellular tight junction by zonula occludens toxins and its eukaryotic analogue zonulin. Ann NY Acad Sci. 2000;915:214–22.
42. Di Pierro M, Lu R, Uzzau S, Wang W, Margaretten K, Pazzani C, Maimone F, Fasano A. Zonula occludens toxin structure–function analysis. Identification of the fragment biologically active on tight junctions and of the zonulin receptor binding domain. J Biol Chem. 2001;276:19160–5.
43. Uzzau S, Lu R, Wang W, Fiore C, Fasano A. Purification and preliminary characterization of the zonula occludens toxin receptor from human (CaCo2) and murine (IEC6) intestinal cell lines. FEMS Microbiol Lett. 2001;194:1–5.
44. Trucksis M, Galen JE, Michalski J, Fasano A, Kaper JB. Accessory cholera enterotoxin (Ace), the third toxin of a *Vibrio cholerae* virulence cassette. Proc Natl Acad Sci USA. 1993;90:5267–71.
45. Trucksis M, Conn TL, Fasano A, Kaper JB. Production of *Vibrio cholerae* accessory cholera enterotoxin (Ace) in the yeast *Pichia pastoris*. Infect Immun. 1997;65:4984–8.
46. Pearson GD, Woods A, Chiang SL, Mekalanos JJ. CTX genetic element encodes a site-specific recombination system and an intestinal colonization factor. Proc Natl Acad Sci USA. 1993;90:3750–4.
47. Taylor DN, Killeen KP, Hack DC, Kenner JR, Coster TS, Beattie DT, Ezzell J, Hyman T, Trofa A, Sjogren MH. Friedlander A, Mekalanos JJ, Sadoff JC. Development of a live, oral, attenuated vaccine against El Tor cholera. J Infect Dis. 1994;170:1518–23.
48. Michalski J, Galen JE, Fasano A, Kaper JB. CVD110, an attenuated *Vibrio cholerae* O1 El Tor live oral vaccine strain. Infect Immun. 1993;61:4462–8.
49. Kaisho T, Akira S. Toll-like receptor function and signaling. J Allergy Clin Immunol. 2006;1176:979–87.
50. Wu Z, Milton D, Nybom P, Sjo A, Magnusson KE. *Vibrio cholerae* hemagglutinin/protease (HA/protease) causes morphological changes in cultured epithelial cells and perturbs their paracellular barrier function. Microb Pathog. 1996;21:111–23.

51. Mel SF, Fullner KJ, Wimer-Mackin S, Lencer WI, Mekalanos JJ. Association of protease activity in *Vibrio cholerae* vaccine strains with decreases in transcellular epithelial resistance of polarized T84 intestinal epithelial cells. Infect Immun. 2000;68:6487–92.
52. Fullner KJ, Lencer WI, Mekalanos JJ. *Vibrio cholerae*-induced cellular responses of polarized T84 intestinal epithelial cells are dependent on production of cholera toxin and the RTX toxin. Infect Immun. 2001;69:6310–7.
53. Figueroa-Arredondo P, Heuser JE, Akopyants NS, Morisaki JH, Giono-Cerezo S, Enriquez-Rincon F, Berg DE. Cell vacuolation caused by *Vibrio cholerae* hemolysin. Infect Immun. 2001;69:1613–24.
54. Sharma C, Thungapathra M, Ghosh A, Mukhopadhyay AK, Basu A, Mitra R, Basu I, Bhattacharya SK, Shimada T, Ramamurthy T, Takeda T, Yamasaki S, Takeda Y, Nair GB. Molecular analysis of non-O1, non-O139 *Vibrio cholerae* associated with an unusual upsurge in the incidence of cholera-like disease in Calcutta, India. J Clin Microbiol. 1998;36:756–63.
55. Rader AE, Murphy JR. Nucleotide sequences and comparison of the hemolysin determinants of *Vibrio cholerae* El Tor RV79(Hly+) and RV79(Hly$^-$) and classical 569B(Hly$^-$). Infect Immun. 1988;56:1414–9.
56. Zhu J, Miller MB, Vance RE, Dziejman M, Bassler BL, Mekalanos JJ. Quorum-sensing regulators control virulence gene expression in *Vibrio cholerae*. Proc Natl Acad Sci USA. 2002;99:3129–34.
57. Zhu J, Mekalanos JJ. Quorum sensing-dependent biofilms enhance colonization in *Vibrio cholerae*. Dev Cell. 2003;5:647–56.
58. Kenner JR, Coster TS, Taylor DN, Barrera-Oro M, Hyman T, Adams J, Beattie D, Killen K, Mekalanos JJ, Sadoff. Peru-15, an improved live attenuated oral vaccine candidate for *Vibrio cholerae* O1. J Infect Dis. 1995;172:1126–9.
59. Cohen MB, Giannella RA, Bean J, Taylor DN, Parker S, Hoeper A, Wowk S, Hawkins J, Kochi SK, Schiff G, Killeen KP. Randomized, controlled human challenge study of the safety, immunogenicity, and protective efficacy of a single dose of Peru-15, a live attenuated oral cholera vaccine. Infect Immun. 2002;70:1965–70.
60. Qadri F, Chowdhury MI, Faruque SM, Salam MA, Ahmed T, Begum YA, Saha A, Alam MS, Zaman K, Seidlein LV, Park E, Killeen KP, Mekalanos JJ, Clemens JD, Sack DA. Randomized, controlled study of the safety and immunogenicity of Peru-15, a live attenuated oral vaccine candidate for cholera, in adult volunteers in Bangladesh. J Infect Dis. 2005;192:573–9.
61. Ghosh A, Saha DR, Hoque KM, Asakura M, Yamasaki S, Koley H, Das SS, Chakrabarti MK, Pal A. Enterotoxigenicity of mature 45-kilodalton and processed 35-kilodalton forms of hemagglutinin protease purified from a cholera toxin gene-negative *Vibrio cholerae* non-O1, non-O139 strain. Infect Immun. 2006;75:2937–46.
62. Lin W, Fullner KJ, Clayton R, Sexton JA, Rogers MB, Calia KE, Calderwood SB, Fraser C, Mekalanos JJ. Identification of a *Vibrio cholerae* RTX toxin gene cluster that is tightly linked to the cholera toxin prophage. Proc Natl Acad Sci USA. 1999;96:1071–6.
63. Fullner KJ, Mekalanos JJ. In vivo covalent cross-linking of cellular actin by the *Vibrio cholerae* RTX toxin. EMBO J. 2000;19:5315–23.
64. Sheahan KL, Cordero CL, Satchell KJ. Identification of a domain within the multifunctional *Vibrio cholerae* RTX toxin that covalently cross-links actin. Proc Natl Acad Sci USA. 2004;101:9798–803.
65. Fullner KJ, Boucher JC, Hanes MA, Haines GK III, Meehan BM, Walchle C, Sansonetti PJ, Mekalanos JJ. The contribution of accessory toxins of *Vibrio cholerae* O1 El Tor to the proinflammatory response in a murine pulmonary cholera model. J Exp Med. 2002;195: 1455–62.
66. Jonson G, Holmgren J, Svennerholm AM. Identification of a mannose-binding pilus on *Vibrio cholerae* El Tor. Microb Pathog. 1991;11:433–41.
67. Attridge SR, Manning PA, Holmgren J, Jonson G. Relative significance of mannose-sensitive hemagglutinin and toxin-coregulated pili in colonization of infant mice by *Vibrio cholerae* El Tor. Infect Immun. 1996;64:3369–73.

68. Tacket CO, Taylor RK, Losonsky G, Lim Y, Nataro JP, Kaper JB, Levine MM. Investigation of the roles of toxin-coregulated pili and mannose-sensitive hemagglutinin pili in the pathogenesis of *Vibrio cholerae* O139 infection. Infect Immun. 1998;66:692–5.
69. Gardel CL, Mekalanos JJ. Alterations in *Vibrio cholerae* motility phenotypes correlate with changes in virulence factor expression. Infect Immun. 1996;64:2246–55.
70. Everiss KD, Hughes KJ, Kovach ME, Peterson KM. The *Vibrio cholerae acfB* colonization determinant encodes an inner membrane protein that is related to a family of signal-transducing proteins. Infect Immun. 1994;62:3289–98.
71. Hung DT, Mekalanos JJ. Bile acids induce cholera toxin expression in *Vibrio cholerae* in a ToxT-independent manner. Proc Natl Acad Sci USA. 2005;102:3028–33.
72. Xu Q, Dziejman M, Mekalanos JJ. Determination of the transcriptome of *Vibrio cholerae* during intraintestinal growth and midexponential phase in vitro. Proc Natl Acad Sci USA. 2003;100:1286–91.
73. Merrell DS, Butler SM, Qadri F, Dolganov NA, Alam A, Cohen MB, Calderwood SB, Schoolnik GK, Camilli A. Host-induced epidemic spread of the cholera bacterium. Nature. 2002;417:642–5.
74. Coster TS, Killeen KP, Waldor MK, Beattie DT, Spriggs DR, Kenner JR, Trofa A, Sadoff JC, Mekalanos JJ, Taylor DN. Safety, immunogenicity, and efficacy of live attenuated *Vibrio cholerae* O139 vaccine prototype. Lancet. 1995;345:949–52.
75. O'Shea YA, Finnan S, Reen FJ, Morrissey JP, O'Gara F, Boyd EF. The *Vibrio* seventh pandemic island-II is a 26.9 kb genomic island present in *Vibrio cholerae* El Tor and O139 serogroup isolates that shows homology to a 43.4 kb genomic island in *V. vulnificus*. Microbiology. 2004;150:4053–63.
76. Park KS, Ono T, Rokuda M, Jang MH, Okada K, Iida T, Honda T. Functional characterization of two type III secretion systems of *Vibrio parahaemolyticus*. Infect Immun. 2004;72: 6659–65.
77. Tam VC, Serruto D, Dziejman M, Brieher W, Mekalanos JJ. A type III secretion system in *Vibrio cholerae* translocates a formin/spire hybrid-like actin nucleator to promote intestinal colonization. Cell Host Microbe. 2007;1:95–107.
78. Pukatzki S, Ma AT, Sturtevant D, Krastins B, Sarracino D, Nelson WC, Heidelberg JF, Mekalanos JJ. Identification of a conserved bacterial protein secretion system in *Vibrio cholerae* using the *Dictyostelium* host model system. Proc Natl Acad Sci USA. 2006;103:1528–33.
79. Das S, Chaudhuri K. Identification of a unique IAHP (IcmF associated homologous proteins) cluster in *Vibrio cholerae* and other proteobacteria through in silico analysis. In Silico Biol. 2003;3:287–300.
80. Mougous JD, Cuff ME, Raunser S, Shen A, Zhou M, Gifford CA, Goodman AL, Joachimiak G, Ordoñez CL, Lory S, Walz T, Joachimiak A, Mekalanos JJ. A virulence locus of *Pseudomonas aeruginosa* encodes a protein secretion apparatus. Science. 2006;312: 1526–30.
81. Faruque SM, Islam MJ, Ahmad QS, Faruque AS, Sack DA, Nair GB, Mekalanos JJ. Self-limiting nature of seasonal cholera epidemics: role of host-mediated amplification of phage. Proc Natl Acad Sci USA. 2005;102:6119–24.
82. Faruque SM, Naser IB, Islam MJ, Faruque AS, Ghosh AN, Nair GB, Sack DA, Mekalanos JJ. Seasonal epidemics of cholera inversely correlate with the prevalence of environmental cholera phages. Proc Natl Acad Sci USA. 2005;102:1702–7.
83. Kirn TJ, Jude BA, Taylor RK. A colonization factor links *Vibrio cholerae* environmental survival and human infection. Nature. 2005;438:863–6.
84. Faruque SM, Biswas K, Udden SM, Ahmad QS, Sack DA, Nair GB, Mekalanos JJ. Transmissibility of cholera: in vivo-formed biofilms and their relationship to infectivity and persistence in the environment. Proc Natl Acad Sci USA. 2006;103:6350–5.
85. Huber KE, Waldor MK. Filamentous phage integration requires the host recombinases XerC and XerD. Nature. 2002;417:656–59.

86. Iida T, Makino K, Nasu H, Yokoyama K, Tagomori K, Hattori A, Okuno T, Shinagwa H, Honda T. Filamentous bacteriophage of Vibrios are integrated into the *dif*-like site of the host chromosome. J Bacteriol 2002;184:4933–35.
87. Val ME, Bouvier M, Campos J, Sherratt D, Cornet F, Mazel D, Barre FX. The single-stranded genome of phage CTX is the form used for integration into the genome of *Vibrio cholerae*. Mol Cell 2005;19:559–66.
88. Hassan F, Kamruzzaman M, Mekalanos JJ, Faruque SM. Satellite phage TLCΦ enables toxigenic conversion by CTX phage through *dif* site alteration. Nature 2010. In press.
89. Rui H, Ritchie JM, Bronson RT, Mekalanos JJ, Zhang Y, Waldor MK. Reactogenicity of live-attenuated *Vibrio cholerae* vaccines is dependent on flagellins. Proc Natl Acad Sci U S A. 2010;107(9):4359–64.
90. Ritchie JM, Rui H, Bronson RT, Waldor MK. Back to the future: studying cholera pathogenesis using infant rabbits. MBio. 2010;1(1). pii: e00047–10.
91. Ma AT, Mekalanos JJ. In vivo actin cross-linking induced by *Vibrio cholerae* type VI secretion system is associated with intestinal inflammation. Proc Natl Acad Sci U S A. 2010;107(9):4365–70.
92. Ma AT, McAuley S, Pukatzki S, Mekalanos JJ. Translocation of a *Vibrio cholerae* type VI secretion effector requires bacterial endocytosis by host cells. Cell Host Microbe. 2009;5(3):234–43.

Chapter 7
Molecular Epidemiology of Toxigenic *Vibrio cholerae*

Shah M. Faruque, G. Balakrish Nair, and Yoshifumi Takeda

Abstract Toxigenic *Vibrio cholerae* of the O1 or O139 serogroups, the causative agents of cholera, undergo frequent genetic changes leading to the origination of diverse clones, which can be differentiated using defined genetic markers. Developments in DNA analysis techniques have introduced several new typing methods that have enabled to study the epidemiology of *V. cholerae* on a larger global perspective. Molecular epidemiological studies have revealed considerable diversity among epidemic strains isolated from different parts of the world and at different times, and can be a basis for tracking the emergence and spread of strains with epidemic potential. Besides diversity in genetic markers, these clones often show difference in their antibiotic resistance profile, phage resistance, and several other phenotypic properties. These studies have further indicated that there has been a continual emergence of new clones of toxigenic *V. cholerae*, which replaced existing clones, possibly through a natural selection involving unidentified genetic and environmental factors as well as immunity of the host population. The continual emergence of new genetic variants and their selective enrichment during cholera outbreaks appear to constitute an essential component of the natural ecosystem for the evolution of epidemic *V. cholerae* strains.

7.1 Introduction

Toxigenic *Vibrio cholerae* belonging to the O1 or O139 serogroups are the causative agents of cholera, an acute dehydrating diarrhea that occurs in epidemic form particularly in South Asia, Africa, and Latin America [1, 2]. The major surface antigen employed in the characterization of *V. cholerae* is the O antigen. A flagellar (H) antigen is also present, but the value of this antigen for species identification is limited due to the presence of common H epitopes among all *Vibrio* species.

S.M. Faruque (✉)
Molecular Genetics Laboratory, International Centre for Diarrhoeal Diseases Research, Dhaka 1212, Bangladesh
e-mail: faruque@icddrb.org

T. Ramamurthy, S.K. Bhattacharya (eds.), *Epidemiological and Molecular Aspects on Cholera*, Infectious Disease, DOI 10.1007/978-1-60327-265-0_7,
© Springer Science+Business Media, LLC 2011

Of more than 200 known O serogroups of *V. cholerae*, strains belonging to two serogroups, namely O1 and O139, have been associated with epidemics of cholera. The remaining serogroups collectively referred to as *V. cholerae* non-O1, non-O139 are not involved in cholera epidemics but occasionally cause gastrointestinal or extra-intestinal illness utilizing similar or distinctly different virulence mechanisms [3–5].

Vibrio cholerae O1 strains have historically been divided into two biotypes, classical and El Tor based on certain biochemical properties and susceptibility to bacteriophages [2]. Each biotype can be further subdivided into two major serotypes "Ogawa" and "Inaba." A third serotype Hikojima has been reported but it is rare and unstable. Strains of serogroup O1 was supposed to include all the strains responsible for epidemic and endemic cholera until late 1992 when a large cholera-like outbreak in India and Bangladesh was caused by a *V. cholerae* non-O1 strain [3, 6]. This organism did not belong to any of the 138 O serogroups of *V. cholerae* described until then but belonged to a new serogroup, which was later designated as O139. Since then, *V. cholerae* O139 has been persisting as a second etiologic agent of cholera. Hence, there are now two serogroups O1 and O139 that have been associated with epidemics of cholera. However, there are also strains of these serogroups which do not produce cholera toxin and are not involved in epidemics. On the other hand, even within toxigenic strains of the O1 and O139 serogroups, there are genetic variations that determine antibiotic resistance, phage resistance, and other phenotypes that may be linked to environmental fitness, as well as the pandemic potential of a pathogenic clone [1, 2, 7–9].

The ability of *V. cholerae* strains to cause severe enteric infection in humans depends primarily on their virulence gene content. Strains capable of causing cholera invariably carry the genes for cholera toxin (CT) and toxin-coregulated pilus (TCP) [10]. In addition to CT and TCP, all pathogenic strains also carry genes for a number of putative accessory virulence-associated factors, e.g., the mannose-sensitive hemagglutinin (MSHA) pilus, the RTX toxin, and a new type IV pilus [11]. However, the roles of the accessory virulence factors in cholera pathogenesis are not well established, and recent studies are beginning to reveal that at least some of these putative virulence factors also play a role in environmental fitness of the pathogen [12, 13]. The epidemiological success of a pandemic strain may depend on its environmental fitness in addition to the ability to produce the major virulence factors.

Epidemiological surveillance of cholera was limited before the 1970s by the lack of suitable typing systems. A variety of molecular techniques can now efficiently distinguish between different clones of *V. cholerae* and this approach has provided unique insights into the epidemiology of cholera on a larger global perspective [9, 14–21]. The purpose of this chapter is to review and summarize available information on the molecular epidemiology of toxigenic *V. cholerae* and mechanisms leading to genetic diversity of epidemic strains. Special emphasis will be paid to compile scientific data obtained from molecular studies on *V. cholerae* and provide an insight into the possible epidemiological significance of increasing genetic diversity and emergence of new clones of pathogenic *V. cholerae*.

7.2 Epidemiology of Cholera: Overview

Seven distinct pandemics of cholera have occurred since 1817 [1, 2]. Except for the seventh pandemic which originated in Indonesia, six of the pandemics arose from the Indian subcontinent, usually from the Ganges Delta region, reaching to other continents. The sixth pandemic and presumably the earlier pandemics were caused by *V. cholerae* O1 of the classical biotype, whereas the current seventh pandemic, which is the most extensive in geographic spread and duration, has been caused by *V. cholerae* O1 of the El Tor biotype. The seventh pandemic began in 1961 from the island of Sulawesi in Indonesia and spread to other islands including Java, Sarawak, and Borneo, and then to the Philippines, Sabah, and Taiwan, thereby affecting the entire southeast Asian archipelago by the end of 1962. During 1963–1969, the pandemic spread to the Asian mainland and affected Malaysia, Thailand, Burma, Cambodia, Vietnam, India, Bangladesh, and Pakistan. Soon after, El Tor cholera reached Pakistan, Afghanistan, Iran, Iraq, and nearby republics within the Soviet Union experienced outbreaks of cholera. By 1970, El Tor cholera invaded the Arabian Peninsula, Syria, and Israel, and thereafter reached the sub-Saharan west Africa in the early 1970s. Several events that further marked the importance of the seventh pandemic include the reemergence of cholera in Latin America in 1991 [22] and explosive outbreak of cholera among Rwandan refugees in Goma, Zaire, which resulted in about 70,000 cases and 12,000 deaths in 1994 [23]. In late 1992, *V. cholerae* O139 emerged as a predominant epidemic strain and indicated the beginning of a possible eighth pandemic. The epidemiological trend of *V. cholerae* O139 remains under close observation.

7.3 Molecular Epidemiological Tools

Epidemiological surveillance of cholera requires differentiating among different clones of *V. cholerae* and hence it was necessary to develop new typing methods. Previously bacteriophages acting on *V. cholerae* (vibriophages) had been used to type strains based on their susceptibility to different phages. *Vibrio cholerae* O1 strains of the same biotype and serotype could be differentiated into 146 phage types using 10 typing phages [24], and a similar phage-typing scheme has been established for the O139 strains [25]. Since the 1990s, several genetic typing schemes based on restriction fragment length polymorphisms (RFLP) of different genes were developed. This included RFLP in conserved rRNA genes (ribotype) as well as genes encoding virulence factors including the *ctxAB* operon (CTX genotype). The observed mutations, or rearrangement of *rrn* operons, allow ribotype diversity which can be used to monitor strains with epidemic potential. A standardized ribotyping scheme for *V. cholerae* O1 [26] and for O139 [27] that can distinguish 7 different ribotypes among classical strains, 20 ribotypes and subtypes among El Tor strains, and 6 distinct ribotypes among O139 strains have been effectively used to study the molecular epidemiology of cholera. Restriction fragment length polymorphism analysis using *ctxB* (99% conserved) gene probe can divide toxigenic O1 serogroup into different genotypes. Genotype 1 was found in classical biotype and El Tor strains from the US Gulf Coast. Genotype 2 was found in El Tor strains from Australia and genotype 3 was found in seventh pandemic strains and the Latin American outbreak.

The introduction of more advanced techniques, including multilocus enzyme electrophoresis (MEE), offered new opportunities to study the molecular epidemiology depending on allelic variations in housekeeping enzymes detected by electrophoretic mobilities. MEE can distinguish between classical and El Tor strains [9, 28] and has grouped the toxigenic El Tor biotype strains of *V. cholerae* O1 into four major clonal groups or electrophoretic types (ET) representing broad geographical areas.

Pulse-field gel electrophoresis (PFGE) is one of the powerful tools in detecting bacterial strain similarity or clonal relatedness using specific restriction enzymes which are species specific, e.g., *Xba*I for *Escherichia coli* and *Salmonella*, *Not*I for *V. cholerae*. The Centers for Disease Control in US (CDC) have developed several standard PFGE protocols for universal use so that the generated profiles can be compared. Recently, PFGE protocol for *V. cholerae* O1 has also been introduced [29].

Multilocus sequence typing (MLST), a method that is based on partial nucleotide sequences of multiple (usually around seven) housekeeping genes, has recently been shown to be a powerful technique for bacterial typing [30]. Housekeeping genes are preferred over virulence-associated genes, because an analysis of mutations (most of which are usually synonymous, given the strong selection against changes of the amino acid sequence in genes coding for proteins required for growth) in such genes is more likely to adequately reflect the phylogeny of strains. The other version of MLST is the detection of variable number tandem repeats (VNTR) in the housekeeping genes. This technique differentiated *V. cholerae* O1 and O139 strains, which were genetically related as detected by using other molecular tools such as RFLP and PFGE [31, 32].

7.4 Molecular Epidemiology of Cholera

The frequently changing relative prevalence of different clones of pathogenic *V. cholerae* has prompted numerous studies of the molecular basis of these changes. The clonal diversity and epidemiological associations of different clones of toxigenic *V. cholerae* have been reviewed by Wachsmuth et al. [9]. Multilocus enzyme electrophoresis (MEE) analysis and analysis of restriction fragment length polymorphism (RFLP) in different genes have enabled the study of *V. cholerae* epidemiology by analyzing strains from various geographical regions [1, 9, 14]. Analysis of strains using MEE has identified four major electrophoretic types, which include the Australian clone (ET1), the Gulf Coast clone (ET2), the seventh pandemic clone (ET3), and Latin American clone (ET4) [15, 16, 32].

RFLP of conserved rRNA genes (ribotype) has been used to separate *V. cholerae* strains into different clonal groups. For example, ribotyping indicates that toxigenic

US Gulf Coast isolates are clonal and that they differ from other seventh pandemic isolates, while analysis of isolates from the 1991 Latin American epidemic showed that these were clearly related to seventh pandemic isolates from other parts of the world.

Ribotyping provided the initial indication that *V. cholerae* O139 strains are likely to be closely related to the O1 El Tor strains [19]. Eventually, further comparative molecular analysis of O1 El Tor strains and the epidemic O139 strains suggested that the O139 strains were derived from the El Tor strains by serotype-specific genetic changes [33]. Microarray-based comparative genomic analysis of *V. cholerae* isolates [34] identified genes that are shared by all pandemic strains, as well as genes specific to seventh pandemic O1 El Tor and O139 strains. The latter genes were contained on two chromosomal "islands" known as *Vibrio* seventh pandemic islands (VSP-1 and VSP-2), and may define genes involved in the epidemiological success of the seventh pandemic clone. None of the non-O1, non-O139 strains carried the complete VSP island, indicating that the O1 and O139 serogroups are different from the rest of the *V. cholerae* serogroups [35]. The detailed molecular epidemiology of cholera in Asian and African countries is described in Chapters 2 and 3, respectively.

7.4.1 Clonal Diversity of Epidemic Strains in Bangladesh and India

Molecular analysis of epidemic isolates of *V. cholerae* in Bangladesh and India revealed clonal diversity among strains isolated during different epidemics [20, 21, 36–42]. These studies demonstrated the transient appearance and disappearance of more than six ribotypes among classical vibrios, at least five ribotypes of El Tor vibrios, and three different ribotypes of *V. cholerae* O139. Different ribotypes often showed different CTX genotypes resulting from differences in copy number of the CTX prophage and adjacent RS elements. Epidemiological data on the emergence and prevalence of *V. cholerae* O139 and its coexistence with the O1 El Tor strains are available primarily from Bangladesh and India [20, 21, 27, 37–42]. Analysis of strains collected since the first detection of the O139 serogroup has demonstrated emergence of new clones of *V. cholerae* O139 and their association with different epidemic outbreaks. Ribotype analysis as well as CTX genotypes has been used to track clonal relationships among outbreak strains. At least seven different ribotypes have been reported among O139 strains isolated until 2000, and since then two new ribotypes have been detected [27, 36]. Within the short span of emergence of O139, the genomic instability of O139 vibrios, at least considering that of the *rrn* operons as a marker, appears to be more than that of *V. cholerae* O1 El Tor strains. Yearly isolation of different ribotypes of *V. cholerae* O139 in Bangladesh and India showed a striking difference in the proportion of strains belonging to different ribotypes, as well as the distribution of the ribotypes between the two neighboring countries. Since 1993, a decline in the proportion of strains belonging to ribotype B-I and an increase in the proportion of strains belonging to ribotype B-II were noticed both in India and in Bangladesh, and subsequently in some districts of Bangladesh, strains belonging to ribotype B-III increased considerably and caused an outbreak in 1997 [43]. This finding demonstrated the dynamic nature of the epidemiology of *V. cholerae* O139 and the different clones involved in epidemics. A huge resurgence of *V. cholerae* O139 was witnessed in September 1996 in Calcutta [39]. Phenotypically, the reemerged *V. cholerae* O139 was different from those that appeared in late 1992 in that the current O139 strains were sensitive to co-trimoxazole. Despite differences in antibiogram, the 1992 O139 and 1996 O139 strains had identical ribotype.

RFLPs in *ctxA* genes and its flanking sequences (CTX genotypes) have also been used to define different clones. A total of 11 distinct CTX genotypes were identified among six ribotypes until 2000 [27]. Although the structural gene for CT is identical in different strains, the observed diversity of CTX genotypes may result from duplication of the CTX prophage. RFLP of CTX prophage indicated that there is a continuous change in the structure and organization of CTX prophage along with emergence of a new type of CTX prophage. The 1992–1993 strains showed two CTX prophages connected by an RS1 element, while the 1996 strains showed three CTX prophages arranged in tandem. These strains exhibited two types of CTX prophages with the first of the three prophages being an El Tor-type CTX prophage and the second and third CTX prophages being a new type of CTX prophage, with difference primarily lying in *rstR* gene which codes for the repressor proteins of CTX prophage. In 1998, the prevalence of two new clones of *V. cholerae* O139 was observed at two cholera-endemic areas including Calcutta and Alleppey, in southern India. Whereas most of the O139 strains from Calcutta have only the El Tor-type CTX prophage, the strains from Alleppey were found to contain the unique arrangement of the CTX prophage of the O139 isolates of 1996 [39]. Thus, two different clones of O139 were prevalent at two regions in India, showing contrasting distribution of different O139 clones in different geographic locations. It is not clear how the CTX genotype diversity can contribute to increased incidence of cholera. Interestingly, the reemerged O139 strains in 2002 in Bangladesh carried the Calcutta-type CTX^{Cal} prophages [11].

In July 1993, while the O139 epidemic was waning in Calcutta, *V. cholerae* O1 reemerged both in Calcutta and in neighboring Dhaka, Bangladesh [20, 37, 40, 44]. The introduction of *V. cholerae* O139 in 1992 and the concurrent disappearance of *V. cholerae* O1 biotype El Tor provided a unique opportunity in that it punctuated the otherwise continuous incidence of *V. cholerae* O1, an event which perhaps had never occurred before. This enabled to identify two groups of *V. cholerae* O1one that prevailed before the genesis of O139 and the other that reappeared after the epidemic of O139 in Calcutta. Macro-restriction analysis of DNA using pulsed-field gel electrophoresis, and comparison with patterns generated by four standard strains of *V. cholerae* O1, namely ET1, which represents the Australian clone, ET2, which represents the US Gulf Coast clone, ET3, which represents the seventh pandemic clone, ET4, which represents the Latin American pandemic clone, suggested that the remerged El Tor strains represented a new ET type. This strain was also found to belong to a new ribotype. Epidemiologically, the new clone of *V. cholerae* O1 had by now prevailed in Calcutta and became the dominant serogroup and the isolation of O139 on the other hand had became few and far apart. By analyzing strains received from different parts of the country, it was established that the new clone had spread into several of the cholera-endemic areas in India and neighboring countries. Extended molecular studies showed that the new clone of *V. cholerae* O1 had further spread into the West African country of Guinea-Bissau, where it was responsible for an epidemic of cholera during 1996 [41].

The distribution and the significance of VSP islands in *V. cholerae* O1 strains collected from different geographical regions have been studied. Peruvian strains that existed between 1991 and 2003 lacked the VSP-2 island (VC0511–VC0515), which is a unique feature and different from the seventh cholera pandemic strains from many Asian regions [45]. Based on the difference in the VSP-2 region and detection of clonal similarity using PFGE, the Peruvian strains seem to have evolved independently and shared its homology with Eurasian and African counterpart strains [45]. The indigenous Australian strains of *V. cholerae* O1 on the other hand did not harbor VSP-1 and VSP-2 islands and presumably have pre-seventh pandemic ancestry [46].

7.4.2 Changing Antibiotic Resistance Among Toxigenic **Vibrio cholerae**

Epidemic strains of *V. cholerae* are known to undergo frequent changes in their antibiotic resistance properties. Although genetic determinants of antibiotic resistance are not directly involved in pathogenesis, alteration in antibiotic sensitivity of epidemic strains can significantly influence the epidemiology of cholera [1]. In 1979, a multiple drug-resistant *V. cholerae* O1 strain appeared in Matlab, a rural sub-district of Bangladesh, causing an outbreak of cholera [47, 48]. Screening of isolates from the outbreak showed that 16.7% of the isolates were resistant to five antibiotics including tetracycline, ampicillin, kanamycin, streptomycin, and trimethoprim–sulfamethoxazole, whereas another 10% of the isolates were resistant to any four of these antibiotics including tetracycline. Epidemiological assessment suggested that the outbreak began from the introduction into the area of a single multiple drug-resistant strain of *V. cholerae* O1. By 1986, the drug resistance pattern changed, and screening of *V. cholerae* O1 isolated from cholera patients in January 1986 in Dhaka showed that none of these isolates was resistant to tetracycline, streptomycin, chloramphenicol, amoxicillin, or nalidixic acid [49]. After almost a decade, reemergence of tetracycline-resistant El Tor strains was observed during the 1991 epidemic in Bangladesh [50]. In October 1995, emergence of nalidixic acid-resistant *V. cholerae* O1 was observed in southern India [51]. The susceptibility of *V. cholerae* O1 strains to certain antibiotics varied in regard to the isolation time and geographical location. Between March 1994 and December 1996, 80–100% of *V. cholerae* O1 isolates in Kenya and south Sudan, and 65–90% of isolates in Somalia were sensitive to tetracycline [52], whereas all isolates in Tanzania and Rwanda were resistant to tetracycline. In Kenya and Somalia, the percentage of isolates resistant to chloramphenicol and co-trimoxazole markedly increased from 15% in 1994 to more than 90% in 1996.

The O139 serogroup of *V. cholerae*, which emerged during 1992–1993, were sensitive to tetracycline [6]. Although the new serogroup showed a trend of increased resistance to trimethoprim–sulfamethoxazole, it was more susceptible to ampicillin and tetracycline than was the O1 serogroup of *V. cholerae*. Waldor and coworkers [53] reported the presence of a self-transmissible transposon-like element (SXT element) encoding resistance to sulfamethoxazole, trimethoprim, and streptomycin in *V. cholerae* O139. The SXT element could be conjugally transferred from *V. cholerae* O139 to *V. cholerae* O1 and *E. coli* strains, where it integrated into recipient chromosomes in a site-specific, *recA*-independent manner. Comparison of the antibiotic resistance patterns between the O139 strains isolated during 1992 and 1993 and those isolated in 1996 and 1997 in India showed that the latter strains were susceptible to co-trimoxazole, unlike the O139 strains from 1992 and 1993. The rapidly changing pattern of antibiotic resistance observed among *V. cholerae* suggests that there is substantial mobility in genetic elements encoding antibiotic resistance in *V. cholerae*.

7.5 Molecular Basis for Clonal Diversity

The origination of different clones of pathogenic *V. cholerae* could be due to acquisition of critical virulence genes and other gene clusters by diverse nonpathogenic progenitor strains, and thus transform into pathogenic strains. Alternatively, an existing pathogenic strain may undergo genetic changes leading to the origination of a new strain. The mechanism by which horizontal transfer of CT genes occurs has been reasonably defined with the discovery of CTXΦ [54], but the mechanism involved in the transfer of TCP pathogenicity island which encodes the major colonization factor is still not clearly known. The O139 strains are presumed to have been derived from a *V. cholerae* O1 El Tor strain by genetic changes in serotype-specific genes [11, 33], involving multiple gene transfer and recombination events. The *V. cholerae* genome has been found to contain a distinctive class of integrons, which are gene expression elements that acquire open reading frames and convert them to functional genes [55]. This permits the bacteria to entrap genes from other microorganisms and thus constitute a mechanism for the clustering and spread of pathogenic genes as well as genes for other biochemical functions.

Microevolution among toxigenic *V. cholerae* may also contribute significantly to the emergence of new strains. For example, several researchers have documented the existence of different *tcpA* alleles in vibrios of various serogroups, including toxigenic and nontoxigenic strains from clinical and environmental sources [56]. Similarly, a recent study of non-O1, non-O139 environmental isolates detected considerable genetic mosaicism in both the CTXΦ and the *tcp* gene cluster [57]. Some of these strains were shown to possess new alleles of the *tcpA*, *toxT*, and *tcpF* genes carried by the TCP pathogenicity island. Different alleles of the CTXΦ prophage repressor *rstR* were also found. Interestingly, some of the non-O1, non-O139 environmental strains possessing novel *tcpA*, *toxT*, and *tcpF* alleles were able to colonize, indicating that non-canonical virulence alleles may still have virulence function. The functional virulence genes may have evolved from ancestral genes which exist as virulence gene homologues in some environmental *V. cholerae* strains. Thus genetic variation among *V. cholerae* strains may be quite extensive, and the fitness of strains in the environment may be favored by different alleles of virulence genes. Genomic microarrays have been recently used to perform comparative analysis of *V. cholerae* isolates [34]. Microarray-based genomic analysis has not yet been applied extensively to environmental strains or non-O1, non-O139 *V. cholerae*. Such studies will offer an opportunity to expand our understanding of the phylogenic relationships between pathogenic and nonpathogenic strains of *V. cholerae*. Although clinical strains could be distinguished from environmental strains by various molecular studies, both types of strains are very similar overall depending upon their geographical origin. These studies seem to suggest that pathogenic *V. cholerae* are probably derivatives of some ancestral environmental strains but the complete pathway to emergence has not yet been fully observed in any geographical locale.

7.6 Influence of Clonal Diversity on the Epidemiology of Cholera

Studies have been conducted to understand the genetic variation in successive isolates of *V. cholerae* O1 and O139 and to discern if these genetic variations lead to the emergence of new clones of *V. cholerae,* with considerably altered pathobiological properties or increased environmental fitness. A pertinent question would be whether changes in the genetic markers in the O1 and O139 strains are associated with other undiscovered discreet changes in the organism that affect its epidemiological characteristics. Although substantial information is available on the epidemiology of *V. cholerae* and the emergence of new epidemic clones, it is not clear what drives the frequent emergence of new clones often associated with epidemics and replacement of existing clones. An important area that needs to be addressed is whether preexisting immunity against one clone of either O1 or O139 can completely provide protection against another emerging clone. From an epidemiological viewpoint, it certainly appears that genetic rearrangement fosters some advantage to the emerging clone. It is also crucial to know whether the observed genetic reassortment in the O1 and O139 strains accompanies other subtle changes in the organism that enable the organism to escape the immune pressure of the host population against previously existing clones of toxigenic *V. cholerae*. There is no cross-protection between *V. cholerae* O1 and O139 in the animal models. The predominantly adult population infected by O139 cholera [6] also provides evidence that O1 strains do not protect efficiently against O139 strains. Considering the extent of longevity of a clone and frequency of appearance and reappearance of different clones in a cholera-endemic area, as well as occurrence of epidemics of cholera with seasonal regularity, it appears that the clonal turnover has greater implications. The continual emergence of new toxigenic strains and their selective enrichment during cholera outbreaks perhaps constitute an essential component of the natural ecosystem for the evolution of epidemic *V. cholerae* strains to ensure its continued existence.

References

1. Faruque SM, Albert MJ, Mekalanos JJ. Epidemiology, genetics, and ecology of toxigenic *Vibrio cholerae*. Microbiol Mol Biol Rev. 1998;62:1301–14.
2. Kaper JB, Morris JG, Levine MM. Cholera. Clin Microbiol Rev. 1995;8:48–86.
3. Ramamurthy T, Bag PK, Pal A, Bhattacharya SK, Shimada T, Takeda T, Karasawa T, Kurasono H, Takeda Y, Nair GB. Virulence patterns of *Vibrio cholerae* non-O1 strains isolated from hospitalised patients with acute diarrhoea in Calcutta, India. J Med Microbiol. 1993;39:310–7.
4. Morris JG. Non-O1 group 1 *Vibrio cholerae* strains not associated with epidemic disease. In: Wachsmuth IK, Blake PA, Olsvik O, editors. *Vibrio cholerae* and cholera: molecular to global perspectives. Washington, DC: ASM Press; 1994. pp. 103–16.
5. Mukhopadhyay AK, Saha PK, Garg S, Bhattacharya SK, Shimada T, Takeda T, Takeda Y, Nair GB. Distribution and virulence of *Vibrio cholerae* belonging to serogroups other than O1 and O139: a nation survey. Epidemiol Infect. 1995;114:65–70.
6. Cholera Working Group. Large epidemic of cholera-like disease in Bangladesh caused by *Vibrio cholerae* O139 synonym Bengal. Lancet. 1993;342:387–90.
7. Guidolin A, Manning PA. Genetics of *Vibrio cholerae* and its bacteriophages. Microbiol Rev. 1987;51:285–98.
8. Takeya K. Lysogeny in El Tor vibrios and prophage typing. In: Barua D, Burrows W, editors. Cholera. Philadelphia: W.B. Saunders; 1974. pp.74–83.
9. Wachsmuth IK, Olsvik Ø, Evins GM, Popovic T. Molecular epidemiology of cholera. In: Wachsmuth IK, Blake PA, Olsvik Ø, editors. *Vibrio cholerae* and cholera: molecular to global perspectives. Washington, DC: American Society for Microbiology; 1994. pp. 357–70.
10. Herrington DA. Hall RH, Losonsky G, Mekalanos JJ, Taylor RK, Levine MM. Toxin, toxin-coregulated pili, and the *toxR* regulon are essential for *Vibrio cholerae* pathogenesis in humans. J Exp Med. 1988;168:1487–92.
11. Faruque SM, Mekalanos JJ. Pathogenicity islands and phages in *Vibrio cholerae* evolution. Trend Microbiol. 2003;11:505–10.
12. Chiavelli DA, Marsh JW, Taylor RK. The mannose-sensitive hemagglutinin of *Vibrio cholerae* promotes adherence to zooplankton. Appl Environ Microbiol. 2001;67:3220–5.
13. Watnick PI, Fullner KJ, Kolter R. A role for the mannose sensitive hemagglutinin in biofilm formation by *Vibrio cholerae* El Tor. J Bacteriol. 1999;181:3606–9.
14. Kaper JB, Bradford HB, Roberts NC, Falkow S. Molecular epidemiology of *Vibrio cholerae* in the US Gulf Coast. J Clin Microbiol. 1982;16:129–34.
15. Chen F, Evins GM, Cook WL, Almeida R, Hargrett-Bean N, Wachsmuth K. Genetic diversity among toxigenic and nontoxigenic *Vibrio cholerae* O1 isolated from the Western Hemisphere. Epidemiol Infect. 1991;107:225–33.
16. Wachsmuth IK, Evins GM, Fields PI, Olsvik Ø, Popovic T, Bopp CA, Wells JG, Carrillo C, Blake PA. The molecular epidemiology of cholera in Latin America. J Infect Dis. 1993;167:621–6.
17. Faruque SM, Albert J. Genetic relation between *Vibrio cholerae* O1 strains in Ecuador and Bangladesh. Lancet. 1992;339:740–1.
18. Faruque, SM, Alim ARMA, Rahman MM, Siddique AK, Sack RB Albert MJ. Clonal relationships among classical *Vibrio cholerae* O1 strains isolated between 1961 and 1992 in Bangladesh. J Clin Microbiol. 1993;31:2513–6.
19. Faruque SM, Alim ARMA, Roy SK, Khan F, Nair GB, Sack RB, Albert MJ. Molecular analysis of rRNA and cholera toxin genes carried by the new epidemic strain of toxigenic *Vibrio cholerae* O139 synonym Bengal. J Clin Microbiol. 1994;32:1050–3.

20. Faruque SM, Ahmed KM, Abdul Alim AR, Qadri F, Siddique AK, Albert MJ. Emergence of a new clone of toxigenic *Vibrio cholerae* O1 biotype El Tor displacing *V. cholerae* O139 Bengal in Bangladesh. J Clin Microbiol. 1997;35:624–30.
21. Faruque SM, Ahmed KM, Siddique AK, Zaman K, Alim AR, Albert MJ. Molecular analysis of toxigenic *Vibrio cholerae* O139 Bengal strains isolated in Bangladesh between 1993 and 1996: evidence for emergence of a new clone of the Bengal vibrios. J Clin Microbiol. 1997;35:2299–306.
22. Levine MM. South America: the return of cholera. Lancet. 1991;338:45–6.
23. Siddique, AK, Salam A, Islam MS, Akram K, Majumdar RN, Zaman K, Fronczak N, Laston S. Why treatment centres failed to prevent cholera deaths among Rwandan refugees in Goma, Zaire. Lancet. 1995;345:359–61.
24. Chattopadhyay DJ, Sarkar, BL, Ansari MQ, Chakraborti BK, Roy MK, Ghose, AN, Pal SC. New phage typing scheme for *Vibrio cholerae* O1 biotype El Tor strains. J Clin Microbiol. 1993;31:1579–85.
25. Chakrabarti AK, Ghosh AN, Nair GB, Niyogi SK, Bhattacharya SK, Sarkar BL. Development and evaluation of a phage typing scheme for *Vibrio cholerae* O139. J Clin Microbiol. 2000;38:44–9.
26. Popovic T, Bopp C, Olsvik O, Wachsmuth K. Epidemiologic application of a standardized ribotype scheme for *Vibrio cholerae* O1. J Clin Microbiol. 1993;31:2474–82.
27. Faruque SM, Saha, MN, Asadulghani, Bag PK, Bhadra RK, Bhattacharya SK, Sack RB, Takeda Y, Nair GB. Genomic diversity among *Vibrio cholerae* O139 strains isolated in Bangladesh and India between 1992 and 1998. FEMS Microbiol Lett. 2000;84:279–84.
28. Freitas FS, Momen H, Salles CA. The zymovars of *Vibrio cholerae*: multilocus enzyme electrophoresis of *Vibrio cholerae*. Mem Inst Oswaldo Cruz. 2002;97:511–6.
29. Cooper KL, Luey CK, Bird M, Terajima J, Nair GB, Kam KM, Arakawa E, Safa A, Cheung DT, Law CP, Watanabe H, Kubota K, Swaminathan B, Ribot EM. Development and validation of a PulseNet standardized pulsed-field gel electrophoresis protocol for subtyping of *Vibrio cholerae*. Foodborne Pathog Dis. 2006;3:51–8.
30. Maiden MC, Bygraves JA, Feil E, Morelli G, Russell JE, Urwin R, Zhang Q, Zhou J, Zurth K, Caugant DA, Feavers IM, Achtman M, and Spratt BG. Multilocus sequence typing: a portable approach to the identification of clones within populations of pathogenic microorganisms. Proc Natl Acad Sci USA. 1998;95:3140–5.
31. Ghosh R, Nair GB, Tang L, Morris JG, Sharma NC, Ballal M, Garg P, Ramamurthy T, Stine OC. Epidemiological study of *Vibrio cholerae* using variable number of tandem repeats. FEMS Microbiol Lett. 2008;288:196–201.
32. Salles CA, Momen H. Identification of *Vibrio cholerae* by enzyme electrophoresis. Trans R Soc Trop Med Hyg. 1991;85:544–7.
33. Bik EM, Bunschoten AE, Gouw RD, Mooi FR. Genesis of the novel epidemic *Vibrio cholerae* O139 strain: evidence for horizontal transfer of genes involved in polysaccharide synthesis. EMBO J. 1995;14:209–16.
34. Dziejman M, Balon E, Boyd D, Fraser CM, Heidelberg JF, Mekalanos JJ. Comparative genomic analysis of *Vibrio cholerae*: genes that correlate with cholera endemic and pandemic disease. Proc Natl Acad Sci USA. 2002;99:1556–61.
35. Rahman MH, Biswas K, Hossain MA, Sack RB, Mekalanos JJ, Faruque SM. Distribution of genes for virulence and ecological fitness among diverse *Vibrio cholerae* populations in a cholera endemic area: tracking the evolution of pathogenic strains. DNA Cell Biol. 2008;27:347–55.
36. Faruque SM, Saha MN, Asadulghani, Sack DA, Sack RB, Takeda Y, Nair GB. The O139 serogroup of *Vibrio cholerae* comprises diverse clones of epidemic and non epidemic strains derived from multiple *V. cholerae* O1 and non-O1 progenitors. J Infect Dis. 2000;182:1161–8.
37. Mukhopadhyay AK, Garg S, Mitra R, Basu A, Dutta D, Bhattacharya SK, Shimada T, Takeda T, Takeda Y, Nair GB. Temporal shifts in traits of *Vibrio cholerae* strains isolated

from hospitalized patients in Calcutta: a 3-year (1993–1995) analysis. J Clin Microbiol. 1996;34:2537–43.
38. Mukhopadhyay AK, Basu A, Garg P, Bag PK, Ghosh A, Bhattacharya SK, Takeda Y, Nair GB. Molecular epidemiology of reemergent *Vibrio cholerae* O139 Bengal in India. J Clin Microbiol. 1998;36:2149–52.
39. Sharma C, Maiti S, Mukhopadhyay AK, Basu A, Basu I, Nair GB, Mukhopadhyaya R, Das B, Kar S, Ghosh RK, Ghosh A. Unique organization of the CTX genetic element in *Vibrio cholerae* O139 strains which reemerged in Calcutta, India, in September, 1996. J Clin Microbiol. 1997;35:3348–50.
40. Sharma C, Nair GB, Mukhopadhyay AK, Bhattacharya SK, Ghosh RK, Ghosh A. Molecular characterization of *Vibrio cholerae* O1 biotype El Tor strains isolated between 1992 and 1995 in Calcutta, India: evidence for the emergence of a new clone of the El Tor biotype. J Infect Dis. 1997;175:1134–41.
41. Sharma C, Ghosh A, Dalsgaard A, Forslund A, Ghosh RK, Bhattacharya SK Nair GB. Molecular evidence that a distinct *Vibrio cholerae* O1 biotype El Tor strain in Calcutta may have spread to the African continent. J Clin Microbiol. 1998;36:843–4.
42. Mitra R, Basu A, Dutta D, Nair GB Takeda Y. Resurgence of *Vibrio cholerae* O139 Bengal with altered antibiogram in Calcutta, India. Lancet. 1996;348:1181.
43. Faruque SM, Siddique AK, Saha MN, Asadulghani, Rahman MM, Zaman K, Albert MJ, Sack DA, Sack RB. Molecular characterization of a new ribotype of Vibrio cholerae O139 Bengal associated with an outbreak of cholera in Bangladesh. J Clin Microbiol. 1999;37:1313–8.
44. Yamasaki S, Nair GB, Bhattacharya SK, Yamamoto S, Kurazono H, Takeda Y. Cryptic appearance of a new clone of *Vibrio cholerae* O1 biotype El Tor in Calcutta, India. Microbiol Immunol. 1997;41:1–6.
45. Nusrin S, Gil AI, Bhuiyan NA, Safa A, Asakura M, Lanata CF, Hall E, Miranda H, Huapaya B, Vargas GC, Luna MA, Sack DA, Yamasaki S, Nair GB. Peruvian *Vibrio cholerae* O1 El Tor strains possesses a distinct region in the Vibrio seventh pandemic island-II that differentiates them from the prototype seventh pandemic El Tor strains. J Med Microbiol. 2009;58:342–54.
46. Safa A, Bhuiyan NA, Murphy D, Bates J, Nusrin S, Kong RY, Chogsanguan M, Chaicumpa W, Nair GB. Multilocus genetic analysis reveals that the Australian strains of *Vibrio cholerae* O1 are similar to the pre-seventh pandemic strains of the El Tor biotype. J Med Microbiol. 2009;58:105–11.
47. Glass RI, Huq I, Alim ARMA, Yunus M. Emergence of multiple antibiotic-resistant *Vibrio cholerae* in Bangladesh. J Infect Dis. 1980;142:939–42.
48. Glass RI, Huq MI, Lee JV, Threlfall EJ, Khan MR, Alim ARMA, Rowe B, Gross RJ. Plasmid-borne multiple drug resistance in *Vibrio cholerae* serogroup O1, biotype El Tor: evidence of a point-source outbreak in Bangladesh. J Infect Dis. 1983;147:204–9.
49. Nakasone N, Iwanaga M, Eeckels R. Characterization of *Vibrio cholerae* O1 recently isolated in Bangladesh. Trans Roy Soc Trop Med Hyg. 1987;81:876–8.
50. Siddique AK, Zaman K, Baqui AH, Akram K, Mutsuddy P, Eusof A, Haider K, Islam S, Sack RB. Cholera epidemics in Bangladesh: 1985–1991. J Diarrhoeal Dis Res. 1992;10:79–86.
51. Jesudason MV, Saaya R. Resistance of *Vibrio cholerae* O1 to nalidixic acid. Indian J Med Res. 1997;105:153–4.
52. Materu SF, Lema OE, Mukunza HM, Adhiambo CG, Carter JY. Antibiotic resistance pattern of *Vibrio cholerae* and *Shigella* causing diarrhea outbreaks in the eastern Africa region: 1994–1996. East Afr Med J. 1997;74:193–7.
53. Waldor MK, Tschape H, Mekalanos JJ. A new type of conjugative transposon encodes resistance to sulfamethoxazole, trimethoprim, and streptomycin in *Vibrio cholerae* O139. J Bacteriol. 1996;178:4157–65.
54. Waldor MK, Mekalanos JJ. Lysogenic conversion by a filamentous bacteriophage encoding cholera toxin. Science. 1996;272:1910–4.
55. Mazel D, Dychinco B, Webb VA, Davies J. A distinctive class of integron in the *Vibrio cholerae* genome. Science. 1998;280:605–8.

56. Boyd EF, Waldor MK. Evolutionary and functional analyses of variants of the toxin-coregulated pilus protein TcpA from toxigenic *Vibrio cholerae* non-O1/non-O139 serogroup isolates. Microbiol. 2002;148:1655–66.
57. Mukhopadhyay AK, Chakraborty S, Takeda Y, Nair GB, Berg DE. Characterization of VPI pathogenicity island and CTXphi prophage in environmental strains of *Vibrio cholerae*. J Bacteriol. 2001;183:4737–46.

Chapter 8
Diversity and Genetic Basis of Polysaccharide Biosynthesis in *Vibrio cholerae*

Shanmuga Sozhamannan and Fitnat H. Yildiz

Abstract *Vibrio cholerae* elaborates three types of polysaccharide structures: lipopolysaccharide (LPS), a component of which is the O-polysaccharide or O-antigen, capsular polysaccharide (CPS) or K-antigen, and "rugose" polysaccharide also known as exopolysaccharide (EPS) or *Vibrio* polysaccharide (VPS). The major protective antigen for *V. cholerae* is the O-antigen. A strain typing scheme based on the somatic O-antigen has been in use for a number of years. There are 206 serogroups identified so far and of these only O1 and O139 are known to cause epidemic/pandemic cholera, although a handful of non-O1/non-O139 strains are known to possess the major virulence factors. The O-antigen diversity is due to the number and composition of monosaccharide components, linkages, addition of non-sugar moieties, modal length of the polysaccharide chain, and biosynthesis mechanisms. The genetic basis of this diversity is just beginning to be understood with the sequencing of a number of gene clusters that encode O-polysaccharide (OPS)/capsule structures. In this review, we summarize our current knowledge on the biochemical composition and structure of some of the O-polysaccharides, genes involved in their biosynthesis, and touch upon the role of horizontal gene transfer in creating this diversity and possible mechanisms that may be operative in this process. We highlight the fact that the distinction between OPS and CPS seems to be less evident in *V. cholerae* than in other species since the genes encoding these structures are shared and map in the same region of the genome. We also describe our current understanding of the genetics and regulation of EPS/VPS synthesis and its role in biofilm formation and environmental survival of *V. cholerae*.

S. Sozhamannan (✉)
Genomics Department, Biological Defense Research Directorate, Naval Medical Research Center, Silver Spring, MD, USA
e-mail: shanmuga.sozhamannan@med.navy.mil

T. Ramamurthy, S.K. Bhattacharya (eds.), *Epidemiological and Molecular Aspects on Cholera*, Infectious Disease, DOI 10.1007/978-1-60327-265-0_8,
© Springer Science+Business Media, LLC 2011

8.1 Introduction

Vibrio cholerae possesses a typical Gram-negative cell envelope structure: an inner and outer membrane composed of lipid bilayers separated by a periplasmic space containing a thin peptidoglycan (murein) skeleton layer. The outer leaflet of the outer membrane also known as the LPS, in turn, is composed of a three-layer structure: lipid A which is also known as endotoxin, core polysaccharide, and O-polysaccharide, OPS [1]. These multiple structures are composed of lipid, sugars, and some peptides. While the inner membrane, periplasm, and outer membrane are integral structural components essential for cell integrity and protection against osmotic stress, the outer leaflet of the outer membrane serves as effective barrier to deleterious hydrophobic compounds such as antibiotics [2]. An unintended consequence of the presence of the surface-exposed OPS structure is the induction of a strong immune response when the bacterium comes into contact with the host and thus serves as an effective immunogen. The OPS is referred as somatic antigen or O-antigen, an antigen located in the cell wall of a bacterium.

The O-antigen contains repeating oligosaccharide units generally composed of 3–6 sugar moieties (O units; often repeated 10–30-fold) [1], giving rise to a characteristic ladder-like structure on western blot analysis of purified O-antigen [3]. The O-antigens are extremely variable in structure. The variability is due to the number and composition of monosaccharide components, linkages, addition of non-sugar moieties, modal length of the polysaccharide chain, biosynthesis mechanisms, and possibly in their genetic basis as well [1, 4]. Thus, bacteria in turn have evolved mechanisms to circumvent or evade immune detection by their host, for example, by antigen switching. There are 206 O-antigens known in *V. cholerae* and they are classified based on the serology of purified O-antigens from various strains [5]. However, interestingly, only a couple of the serogroup strains (O1 and O139) possess epidemic and pandemic potential although several non-O1/non-O139 possess the known virulence factors such as cholera toxin (CTX) and toxin-coregulated pilus (TCP) [6–9] suggesting a role for O-antigen in epidemicity. Of the 206 O-antigens, the chemical structure and the genetic basis of O-antigen biosynthesis of only a few serogroups are known.

Some strains of *V. cholerae* produce a capsule, also known as K-antigen, which is a polysaccharide secreted outside the O-antigen and is loosely attached to the cell wall. The capsule lacks the characteristic ladder-like display of repeating sugar units seen, in general, with O-polysaccharide, instead it displays a high molecular weight band in Western blot analysis. More than 85% of non-O1 *V. cholerae* isolated from patients with septicemia or diarrhea or from environment have a capsule [10]. However, structures and the genetics of CPS in *V. cholerae* are poorly understood. Only two serogroups, O139 and O31, known to possess CPS have been genetically characterized.

The epidemic strain O139 Bengal was shown to possess a *hybrid* structure: a shortened O-antigen substituted with a capsule and its biosynthetic genes mapping to the LPS region on chromosome I [3, 11, 12]. Recently, a similar arrangement was shown to be present in a non-O1/non-O139/non-epidemic, diarrhea causing strain, NRT36S (serogroup O31) [13]; i.e., the O31 capsule biosynthesis cluster was mapped to the LPS biosynthesis region. Thus, in contrast to *Escherichia coli*, evidence for a distinct genetic region required for capsule biosynthesis is yet to be found in *V. cholerae*. Hence, in this review, the capsular biosynthesis will be discussed only in the context of O-antigen biosynthesis in O139 and O31 serogroups.

Epidemic *V. cholerae* strains are known to produce yet another exopolysaccharide (EPS), also known as rugose polysaccharide or *Vibrio* polysaccharide (VPS), in response to harsh environmental conditions such as chlorine treatment [14–17] The rugose phenotype, as originally described by Bruce White in 1938, is characterized by wrinkled colony morphology associated with the secretion of copious amounts of extracellular polysaccharide [18, 19]. Extensive work in the recent past has shed light on identification of VPS biosynthesis genes, its chemical composition, regulation of biosynthesis, and ecological significance of rugose phenotype.

The three complex surface structures (LPS, CPS, and EPS/VPS) not only protect the bacterium from harsh environmental conditions such as chemicals, detrimental hydrophilic compounds, and antibiotics entering the cells, phagocytosis, complement and phage-mediated killing, but also enable them to persist in unfavorable environments by biofilm formation [2].

The focus of this review is on the diversity of the polysaccharide structures such as LPS, CPS, and EPS/VPS that come into contact with the external environment or host cells. More specifically, we will elaborate on the genetics of lipid A, core, O-antigen and its diversity and review evidence for horizontal transfer of O-antigen biosynthesis clusters and the recent advances in our understanding of rugose polysaccharide biosynthesis, regulation, and its role in biofilm formation. The reader is also referred to some excellent earlier reviews dealing with similar and other aspects of polysaccharide structures in *V. cholerae* [3, 20–24].

8.2 Lipopolysaccharide (LPS)

8.2.1 Lipid A

8.2.1.1 Composition and Structure of Lipid A

The biochemical nature of the lipid A part of the LPS has been determined. The backbone structure is P-GlcN (β-1–6) GlcN-1-pp-Etn [3]. It consists of a β-1,6-linked D-glucosamine oligosaccharide substituted with a phosphate group ester bound to the nonreducing glucosamine residue and a pyrophosphorylethanolamine (PP-Etn) linked to C-1 of the reducing glucosamine residue [25, 26]. The glucosamine residues are further substituted at their hydroxyl and amino groups by a number of fatty acids [27]. Three of these fatty acids, tetradecanoic, hexadecanoic, and 3-hydroxydodecanoic acid, are involved in ester linkages, whereas 3-hydroxytetradecanoic acid (3-hydroxymyristic acid) is involved in an amide linkage. The fatty acids found in lipid A occur in all serogroups [25]. The

V. cholerae lipid A is similar to lipid A from other species except in containing D-3-hydroxydodecanoic acid, which is 3-*O*-acylated by D-3-hydroxydodecanoic acid. As in *Salmonella* spp., the lipid A is linked to the KDO [3].

8.2.1.2 Genetics of Lipid A Biosynthesis

All the genes involved in lipid A biosynthesis have not been deciphered, although based on annotation of the *V. cholerae* genome sequence, two genes with putative lipid A biosynthesis functions (VC0212 and VC0213) have been located very close to the OPS biosynthesis cluster. They are apparently present and conserved in all the serogroups sequenced so far [28, TIGR-CMR database].

8.2.2 Core Oligosaccharide

8.2.2.1 Composition and Structure of Core Oligosaccharide

The core oligosaccharide of *V. cholerae* contains glucose, heptose, fructose, ethanolamine phosphate, and KDO [29]. The remaining sugars found in the core region of the LPS form a backbone substituted at various points by sugars such as fructose, heptose, ethanolamine, and *N*-acetyl glucosamine. KDO seems to serve as the linker between the core and the lipid A. Cloned *V. cholerae* O-antigen gene cluster (*rfb/wbe*) failed to display the *V. cholerae* O-antigen on the *E. coli* core lacking glucose, suggesting that the *V. cholerae* core contains glucose [30].

The structure of core OS has been resolved for two O1 strains (classical; smooth and rough), two O139 (encapsulated and nonencapsulated) strains, two O22 strains, and one non-O1, non-O139 isolate, H11. The O1, O139, and O22 core structures are very similar, while the structure of the non-O1, non-O139 isolate H11 differs significantly in side branches (see [31] and references therein for more details).

8.2.2.2 Genetics of Core Biosynthesis

An extensive study by Nesper et al. has deciphered the genetics of *V. cholerae* core biosynthesis [31]. The core polysaccharide biosynthesis cluster has been designated as *wav* cluster in accordance with the nomenclature proposed by Reeves et al. [1]. The cluster is located on chromosome I of *V. cholerae* between ORFs VC0223–VC0240 adjacent to the O-antigen cluster. The left junction of this cluster is VCO222 (*kdtB* or coenzyme A biosynthesis gene *coaD*) and the right junction is *rfaD* or VCO240 (*gmhD*).

The *V. cholerae* O1 El Tor *wav* genes have been characterized by isolation of mutants and subsequent complementation of the mutants by wild-type gene for the following phenotypes: phage resistance (the putative receptor of the phage, K139, is the O-antigen), rough phenotype, and LPS profile [32, 33]. The *wav* cluster genes have been assigned functions based on homology and several gene functions have been verified by genetic analysis; i.e., isolation of mutants of *waaL*, *waaF*, and *wavB* and subsequent complementation by the respective wild-type alleles.

The *wav* cluster has been further subdivided into four clusters *wav1* through *wav4*. The genetic organization of the putative *wav* cluster has been deciphered from a number of serogroup strains [31]. These investigators have systematically analyzed the *wav* cluster in 38 different strains using southern hybridization, PCR, and targeted sequencing of the region. Of these 38 isolates, 17 were CT^+TCP^+ (the two major virulence factors of *V. cholerae* that are associated with clinical cholera: cholera toxin, CT, and toxin-coregulated pilus, TCP), 4 human CT^-TCP^- strains, 1 human CT^+TCP^+, 2 environmental CT^+TCP^+, 13 CT^-TCP^- environmental isolates, and 1 CT^-TCP^- strain of unknown origin. The major highlights of this study are as follows:

(a) At least 10 genes are conserved in all the strains tested. They are interspersed with other genes of unknown or putative functions making up the core region from VC0223 to VC0240. (b) O1 El Tor-type I *wav* cluster is highly prevalent among epidemic strains. This is consistent with the idea that the epidemic strains are of clonal origin implicating a role for O-antigen in virulence and epidemicity. (c) All epidemic strains and two O1 environmental isolates (one each of CT^+TCP^+/CT^-TCP^-) and two clinical non-O1/non-O139 CT^+TCP^+ appeared to be identical to the type I cluster found in epidemic strains. (d) The *wav* clusters of 19 environmental isolates have been grouped into four different types (type 2–5). The type 2 cluster was found in one environmental isolate. Comparison of *waaL* enzyme in type 1 and type 2 isolates revealed a diverse genetic relationship among these isolates although functional similarity is evident. Type 3 was the predominant and was found in 9 isolates. Types 4 and 5 were represented by 4 and 5 isolates, respectively. The type 5 appears to be the most evolutionarily unrelated of the *wav* clusters. This study revealed extensive horizontal gene transfer in generating genetic diversity in *wav* cluster in spite of limited structural diversity of the core OS. The differences in the five *wav* gene cluster types have been proposed to be located in side branches of the core OS [31].

8.2.3 O-Polysaccharide (O-Antigen)

The genes responsible for the synthesis of O-antigens are clustered on chromosome I of *V. cholerae* between ORFs VC0240 and VC0264 [22]. The chemical structure and the genetic basis of several OPSs have been determined. Structures of the O-specific polysaccharides of *V. cholerae* lipopolysaccharides have been found to vary significantly from serogroup to serogroup. Structural studies have been performed on the OPS of serogroups O1–O3, O5, O8–O10, O21, O22, O76, O139, O144, and O155, and a number of unique sugars and non-carbohydrate groups have been identified as OPS components [34, 35]. Recent studies on the structures of the OPS of *V. cholerae* O6 and O31 indicate that they are strikingly similar [36, 13]. The DNA sequence of O-antigen clusters of 11 serogroups has been determined.

8.2.3.1 Serogroup O1

All seven pandemics of cholera are believed to have been caused by strains of the O1 serogroup. The O1 serogroup is further divided into two biotypes: classical and El Tor, the latter of which has unique characteristic features: El Tor hemolysis, agglutination of chicken erythrocytes, and resistance to classic strain-specific phage (Mukerjee's group IV). The sixth and seventh pandemics were caused by O1 classical strain and O1 El Tor strains, respectively. The O1 strains of both biotypes are further subdivided into three serotypes: Inaba, Ogawa, and Hikojima. This grouping is based on the composition of three antigenic forms in O1 antigen: the three serotypes have in common an antigenic factor referred to as the A antigen. In addition, there are two specific antigens, B and C, which are expressed to varying degrees on the different serotypes. Inaba express only C, while Ogawa strains express both B and C, although C is present in a much reduced amount compared to the amount in Inaba. The third serotype, Hikojima, is regarded as rare and unstable and supposed to express all three antigens at very high amounts [37]. The nature and the genetic basis of this serotype are not clear. It has been suggested that these strains may be segregating diploids or undergo high-frequency serotype conversion [3, 37, 38].

Composition and Structure of O1 O-Antigen

The O-antigen of the O1 serogroup consists of a homopolymer of 4-amino-4,6-dideoxy-mannose (perosamine) (17–18 units) which is substituted with 3-deoxy-L-glycero-tetronic acid (tetronate). The tetronate substitution is common to both Inaba and Ogawa serotypes and has been proposed to correspond to the A antigen which is common to both serotypes. The LPS also contains another sugar, quinovosamine, found at ratio of 1:20 compared to perosamine. It has been suggested that quinovosamine may cap the LPS core since it is also found in O139 that has a different O-antigen composition [21]. Thus, it appears that part of the core may be encoded by O-antigen cluster [21, 39, 40].

Genetics of O1 O-Antigen Biosynthesis

The genetic region required for the synthesis of the O-antigen has been localized between the *gmhD* (VC0240) and the *rjg* (VC0264) genes (Fig. 8.1). The O1 *wbe* region was originally located on a *Sac*I fragment by heterologous complementation in *E. coli* and it did not include the entire region necessary for O-antigen biosynthesis in *V. cholerae* [41, 42]. Subsequent work identified several additional genes further downstream of this region [43]. The O1 *wbe* cluster has been subdivided into five regions according to the functions: (1) perosamine biosynthesis, (2) O-antigen transport, (3) tetronate biosynthesis, (4) O-antigen modification, and (5) additional genes required for O-antigen biosynthesis in *V. cholerae*. The leftmost gene in the cluster, *gmhD*, is separated from the rest by jump-start sequences [44], which are important regulatory sequences found in LPS biosynthesis regions. There

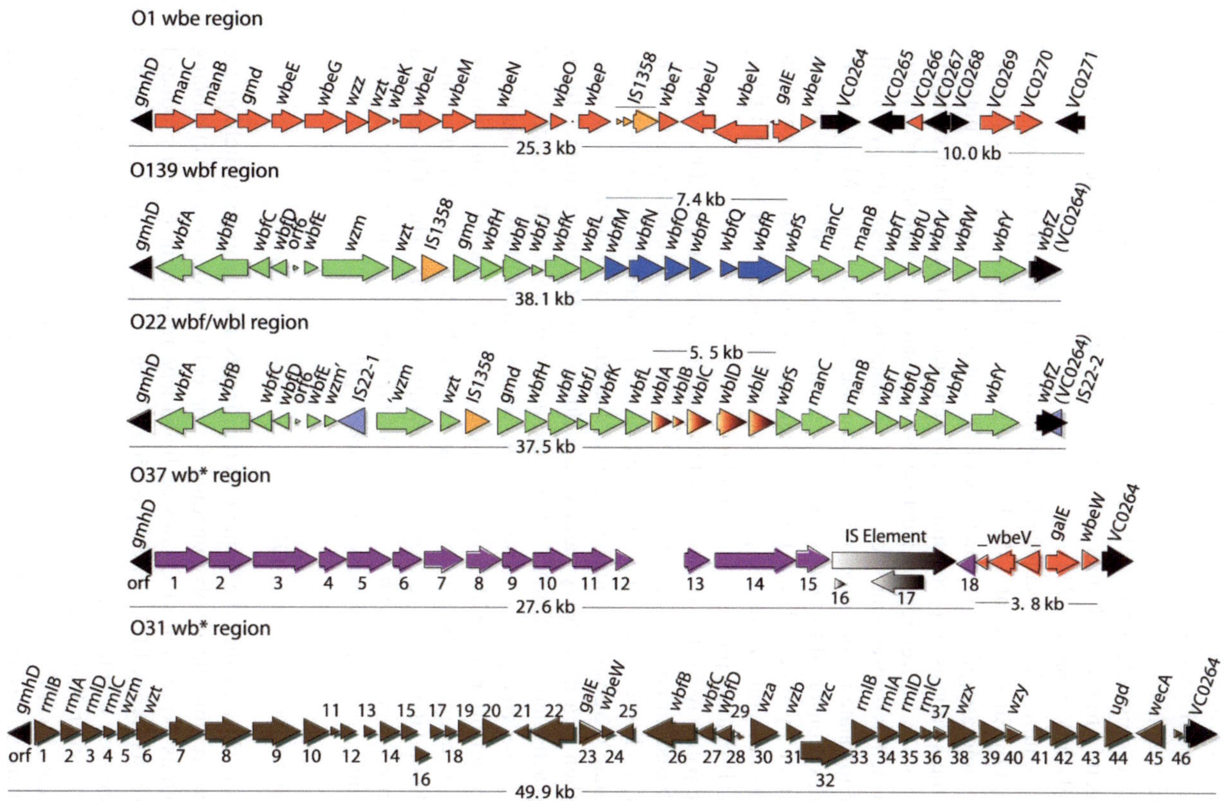

Fig. 8.1 Genetic organization of OPS/CPS biosynthesis (*wb**) regions of *V. cholerae* serogroups O1, O139, O22, O37 and O31. The ORFs and putative genes, if known, and their orientation are indicated by block arrows (not drawn to scale). Gene similarities are indicated by the same color arrows in different serogroups. ORFs/genes unique to a serogroup are represented by unique colors. Similarities at the DNA level only are indicated, while similarities at protein/functional level, are much more prevalent in *wb** regions. The left and right junctions of the *wb** region are *gmhD* and *rjg* (VC0264) respectively, although genes downstream of VC0264 (VC0266, VC0269, and VC0270) have been implicated in O-antigen biosynthesis (49)

are several transcriptional units: the perosamine biosynthesis, O-antigen transport, and tetronate biosynthesis form a single unit separated by a defective transposon *IS*1358, which is followed by *wbeT,* the gene responsible for serotype conversion. The O-antigen modification leading to the expression of B-antigen in the Ogawa serotype is determined by *wbeT* gene product which is in a separate transcriptional unit to the preceding *wbe* genes. The genes downstream of *wbeT* are organized as two divergently transcribed units and these genes are needed for the O-antigen expression only in *V. cholerae* [21].

The genes required for perosamine biosynthesis *manC*, *manB*, *gmd,* and *wbeE* were described by Stroeher et al. [45]. Homology comparisons have led to a better understanding and assigning the functions of O1 antigen export genes. A putative pathway for the synthesis of perosamine has been proposed [45]. The first step in this pathway is the conversion of fructose-6-phosphate to mannose-6-phosphate by ManC, which is subsequently converted to mannose-1-phosphate by ManB. Mannose-1-phosphate is converted to GDP-mannose by the action of ManC. GDP-mannose is next converted to GDP-4-keto-6-deoxymannose by WbeD and then to GDP-perosamine by WbeE. The export of O-antigen involves a pair of genes: the carrier or pore integral membrane protein, encoded by *wzm,* and the energizing partner, an ATP-binding protein encoded by *wzt*.

A putative pathway for the synthesis of tetronate has been proposed by Morona et al. solely based on homologies of the gene products involved, i.e., genes *wbeK,* wbeL, *wbeN,* and *wbeO* have been assigned to the tetronate biosynthesis. In the first step, malate (Krebs cycle intermediate) is converted to an aldehyde by *wbeN* protein and then to a di-hydroxy carboxylic acid by *wbeM* gene product. This is subsequently activated to a Co-A form by the action of *wbeL.* The substrate for this reaction is 3-*deoxy-glycero*-tetronyl: CoA, which is then likely to be activated to the acyl carrier protein (ACP) form by an enzyme such as acetyl-CoA transcyclase, which is thought to be a housekeeping enzyme from the general metabolic pool. *WbeK* has all the conserved features of a number of ACPs. The ACP-activated precursor can be condensed with a molecule of GDP-perosamine to give rise to a complete O-antigen subunit by the action of *wbeO* [46].

Serotype Switching in *V. cholerae* O1

Manning et al. showed that the genetic determinants for serotype switching are within the *wbe* region and involve unequal reciprocal interconversion; (1) cloning and heterologous expression of Ogawa or Inaba *wbe* regions in *E. coli* produced the respective serotypes [41]. (2) Sequencing of the entire regions from Ogawa and Inaba revealed a single bp deletion in the *wbeT* gene [47]. Further analysis of *wbeT* from other Inaba and Ogawa strains revealed that Inaba strains have a mutated/truncated *wbeT*. The conversion of Inaba to Ogawa was possible by introducing the Ogawa *wbeT* into Inaba and the converse was possible only by mutating the *wbeT* in Ogawa. This observation also explains why the conversion from Ogawa to Inaba occurs at much higher frequency than the converse since this conversion only requires a mutation in the *wbeT* gene, whereas the converse requires a specific base correction to restore the ORF [47]. The precise modification giving rise to the B antigen has now been determined to be the incorporation of a methyl group onto the perosamine backbone of the O-antigen [48]. However, it is not clear how this modification occurs or the role of *wbeT* in methylation since it does not show any similarities to any known methylase [21]. Although genetic evidence shows the involvement of *wbeT* in O-antigen modification, it is not vital for O-antigen synthesis and it appears to be associated with a defective transposon IS*1358*d1 and is in a separate transcriptional unit. It has been proposed that the IS*1358* defines the end of the actual *wbe* region and the genes *wbeT*, *wbeU*, *wbeV*, *galE,* and *wbeW* are separate from the rest of the operon [21, 43]. Recent study extended the O1 *wbe* region (32 kb) further downstream of VC0264, placing the right junction within or downstream of VC0271 (Fig. 8.1) [49].

8.2.3.2 Serogroup O139

The Discovery of *V. cholerae* O139 Bengal and Subsequent Studies

In 1992, *V. cholerae* belonging to serogroup O139 emerged as an epidemic strain in the Indian subcontinent [50–52]. The major findings ensuing from a number of studies can be summarized as follows: (1) Clinically, the O139 strains cause a disease similar to the one caused by the O1 strains. (2) The O139 strains are genetically similar to Asian seventh pandemic O1 El Tor strains and possess the known virulence factors. (3) The O139 strains do not possess the typical O1 LPS but a hybrid LPS/CPS. (4) The O139 strains are immunologically distinct from O1 El Tor strains, as evidenced by the susceptibility of the individuals preexposed to O1 strains by vaccination or infection.

The genetic differences accounting for the phenotypically distinct surface polysaccharide of O1 El Tor and O139 Bengal have been determined. The first evidence of the genetic change in the O-antigen region was provided by the work of Waldor et al. [53]. These investigators showed that *V. cholerae* O139 was serum resistant (an indication of the presence of a capsule) and exhibited three electrophoretic forms of the O-antigen corresponding to both LPS and CPS. A transposon insertion abolished the production of both of these materials suggesting common genes involved in their biosynthesis. Further, Manning et al. [3] showed that the O139 genome lacked the O1 specific *wbe* genes except for the *gmhD*, IS*1358,* and *rjg* (VC0264).

Composition and Structure of O139 O-antigen

Serogroup O139 strains do not possess a typical LPS but rather a shortened LPS consisting of a single subunit in its OPS form attached on the core OS and a capsule. Colony morphology of O139 strain was described as semi-rough [3]. Silver staining analysis, following SDS-PAGE of the O139 LPS, revealed marked differences compared to O1. *V. cholerae* O139 LPS has the core and lacked a smear and diffuse band-like pattern characteristic of the O1 LPS but instead had very high molecular weight forms which are consistent with the capsular structure [53]. *V. cholerae* O139 LPS contained only one O-antigen unit on a more completely substituted core.

Accordingly, these strains exhibit two colony morphologies: opaque and translucent, typical of capsulated strains. The capsulated forms are resistant to serum killing [54]. Analysis of the sugar composition of LPS of O139 revealed striking absence of perosamine found in O1 strains and is composed of a phosphorylated hexasaccharide consisting of colitose, galactouronic acid, quinovosamine, galactose, and glucosamine residues [55, 56].

Three different studies looked at the monosaccharide content and structure of the O139 CPS. This macromolecule contains six different sugars: D-galactose, 3,6-dideoxy-xylo-hexose (colitose), 2-acetamido-2-dexoy-D-glucose, 2-acetamido-2,6-didexoy-D-glucose, -acetamido-2,6-didexoy-D-glucose (*N*-acetyl-D-quinovosamine), D-galacturonic acid and phosphate [57–59]. The CPS of O139 contains a repeating unit of six sugar residues, which included one residue each of *N*-acetylglucosamine (GlcNac), *N*-acetylquinovosamine (QuiNAc), galacturonic acid (GalA), galactose, and two residues of 3,6-didexoy-xylo-hexose (Xylhex) [57]. Knirel et al. further refined the structure [59] and proposed a similar structure containing D-galactose 4,6-cyclophosphate. Recently, the hexasaccharide repeating structure of the O139 CPS was isolated using a polysaccharide lyase of bacteriophage JA1. It specifically cleaves at a single position of the 4-linked galactouronic acid producing an unsaturated sugar product. This structure was found to be a tetrasaccharide epitope homologous to the human Lewis blood group antigen and its conformation is identical or similar to that of the intact CPS [60].

Genetics of O139 *wbf* Region

The genes responsible for the synthesis of O139 O-antigen are present in a cluster designated *wb** (*rfb*) region (Fig. 8.1). It was shown that a large portion of DNA corresponding to the *wbe* (*rfb*) region of O1 strains is missing in O139 strains and that O139 has acquired a unique DNA region [61–63]. Specifically, it was demonstrated that the serogroup O139 resulted from a 22-kb deletion of the *wbe* (*rfb*) region of O1 and replacement with a 35-kb *wbf* region encoding the O139 O-antigen [62]. Several groups sequenced the O139-specific *wbf* (*rfb*) region. The 14.363-kb sequence of the left part of the *wbf* region, *gmhD* to *gmd* (originally designated *otn*), was reported by Bik et al. [61], the 12.938-kb right part of the *wbf* region, *wbfQ* to *wbfX*, was reported by Comstock et al. [60], and the intervening region was reported by Stroeher et al. [63]. The *wbf* region has complex organization and independent transcriptional units. Based on mutational analyses from different studies and homology searches, Stroeher et al. have assigned functions for many of the genes in this cluster [21]. The striking feature of this cluster is the absence of discrete pathway-associated clusters as seen in O1 *wbe* region. Mutational analyses showed that O139 region contains genes involved in both O-antigen and CPS production as well as genes specific for either the capsule or the O-antigen production.

The region described by Bik et al. [61] has two genes that are involved in the capsule biosynthesis but not O-antigen biosynthesis: *wzm* (capsule transport) and *wzz* (chain length determination). The gene products of other genes located in the region, *wbfA*, *wbfB*, *wbfC*, *wbfD*, and *wbfE*, do not show any resemblance to polysaccharide biosynthesis genes.

Comstock et al. [62] defined the genes on the right end of the cluster and identified genes with similarities to polysaccharide biosynthesis. A gene downstream of IS*1358*, *gmd*, is similar to oxidoreductase of O1 strains. There are genes with redundancies in functions; for example, two possible galactosyl transferases of different classes, and a number of UDP-galactose 4-epimerases involved in the synthesis of UDP- galactose. *wbfU* shows homology (68% identity) to O1 *wbeW* galactosyl transferase and it has been suggested that in combination with the *galE* homologs (*wbfT* and *wbfW*) they produce UDP-galactose which is used as a substrate by one of the putative galactosyl transferases (*wbfS* and *wbfU*) and transferred to the lipid carrier, bactoprenol.

The regions encompassing *wbfH*, *wbfI*, and *wbfJ* are most likely involved in colitose biosynthesis since they show extensive similarity to *E. coli* O111 *rfb* cluster genes [64] which also produce colitose. *ManB* and *ManC* have assigned functions in the synthesis of GDP-4-keto-6-deoxymannose from fructose-6-phosphate similar to that described for *V. cholerae* O1 ManB and ManC. ManC is likely to have phosphomannoisomerase (PMI) and guanosine 5′ diphospho-d-mannose-pyrophosphorylase (GMP) activities whereas ManB has the phosphomannomutase (PMM) activity. Interestingly, the *gmd* homolog of *V. cholerae* O1 found in *V. cholerae* O139 also plays a role in the conversion of GDP mannose to GDP-4-keto-6-deoxymannose. Thus, GDP-4-keto-6-deoxymannose is required for the biosynthesis of perosamine in O1 and colitose in O139 [21].

V. cholerae O139 has a *wzy* homolog (*wbfQ*/*wzy*). The Wzy protein is involved in O-antigen polymerization and is not usually found in bacterial strains which have a homopolymeric O-antigen as in *V. cholerae* O1. Furthermore, in *V. cholerae* O139 there would not appear to be a need for a *wzy* homolog, since only a single O-antigen subunit sugar is substituted on the core. Thus, the *wzy* found in *V. cholerae* O139 is either non-functional or involved in polymerization of the capsule which is thought to be composed of a material identical to the O-antigen but not linked to the lipid A core [53]. Mutational analysis indicates that *wzy* homolog is solely involved in capsule biosynthesis. It is possible that the *wzz* product of O139 *wbf* cluster interacts with *wzy* product in capsule biosynthesis. Another interesting aspect is that the O139 export system has only the export protein *wzm* (transmembrane protein) and not the energizing partner *wzt* as seen in *V. cholerae* O1. Mutations in *wzm* eliminate capsule production and so one of the as yet unassigned proteins might function as the energizing partner, Wzt, in *V. cholerae* O139 [21].

8.2.3.3 Serogroup O22

Composition and Structure of O22 O-antigen

The O-antigen polysaccharides of *V. cholerae* O22 and O139 are similar in chemical composition and structure and differ only in the presence of an *O*-acetylated β-Gal-4, 6 P in the latter and in the anomeric configuration of GlcNAc. *V. cholerae*

O22 lacks a CPS but possesses an OPS similar to that of O139 except for some minor differences. Both serogroups contain the same trisaccharide, GlcNAc-GalA-QuiNAc, and in both cases the *N*-acetylglucosamine is di-substituted at the 3- and 4-positions. Further, the rare terminal 3,6 dideoxy-L-xylo hexose (colitose) residues are present in both serogroups. The *N*-acetyl glucosamine of O22 has the α-configuration instead of the β seen in O139. The substituent at the 3-position is also different, a galactouronic acid in O22 and a cyclic phosphorylated galactose in O139. The phosphate group is absent in O22. Despite these differences, the similarity in structure correlates with serological cross reactivity and genetic relatedness of the O-antigen biosynthesis regions [22, 65, 66].

Genetics of O22 *wb** Region

As indicated above the chemical constituents and structure of O22 O-antigen is very similar to that of O139 suggesting that the O22 O-antigen genes might be similar to O139 *wbf* genes. This possibility was tested by Yamasaki et al. [67]. When DNA fragments derived from the O139 O-antigen biosynthesis region were used as probes, the entire O139 O-antigen biosynthesis gene region could be divided into five classes, designated as I–V based on Southern hybridization patterns of the probes against reference strains of *V. cholerae* representing serogroups O1–O193. Class IV was specific to O139 serogroup, while classes I–III and class V were homologous to varying extents to some of the non-O1, non-O139 serogroups. Interestingly, the regions other than class IV were also conserved in the O22 serogroup. High-fidelity PCR analysis was employed to determine whether a simple deletion or substitution could account for the difference in class IV between O139 and O22. The class IV region is approximately 8 kb in size in O139. Using primers flanking the class IV region, a product of approximately 15 kb was amplified when O139 DNA was used as the template, while a product of approximately 12.5 kb was amplified when O22 DNA was used as the template. A complete deletion of 8 kb would have resulted in a 7 kb fragment in O22, indicating that a substitution with a 5.5 kb fragment but not a deletion could account for the difference in class IV region between O22 and O139 serogroups. In order to precisely compare the genes responsible for O-antigen biosynthesis of O139 and O22, the region responsible for O-antigen biosynthesis of O22 serogroup was cloned and analyzed. In concurrence with the results of the hybridization test, all regions were well conserved in O22 and O139 serogroups, although *wbfA* and the five or six genes comprising class IV in O22 and O139 serogroups, respectively, were exceptions. Again the genes in class IV in O22 were confirmed to be specific to O22 among the 155 'O' serogroups of *V. cholerae*. These data suggest that the gene clusters responsible for O139 O-antigen biosynthesis are most similar to those of O22 and genes within class IV of O139 and O22 define the unique O antigen of O139 or O22 (Fig. 8.1) [67].

Further confirmation of the hybridization data was obtained by sequencing the O-antigen biosynthesis regions in O22 and O139 strains which corresponded to 36.6 and 35.9 kb in length respectively (Fig. 8.1). The major highlights of the sequence analysis were as follows: (1) Presence of IS elements in *wbfF* and *wbfZ*, possibly interrupting the genes. (2) All genes of *V. cholerae* O139 *wbf* cluster, including the hypothetical *wbfF* and *wbfZ* found in O22 serogroup, with the exception of *wbfA* and *wblA* to *wblE*, have very high homology to corresponding genes found in O139 serogroup. (3) The predicted protein encoded by *gmhD* in O22 is highly homologous to that encoded by *gmhD* in O139 whose possible function is LPS core synthesis. (4) Although the similarity is very low (37%), the predicted protein encoded by *wbfA* in O22 was homologous to that encoded by that in O139 whose function is unknown. (5) *wbfB* to *wbfK* of O22 encode proteins which are almost identical to those of *wbfB* to *wbfK* in O139. (6) Whereas the predicted protein encoded by *wbfL* in O22 was about 61% identical to that of O139 *wbfL* in O139, the predicted proteins encoded by *wblA* to *wblE* in O22 have no homology to those found in O139 serogroup, except for WblC, which was 44% identical to WbfO. (7) *wblA* encodes a protein consisting of 281 aa residues which is homologous to α-(1,2)-FT of human with 29.2% identity in 176 aa overlap whose possible function is β-D-galactoside 2-α-L-fucosyl transferase. *wblB* encodes a protein consisting of 161 aa residues which has homology to LacA of *E. coli* with 39.6% identity in a 91 aa overlap whose possible function is galactoside-*O*-acetyl transferase. *wblC* encodes a protein consisting of 340 aa residues which is homologous to WbfO with 43.8% identity in 333 aa overlap or to LgtD homolog of *Haemophilus influenzae* with 31.8% identity in 169 aa overlap whose possible function is glycosyl transferase. *wblD* encodes a protein consisting of 398 aa residues which is homologous to NADH dehydrogenase of *Trypanosoma brucei* with 18.9% identity in 358 aa overlap whose possible function is NADH dehydrogenase. *wblE* encodes a protein consisting of 359 aa residues which is homologous to RfaG of *E. coli*, with 24.2% identity in 157 aa overlap whose possible function is glucosyltransferase. The products of *wbfS* to *wbfZ* in O22 were again highly homologous to those of *wbfS* to *wbfZ* in O139. (8) JUMP start sequences characteristic of O-antigen clusters were found between *orf6* and *wbfE* [44, 67].

8.2.3.4 Serogroup O37

Composition and Structure of O37 O-antigen

Neither the biochemical constituents nor the structure of O37 O-antigen is known.

Genetics of O37 *wb** Region

Serogroup O37 strains were implicated in localized cholera outbreaks in Sudan and Czechoslovakia in 1962 [68, 69]. The O37 strains were shown to contain virulence gene clusters, CTXϕ and TCP island [6]. The genes responsible for the synthesis of O37 O-antigen were isolated and sequenced. This was accomplished by PCR using primers anchored at the left in *gmhD* and on the right at *rjg* (VC0264). The intervening sequence was PCR amplified, cloned, and sequenced. The entire region was found to be 27,552 bp and the O37 specific O-antigen region was 23,388 bp (Fig. 8.1). The O37 *wb** cluster is unique, since no significant DNA similarity was found between the O37-specific sequences and the previously published *V. cholerae* *wb** cluster sequences except at the right junction. Twenty-three ORFs were identified in the O37 *wb** region, and as expected, many of these *orfs* (*orf-1* to *-13* and *orf-18*) encode enzymes involved in polysaccharide biosynthesis. *orf-14* and *orf-15* encode hypothetical proteins of unknown functions. *orf-14* has very weak similarity (26% identical and 46% positive in a 64-amino acid region) to *yhfO* (which encodes a hypothetical protein) of *Bacillus subtilis*. A 1,549-bp promoter region separates *gmhD* and *orf-1*, and this region contains a putative promoter and *ops* elements found in known polysaccharide biosynthesis regions. As seen in other *wb** regions, there is an IS element in the interval between the regions that is unique to O37 and the right junction, and this element, which appears to contain an insertion of another fragment encoding a transposase, spans 3,580 bp and has an 18-bp inverted repeat at its ends. The original IS element is 1,054 bp long and is virtually identical (95% identical at the DNA level; the transposase is 90% identical [278 of 306 amino acids]) to the IS elements found in *Vibrio parahaemolyticus* (ISV-3L, ISV-5R, ISV-5L, ISV-4R). Interestingly, the *tnp* gene of the *V. cholerae* element is interrupted after the 76 amino acid residue by a 2,527-bp DNA fragment. This insert encodes an ORF (*orf-17*) that is transcribed in the opposite direction and has extensive homology to transposases of various IS elements found in other bacterial species. The region downstream of the IS element has three ORFs almost identical to the O1 *wbe* cluster: *wbeV*, *galE*, and *wbeW* followed by *rjg* (VC0264). The *wbeV* gene in the O37 strain appears to consist of three small ORFs [6].

8.2.3.5 Serogroup O31

Composition and Structure of O31 O-antigen

Recently, the capsule structure and genetic basis for capsular biosynthesis were elucidated for a *V. cholerae* serogroup O31 strain NRT36S. It is a human pathogen that produces a heat-stable enterotoxin (NAG-ST) [13]. Carbohydrate analysis of the O-antigen of O31 strain indicated the presence of L-rhamnose (Rha), D-glucosamine (GlcNAc), D-glucuronic acid (GlcA), and D-galactose. The structure of the capsular (CPS) polysaccharide was determined by high-resolution NMR spectroscopy and shown to be a complex structure with four residues in the repeating subunit. The structure proposed for *V. cholerae* O31 CPS is very similar to that proposed for O6 O-antigen except for minor differences in the degree and positions of O-acetylation [36].

Genetics of O31 *wb** Region

A combination of genetic and whole genome sequencing approaches has been used to decipher the *V. cholerae* O31-specific capsular region. The entire region between *gmhD* and *rjg* (VC0264) was determined to be 49,916 bp in length (Fig. 8.1). This region also contained the transposon insertions resulting in translucent phenotype suggesting that it is the CPS region as well as the O-antigen region. It appears that there is extensive redundancy in functions. Of the 46 potential ORFs, 12 are glycosyltransferases, 16 genes are involved in pathways for synthesis of nucleotide precursor of exopolysaccharide, and 6 other genes are involved in polysaccharide processing and transport. The other 12 ORFs are of unknown function. A jump start sequence was located just downstream of *gmhD*. The transposon insertions resulting in translucent phenotype were located in *orf5* (*wzm*-an ABC transporter gene), *orf8* (a glycosyl transferase gene), *orf23* (galE-UDP-glucose epimerase), and *orf43* (rhamnosyltransferase). Complementation of the *galE* mutants resulted in formation of opaque colonies which were resistant to serum killing [13]. These authors have proposed putative functions for many of the ORFs based on the proposed biochemical structure of the O31 capsule. Some of the highlights of these analyses are as follows:

Glycosyltransferases: *Orf45, wecA* homolog [70], is an undecaprenylphosphate *N*-acetylglucosamine 1-phosphate transferase gene. WecA is the putative initial transferase to catalyze the transfer of *N*-acetylglucosamine 1-phosphate to undecaprenylphosphate in the capsule polysaccharide biosynthesis pathway.

Synthesis genes: The structural data for *V. cholerae* NRT36S indicate that the capsule contains one residue each of rhamnose and glucuronic acid and two *N*-acetyl glucosamine residues; genes required for the synthesis of these moieties are present in the CPS region. There are two sets of L-rhamnose biosynthesis genes, homologous to *rmlB*, *rmlA*, *rmlC*, and *rmlD*, that are almost identical; they are *orf* 1–4 and *orf* 33–36 in the same order as *rml* genes. *Orf43* may be the rhamnosyltransferase that catalyzes the addition of rhamnose to the CPS backbone. The disruption of *orf43* by a transposon insertion resulted in the loss of the capsule. This observation is consistent with the presence of rhamnose in the repeating polysaccharide backbone of the capsule. *Orf23* is similar to *galE* gene; its product UDP-glucose 4-epimerase catalyzes the conversion of UDP-glucose to UDP-galactose. Disruption of *galE* gene in mutant TR3 caused the loss of the capsule. *Orf24* (*wbeW*) transfers galactose to the capsule complex. *Orf41*, a sugar O-acetyltransferase homolog could be involved in the observed O-acetylation of the capsule. *Orfs 11, 12, 22, 37, 41,* and *44* are also putative pathway genes for the synthesis of nucleotide sugar precursors, but their precise functions are not clear at this time.

Translocation and Processing Genes: Some of the translocation- and processing-associated genes identified are *wza, wzb, wzc, wzm, wzt, wzx* (*orf38*), *wzy* (*orf40*). The disruption of an ABC transporter system integral membrane protein gene, *wzm,* by transposon insertion had significantly reduced the amount of capsule and resulted in translucent colonies that were susceptible to serum killing. Presence of *wzt*, which is another component of the ABC transporter system in the CPS region, suggested that the processing and translocation of the capsule in *V. cholerae* NRT36S involve an ABC transporter system.

8.3 Horizontal Gene Transfer (HGT) in Creation of O-Antigen Diversity

There is overwhelming evidence for the role of horizontal gene transfer (HGT) in creating new pathogens by generating diversity in the O-antigen/capsule regions. In the case of O-antigen/capsular biosynthesis regions, at least two types of such events occur. (1) "En bloc" HGT of O-antigen clusters from one serogroup to another resulting in change of O-antigens in completely different genetic backgrounds. The prime example for such an event is the emergence of *V. cholerae* O139 Bengal which appears to be the result of a horizontal transfer event of the entire *wbf* cluster, probably from a nonpathogenic *V. cholerae* strain such as Arg-3 [71] to a *V. cholerae* O1 El Tor strain thereby creating a new epidemic strain [72, 73]. (2) "En morceaux" or "Mosaic" HGT of O-antigen genes/fragments between highly conserved O-antigen regions that results in O-antigen switching. The best example of this type of HGT is seen in *V. cholerae* serogroups O22/O139 which have serogroup-specific fragments embedded within highly conserved O-antigen genes. Serogroup changes in non-pathogenic *V. cholerae* may not have serious consequences and may go unnoticed, whereas in pathogenic backgrounds, such as O1, antigen conversion may result in novel pathogens (e.g., *V. cholerae* O139 Bengal, which causes cholera epidemics even in populations immune to *V. cholerae* O1).

The first evidence for HGT events, stemming from phylogenetic studies of *V. cholerae* [74], indicated that exchange of *wb** regions may be quite common in this species. Mooi and Bik used IS*1004*-based fingerprinting analysis for *V. cholerae* strain typing and identified several strains with identical or related fingerprint patterns but expressed different serotypes. Conversely, strains with different fingerprint patterns with identical serotypes were also found. These observations led to the idea that the gene clusters coding for distinct O antigens may be transferred horizontally between *V. cholerae* strains. Two examples of non-O1 strains with a fingerprint resembling that of epidemic O1 strains were found; they were the O139 Bengal strain and an O37 strain. The O139 Bengal strain is closely related to the El Tor biotype [74], later confirmed by comparative genome hybridization analysis [75]. The O37 strain was responsible for a large cholera outbreak in Sudan in 1968 and was classified as a noncholera *Vibrio* [69]. These authors showed that the O37 Sudan strain is genetically closely related to classical O1 strains, a conclusion later confirmed by additional analyses and direct sequencing of O37 *wb** cluster [6].

Additional non-O1 and non-O139 strains (O27, O37, O53, and O65) have been identified that further expand the repertoire of *V. cholerae* strains with epidemic potential and underscore the idea that the emergence of the O139 serogroup was not a unique event [6]. The genetic relatedness of the four non-O1 strains, based on IS*1004* fingerprinting and RFLP analyses, was supported by an extensive multilocus sequence-typing analysis in which these four strains always clustered with the epidemic strains and had genomic regions very similar to those of O1 strains. Interestingly, the four strains have diverged further by acquiring different virulence cassettes or parts of cassettes. For example, the O37 serogroup strain with a classical background and a classical VPI acquired an El Tor CTXϕ, the O53 and O65 strains with an El Tor background acquired a preclassical CTXϕ (without the *ctx* genes), and the O27 strain with an El Tor background acquired a novel CTXϕ and a VPI cluster. Thus, it appears that the core genome, *wb** cluster, VPI, and CTX regions have evolved as independent units by HGT. Further, it was shown that the O53 and O65 strains are very similar to each other indicating that these strains may have arisen by "en morceaux" or "mosaic" HGT in their *wb** regions.

The DNA sequence of the O-antigen/capsule biosynthesis regions has been determined for 19 *V. cholerae* strains (TIGR-CMR database). Of these, six belong to serogroup O1, three O37, two O139 and one each of O12, O14, O22, O31, O39, O135, O141 and one unknown serogroup. A multigenome BLAST comparison of the *wb** regions using O1 *wbe* cluster as the reference is shown in Fig. 8.2. The major highlights of this analysis can be summarized as follows: (1) All the O1 genomes are highly conserved except for minor differences in one of the strains (2740–80) in the *wb** region in the core region of strain, NCTC8457. (2) All non-O1 serogroups have the junction genes VC0241 and VC0264 conserved and the intervening genes different and serogroup specific. (3) The LPS core region seems to be conserved among the serogroups of O1, O37, and O139 and more diverged in non-O1/non-O139 serogroups supporting earlier conclusions [31].

Thus, conservation of the genetic organization of the *wb** regions, i.e., conserved regions separated by divergent serogroup specific segments, provides ample opportunity for genetic exchange of O-antigen regions "en bloc" or "en morceaux" between different serogroups creating novel pathogens via homologous recombination between the conserved regions (Fig. 8.1).

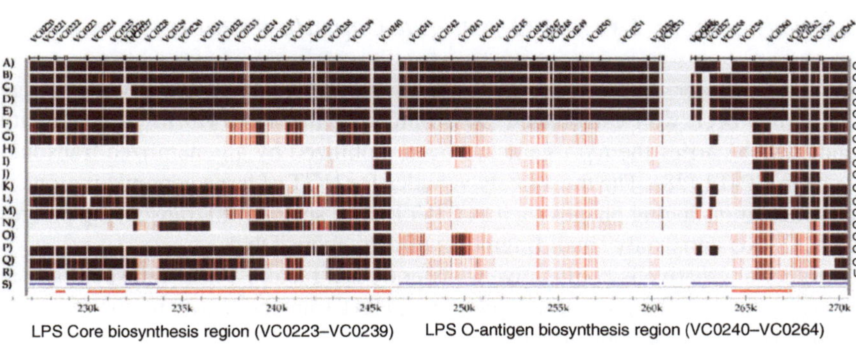

Fig. 8.2 Comparative genome analysis of the LPS core and O-antigen biosynthesis regions in various serogroups of *Vibrio cholerae*. Genome BLAST atlas (linear) of O-antigen biosynthesis region in *V. cholerae* genome. *V. cholerae* O1 strain N16961 (Accession AE003852/AE003853) was used as a reference strain and compared to a set of other publicly available *V. cholerae* strains. ORFs showing conservation are indicated by the horizontal *dark bars*. The *light bars* represent protein residues and the *gaps* indicate ORFs that are present in the reference strain but absent in the other sequences. Hits are defined by the "50/50" rule (at least 50% of the length of the protein, and a minimum of 50% sequence identity). The sequences of the strains (serogroup) and accession numbers are from top to bottom: (A) 2740–80 (O1) AAUT00000000; (B) O395 (O1) AAKG00000000; (C) NCTC8457 (O1) AAWD00000000; (D) B33 (O1) AAWE00000000; (E) MAK757 (O1) AAUS00000000; (F) 1587 (O12) AAUR00000000; (G) MZO-2 (O14) AAWF00000000; (H) 169–98 (O22) AB012957; (I) NRT36S (O31) DQ915177; (J) 1322–69 (O37) AF390573; (K) V52 (O37) AAKJ00000000; (L) MZO-3 (O37) AAUU00000000; (M) AM19226 (O39) AATY00000000; (N) RC385 (O135) AAKH00000000; (O) MO45 (O139) AB012956; (P) MO10 (O139) AAKF00000000; (Q) V51 (O141) AAKI00000000; (R) 623–39 (unknown) AAWG00000000

8.3.1 Mechanism of HGT-Evidence for Homologous Recombination Events

Closer examination of the sequences at the junctions of the O-antigen clusters indicates that O37 and O139 strains resulted from an exchange of the O1 *wbe* cluster with a *wb** O37 cluster and *wbf* cluster, respectively, via homologous recombination [6, 11]. However, the homology breakpoints in O37 are different from the breakpoints of the O139 junctions. Unlike O139, in which the entire *wbe* region has been replaced by the *wbf* region, in O37 some of the O1 *wbe* genes closer to the right junction have been retained, which provides further support for the idea that the O37 strain emerged from an O1 strain by O-antigen conversion. Although the precise crossover points in this recombination event cannot be predicted from the sequence, evidence for recombination is obvious from the breakpoints. There is also some evidence that the *wb** junctions are hyper-recombinogenic due to the presence of chi-like sequences at the junctions of *wb** clusters in *E. coli* and *Klebsiella* spp. [76]. A role for IS elements in HGT has also been proposed [73]. IS elements may be involved in transposing O-antigen genes into *wb** regions or they may provide regions of homology in transferring non-homologous segments [49, 73].

8.3.2 Vehicles for HGT of O-Antigen/Capsule Regions

Horizontal transfer of virulence genes such as cholera toxin gene via a phage (CTXϕ) has been well established [77]. Similarly, the VPI (*Vibrio* pathogenicity island) which carries the *tcpA* gene has been mobilized by a generalized transducing phage CP-T1 [78]. The most likely vehicles for HGT of large genomic regions is generalized transducing phages such as CP-T1 since specialized transducing phages such as CTXϕ carry only small fragments of extraneous DNA.

Another alternative is conjugative transfer by sex factors. Although such factors have been shown to be present in *V. cholerae*, HGT transfer of *wb** regions by such elements is yet to be demonstrated.

A third possibility is DNA transformation. Recent studies support the idea of HGT of O-antigen/capsule biosynthesis regions by transformation in *V. cholerae* [49, 79]. It was shown that chitin, a polymer of *N*-acetylglucosamine, not only serves as a surface for biofilm formation in aquatic habitats and as a nutrient source, but also induces competence for natural transformation. In the initial report of this phenomenon, competence was experimentally demonstrated by showing that it could mediate the acquisition of genes conferring antibiotic resistance during growth of *V. cholerae* strain on crab shell fragment immersed in seawater [79]. More recently, serogroup conversion was demonstrated by chitin-induced natural transformation [49]. These authors demonstrated serogroup conversion of an O1 strain into O139 and O37 by transformation. Serogroup conversion was shown to have occurred as a single-step exchange of large fragments of DNA. The precise crossover points were localized to regions of homology common to *V. cholerae* serogroups that flank serogroup-specific sequences. The successful serogroup conversion of an O1 strain by O139 or O37 genomic DNA demonstrated that chitin-induced natural transformation might be a common mechanism for serogroup conversion in aquatic habitats [49].

8.4 Exopolysaccharide (EPS)/*Vibrio* Polysaccharide (VPS)/Rugose Polysaccharide

8.4.1 V. cholerae *Generates Colonial Variants Termed Smooth and Rugose*

Growth of many bacteria can result in populations that contain genotypically and phenotypically diverse variants that can arise from programmed genetic rearrangements, adaptive mutations, or variable expression patterns [80–82], *V. cholerae* has the ability to switch colonial morphology types between smooth and rugose [15, 17, 18, 83] (Fig. 8.3).

The capacity of *V. cholerae* to switch from the smooth and translucent colonies to corrugated and opaque colonies (also known as rugose) was first documented by White in 1938 [18]. Smooth to rugose conversion can occur spontaneously under diverse conditions, including in response to carbon limitation, growth in biofilms, and treatment with bactericidal agents [15, 17, 83, 84]. It has also been shown that in some strains of *V. cholerae* lacking *hapR* (the master regulator of quorum sensing) [85–87], *flaA* (a major flagellin subunit) [88, 89] or *cytR* [90] (a regulator of nucleoside uptake and catabolism) can lead to rugose colony formation. This indicates that smooth to rugose switch is controlled by multiple mechanisms.

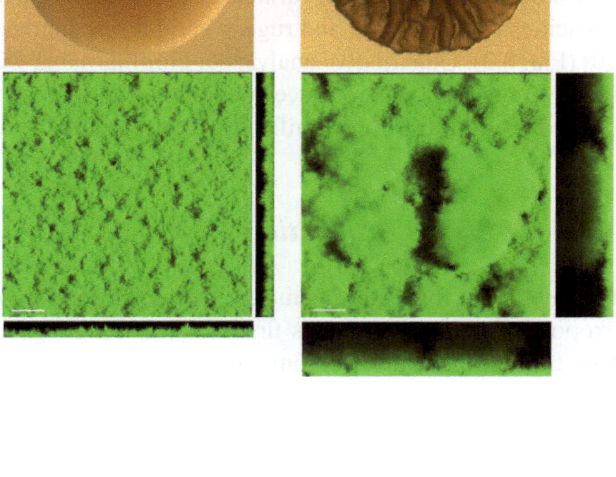

Fig. 8.3 Colony morphology and biofilm formation phenotypes of smooth and rugose phase variants. Colony morphology of the smooth and rugose variants grown on LB Agar plates at 30°C for 48 h (*top panels*). Biofilm structures of the variants formed 24 h after inoculation in once-through flow cell. Images were acquired with CSLM and top-down view of biofilms are shown. Bar= 30 μm (*bottom panels*)

The smooth to rugose switch appears to take place in natural environments, as rugose variants have been isolated from environmental biofilm samples collected in Bangladesh [91]. Rugose variants of *V. cholerae* have also been isolated from stools of cholera patients (Dr. Islam, International Center for Diarrheal Disease Research, Dhaka, Bangladesh, and Dr. Schoolnik, Stanford University, personal communication). Finally, as by human volunteer studies, the rugose variant can infect humans and can cause disease cholera [15].

8.4.2 The Rugose Variant Has an Increased Capacity to Survive Biocides and Environmental Stresses

Both Rice et al. and Morris et al. have shown that a switch to the rugose colonial form is associated with increased survival in chlorinated water [14, 15]. The rugose variant also exhibits increased resistance to osmotic and oxidative stress compared to the smooth variant [17, 83]. Recent studies have shown that production of smooth and rugose variants is a defensive strategy adapted by *V. cholerae* against predation by protozoa and phages, a process considered to be major cause of bacterial mortality in natural aquatic habitats [92, 93]. Thus, switching from smooth to rugose enhances proliferation rates and environmental persistence of *V. cholerae.*

8.4.3 The Rugose Variant Has Increased Capacity to Form Biofilms

Biofilms are surface-attached microbial communities encased in extra-polymeric matrix composed of exopolysaccharides, proteins, and nucleic acids [94]. In natural settings, the biofilm mode of growth facilitates environmental survival by protecting its members from environmental stresses, predators, such as protozoa and viruses, and antimicrobial agents [95–97]. Biofilm formation is an important part of *V. cholerae*'s life cycle and modulates environmental survival, transmission, and infectivity.

Both smooth and rugose variants can form biofilms; however, the biofilm forming capacities of the smooth and rugose variants are markedly different [17, 83, 85, 87, 90] (Fig. 8.3). Quantitative analysis of biofilms using COMSTAT [98] demonstrated that average thickness, surface roughness, substratum coverage, and total biomass of rugose biofilms differ significantly compared to those of the smooth variant [86].

8.4.4 EPS/VPS Production

In *V. cholerae* formation of mature biofilms requires VPS (*Vibrio* poly*s*accharide) exopolysaccharide. VPS was first isolated from rugose strains of *V. cholerae* [17, 83]. Indeed, corrugated colony development is mainly due to increased levels of

VPS [83, 99]. Initial chemical analysis of VPS isolated from different strains revealed that the glycosyl composition of the VPS varies among different isolates [17, 83]. The VPS produced by the *V. cholerae* O1 El Tor A1552 strain has equal amounts of glucose and galactose, with smaller amounts of *N*-acetylglucosamine and mannose [83]. In contrast, VPS of *V. cholerae* O1 El Tor strain (TSI-4) lacked glucose and contained *N*-acetylglucosamine, D-mannose, 6-deoxy-D-galactose, and D-galactose [17]. Importance of differences in VPS glycosyl composition to biofilm structure and function is not known.

8.4.5 VPS Biosynthesis and Analysis of the Functions of the vps Genes

The genes required for VPS synthesis (*vps* genes) are clustered in two regions on the large chromosome of *V. cholerae* O1 El Tor [*vps*U (VC0916), *vpsA-K*, VC0917-27 (*vps*-I cluster); *vpsL-Q*, VC0934-9 (*vps*-II cluster)] (Fig. 8.4). We can make limited predictions about possible functions of these genes. The proteins encoded by VC0917, VC0918, and VC0927 may be involved in production of nucleotide sugar precursors required for synthesis of VPS. VC0920, VC0925, and VC0934 encode glycosyltransferases and may be involved in initiation and/or elongation of possible VPS subunits. Thus, the mutants of these genes may not be able to produce VPS, because they are unable to initiate/elongate VPS subunits. VC0921, VC0924, and VC0937 appear to encode proteins required for polymerization and export of VPS. We anticipate that these mutants may produce lipid-linked subunits of VPS, but not full-length VPS. If VPS transport takes place after complete polymerization in the cytoplasm or periplasm, VPS could accumulate in the cytoplasm or periplasm. Alternatively, unpolymerized/partially polymerized VPS may be secreted. VC0919 and VC0923 encode serine acetyltransferase-related proteins that could be involved in addition of an acetyl group to VPS. It is yet to be determined whether the VPS contains the acetyl, succinyl, and pyruval constituents. If the VPS has acetyl modifications, these mutants could produce VPS without acetyl modifications. VC0922, VC0926, VC0935, VC0936, VC0938, and VC0939 encode hypothetical proteins and no function has been assigned. Besides *vps* genes, other genes necessary for production of nucleotide sugar precursors such as *galU*, encoding for UDP-glucose pyrophosphorylase and *galE*, encoding UDP-galactose epimerase are also required for synthesis of VPS and in turn biofilm formation [33].

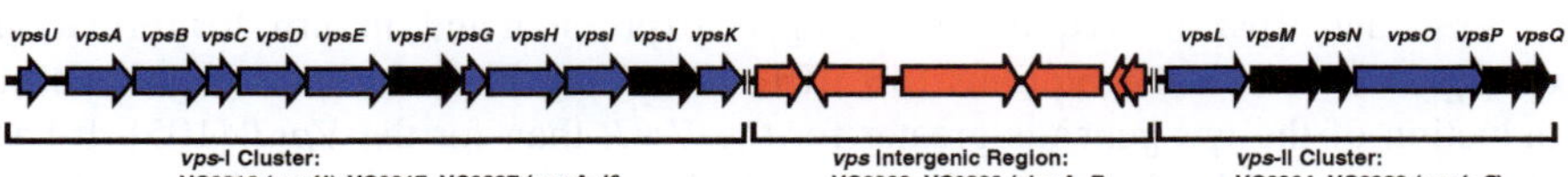

Fig. 8.4 Genetic organization of the *vps* region. *Blue boxes* in *vps*-I and *vps*-II clusters represent genes homologous to EPS biosynthesis from other organisms. *Black boxes* represent genes of unknown functions

8.4.6 VPS Cluster Is Unique to V. cholerae

A multigenome BLAST analysis of the VPS region was done using the genome sequence of *V. cholerae* O1 El Tor strain N16961 as the reference and compared to 28 other *Vibrio* genomes. This set included 15 *V. cholerae* genomes and 13 genomes representing other *Vibrio* species (T. Binneweis and D. Ussery, unpublished:http://www.cbs.dtu.dk/services/GenomeAtlas/). The significant finding of this analysis is that the VPS genes are highly conserved and unique to *V. cholerae*. Moreover, they are absent in other *Vibrio* species (Fig. 8.5). However, this observation does not preclude the existence of rugose/smooth switching and other rugose polysaccharide biosynthesis systems in *Vibrios* other than *V. cholerae*.

8.4.7 Regulators of VPS Production

VPS production in *V. cholerae* is controlled by a complex regulatory network involving a set of positive and negative transcriptional regulators as well as alternative sigma factors. A model of regulation of *vps* genes is presented in Fig. 8.6.

VPS production is positively regulated by VpsR (*Vibrio* polysaccharide regulator) which was identified by transposon mutagenesis using rugose to smooth conversion as a screen [99, 100]. VpsR exhibits homology to the NtrC subclass of response regulators of the two-component regulatory system. Disruption of *vpsR* in the rugose variant results in smooth colonial morphology, prevents expression of *vps* genes and production of VPS, and abolishes formation of mature biofilms [100]. VpsR belongs to the NtrC subclass of response regulators which acts in concert with alternative sigma factor RpoN suggesting that *vps* genes might require both VpsR and RpoN for their expression. It has been shown that deletion of *rpoN* significantly reduces the expression of *vps* genes [86].

The second positive regulator of the *vps* genes is VpsT [86, 101] that shows similarity to proteins belonging to the UhpA (FixJ) family of transcriptional response regulators. VpsT is homologous to the transcriptional regulators CsgD *E. coli* and *Salmonella enterica* [101]. CsgD positively regulates expression of the genes necessary for production of extracellular matrix components, cellulose and curli fimbriae, which are important for the development of wrinkled colonies and biofilm formation in these bacteria [102–104]. Disruption of *vpsT* in the rugose variant results in reduction in *vps* gene expression, yields phenotypically smooth colonies, and reduces the biofilm-forming capacity [101, 105].

VpsT and VpsR are required for *vps* gene expression and, in turn, for formation of rugose colonies and biofilms [86, 101, 102, 106]. However, the magnitude of induction of the *vps* genes is greater for the *VpsR* than for the *VpsT* [105]. It has also been shown that VpsR and VpsT positively regulate their own and each other's expression [101] indicating presence of a regulatory feedback loop.

Negative regulators of rugosity and biofilm formation have also been identified. In *V. cholerae*, quorum-sensing systems (QS) signal transduction system

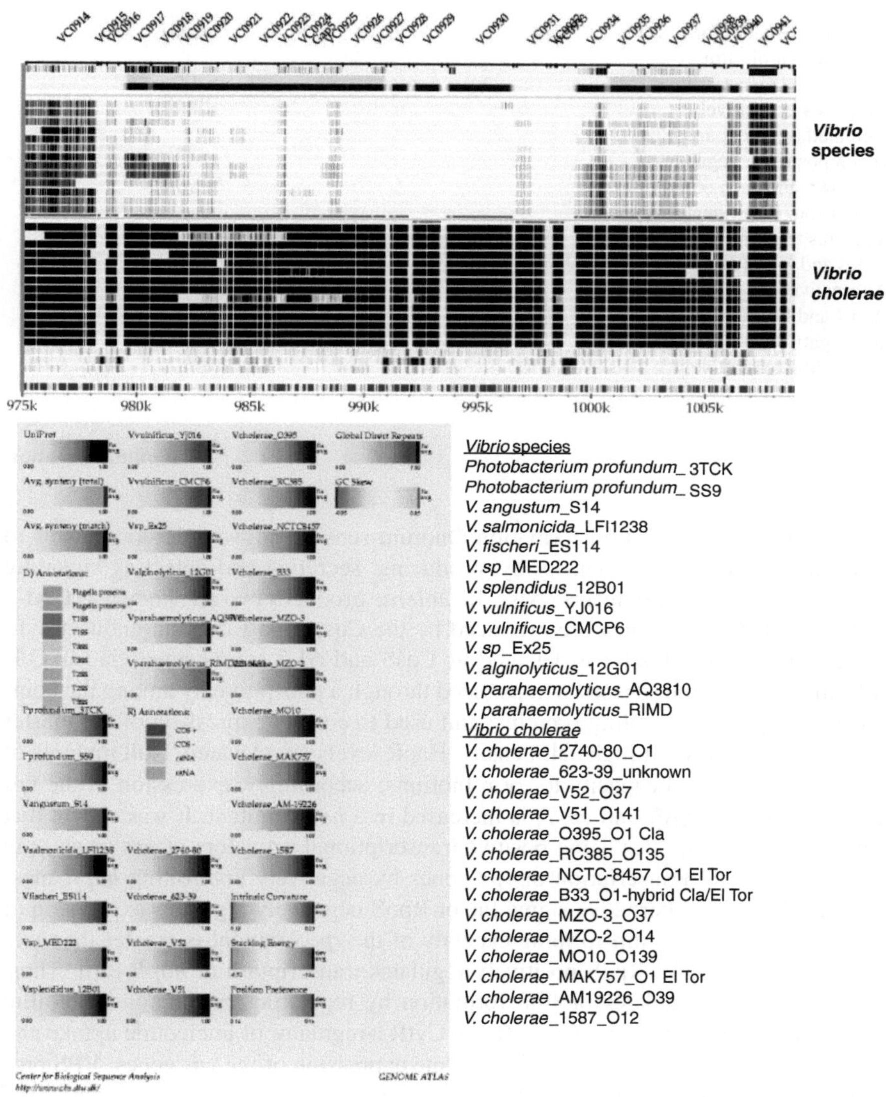

Fig. 8.5 Comparative genome analysis of the VPS region in various *V. cholerae* and other *Vibrio* species strains. Genome BLAST atlas (linear) of VPS region in *V. cholerae* genome. *V. cholerae* O1 strain N16961 (Accession AE003852/AE003853) was used as a reference strain and compared to a set of 28 other publicly available *V. cholerae* and other *Vibrio* species strains. The region includes ORFs VC0194 to VC0941. ORFs showing conservation are indicated by the *horizontal dark bars*. The *light bars* represent protein residues and the *gaps indicate* ORFs that are present in the reference strain but absent in the other sequences. Hits are defined by the "50/50" rule (at least 50% of the length of the protein, and a minimum of 50% sequence identity). The strain designations and the serogroup if known are indicated

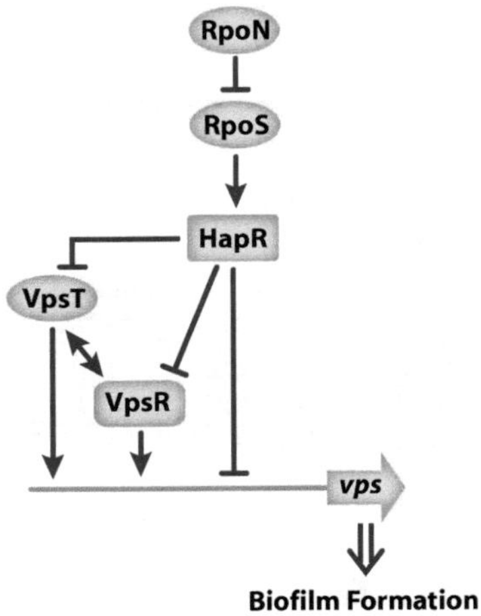

Fig. 8.6 A model for regulation of biofilm formation. Transcription of *vps* genes is positively regulated by VpsR and VpsT which positively regulate their own and each other's expression. HapR negatively regulates transcription of *vps* genes and VpsR and VpsT. Alternative sigma factors, RpoN and RpoS positively and negatively regulate *vps* expression, respectively

negatively regulates biofilm formation Quorum sensing allows microorganisms to evaluate their population density by, producing, secreting and detecting signaling molecules termed autoinducers (AIs). *V. cholerae* produces two AIs known as CAI-1 and AI-2 [106–108] which are produced by the CqsA and LuxS autoinducer synthases, respectively. CAI-1 is detected by CqsS and AI-2 are detected via LuxQ/P. Information from the sensors is transduced through a phosphorelay among the components of the QS signaling proteins; and used to control expression of the master QS regulator HapR. At high cell density, HapR levels are elevated. Null mutants of *hapR* have increased capacity to form biofilms; accordingly, expression of the *vps* genes, as well as *vpsR* and *vpsT*, is increased in a *hapR* mutant. It was shown that VpsT and VpsR, besides being positive transcriptional activators of *vps* genes, can also contribute to transcription of *vps* genes by negatively controlling *hapR* message levels [105]. Alternative sigma factor RpoS negatively regulates expression of *vps* genes and the biofilm-forming capacity of the *rpoS* mutant is greater than that of the wild type [86]. RpoS positively regulates transcription of *hapR* [86]. Thus, RpoS negatively regulates biofilm formation by regulating production of biofilm regulators.In addition to HapR and Rpos, CytR a regulator of nucleotide uptake and catabolism was shown to negatively regulate expression of the *vps* genes, VPS production, and biofilm formation in *V. cholerae* O139 [90]. In-depth understanding of wiring of the regulatory circuitry controlling biofilm formation and elucidation of specific mechanisms by which regulatory proteins discussed above controls biofilm formation awaits further investigation.

8.4.8 Type II Secretion System in VPS Export

Using a Tn*PhoA* mutagenesis approach, Ali et al. identified a number of stable mutants of a *V. cholerae* rugose strain that turned smooth [99]. In a later study, it was shown that the extracellular protein secretion system (gene designated *eps*) is involved directly or indirectly in the production of rugose polysaccharide [109]. These investigators identified a Tn*phoA* insertion in the *epsD* gene of the *eps* operon that abolished the production of rugose polysaccharide, reduced the secretion of cholera toxin and hemolysin, and resulted in a nonmotile phenotype. Further, they constructed defined *epsD* and *epsE* mutants which affected these phenotypes and restored the wild-type phenotypes by complementation of these defects by plasmid clones of the respective wild-type genes. These results suggested a major role for the *eps* system in pathogenesis and environmental survival of *V. cholerae* [109]. It is surprising that the *eps* system, a type II secretion system, involved in transporting proteins across the outer membrane, can also effect polysaccharide secretion. However, it is not clear whether mutations in this pathway cause loss of rugosity because they prevent VPS secretion or because they prevent the secretion of protein(s) involved in the transport/assembly of VPS. Hence, this area is still open for investigation and interpretation.

8.5 Concluding Remarks

There are significant gaps in our understanding of the genetic basis of polysaccharide biosynthesis in general and O-antigen diversity in particular. Currently, DNA sequences of only 11 of the 206 *V. cholerae* serogroups are available and functional genetic analysis of O-antigen biosynthesis of even fewer serogroups has been performed. Recent advances in rapid and economical sequencing technologies should enable us to look at the vastly unexplored territory of genetic diversity in O-antigen biosynthesis [110]. Sequencing of specific genomic regions rather than the entire genomes would allow examining genetic diversity in hundreds of strains, especially in *V. cholerae* where majority of the gene content is conserved among different serogroups. The level of conservation among the 15 *V. cholerae* genomes sequenced so far is around 90–95% (T. Binneweis and D. Ussery, unpublished). Earlier, comparative genome analysis estimated a $\sim$1% difference between epidemic and environmental strains compared to an O1 El Tor reference strain [75]. Also, a catalog of well-defined serogroups, their origin, geographic distribution, epidemiology, and associated virulence phenotypes is available for *V. cholerae* [5]. Sequencing polysaccharide biosynthesis regions from these strains will shed light on the mechanisms underlying the generation of diversity and the role of HGT in this process.

An additional question involves the contribution of LPS/CPS to the epidemic potential of *V. cholerae*. In other words, it appears only O1 serogroup strains (*V. cholerae* O139 has an O1 genetic background) have the epidemic potential although many other serogroups apparently carry the major virulence factors and cause cholera-like illness. A comparative genome hybridization study, using ten O1, O139, and environmental isolates, has identified many genes associated with epidemicity of *V. cholerae*; more specifically, genes involved in displacement of the preexisting classical O1 strains in South Asia and establishment of endemic disease in previously cholera-free locations [75]. Such an approach, however, fails to reveal the genes present in the test strains and absent in the reference strain. Whole genome sequencing combined with SNP/indel association studies using an expanded set of carefully selected *V. cholerae* strains isolated from known outbreaks, epidemics/pandemics, other clinical settings, and the environment may address the problem associated with CGH and further expand our understanding of the genetic factors contributing to epidemic potential of *V. cholerae*. However, the role of environment in epidemicity of *V. cholerae* cannot be overlooked. Faruque et al. have shown that seasonal epidemics of cholera inversely correlate with the prevalence of environmental cholera phages [111]. The serogroup specificity of the prevalent phage during an interepidemic period determines the successful serogroup in the following epidemic. It may be that the non-O1/non-O139 pathogenic strains never reach high enough concentrations in the environment to cause an epidemic due to the lytic activity of specific phages.

Finally, the various intertwined regulatory circuits involved in rugose polysaccharide synthesis are just beginning to be unraveled. A more thorough understanding of the regulation will broaden our understanding of the significance of rugose polysaccharide in biofilm formation and environmental survival of *V. cholerae*.

Acknowledgments The authors thank Tim T. Binnewies and David W. Ussery of Center for Biological Sequence Analysis, BioCentrum-DTU-Denmark (web page: http://www.cbs.dtu.dk/services/GenomeAtlas/) for BLAST analysis of unfinished and finished Vibrio genomes and generating the BLAST atlas in Figs. 8.2 and 8.5. We also would like to thank TIGR for giving access to the sequences of unfinished genomes. S.S. is supported by funding from Defense Threat Reduction Agency, Department of Defense of the US Government and F.Y. is supported by funding from NIH RO1 grant # AI055987. The views expressed in this chapter are those of the author and do not necessarily reflect the official policy or position of the Department of the Navy, Department of Defense, or the U.S. Government. The authors thank ASM Press for use of certain text extracts from Chapter 6 of the book entitled *Vibrio cholerae and Cholera: Molecular to Global Perspectives* and Joachim Reidl for comments on the manuscript.

References

1. Reeves PR, Hobbs M, Valvano MA, Skurnik M, Whitfield C, Coplin D, Kido N, Klena J, Maskell D, Raetz CRH, Rick PD. Bacterial polysaccharide synthesis and gene nomenclature. Trend Microbiol. 1996;4:495–503.
2. Schaechter M, Ingraham JL, Neidhart FC. Prokaryotic cell structure and function: envelopes and appendages. In: Microbe. Washington, DC: ASM; 2005. pp. 19–37.
3. Manning PA, Stroeher UH, Morona R. Molecular basis of O-antigen biosynthesis in *Vibrio cholerae* O1: Ogawa-Inaba switching. In: Wachsmuth IK, Blake PA, Olsvik O, editors. *Vibrio cholerae* and cholera: molecular to global perspectives. Washington, DC: ASM; 1994. pp. 77–94.
4. Kenne L, Lindberg B. Bacterial lipopolysaccharides. In: Aspinell GO, editor. The Polysaccharides, vol. 2. New York: Academic; 1983. pp. 287–363.
5. Shimada T, Arakawa E, Itoh K, Okitsu T, Matsushima A, Asai Y, Yamai S, Nakazato T, Nair GB, Albert MJ, Takeda Y. Extended serotyping scheme for *Vibriocholerae*. Curr Microbiol. 1994; 28: 175–8.
6. Li M, Shimada T, Morris JG Jr., Sulakvelidze A, Sozhamannan S. 2002. Evidence for the emergence of non-O1 and non-O139 *Vibrio cholerae* strains with pathogenic potential by exchange of O-antigen biosynthesis regions. Infect Immun. 2002;70:2441–53.

7. Boyd EF, Waldor MK. Evolutionary and functional analyses of variants of the toxin- coregulated pilus protein TcpA from toxigenic *Vibrio cholerae* non-Ol/non-O139 serogroup isolates. Microbiology. 2002;148:1655–66.
8. Mukhopadhyay AK, Chakraborty S, Takeda Y, Nair GB, Berg DE. Characterization of VPI pathogenicity island and CTXϕ prophage in environmental strains of *Vibrio cholerae*. J Bacteriol. 2001;183:4737–46.
9. Li M, Kotetishvili M, Chen Y, Sozhamannan S. Comparative genomic analyses of the vibrio pathogenicity island and cholera toxin prophage regions in nonepidemic serogroup strains of *Vibrio cholerae*. Appl Environ Microbiol. 2003;69:1728–38.
10. Johnson JA, Joseph A, Panigrahi P, Morris JG. Frequency of encapsulated versus unencapsulated strains of non-O1 *Vibrio cholerae* isolated from patients with septicemia or diarrhea, or from environmental strains. In: American Society of Microbiology annual meeting. New Orleans: Lousiana; 1992.
11. Comstock LE, Johnson JA, Michalski JM, Morris JG, Kaper JB. Cloning and sequence of a region encoding a surface polysaccharide of *Vibrio cholerae* O139 and characterization of the insertion site in the chromosome of *Vibrio cholerae* O1. Mol Microbiol. 1996;19:815–26.
12. Comstock LE, Maneval D, Panigrahi P, Joseph A, Levine MM, Kaper JB, Morris JG, Johnson JA. The capsule and O antigen in *Vibrio cholerae* O139 Bengal are associated with a genetic region not present in *Vibrio cholerae* O1. Infect Immun. 1995;63:317–23.
13. Chen Y, Bystricky P, Adeyeye J, Panigrahi P, Ali A, Johnson JA, Bush CA, Morris JG, Stine OC. The capsule polysaccharide structure and biogenesis for non-O1 *Vibrio cholerae* NRT36S: genes are embedded in the LPS region. BMC Microbiol. 2007;7:1–15.
14. Rice EW, Johnson CJ, Clark RM, Fox KR, Reasoner OJ, Dunnigan ME, Panigraghi P, Johnson JA, Morris JG Jr. Chlorine and survival of "rugose" *Vibrio cholerae*. Lancet. 1992;340:740.
15. Morris JG Jr, Sztein MB, Rice EW, Nataro JP, Losonsky GA, Panigrahi P, Tacket CO, Johnson JA. 1996. *Vibrio cholerae* O1 can assume a chlorine-resistant rugose survival form that is virulent for humans. J Infect Dis. 1996;174:1364–8.
16. Rice EW, Johnson CH, Clark RM, Fox KR, Reasoner DJ, Dunnigan ME, Panigrahi P, Johnson JA, Morris JG Jr. 1993. *Vibrio cholerae* O1 can assume a "rugose" survival form that resists killing by chlorine, yet retains virulence. Int J Environ Health Res. 1993;3:89–98.
17. Wai SN, Mizunoe Y, Takade A, Kawabata SI, Yoshida SI. *Vibrio cholerae* O1 strain TSI-4 produces the exopolysaccharide materials that determine colony morphology, stress resistance, and biofilm formation. Appl Environ Microbiol. 1998;64:3648–55.
18. White PB. The rugose variant of vibrios. J Pathol Bacteriol. 1938;46:1–6.
19. White PB. The characteristic hapten and antigen of rugose races of cholera and El Tor vibrios. J Pathol Bacteriol. 1940;50:160–4.
20. Morris JG, and the cholera laboratory task force. *Vibrio cholerae* O139 Bengal. In: Wachsmuth IK, Blake PA, Olsvik O, editors. *Vibrio cholerae* and cholera: Molecular to Global Perspectives. Washington, DC: ASM; 1994. pp. 95–102.
21. Stroeher UH, Jedani KE, Manning PA. Genetic organization of the regions associated with surface polysaccharide synthesis in *Vibrio cholerae* O1, O139 and *Vibrio anguillarum* O1 and O2: a review. Gene. 1998;223:269–82.
22. Chatterjee SN, Chaudhuri K. Lipopolysaccharides of *Vibrio cholerae* I. Physical and chemical characterization. Biochimica et Biophysica Acta. 2003;1639:65–79.
23. Chatterjee SN, Chaudhuri K. Lipopolysaccharides of *Vibrio cholerae* II. Genetics of biosynthesis. Biochimica et Biophysica Acta. 2003;1690:93–109.
24. Chatterjee SN, Chaudhuri K. Lipopolysaccharides of *Vibrio cholerae* III. Biological functions. Biochimica et Biophysica Acta. 2003;1762:1–16.
25. Broady KW, Rietschel E, Luderitz O. The chemical structure of the lipid A component of lipopolysaccharides from *Vibrio cholerae*. Eur J Biochem. 1981;115:463–8.
26. Raziuddin S. Studies of the polysaccharide fraction from the cell wall lipopolysaccharide (O-antigen) of *Vibrio cholerae*. Indian J Biochem Biophys. 1997;14:262–3.

27. Armstrong JL, Redmond JW. The fatty acids present in the lipopolysaccharide of *Vibrio cholerae* 569B (Inaba). Biochem Biophys Acta. 1973;348:302–5.
28. Heidelberg JF, Eisen JA, Nelson WC, Clayton RA, Gwinn ML, Dodson RJ, Haft DH, Hickey EK, Peterson JD, Umayam L, Gill SR, Nelson KE, Read TD, Tettelin H, Richardson D, Ermolaeva MD, Vamathevan J, Bass S, Qin H, Dragoi I, Sellers P, McDonald L, Utterback T, Fleishmann RD, Nierman WC, White O, Salzberg SL, Smith HO, Colwell RR, Mekalanos JJ, Venter JC, Fraser CM. DNA sequence of both chromosomes of the cholera pathogen *Vibrio cholerae*. Nature. 2006;406:477–83.
29. Brade H. Occurance of 2-keto-deoxyoctonic 5-phosphate in lipopolysaccharides of *Vibrio cholerae* ogawa and Inaba. J Bacteriol. 1985;161:795–8.
30. Morona R, Brown MH, Yeadon J, Heuzenroeder MW, PA Manning. Effect of lipopolysaccharide core synthesis mutations on the production of *Vibrio cholerae* O-antigen in Escherichia coli K-12. FEMS Mirobiol Lett. 1991;82:279–86.
31. Nesper J, KraibA, Schild S, Blab J, Klose KE, Bockemuhl J, J Reidl. Comparative and genetic analyses of the putative *Vibrio cholerae* lipopolysaccharide core oligosaccharide biosynthesis (*wav*) gene cluster. Infect Immun. 2002;70:2419–33.
32. Nesper J, Kapfhammer D, Klose KE, Merkert H, Reidl J. 2000. Characterization of *Vibrio cholerae* O1 antigen as a bacteriophage K139 receptor and identification of IS *1004* insertions aborting O1-antigen biosynthesis. J Bacteriol. 2000;182:5097–104.
33. Nesper J, Lauriano CM, Klose KE, Kapfhammer D, Kraiss A, and J. Reidl. Characterization of *Vibrio cholerae* O1 El Tor *galU* and *galE* mutants: influence on lipopolysaccharide structure, colonization, and biofilm formation. Infect Immun. 2001;69:435–45.
34. Jansson PE. The chemistry of O-polysaccharide chains in bacterial lipopolysaccharides. In: Brade H, Opal SM, Vogel SN, Morrison DC, editors. Endotoxin in health and disease. New York: Marcel Dekker; 1999. pp. 155–78.
35. Kocharova NA, Perepelov AV, Zatonsky GV, Shashkov AS, Knirel YA, Jansson PE, Weintraub A. Structural studies of the O-specific polysaccharide of *Vibrio cholerae* O:2. Carbohydr Res. 2001;330:83–92.
36. Bergstrom N, Nair GB, Weintraub A, Jansson PE. Structure of the O-polysaccharide from the lipopolysaccharide from *Vibrio cholerae* O6. Carbohydr Res. 2002;337:813–7.
37. Sakazaki R. 1994. Bacteriology of vibrio and related organisms. In: Wachsmuth IK, Blake PA, Olsvik O, editors. *Vibrio cholerae* and cholera: Molecular to Global Perspectives. Washington, DC: ASM; 1994. pp. 37–55.
38. Bhaskaran, K, Gorrill RH. A study of antigenic variation in *Vibrio cholerae*. J Gen Microbiol. 1957;16:721–9.
39. Kenne L, Lindberg B, Unger P, Holme T, Holmgren J. Structural studies of the *Vibrio cholerae* O-antigen. Carbohydr. Res. 1982;100:341–9.
40. Redmond JW. The structure on the O-antigenic side chain of the lipopolysaccharide of *Vibrio cholerae* 569B (Inaba). Biochem. Biophys. Acta. 1979;584:346–52.
41. Manning PA, Heuzenroeder MW, Yeadon J, Leavesley DI, Reeves PR, D. Rowley D. Molecular cloning and expression in *Escherichia coli* K-12 of the O-antigen of the Ogawa and Inaba serotypes of the lipopolysaccharide of *Vibrio cholerae* O1 and their potential for vaccine development. Infect. Immun. 1986;53:272–7.
42. Ward HM, Morelli G, Kamke M, Morona R, Yeadon J, Hackett JA, Manning PA. A physical map of the chromosomal region determining O-antigen biosynthesis in *Vibrio cholerae* O1. Gene. 1987;55:197–204.
43. Fallarino A, Mavrangelos C, Stroeher UH, Manning PA. Identification of additional genes required for O-antigen biosynthesis in *Vibrio cholerae* O1. J Bacteriol. 1997;179:2147–53.
44. Hobbs M, Reeves PR. The JUMPstart sequence: a 39 bp element common to several polysaccharide gene clusters. Mol Microbiol. 1994;12:855–6.
45. Stroeher UH, Karageorgos LE, Brown MH, Morona R, Manning PA. A putative pathway for perosamine biosynthesis is the first function encoded within the *rfb* region of *Vibrio cholerae* O1. Gene. 1995;166:33–42.

46. Morona R, Stroeher UH, Karageorgos lE, Brown MH, Manning PA. A putative pathway for biosynthesis of the O-antigen component, *3-deoxy-L-glycero-tetronic* acid, based on the sequence of the *Vibrio cholerae* O1 *rfb* region. Gene. 1995;166:19–31.
47. Stroeher UH, Karageorgos LE, Morona R, Manning PA. Serotype conversion in *Vibrio cholerae* O1. Proc Natl Acad Sci USA. 1992;89:2566–70.
48. Hisatsune K, Kondo S, Isshiki Y, Iguchi T, Haishima Y. Occurance of 2-O-methyl-N-(3-deoxy-L-glycero-tetronyl)-D-perosamine (4-amino-4, 6-didexoy-D-mannose-pyranose) in lipopolysaccharide from the Ogawa but not from Inaba forms of O1 *Vibrio cholerae*. Biochem Biophys Res Commun. 1993;190:302–7.
49. Blokesch M, Schoolnik GK. Serogroup conversion of *Vibrio cholerae* in aquatic reservoirs. Plos Pathogens. 2007;3:0733–42.
50. Albert MJ, Siddique AK, Islam MS, Faruque ASG, Ansaruzzaman M, Faruque MSM, Sack RB. Large outbreak of clinical cholera due to *Vibrio cholerae* non-O1 in Bangladesh. Lancet. 1993;341:704.
51. Albert MJ, Ansaruzzaman M, Bardhan PK, Faruque ASG, Faruque SM, Islam MS, Mahalanabis D, Sack RB, M. Salam MS, Siddique AK, Yunus MD, Zaman K. Large epidemic of cholera-like disease in Bangladesh caused by *Vibrio cholerae* O139 synonym Bengal. Lancet. 1993;342:387–90.
52. Ramamurthy T, Garg S, Sharma R, Bhattacharya SK, Nair GB, Shimada T, Takeda T, Karasawa T, Kurazano H, Pal A, Takeda Y. 1993. Emergence of novel strain of *Vibrio cholerae* with epidemic potential in southern and eastern India. Lancet. 1993;341:703–4.
53. Waldor MK, Colwell R, Mekalanos JJ. The *Vibrio cholerae* O139 serogroup antigen includes O-polysaccharide capsule and lipopolysaccharide virulence determinant. Proc Natl Acad Sci USA. 1994;91:11388–92.
54. Johnson JA, Salles CA, Panigrahi P, Albert MJ, Wright AC, Johnson RJ, Morris Jr JG. *Vibrio cholerae* O139 synonym Bengal is related to *Vibrio cholerae* El Tor but has important differences. Infec Immun. 1994;62:2108–10.
55. Cox AD, Brisson JR, Varma V, Perry MB. Structural analysis of the lipopolysaccharide from *Vibrio cholerae* O139. Carbohydr Res. 1996;290:43–58.
56. Knirel YA, Widmalm G, Senchenkova SN, Jansson PE, Weintraub A. Structural studies on the short-chain lipopolysaccharide of *Vibrio cholerae* O139 Bengal. Eur Biochem. 1997;247:402–10.
57. Weintraub A, Widmalm G, Jansson PE, Jansson M, Hultenby K, Albert MJ. *Vibrio cholerae* O139 Bengal possesses a capsular polysaccharide which may confer increased virulence. Microb. Pathoge. 1994;16:235–41.
58. Preston LM, Xu Q, Johnson JA, Joseph A, Maneval DR, Husain K, Reddy GP, Bush CA, Morris JG Jr. 1995. Preliminary structure determination of the capsular polysaccharide of *Vibrio cholerae* O139 Bengal AI-1837. J. Bacteriol. 1995;177:835–8.
59. Knirel YA, Pardes L, Jansson PE, Weintraub A, Widmalm G, Albert MJ. Structure of the capsular polysaccharide of *Vibrio cholerae* O139 Bengal containing D-galactose 4,6-cyclophosphate. Eur J Biochem. 1995;232:391–6.
60. Adeyeye J, Azurmundi HF, Stroop CJM, Sozhamannan S, Williams AL, Adetumbi AM, Johnson JA, Bush CA. Conformation of the hexasaccharide repeating subunit from the *V. cholerae* O139 capsular polysaccharide. Biochemistry. 2003;42:3979–88.
61. Bik EM, Bunschoten AE, Gouw RD, Mooi FR. Genesis of the novel epidemic *Vibrio cholerae* O139 strain: evidence for horizontal transfer of genes involved in polysaccharide synthesis. EMBO J. 1995;14:209–16.
62. Comstock LE, Johnson JA, Michalski JM, Morris JG Jr, Kaper JB. Cloning and sequence of a region encoding a surface polysaccharide of *Vibrio cholerae* O139 and characterization of the insertion site in the chromosome of *Vibrio cholerae* O1. Mol Microbiol. 1996;19: 815–26.
63. Stroeher UH, Parasivam G, Dredge BK, Manning PA. Novel *Vibrio cholerae* O139 genes involved in lipopolysaccharide biosynthesis. J Bacteriol. 1997;179:2740–7.

64. Bastin DA, Reeves PR. Sequence analysis of the O-antigen gene (*rfb*) cluster of *Escherichia coli* O111. Gene. 1995;164:17–23.
65. Cox AD, Brisson JR, Thibault P, Perry MB. Structural analysis of the lipopolysaccharide from *Vibrio cholerae* serotype O22. Carbohyd Res. 1997;304:191–208.
66. Knirel, YA, Senchenkova SN, Jansson PE, Weintraub A. More on the structure of *Vibrio cholerae* O22 lipopolysaccharide. Carbohydr Res. 1998;310:117–9.
67. Yamasaki S, Shimizu T, Hoshino K, Ho S-T, Shimada T, Nair GB, Takeda Y. The genes responsible for O-antigen synthesis of *Vibrio cholerae* O139 are closely related to those of *Vibrio cholerae* O22. Gene. 1999;237:321–32.
68. Aldova E, Laznickova K, Stepankova E, Lietava J. Isolation of nonagglutinable vibrios from an enteritis outbreak in Czechoslovakia. J Infect Dis. 1968;118:25–31.
69. Kamal AM. Outbreak of gastro-enteritis by non-agglutinable (NAG) vibrios in the republic of the Sudan. J Egypt Public Health Assoc. 1971;XLVI:125–59.
70. Raetz CR, Whitfield C. Lipopolysaccharide endotoxins. Annu Rev Biochem 2002;71: 635–700.
71. Rivas M, Toma C, Miliwebsky E, Caffer MI, Galas M, Varela P, Tous M, Bru AM, Binsztein N. Cholera isolates in relation to the "eighth pandemic". Lancet. 1993;342:926–7.
72. Mooi FR, Bik EM. The evolution of epidemic *Vibrio cholerae* strains. Trends Microbiol. 1997;4:161–5.
73. Stroeher UH, Manning PA. *Vibrio cholerae* serotype O139: swapping genes for surface polysaccharide biosynthesis. Trends Microbiol. 1997;5:178–80.
74. Bik EM, Gouw RD, Mooi FR. DNA fingerprinting of *Vibrio cholerae* strains with a novel insertion sequence element: a tool to identify epidemic strains. J Clin Microbiol. 1996;34:1453–61.
75. Dziejman M, Balon E, Boyd D, Fraser CM, Heidelberg JF, Mekalanos JJ. Comparative genomic analysis of *Vibrio cholerae*: Genes that correlate with cholera endemic and pandemic disease. Proc Natl Acad Sci USA 2002;99:1556–61.
76. Sugiyama T, Kido N, Kato Y, Koide N, Yoshida T, Yokochi T. Evolutionary relationship among *rfb* gene clusters synthesizing mannose homo polymer as O-specific polysaccharides in *Esherichia coli* and *Klebsiella*. Gene. 1997;198:111–3.
77. Waldor MK, Mekalanos JJ. Lysogenic conversion by a filamentous phage encoding cholera toxin. Science. 1996;272:1910–4.
78. O'Shea YA, Boyd EF. Mobilization of the Vibrio pathogenicity island between *Vibrio cholerae* isolates mediated by CP-T1 generalized transduction. FEMS Microbiol Lett. 2002;214:153–7.
79. Meibom KL, Blokesch M, Dolganov NA, Wu CY, Schoolnik GK. Chitin induces natural competence in *Vibrio cholerae*. Science. 2005;310:1824–7.
80. Dybvig, K. DNA rearrangements and phenotypic switching in prokaryotes. Mol Microbiol. 1993;10:465–71.
81. Henderson IR, Owen P, Nataro JP. 1999. Molecular switches-the ON and OFF of bacterial phase variation. Mol Microbiol. 1999;33:919–32.
82. Smits WK, Kuipers OP, Veening JW. Phenotypic variation in bacteria: the role of feedback regulation. Nat Rev Microbiol. 2006;4:259–71.
83. Yildiz FH, Schoolnik GK. *Vibrio cholerae* O1 El Tor: identification of a gene cluster required for the rugose colony type, exopolysaccharide production, chlorine resistance, and biofilm formation. Proc Natl Acad Sci USA. 1999;96:4028–33.
84. Ali A, Rashid MH, Karaolis DK. High-frequency rugose exopolysaccharide production by *Vibrio cholerae*. Appl Environ Microbiol. 2002;68:5773–8.
85. Hammer BK, Bassler BL. Quorum sensing controls biofilm formation in *Vibrio cholerae*. Mol Microbiol. 2003;50:101–4.
86. Yildiz FH, Liu XS, Heydorn A, and Schoolnik GK. Molecular analysis of rugosity in a *Vibrio cholerae* O1 El Tor phase variant. Mol Microbiol. 2004;53:497–515.

87. Zhu J, Mekalanos JJ. Quorum sensing-dependent biofilms enhance colonization in *Vibrio cholerae*. Dev Cell. 2003;5:647–56.
88. Lauriano CM, Ghosh C, Correa NE, and Klose KE. The sodium-driven flagellar motor controls exopolysaccharide expression in *Vibrio cholerae*. J Bacteriol. 2004;186:4864–74.
89. Watnick PI, Lauriano CM, Klose KE, Croal L, Kolter R. The absence of a flagellum leads to altered colony morphology, biofilm development and virulence in *Vibrio cholerae* O139. Mol Microbiol. 2001;39:223–35.
90. Haugo AJ, Watnick PI. *Vibrio cholerae* CytR is a repressor of biofilm development. Mol Microbiol. 2002;45:471–83.
91. Islam MS, Jahid MI, Rahman MM, Rahman MZ, Islam MS, Kabir MS, Sack DA, Schoolnik GK. 2007. Biofilm Acts as a Microenvironment for Plankton-Associated *Vibrio cholerae* in the Aquatic Environment of Bangladesh. Microbiol Immunol. 2007;51:369–79.
92. Beyhan S, Yildiz FH. Smooth to rugose phase variation in *Vibrio cholerae* can be mediated by a single nucleotide change that targets c-di-GMP signalling pathway. Mol Microbiol. 2007;63:995–1007.
93. Matz C, McDougald D, Moreno AM, Yung PY, Yildiz FH, Kjelleberg S. Biofilm formation and phenotypic variation enhance predation-driven persistence of *Vibrio cholerae*. Proc Natl Acad Sci USA. 2005;102:16819–24.
94. Costerton JW, Lewandowski Z, Caldwell DE, Korber DR, Lappin-Scott HM. Microbial biofilms. Annu Rev Microbiol. 1995;49:711–45.
95. Davey ME, O'Toole GA. Microbial biofilms: from ecology to molecular genetics. Microbiol Mol Biol Rev. 2000;64:847–67.
96. Donlan RM, Costerton JW. Biofilms: survival mechanisms of clinically relevant microorganisms. Clin Microbiol Rev. 2002;15:167–93.
97. Parsek MR, Singh PK. Bacterial biofilms: an emerging link to disease pathogenesis. Annu Rev Microbiol. 2003;57:677–701.
98. Heydorn A, Nielsen AT, Hentzer M, Sternberg MC, Givskov M, Ersboll BK, Molin S. Quantification of biofilm structures by the novel computer program COMSTAT. Microbiology. 2000;146:2395–407.
99. Ali A, Mahmud ZH, Morris JG Jr, Sozhamannan S, Johnson JA. Sequence analysis of TnphoA insertion sites in *Vibrio cholerae* mutants defective in rugose polysaccharide production. Infect Immun. 2000;68:6857–64.
100. Yildiz FH, Dolganov NA, and Schoolnik GK. 2001. VpsR, a Member of the response regulators of the two-component regulatory systems is required for expression of vps biosynthesis genes and EPS(ETr)-associated phenotypes in *Vibrio cholerae* O1 El Tor. J Bacteriol. 2001;183:1716–26.
101. Casper-Lindley C, Yildiz FH. VpsT is a transcriptional regulator required for expression of vps biosynthesis genes and the development of rugose colonial morphology in *Vibrio cholerae* O1 El Tor. J Bacteriol. 2004;186:1574–8.
102. Romling U, Sierralta WD, Eriksson K, and Normark S. Multicellular and aggregative behaviour of *Salmonella typhimurium* strains is controlled by mutations in the *agfD* promoter. Mol Microbiol. 1998;28:249–64.
103. Romling U, Rohde M, Olsen A, Normark S, and Reinkoster J. 2000. AgfD, the checkpoint of multicellular and aggregative behaviour in *Salmonella typhimurium* regulates at least two independent pathways. Mol Microbiol. 2000;36:10–23.
104. Uhlich GA, Keen JE, Elder RO. Mutations in the *csgD* promoter associated with variations in curli expression in certain strains of *Escherichia coli* O157:H7. Appl Environ Microbiol. 2001;67:2367–70.
105. Beyhan S, Bilecen K, Salama SR, Casper-Lindley C, Yildiz FH. 2007. Regulation of rugosity and biofilm formation in *Vibrio cholerae*: comparison of VpsT and VpsR regulons and epistasis analysis of vpsT, vpsR, and hapR. J Bacteriol. 2007;189:388–402.
106. Miller MB, Bassler BL. Quorum sensing in bacteria. Annu Rev Microbiol. 2001;55: 165–99.

107. Miller MB, Skorupski K, Lenz DH, Taylor RK, Bassler BL. Parallel quorum sensing systems converge to regulate virulence in *Vibrio cholerae.* Cell. 2002;110:303–14.
108. Zhu J, Miller MB, Vance RE, Dziejman M, Bassler BL, Mekalanos JJ. 2002. Quorum-sensing regulators control virulence gene expression in *Vibrio cholerae*. Proc Natl Acad Sci USA. 2002;99:3129–34.
109. Ali A, Johnson JA, Franco AA, Metzger DJ, Connell TD, Morris JG Jr, Sozhamannan S. Mutations in the extracellular protein secretion pathway genes (*eps*) interfere with rugose polysaccharide production in and motility of *Vibrio cholerae*. Infect Immun. 2000;68: 1967–74.
110. Margulies M, Egholm M, Altman WE, Attiya S, Bader JS, Bemben LA, Berka J, Braverman MS, Chen YJ, Chen Z, Dewell SB, Du L, Fierro JM, Gomes XV, Godwin BC, He W, Helgesen S, Ho CH, Irzyk GP, Jando SC, Alenquer ML, Jarvie TP, Jirage KB, Kim JB, Knight JR, Lanza JR, Leamon JH, Lefkowitz SM, Lei M, Li J, Lohman KL, Lu H, Makhijani VB, McDade KE, McKenna MP, Myers EW, Nickerson E, Nobile JR, Plant R, Puc BP, Ronan MT, Roth GT, Sarkis GJ, Simons JF, Simpson JW, Srinivasan M, Tartaro KR, Tomasz A, Vogt KA, Volkmer GA, Wang SH, Wang Y, Weiner MP, Yu P, Begley RF, Rothberg JM. Genome sequencing in microfabricated high-density picolitre reactors. Nature. 2005;437:376–80.
111. Faruque SM, Naser IB, Islam MJ, Faruque AS, Ghosh AN, Nair GB, Sack DA, Mekalanos JJ. Seasonal epidemics of cholera inversely correlate with the prevalence of environmental cholera phages. Proc Natl Acad Sci USA. 2005;102:1702–7.

Chapter 9
Significance of the SXT/R391 Family of Integrating Conjugative Elements in *Vibrio cholerae*

Vincent Burrus

Abstract Integrating conjugative elements (ICEs) are self-transmissible mobile elements that transfer between bacteria via conjugation and integrate into the chromosome. SXT^{MO10} is an ICE that was initially discovered in a 1992 *Vibrio cholerae* O139 clinical isolate from India. SXT^{MO10} and related ICEs became prevalent in Asian *V. cholerae* populations in the 1990s and are now present in most clinical and environmental *V. cholerae* isolates from Asia and Africa, playing an important role in the spread of antibiotic resistance genes in this pathogen. More than 35 SXT^{MO10}-related elements have now been identified worldwide in environmental and clinical isolates of at least nine *Vibrio*-like species. The SXT^{MO10}-related ICEs have a highly conserved overall genome organization, disrupted by ICE-specific regions conferring ICE-specific properties. The appearance of SXT^{MO10}-related ICEs in *V. cholerae* seems to be correlated with an increased use of antibiotics to treat cholera. However, discovery of ICEs that do not confer any drug resistance suggests that these mobile elements could confer other advantages to the bacteria in the environment.

9.1 Introduction

The gram-negative bacterium *Vibrio cholerae*, causative agent of the serious epidemic disease cholera, has been shown in the past to be susceptible to most antibiotics. Yet, oral antibiotics as sole treatment against this pathogen are not effective to avoid death of patients. The rapid loss of fluids such as profuse watery diarrhea and vomiting that occurs during infection by *V. cholerae* leads within a few hours of onset to severe dehydration of patients, making oral antibiotics useless. Case-fatality rate for severe cholera can reach up to 50% if dehydration is left untreated. Fortunately, without any oral antibiotic intake, rehydration by appropriate

V. Burrus (✉)
Département de biologie, Université de Sherbrooke, Sherbrooke, Québec, Canada, J1K 2R1
e-mail: Vincent.Burrus@USherbrooke.ca

T. Ramamurthy, S.K. Bhattacharya (eds.), *Epidemiological and Molecular Aspects on Cholera*, Infectious Disease, DOI 10.1007/978-1-60327-265-0_9,
© Springer Science+Business Media, LLC 2011

and prompt fluid and electrolyte replacement is a very effective therapy; patients generally recover in about 4–5 days, and fatality rate drops drastically so that virtually all cholera deaths can be avoided.

Nevertheless, in addition to intravenous or oral rehydration solutes, patients with severe diarrhea are also given an oral antibiotic including doxycycline, cotrimoxazole (the association of trimethoprim and sulfamethoxazole), erythromycin or quinolones such as ciprofloxacin [1]. While not mandatory, antibiotic treatment as a complement to rehydration has been shown to shorten the recovery time to about 2–3 days and lessen the symptomatic diarrhea.

Inevitably, drug resistance came along with antibiotic use and abuse. Shortly after the introduction of the first antibiotic treatments of cholera cases in countries where cholera is endemic, antibiotic-resistant strains of *V. cholerae* emerged. This is perfectly exemplified by the emergence of tetracycline, chloramphenicol, doxycycline, and cotrimoxazole-resistant strains of *V. cholerae* O1 El Tor in Zambia during cholera outbreaks in the early 1990s. Continued use of antibiotics for therapy and prophylaxis after two cholera outbreaks during 1990–1991 led to a dramatic rise of antibiotic-resistant strains in subsequent outbreaks after 1992 [2]. While widespread antibiotic resistance in *V. cholerae* was still uncommon three decades ago, reports of multiple antibiotic-resistant clinical and environmental isolates of *V. cholerae* are now frequent in countries where cholera is endemic [2–5].

Interestingly, unlike other pathogenic microorganisms, *V. cholerae* does not seem to retain antibiotic-resistant phenotype for extended period of time once the selective agent has been eliminated. On the contrary, the distribution of antibiotic resistance patterns varies with time and location in endemic areas depending on the antibiotics that are used at a given time in a given region. Again in Zambia, a significant drop in tetracycline- and chloramphenicol-resistant isolates occurred after 1993 when tetracycline was replaced with erythromycin in treating cholera [2]. Whether the isolates responsible for earlier and recent epidemics are of the same clonal origin still remains an open question.

Three types of mobile genetic elements usually account for resistance to antibiotics in *V. cholerae*: (i) plasmids, which for most are large and self-transmissible by conjugation, (ii) integrons, which are chromosomal- or plasmid-borne gene capture and expression systems, and (iii) integrating conjugative elements (now commonly referred to as ICEs), which are chromosomal self-transmissible mobile genetic elements. As such, ICEs are important mediators of horizontal gene transfer among prokaryotes, perhaps as prevalent as conjugative plasmids [6].

To date, ICEs have been described not only in low G+C gram-positive bacteria and the *Bacteroides* group (as conjugative transposons) but also in other major bacterial subdivisions, including the actinomycetales, the Rhizobiaceae (α-proteobacteria), the Burkholderiaceae (β-proteobacteria), and in a large number of γ-proteobacteria such as *V. cholerae* [7]. ICEs have features that are reminiscent of both prophages and conjugative plasmids. Like many temperate bacteriophages, ICEs integrate into and replicate with the host chromosome. An obvious consequence of this behavior is that, like lambdoid prophages, ICEs are inherited vertically along with the rest of the genome when the bacterial cells divide. Such

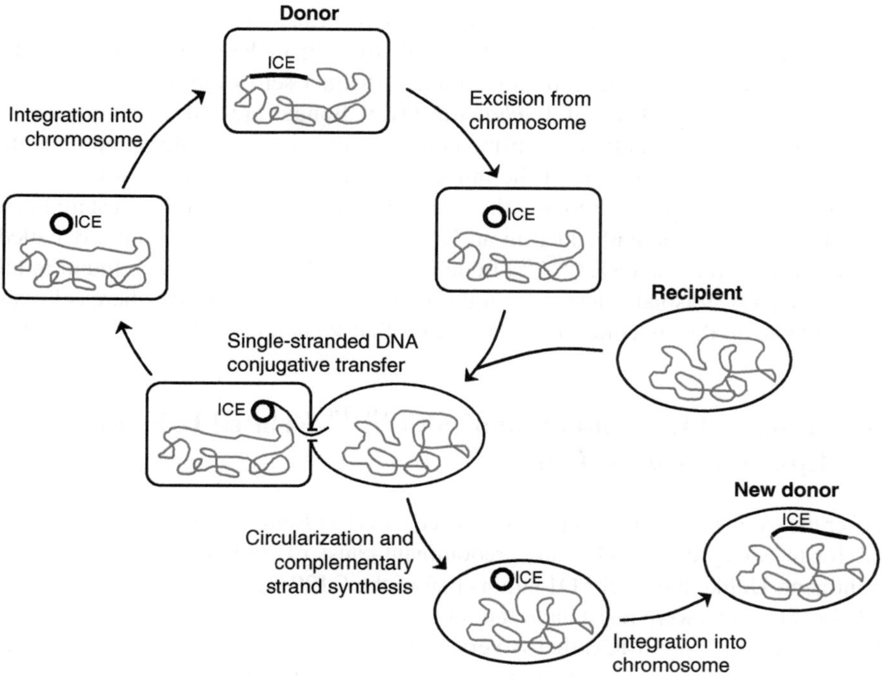

Fig. 9.1 Schematic representation of the life cycle of ICEs

a mechanism of inheritance likely replaces autonomous replication since, unlike plasmids, most if not all ICEs are devoid of an autonomous vegetative replication system. Similar to conjugative plasmids, ICEs transfer via conjugation as shown in a general overview of the life cycle of ICEs (Fig. 9.1). This review is an overview of the ICEs of the SXT/R391 family to which belong all the ICEs that have been described to date in *V. cholerae*.

9.2 Discovery of SXT^{MO10} in a Novel Epidemic-Causing Serogroup of *Vibrio cholerae*

In late 1992 and early 1993, a heretofore unknown serogroup of *V. cholerae* emerged in India and Bangladesh as the first non-O1 *V. cholerae* serogroup to give rise to epidemic cholera [8]. During this period, this new serogroup of *V. cholerae*, designated O139 Bengal, rapidly spread throughout Asia and largely replaced the endemic El Tor O1 strains of *V. cholerae* to become the principal clinical and environmental isolate of *V. cholerae* on most of the Indian subcontinent [9]. The initial characterization of *V. cholerae* O139 revealed that this strain was closely related to the El Tor biotype of *V. cholerae* O1 and suggested that *V. cholerae* O139 was in fact a derivative of an

El Tor strain of *V. cholerae* O1 [9–16]. Yet, two principal features distinguished the newly emerged *V. cholerae* O139 Bengal from the endemic *V. cholerae* O1 El Tor: the novel O139 serogroup antigen and a distinct set of antibiotic resistances [17]. The clinical isolates of *V. cholerae* O139 Bengal are indeed characteristically resistant to the antibiotics trimethoprim, sulphamethoxazole, streptomycin, and chloramphenicol. In one of the initial *V. cholerae* O139 clinical isolates from Madras, India, the genes encoding these resistances were found to be clustered in a composite transposon embedded in an ICE [18, 19]. This ICE was initially called SXT and later renamed SXT^{MO10} (Table 9.1; Fig. 9.2). SXT^{MO10} was found in all other early O139 isolates tested from the Indian subcontinent [18], whereas it was absent from all the endemic El Tor O1 strains that *V. cholerae* O139 replaced [20].

9.3 Apparent Dissemination of SXT^{MO10}-Related ICEs in Epidemic *Vibrio cholerae*

After the extensive cholera outbreaks caused by O139 Bengal strains, *V. cholerae* O1 El Tor re-emerged in 1994 as the predominant cause of cholera on the Indian subcontinent. Unlike the El Tor O1 strains before the O139 outbreak, these re-emerged El Tor O1 strains were resistant to the same four antibiotics as the initial *V. cholerae* O139 Bengal [21]. An ICE that is very similar but not identical to SXT^{MO10} was found to carry the genes encoding these resistances [18–20]. This ICE was originally designated as SXT^{ET} and is now referred to as ICE*Vch*Ind1 (El Tor O1 1994 isolate CO943 from India) and ICE*Vch*Ban1 (El Tor O1 1998 isolate 1811/98 and El Tor O1 1999 isolate C10488 from Bangladesh) [19, 22]. Surprisingly, although ICE*Vch*Ind1 and ICE*Vch*Ban1 were isolated more than 6 years after SXT^{MO10}, comparative DNA sequence analysis of the antibiotic resistance gene clusters of these three ICEs suggests that SXT^{MO10} is not an immediate precursor of two others (Fig. 9.2); instead SXT^{MO10} and ICE*Vch*Ind1, and ICE*Vch*Ban1 on the other hand, seem to have arisen and diverged from a common ancestor similar to ICE*Vch*Ind4, a *V. cholerae* O139-derivated ICE that does not confer resistance to trimethoprim [19] (Table 9.1; Fig. 9.2).

Initial characterization of ICE*Vch*SL1, an ICE found in a 1994 environmental *V. cholerae* O139 strain from Sri Lanka, suggested that this ICE was very closely related, if not identical, to ICE*Vch*Ind1 and ICE*Vch*Ban1 [19] (Fig. 9.2). Such a finding suggested that ICE*Vch*Ind1 and ICE*Vch*Ban1 were not restricted to El Tor O1 *V. cholerae* but instead either moved between O1 and O139 strains or had been independently acquired by strains of both serogroups. However, further characterization based on the analysis of the entry exclusion group-encoding gene *eex* (see below and Table 9.1), ICE*Vch*Ind1 and ICE*Vch*Ban1 from O1 El Tor appears to be identical, whereas ICE*Vch*SL1 from O139, while closely related, is a distinct ICE [23]. This finding argues against promiscuous exchanges of ICE*Vch*Ind1 between O1 and O139 strains and rather suggests that strains of both serogroups may have independently acquired two different lineages of SXT^{MO10} -related ICEs.

Table 9.1 SXT/R391 ICE family members

	Origin						
ICE name	Host[a]	Location	Year	Type[b]	Drug(s) to which resistance is conferred[c]	*eex* allele[d]	References
SXT^{MO10}	*V. cholerae* O139 MO10	India	1992	C	Su, Tm, Cm, Sm	*S*	[18, 19, 23]
ICE*Vch*Ind4	*V. cholerae* O139 AS207	India	1997	C	Su, Cm, Sm	*S*	[19, 23]
ICE*Vch*Ban7	*V. cholerae* O139 2125/98	Bangladesh	1998	C	None	*S*	[19, 23]
ICE*Vch*HKo1	*V. cholerae* O139	Hong Kong	1993	C	None	n.a.	[19]
ICE*Vch*SL1	*V. cholerae* O139 E712	Sri Lanka	1994	E	Su, Tm, Cm, Sm	*S5*	[19, 23]
ICE*Vch*Ban1	*V. cholerae* O1	Bangladesh	1998	C	Su, Tm, Cm, Sm	*R1*	[19, 23]
ICE*Vch*Ban2	*V. cholerae* O1	Bangladesh	2005	C	Su, Tm, Tc	*R1*	[23]
ICE*Vch*Ban3	*V. cholerae* O1	Bangladesh	2005	C	Su, Tm, Cm	*R1*	[23]
ICE*Vch*Ban4	*V. cholerae* O1	Bangladesh	1998	C	Su, Tm	*R1*	[23]
ICE*Vch*Ban5	*V. cholerae* O1	Bangladesh	1998	C	Su, Tm	*R1*	[23]
ICE*Vch*Ban6	*V. cholerae* O1	Bangladesh	1998	C	Su, Tm	*R1*	[23]
ICE*Vch*Ind1	*V. cholerae* O1	India	1994	C	Su, Tm, Cm, Sm	*R1*	[19, 23]
ICE*Vch*Ind2	*V. cholerae* O1	India	1994	C	Su, Tm	*R1*	[23]
ICE*Vch*Ind3	*V. cholerae* O1	India	1994	C	Su, Tm	*R1*	[23]
pJY1[e]	*V. cholerae* O1	Philippines	1973	C	Su, Cm, Sm	n.a.	[40]
ICE*Vch*Lao1	*V. cholerae* O1	Lao PDR	1998	C	Su, Tm, Cm, Sm, Tc	n.a.	[29, 78]
ICE*Vch*Vie0	*V. cholerae* O1	Vietnam	1990	C	None	n.a.	[27]
ICE*Vch*Vie1	*V. cholerae* O1	Vietnam	2000	C	Su, Cm, Sm, Tc	n.a.	[28]
ICE*Vch*Alg1	*V. cholerae* O1	Algeria	1994	C	Su, Tm, Cm, Sm, Tc	n.a.	[36]
ICE*Vch*Saf1	*V. cholerae* O1	South Africa	1998	C	Tc	n.a.	[5]
ICE*Vch*Ang1	*V. cholerae* O1	Angola	1992	C	None	n.a.	[4]
ICE*Vch*Mex1	*V. cholerae* non-O1/O139	Mexico	2001	E	None	*R3*	[33]
ICE*Vch*Ind5	*V. cholerae* O1	India	1994	C	Su, Tm, Cm, Sm	*R1*	V. Burrus
ICE*Vch*Ind6	*V. cholerae* O2	India	1998	E	n.a.	n.a.	[3]
ICE*Vch*Ind7	*V. cholerae* O130	India	1998	E	n.a.	n.a.	[3]
ICE*Vch*Ind8	*V. cholerae* O144	India	1998	E	n.a.	n.a.	[3]

Table 9.1 (continued)

	Origin						
ICE name	Host[a]	Location	Year	Type[b]	Drug(s) to which resistance is conferred[c]	*eex* allele[d]	References
ICE*Vch*Ind9	*V. cholerae* O150	India	1998	E	n.a.	n.a.	[3]
ICE*Vch*Ind10	*Vibrio cholerae* O151	India	1998	E	n.a.	n.a.	[3]
ICE*Vfl*Ind1	*Vibrio fluvialis*	India	2002	C	Su, Tm	*S1*	[23, 32]
ICE*Pda*Spa1	*P. damselae*	Spain	2001	F	Tc	*S3*	[23, 34, 35]
R997[e]	*P. mirabilis*	India	1977	C	Su, Sm, Ap	*S2*	[23, 79]
R705[e]	*P. vulgaris*	South Africa	1975	n.a.	Kn, Hg	n.a.	[80]
R391[e]	*Pr. rettgeri*	South Africa	1967	C	Kn, Hg	*R*	[23, 42, 81]
R748[e]	*Providencia* spp.	South Africa	1974	n.a.	Kn, Hg	n.a.	[82]
ICE*Pal*Ban1	*Pr. alcalifaciens*	Bangladesh	1999	C	Su, Tm	*R2*	[19, 23]
ICE*Spu*PO1	*S. putrefaciens*	Pacific Ocean marine sediment	n.a.	E	Putative multi-drug efflux	*S4*	[23, 43]
pMERPH[e]	*S. putrefaciens*	United Kingdom river sediment	1990	E	Hg	*R4*	[23, 83]

[a]When relevant, the strain is indicated
[b]Type of sample: C, clinical isolate; E, environmental isolate; F, isolate from diseased fish
[c]Ap, ampicillin; Cm, chloramphenicol; Kn, kanamycin; Sm, streptomycin; Su, sulfamethoxazole; Tc, tetracycline; Tm, trimethoprim; Hg, mercury
[d]Variants of entry exclusion determinant *S* and *R* determine the S and R exclusion groups, respectively
[e]Members of the now obsolete IncJ element family. Other members are R392 and R397, which are likely clones of R391, R706, which is likely a clone of R705, and R749, likely clone of R748
n.a., data not available

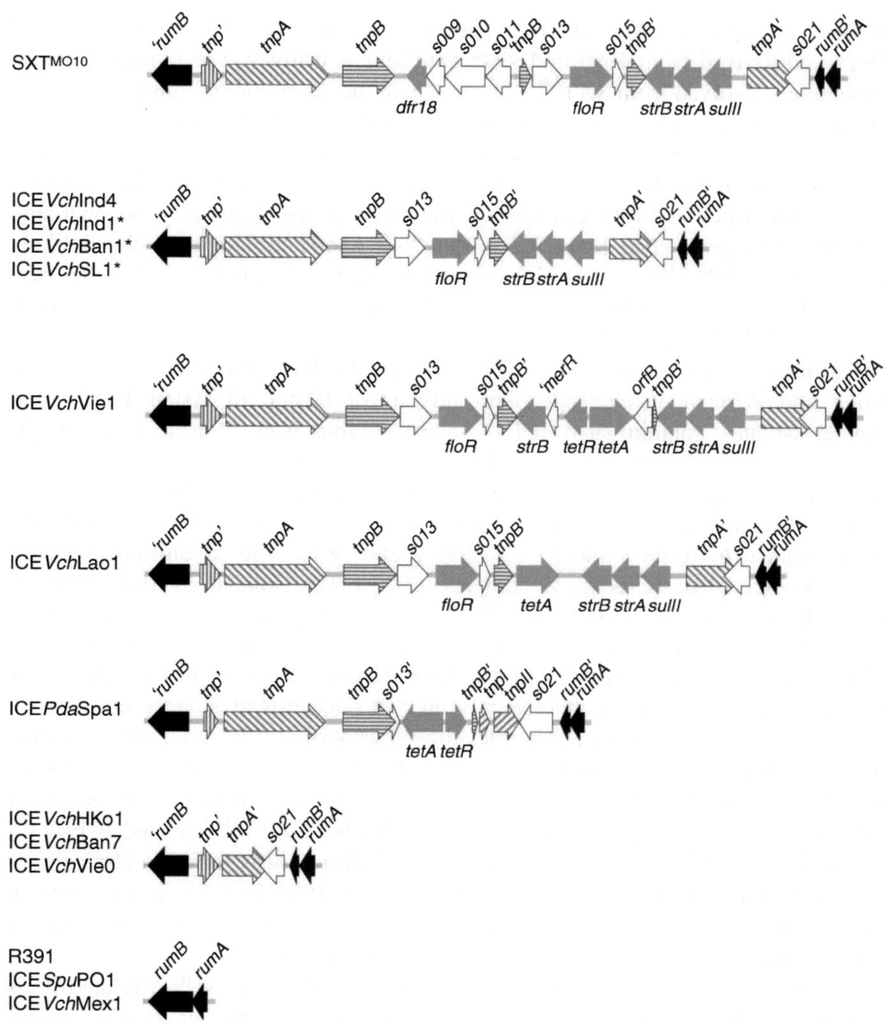

Fig. 9.2 Organization of the antibiotic resistance gene cluster inserted in the *rumB* gene in various SXT/R391 ICE members. The genes mediating resistance to antibiotics, *dfr18* (trimethoprim), *floR* (chloramphenicol), *strAB* (streptomycin), *sulII* (sulfamethoxazole), and *tetAR* (tetracycline), are represented by *gray arrows*. The genes with similarity to transposases, *tnp*, *tnpA*, *tnpB*, *tnpI*, and *tnpII*, are represented by *hatched arrows*. The genes encoding hypothetical proteins of unknown function are shown in *white*. The genes *rumA* and *rumB* are in *black*. * In ICE*Vch*Ind1, ICE*Vch*Ban1, and ICE*Vch*SL1 resistance to trimethoprim is conferred by *dfrA1*, which is located within a class 4 integron integrated in an insertion hotspot between *s073* and *traF*, ~70-kb away from *rumB*

Vibrio cholerae O139 still causes cholera in parts of Asia [24–26] but, unlike initial O139 Bengal isolates, more recent O139 isolates harbor SXT^{MO10}-related ICEs that do not necessarily confer resistance to antibiotics. Comparative DNA sequence analysis suggests that ICE*Vch*HKo1 and ICE*Vch*Ban7, two ICEs formerly called SXT^S and derived from 1993 and 1998 antibiotic-sensitive O139 clinical isolates, respectively, evolved from SXT^{MO10} from which the antibiotic resistance gene cluster has been deleted [19] (Table 9.1; Fig. 9.2). A similar observation has been made by Bani et al. [27] about ICE*Vch*Vie0, which carries the exact same deletion as ICE*Vch*HKo1 and ICE*Vch*Ban7. This is a disturbing observation as ICE*Vch*Vie0 has been identified in a 1990 clinical strains of *V. cholerae* O1 El Tor from Vietnam, a pre-O139 Bengal isolate. Since ICE*Vch*Vie0's deleted antibiotic gene cluster likely derives from SXT^{MO10}'s antibiotic gene cluster, this finding indicates that the presence in *V. cholerae* of ICEs such as SXT^{MO10} largely predates the emergence of *V. cholerae* O139 Bengal in 1992.

During the past decade, SXT^{MO10}-related ICEs have become widespread in Asian and African *V. cholerae*. Currently, most if not all Asian *V. cholerae* clinical isolates contain such ICEs [28–30] (Table 9.1). Dalsgaard et al. [5] reported SXT^{MO10}-related ICEs in 1998 *V. cholerae* O1 El Tor clinical isolates from Mozambique and South Africa. Interestingly, these ICEs, like other recent SXT^{MO10}-related ICEs from Bangladesh, Vietnam, and Lao PDR, appear to mediate resistance to tetracycline [5, 28, 29] (Table 9.1). The ubiquitous presence of ICE*Vch*Ind1-like ICEs in *V. cholerae* O1 El Tor isolated in 1997–1998 in Mozambique, Swaziland and Zimbabwe has been reported by Ceccarelli et al. [31]. Similarly, Mwansa et al. [2] reported SXT^{MO10}-related ICEs in all the 2003–2004 isolates of *V .cholerae* O1 El Tor from Zambia, whereas all the 1996–1997 isolates were devoid of such elements and carried instead a large conjugative plasmid conferring resistance to tetracycline. Furthermore, in Asia, non-toxigenic environmental *V. cholerae* isolates of a variety of serogroups have been reported to contain SXT^{MO10}-related ICEs [3, 30] (Table 9.1). Besides *V. cholerae*, other species of *Vibrio* in Asia have been shown to harbor SXT^{MO10}-related ICEs [32] (Table 9.1). An ICE*Vch*Ind1-like element, ICE*Pal*Ban1, was found in clinical isolates of *Providencia alcalifaciens* in Bangladesh in 1999 [19], suggesting that, in the environment as in the laboratory [18], the host range of SXT^{MO10}-related ICEs is not limited to microorganisms belonging to the genus *Vibrio*.

Yet, Asia and Africa are not the only continents where SXT^{MO10}-related ICEs have been detected. An SXT^{MO10}-related ICE that does not encode resistance to antibiotics has been detected in an environmental *V. cholerae* isolate from Mexico [33] (Table 9.1). This first New World SXT^{MO10}-related ICE, designated ICE*Vch*Mex1, appears to be quite distinct from SXT^{MO10}, ICE*Vch*Ind1 or even ICE*Vch*HKo1, and it is likely that ICE*Vch*Mex1 has evolved independently from these Asian ICEs for a significant period of time [33] (Fig. 9.2). The El Tor O1 strain of *V. cholerae* that is giving rise to the ongoing cholera epidemic in Latin America is highly related to pre-O139 Asian El Tor O1 strains that lacked SXT^{MO10}. The existence of an SXT^{MO10}-related ICE in a Mexican environmental isolate of *V. cholerae* raises the possibility that Latin American pathogenic *V. cholerae* may acquire this ICE or a related one. Furthermore, ICE*Pda*Spa1, an SXT^{MO10}-related ICE conferring resistance to tetracycline, has recently been isolated in Spain from the *Vibrio*-related fish pathogen *Photobacterium damselae* subsp. *piscicida* [34, 35] (Table 9.1).

In all the known instances of apparent spread of SXT^{MO10}-related ICEs between various bacterial species and genera, it is not clear which species (or genera) is the donor and which is the recipient. For example, it is possible that *P. alcalifaciens* could have donated ICE*Pal*Ban1 to El Tor O1 *V. cholerae*. Similarly, as discussed below in more detail, the African SXT^{MO10}-related ICEs may have been the source of the Asian *V. cholerae*-derived ICEs.

9.4 The Origin of SXT^{MO10}-Related ICEs

The origins of SXT^{MO10} and the closely related *V. cholerae*-derived ICEs mentioned above were obscure in 1996 when these ICEs were first described in *V. cholerae* isolates on the Indian subcontinent [18]. The widespread appearance of SXT^{MO10}-related ICEs in Asian *V. cholerae* populations in the early 1990s was geographically and temporally correlated with the emergence of *V. cholerae* O139 Bengal. Although it is clear that SXT^{MO10}-related ICEs were not common in toxigenic Asian *V. cholerae* O1 strains prior to the emergence of *V. cholerae* O139 Bengal [20], it became evident that ICEs similar to SXT^{MO10} were present in Asian *V. cholerae* populations prior to the appearance of epidemic *V. cholerae* O139. This is perfectly exemplified by ICE*Vch*Vie0 identified in a 1990 pre-O139 Vietnamese strain of *V. cholerae* O1 El Tor [27]. Even earlier, such an ICE has been detected in a 1987 non-O1/O139 environmental strain of *V. cholerae* from India (strain VO222 from Varanasi sewage) [30].

SXT^{MO10}-related ICEs also appear to have been present in African vibrios before and during the emergence of *V. cholerae* O139 Bengal. Recently Ceccarelli et al. [4] found SXT^{MO10}-related ICEs that do not mediate resistance to antibiotics in 1992 *V. cholerae* O1 clinical isolates from Angola (Table 9.1). A survey of Algerian *V. cholerae* O1 El Tor clinical isolates from 1994 suggests the presence of SXT^{MO10} -related ICEs mediating trimethoprim resistance [36] (Table 9.1). Thus, although SXT^{MO10}-related ICEs could not be detected in *V. cholerae* isolates from Zambia or Rwanda in the early 1990s [2, 31], they still may have been fairly widespread in African *V. cholerae* during that period. However, our knowledge of the distribution of *V. cholerae* ICEs in Africa remains limited for now.

The identification of SXT^{MO10}-related ICEs in vibrios from Asia, Africa, Europe and Central America likely reflects a long-term association of this group of ICEs with vibrios, as well as their spread. However, vibrios may not be the primary host of SXT^{MO10}-related ICEs. Hochhut et al. [37] discovered that SXT^{MO10} is genetically and functionally related to R391, a mobile element derived from a South African *Providencia rettgeri* (formerly known as *Proteus rettgeri*) strain isolated in 1967. R391 is the prototypical member of the IncJ element family (Table 9.1). IncJ elements were initially thought to be R factors, i.e., conjugative plasmids mediating drug resistance, that were isolated from *Proteus mirabilis*, *Proteus vulgaris*, *P. rettgeri*, *Shewanella putrefaciens* (formerly known as *Pseudomonas putrefaciens*) and *V. cholerae* (Table 9.1). Standard incompatibility testing of the members of the IncJ family showed that they co-existed stably with R factors of all known incompatibility groups but not with other IncJ elements. Based on considerable genetic and functional data [37–39], it is now clear that IncJ elements were misclassified as plasmids rather than ICEs. SXT^{MO10}-related ICEs and IncJ elements have now been unambiguously grouped within a unique family of ICEs recently designed as the SXT/R391 family [22].

Interestingly, the IncJ element pJY1 was identified in a *V. cholerae* O1 clinical isolate in the Philippines nearly 20 years before the emergence of *V. cholerae* O139 Bengal [40] (Table 9.1). Moreover, like SXT^{MO10} and ICE*Vch*Ind1, pJY1 mediates resistance to sulfamethoxazole, chloramphenicol and streptomycin (trimethoprim was not tested) and therefore could be an immediate precursor of these two ICEs. Unfortunately, pJY1 is no longer available, so examination of the genetic relationship between SXT^{MO10}, ICE*Vch*Ind1, and pJY1 is not possible.

Altogether, these data suggest that SXT/R391 family members have been present in several genera of γ-proteobacteria for extended periods. It is important to point out that our knowledge of the distribution and history of SXT/R391 ICEs is limited and biased. Till date, most surveys for such elements have been carried out in antibiotic-resistant clinical isolates. However, it is very likely that more comprehensive analyses would detect new members of the SXT/R391 family in many pathogenic and non-pathogenic γ-proteobacteria species worldwide.

9.5 SXT/R391 Family Members Possess a Large Core Set of Conserved Genes

Comparison of the complete DNA sequences of the genomes of SXT^{MO10} (99.5 kb) and R391 (89 kb) revealed that these ICEs share a conserved core of genes that mediates the essential functions of the elements, including their regulation, excision/integration and conjugative transfer [38, 41, 42]. Additional data provided by the recent sequencing of ICE*Spu*PO1 (110 kb) and ICE*Pda*Spa1 (103 kb) showed that altogether, these four ICEs share more than 90% identity at the nucleotide level over approximately 60 kb of sequences that constitute their functional 'backbone' [34, 43] (Fig. 9.3). A significant part of the conserved sequences are composed of genes of unknown function. Deletion of many of these genes did not impair SXT^{MO10} transfer indicating that they are not critical to the mobility of the SXT/R391 family members [41]. Furthermore, many of these genes are not present in the functional ICE*Vch*Mex1 providing additional evidence that they are not part of the minimal SXT/R391-type ICE [33] (Fig. 9.3). The *rumAB* operon, which encodes a proficient error-prone reparation system [44, 45] , appears to be part of the conserved 'backbone' as it is conserved in all the SXT/R391 ICEs that have been tested. Yet, in SXT^{MO10} and related ICEs from most if not all clinical strains, the *rumB* gene, which encodes the error-prone polymerase IV subunit [44], is disrupted by the antibiotic resistance gene cluster or remnants of this cluster that no longer confer any resistance (Fig. 9.2). On the contrary, environmental strains appear to have an intact *rumB*. This observation raises the possibility that a functional *rumAB*

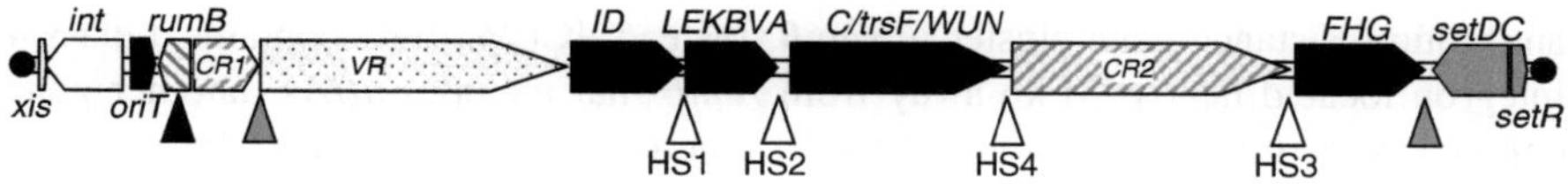

Fig. 9.3 Schematic representation of SXT/R391 ICE genome structure. *Arrows* represent genes or clusters of genes that are likely to be organized in operon. In each cluster, only the most relevant genes or sequences are indicated. *White arrows* indicate the region involved in maintenance (*xis* and *int*). *Black arrows* represent the five *tra* regions involved in conjugative transfer, i.e., the mating apparatus formation and DNA processing function. *Gray arrows* show the region involved in regulation (*setDCR*). *Hatched arrows* indicate conserved regions of known (*rumAB*) or unknown (CR1 and CR2) function. *Dotted arrows* represent a region that is present in SXT^{MO10}, R391, ICE*Pda*Spa1, and ICE*Spu*PO1 but not in ICE*Vch*Mex1 (VR). *White triangles* indicate the position of hotspots for insertions into the conserved 'backbone' (HS1–HS4). The *gray triangles* indicate the insertion sites of the kanamycin resistance-conferring transposon (between CR1 and VR) and mercury resistance-conferring genes (between *traG* and *eex*) in R391. The *black triangle* indicates the position of the antibiotic resistance gene cluster in *rumB* in SXT^{MO10} and many other SXT/R391 ICEs (Fig. 9.2). The *black circles* represent the *attL* and *attR* recombination sites resulting from the integration into the -chromosome

system may be detrimental while colonizing the host, whereas it may be favorable to the survival of *V. cholerae* and related pathogens in the environment.

9.6 SXT/R391 Family Members Also Possess Variable Regions

Besides their shared 'backbone', SXT/R391 ICEs also carry diverse variable regions that confer ICE-specific properties such as antibiotic resistance. Antibiotic resistance genes are often associated with dynamic structures like insertions sequences, transposons and integrons. In SXT^{MO10}, the genes encoding the resistance to cotrimoxazole, streptomycin and chloramphenicol are clustered together within a ~17-kb structure that resembles a composite transposon inserted into *rumB* [19] (Fig. 9.2). Comparison of the *rumB* locus of a diverse set of SXT/R391 ICEs reveals the considerable plasticity of the antibiotic resistance gene cluster (Fig. 9.2). All of these clusters seem to have diverged from a common ancestor resulting from *rumB* disruption by a composite transposon or insertion sequence. At this time, it is not possible to determine whether the common ancestor structure was initially large or small. However, analysis of the current antibiotic resistance clusters indicates that, in some strains, they underwent reduction when the genes mediating resistance were not under selection. For instance, such reduction event likely occurred in ICE*Vch*Vie0 [27] (Fig. 9.2). In contrast, the antibiotic gene cluster may undergo addition of antibiotic resistance genes mediated by transposition of insertion sequences or homologous recombination, as it likely happened in SXT^{MO10} (addition of *dfr18*) [19] or ICE*Vch*Vie1 (addition of *tetAR*) [28] when compared to ICE*Vch*Ind4 (Fig. 9.2).

Kanamycin resistance in R391 is also mediated by a transposon that is inserted a few kilobase away upstream of the intact *rumAB* operon [42] (Fig. 9.3). In ICE*Vch*Ind1, the gene *dfr18* mediating trimethoprim resistance is absent from the antibiotic resistance gene cluster in *rumB*. Instead, ICE*Vch*Ind1 contains a class 4 integron located nearly 70 kb away from *rumB* that includes *dfrA1*, another gene conferring resistance to trimethoprim [19]. Therefore, there appears to be a significant flux of genes mediating resistance to antibiotics in the SXT/R391 family. There are also diverse sets of genes inserted in intergenic regions of many SXT/R391 ICEs. The mechanism of insertion of these gene clusters into the conserved 'backbone' has not yet been discovered and no signature of transposable elements or sequence repeats has been found. Most of these insertions, which contain totally unrelated sequences, are localized in four sites that appear to correspond to insertion 'hotspots' identified by comparative sequence analysis of SXT^{MO10} and R391 genomes [38]. The recently described ICE*Spu*PO1 and ICE*Pda*Spa1 also contain ICE-specific DNA in three out of four of these hotposts [34, 43] (Fig. 9.3). These hotspots are located between *s043* and *traL* (Hotspot 1), *traA* and *s054* (Hotspot 2), *s072*/*s073* and *traF* (Hotspot 3), and *traA* and *s063* (Hotspot 4) [22, 34, 38]. Insertions at these locations separate the highly conserved region that encodes the conjugative transfer functions in four clusters without disturbing the stability and mobility of the SXT/R391 ICEs. In most cases, the function of the genes or group of genes that are inserted in these hotspots is unknown. In ICE*Vch*Ind1, the class 4 integron coding for the trimethoprim resistance is inserted in Hotspot 3 [19]. The insert in Hotspot 2 of ICE*Spu*PO1 carries genes encoding a putative cation efflux protein as well as a putative broad specificity efflux pump for drugs, solvent, and other substrates including antibiotics [34]. Acquisition of an ICE with such potential multiply drug-resistant phenotype by pathogens such as *V. cholerae* could have dreadful consequences on antibiotic therapies.

9.7 Conjugative Transfer and Regulation of the ICEs of the SXT/R391 Family

Based on the similarity of the members of the SXT/R391 family and the apparent conservation of a genetic 'backbone' that ensures the functions essential for conjugative transfer, it is thought that these ICEs share the same life cycle. Therefore, an overview of the steps that result in the transfer of the SXT/R391 ICEs from a donor cell to a recipient cell is presented here using SXT^{MO10} as a model, since the most remarkable progress in functional characterization have been made while studying this ICE.

SXT^{MO10} transfer takes place in three key steps: (i) excision from the host chromosome and formation of a circular intermediate, (ii) conjugative transfer of this substrate to a new host, and (iii) integration of the transferred molecule into the chromosome of the new host (Fig. 9.1).

9.7.1 Excision and Integration

The mechanisms of integration and excision of SXT^{MO10} are reminiscent of those of lambdoid phages. SXT^{MO10} integrates site specifically into the 5′-end of *prfC*, the gene coding for peptide chain release factor 3 (RF3) found on *V. cholerae* chromosome I [46]. Integration of SXT^{MO10} disrupts the 5′-end of *prfC* but provides a novel promoter and 5′ coding sequence that enables the expression of a functional RF3 protein. The integration event is a *recA*-independent process that requires an SXT^{MO10}-encoded tyrosine recombinase Int, which belongs to the λ family of site-specific recombinases. Int catalyzes site-specific recombination between 17-bp nearly identical element (*attP*) and chromosomal sequences (*attB*) [46].

Like λ, integrated SXT^{MO10} excises from the chromosome to form a covalently closed circular extrachromosomal molecule. The excised molecule results from recombination between short sequences, *attL* and *attR*, that are nearly identical and located on each side of the integrated ICE. The excision event regenerates *attP* on the excised circular SXT^{MO10} and *attB* on the chromosome. Int is required but not sufficient to promote the efficient recombination between *attL* and *attR*. Another protein Xis that acts as a recombination directionality factor is also needed for efficient SXT^{MO10} excision [47]. The R391 *xis* gene, *orf4*, is virtually identical to the SXT^{MO10} *xis* and has been shown to have the same function, suggesting that all the ICEs of the SXT/R391 family excise via a similar, if not identical, mechanism [47]. This result has also been confirmed by O'Halloran et al. [48], who confusingly renamed R391 *xis* as *jef*. The discovery of SXT^{MO10} and R391 Xis led to the identification of a previously unrecognized family of putative recombination directionality factors, which are known to be extremely divergent proteins [47, 49]. Sequence comparison of *xis* genes from seven SXT/R391 ICEs suggests that all these ICEs harbor very similar *xis* genes sharing at least 97% identity, encoding identical proteins and grouping into a novel subfamily of recombination directionality factors [48]. Interestingly, unlike λ for which host-encoded nucleoid protein IHF is required for integration into, and IHF and Fis are required for excision from *Escherichia coli* chromosome, neither protein is required for efficient integration or excision of SXT^{MO10} in *V. cholerae* [50]. However, as discussed below, IHF is a critical factor for SXT^{MO10} transfer to and from *V. cholerae*.

9.7.2 Conjugative Transfer

Once excised, the circular SXT^{MO10} molecule appears to be the substrate for conjugative transfer. Indeed, virtually no transfer of SXT^{MO10} is observed from donors that cannot generate the excised circular intermediate [46, 47]. It is thought that SXT^{MO10} transfers to the recipient cell as a single-strand based on analogy to most conjugative plasmids, significant homology of the predicted SXT^{MO10} Tra proteins that constitute its conjugative apparatus with the Tra proteins encoded by plasmids pCAR1 from *Pseudomonas resinovorans* [51], Rts1 from *P. vulgaris* [52], R478 from *Serratia marcescens* [53] and F from *E. coli* [41] as well as the presence of a TraI (nickase) orthologue. Like F in an *Hfr* donor strain, SXT^{MO10} is able to mobilize chromosomal DNA located 3′ of its integration site [54], a behavior that could have an important impact on the plasticity of the genome of pathogens such as *V. cholerae*.

The SXT^{MO10} *tra* genes are well conserved among ICEs of the SXT/R391 family: the *tra* genes of SXT^{MO10}, R391 from *P. rettgeri*, ICE*Spu*PO1 from *S. putrefaciens* and ICE*Pda*Spa1 from *P. damselae* subsp. *piscicida* have identical organization and encode nearly identical amino acid sequences (>94% identity) [34, 38, 43]. Also, the available sequences of R997 and pMERPH *tra* genes are very similar to those in SXT^{MO10} and R391 [39]. The *tra* genes are arranged in four clusters that span more than 25 kb [34, 41–43] (Fig. 9.3). The first gene cluster appears to contain the sequences necessary for transfer initiation and DNA processing, namely *traI*, which encodes a putative nickase (or relaxase), *traD*, which encodes a putative coupling protein, and a 560-bp downstream sequence that was originally identified as the origin of transfer (*oriT*) [38]. However, the *oriT* of all SXT/R391 ICEs was later on undoubtedly relocated in the *s003-rumB* intergenic region as a 299-bp sequence (38a). Additionally, a small new gene, *mobI*, located downstream of *oriT* was identified and shown to be absolutely required for conjugative transfer of these ICEs but not for the SXT^{MO10}-mediated mobilisation of pCloDF13, suggesting that the protein encoded by *mobI* is not part of the mating pore but instead likely recognizes *oriT* (38a). The mating pair formation functions seem to be distributed in three gene clusters: (i) *traLEKBVA*, (ii) *traC/trsF/traWUN*, and (iii) *traFHG* (38). The mechanism allowing simultaneous excision of SXT^{MO10} and synchronous expression of these four gene clusters prior to conjugative transfer is currently not well understood.

9.7.3 Conjugative Transfer Entry Exclusion

Similar to F and other conjugative plasmids, SXT/R391 family members carry genes for an entry exclusion system mediated by two inner membrane proteins, TraG in the donor cell and Eex in the recipient cell [23, 55, 56]. Entry exclusion systems function to specifically inhibit redundant conjugative transfers between donor cells that carry the same element. Marrero and Waldor [55] showed that even though SXT^{MO10} and R391 have nearly identical conjugative transfer genes, these ICEs do not exclude each other; cells harboring SXT^{MO10} inhibit the acquisition of SXT^{MO10} but not R391 and vice versa. Eex variants EexS and EexR, and TraG variants $TraG_S$ and $TraG_R$ encoded by SXT^{MO10} and R391, respectively, mediate element-specific exclusion activity. *traG* and *eex* genes of 21 ICEs belonging to the SXT/R391 family have recently been characterized both genetically and functionally [23]. Unexpectedly, both genes segregate unambiguously into two exclusion groups, called S and R (Table 9.1). In each ICEs, the *traG* gene belonging to one of the two exclusion groups pairs with the *eex* gene of the same exclusion group, likely reflecting the functional interaction between TraG and Eex that mediate exclusion. Consistent with earlier finding for SXT^{MO10} and R391 [55], ICEs harboring an exclusion system of the S group exclude acquisition of ICEs belonging to the same exclusion group but not the ones belonging to the R group and vice versa [23].

Interestingly, while the sample size of tested ICEs was relatively small, Marrero and Waldor [23] found a strict correlation between the exclusion group to which the

ICEs belong and the serogroup of *V. cholerae* strains from which the ICEs were originally isolated. Hence, ICEs from *V. cholerae* O139 isolates belong to the S group, whereas ICEs from *V. cholerae* O1 El Tor isolates belong to the R group. Such finding suggests that more than one SXT/R391-related ICE may have been acquired by pathogenic *V. cholerae* in relatively recent times, as was also suggested earlier by Hochhut et al. [19]. It also suggests that ICEs of the SXT/R391 family might not transfer frequently, if ever, between *V. cholerae* strains in the environment or in the human intestine. This hypothesis is supported by the very low frequency of transfer of SXT^{MO10} from *V. cholerae* in laboratory conditions that remains below 2×10^{-7} exconjugant/recipient, whereas it reaches up to 1×10^{-4} exconjugant/recipient from *E. coli* donors [18]. One could wonder what could be the benefits for ICEs of the SXT/R391 family of bearing an entry exclusion system when the transfer rate is already so low. A possible answer is that *V. cholerae* might not be the primary host of SXT^{MO10} and that these ICEs might not be well adapted to this bacterium.

9.7.4 Regulation

Initial characterization of SXT^{MO10} properties reported that its conjugative transfer requires *recA* [18]. This finding was conflicting with other observations showing that SXT^{MO10} integration was not mediated by RecA but instead was a site-specific recombination event mediated by the SXT^{MO10}-encoded tyrosine recombinase Int [46]. In fact, subsequent studies revealed that the molecular pathway that controls SXT^{MO10} transfer is similar to the pathway that governs the lytic development of phage λ upon activation of the SOS response to DNA damages. An SXT^{MO10} -encoded repressor, SetR, that bears significant similarity to λ cI repressor appears to be at the top of a regulatory cascade controlling SXT^{MO10} excision and transfer. SetR binds to an intergenic region between the divergently transcribed *s086* and *setR* genes [57]. It represses expression from a promoter upstream of *s086*, which appears to be the first gene of an operon that includes *setC* and *setD*, two genes that encode the key activators of SXT^{MO10} excision and transfer [41]. *SetC* and *SetD* activate transcription of both *int* and the four *tra* operons described above. Activation of transcription of R391 *xis* has recently been shown to be *setDC*-dependent as well [47, 48].

Like the λ cI repressor, SetR contains both a helix-turn-helix DNA-binding motif and a protease motif [58]. A marked increase in SXT^{MO10} transfer is observed when donor cells are grown in the presence of DNA-damaging agents that induce the SOS response such as mitomycin C or ciprofloxacin, a widely used fluoroquinolone antibiotic [58]. After DNA damage has stimulated the coprotease activity of RecA, activated RecA appears to promote the autoproteolysis and inactivation of SetR [58]. Depletion of the intracellular pool of SetR alleviates repression of *setC* and *setD* expression, increasing the expression of the genes necessary for SXT^{MO10} transfer. The mechanisms of regulation of *setR* expression are incompletely understood [57], especially in the context of the post-transfer restoration of repression in donor cells and the establishment of SXT^{MO10} in fresh exconjugant cells.

Detection of nearly identical *setR*, *setC*, and *setD* in all characterized SXT/R391 ICEs suggests that the basic scheme governing regulation of SXT^{MO10} transfer applies to all of them. In fact, the SOS response has been shown to induce the transfer of several SXT-related ICEs [33, 59]. Yet, some variations in the regulation of transfer of these ICEs exist. For example, R391 can transfer at a lower frequency in a *recA*-independent fashion [59, 60]. Moreover, significant differences in the basal transfer rate have been reported for several SXT/R391 ICEs [59]. In addition to RecA, another host factor has been recently shown to influence SXT^{MO10} transfer, yielding new understanding of how host cell physiology governs ICE transfer. McLeod et al. [50] showed that the host-encoded nucleoid protein IHF is required for efficient transfer of SXT^{MO10} from and to *V. cholerae* but not from and to *E. coli*. This requirement is not correlated with impaired integration or excision of SXT^{MO10}, instead it is as if *V. cholerae* IHF mutant has a general defect in conjugation since efficient transfer of the broad host range conjugative plasmid RP4, which does not integrate into the chromosome, is also IHF dependent in *V. cholerae* but not in *E. coli* [50]. The precise role of *V. cholerae* IHF in the process of conjugative transfer of SXT^{MO10}-related ICEs has yet to be determined.

9.8 Do ICEs of the SXT/R391 Family Mobilize *Vibrio cholerae* Virulence Determinants?

Publication in 2000 by Heidelberg et al. [61] of the whole DNA sequence of both chromosomes of *V. cholerae* O1 El Tor N16961 has been an important step for studying how horizontal gene transfer has shaped this pathogen's genome. Comparative genomics with currently sequenced whole genome of many other environmental and clinical strains of *V. cholerae* is likely to provide additional data to investigate further the extent and consequences of these gene exchanges and the mechanisms that drive them. In several pathogens such as *Staphylococcus aureus* or *Yersinia* spp. we now begin to understand how virulence determinants or so-called pathogenicity island moves from a virulent strain to an avirulent one and transforms it into a new pathogen [62–64].

Important discoveries have been made in this field in *V. cholerae* as well. Indeed CTX and VPI are two main genetic determinants required in *V. cholerae* strains to cause cholera (Fig. 9.4). The *CTX* gene cluster encodes the cholera toxin (CT) responsible for the profuse and severe secretory diarrhea typical of the disease. Waldor and Mekalanos [65] discovered in 1996 that CTX is the genome of a filamentous phage, CTXϕ, able to convert a non-toxigenic *V. cholerae* strain into a CT-producing strain. Besides CT, CTXϕ also encodes the accessory cholera enterotoxin (Ace), which encodes a minor coat component, and the zonula occludens toxin (Zot), which is thought to be involved in CTXϕ assembly and secretion [66, 67]. The large pathogenicity island VPI is a key virulence determinant in *V. cholerae* as it carries the toxin-coregulated pilus (TCP) gene cluster that encodes a type-IV pilus ensuring two critical functions for pathogenicity. Firstly, TCP acts as a CTXϕ

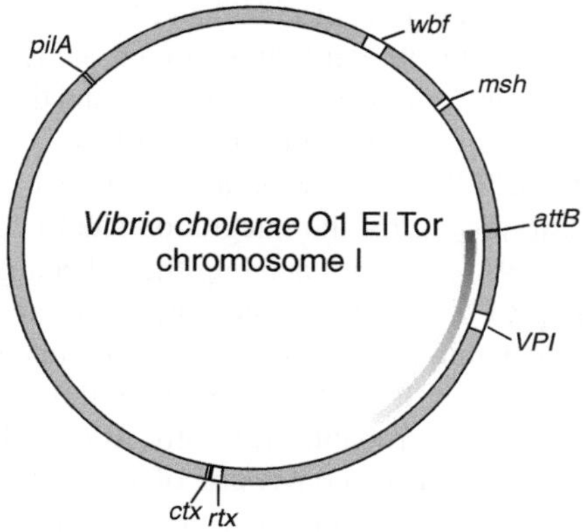

Fig. 9.4 Relative position of pathogenicity determinants on *V. cholerae* O1 El Tor N16961 chromosome I. The clusters of genes encoding toxin of colonization factors are represented by *white boxes*. *attB* indicates the position of the 5′-end of *prfC*, the integration site of SXT^{MO10}, and related ICEs. *wbf* indicates the position of the locus encoding O1 antigen; Wbe, which encodes the O139 antigen, locates at the same relative position in *V. cholerae* O139 strains. *msh* indicates the position of the locus encoding a maltose-sensitive hemagglutinin, a type-IV pilus that plays a role in biofilm formation. The positions of *ctx*, *rtx*, and *VPI* are also indicated. The region of chromosome I that SXT^{MO10} would be able to mobilize in an Hfr manner according to data obtained in *E. coli* is represented by a gradient bar spanning 500 kbp

receptor, therefore it is required for lysogenic conversion of *V. cholerae* into a toxigenic strain [65]. Secondly, it is an essential intestinal colonization factor that might cause the bacteria to aggregate, protecting them from being exposed to and killed by host factors in the intestine [68–70]. VPI is also thought to be the genome of another filamentous phage VPIϕ that seems to use TcpA, the major pilin subunit of TCP, as a coat protein to produce infectious particles [71].

Other toxin and colonization factor encoding genes of *V. cholerae* are known to be exchanged between strains based on their G+C content and trinucleotide composition. For instance, the RTX toxin gene cluster, the *pilA* gene cluster that encodes the prepilin peptidase PilD involved in processing of protein complexes with type-IV prepilin-like signal sequences such as TCP and the region that encodes the O antigen are thought to be horizontally transmitted within *V. cholerae* populations [16, 61, 72] (Fig. 9.4). Until recently, no mechanism that could explain genetic exchange or acquisition of these virulence determinants has been proposed. Therefore, it is very tempting to speculate that the ICEs of the SXT/R391 family could be involved in the mobilization of one or several of these gene clusters. Such hypothesis is relevant since SXT^{MO10} has been found to mobilize from *E. coli* not only plasmids in *trans* but also chromosomal DNA in an Hfr-like manner [54]. Moreover, ICE*Pda*Spa1 has recently been shown to mobilize in *cis* the non-conjugative virulence plasmid pPHDP10 by forming a cointegrate prior to transfer [34]. pPHDP10 encodes the toxin AIP56, an important *P. damselae* virulence factor that induces apoptosis in fish macrophages and neutrophils.

Early work prior to sequencing of SXT^{MO10} and *V. cholerae* genomes suggested that the region encoding the O139 serogroup antigen, *wbe*, was not carried by or closely linked to SXT^{MO10} in *V. cholerae* O139 Bengal and that transfer of SXT^{MO10} does not generally lead to acquisition of the O139 antigen [18]. This was confirmed later by the analysis of SXT^{MO10} and *V. cholerae* N16961 genomes [41, 61]. As noticed by Hochhut et al. [54], *attB*, the integration site of SXT^{MO10}, maps about 460 kbp 3′ of *wbf*, the DNA segment encoding the O1 antigen on chromosome I of *V. cholerae*, whereas SXT^{MO10} seems to be capable of mobilizing only chromosomal DNA that extends within about 500 kbp 3′ of its integration site (Fig. 9.4). Therefore, replacement of *wbf* in *V. cholerae* O1 El Tor through SXT^{MO10}-mediated acquisition of *wbe* is highly improbable. Recently, Blokesch and Schoolnik [73] demonstrated O1-to-O139 serogroup conversion of *V. cholerae* on crab shell mediated by chitin-induced natural transformation confirming that acquisition of *wbe* by the ancestor of *V. cholerae* O139 Bengal did not require SXT^{MO10}. Therefore, the emergence of the multiply drug-resistant *V. cholerae* O139 Bengal in India in 1992 as the first non-O1 serogroup to give rise to epidemic cholera likely occurred as a multi-step process.

In fact, most of the genes encoding virulence or colonization factors are located out of the range of DNA sequences that the ICEs of the SXT/R391 family are capable of mobilizing in an Hfr-manner; they map either 5′ or more than 500 kbp 3′ of *attB*[54] (Fig. 9.4). Other virulence determinants mapping on chromosome II, such as *hap*, which encodes the hemagglutinin protease, *hlyA*, which encodes a secreted haemolysin, or the super-integron, are not physically linked to *attB* located on chromosome I [61]. As a consequence, their mobilization in an Hfr manner by SXT^{MO10} or related ICEs seems to be impossible.

As depicted in Fig. 9.3, only the pathogenicity island VPI is in a position that is favorable to its mobilization by ICEs of the SXT/R391 family as it maps about 200 kbp 3′ of *attB*. Hypothesizing that SXT^{MO10} or related ICEs could mobilize VPI in *trans* is not too far-fetched even though results published by Karaolis et al. [71] suggest that VPI can move as a filamentous phage. Indeed the real nature of VPI is not completely clear and whether or not VPIϕ is a genuine filamentous phage is still subject to debate [67, 74, 75]. Furthermore, the number of TCP^+ *V. cholerae* strains able to produce VPIϕ-transducing particles and the range of TCP^- *V. cholerae* strains capable of acting as recipients of VPIϕ transduction seem to be very limited. In fact, only the TCP^+ strain *V. cholerae* O1 El Tor N16961 has been shown to produce VPIϕ-transducing particles and only one TCP^- *V. cholerae* strain, an O10 serogroup isolate, has been shown to be a suitable recipient for VPIϕ transduction [71]. In addition, distribution of TCP has been primarily restricted to strains of the O1 and O139 serogroups [76, 77]. Whether this limited distribution is a consequence of the lack of genes encoding a VPIϕ receptor in most of the non-O1/O139 strains or of a specific advantage in the human intestine or in the environment of TCP^+ O1/O139 strains over TCP^+ non-O1/O139 *V. cholerae* strains still remains a mystery.

Mobilization of the VPI gene cluster by SXT^{MO10} or related ICEs could alleviate some of the barriers specific to VPIϕ transduction that have been discussed above, drawing attention to the potential for non-O1/non-O139 serogroups to acquire TCP.

Such TCP$^+$ exconjugants becoming proficient for CTXϕ transduction could ultimately turn into new toxigenic *V. cholerae* strains that could potentially cause epidemics and even pandemics.

Such a pessimistic scenario has not been reported to date, but unfortunately, it remains plausible. One of the reasons for which an explosive emergence of toxigenic non-O1/O139 *V. cholerae* strain has not occurred is that the presence of ICEs in *V. cholerae* appears to be a relatively recent event. Indeed, currently no convincing evidence indicates that *V. cholerae* is the primary host of the ICEs of the SXT/R391 family. The oldest isolate of *V. cholerae* containing a genuine SXT/R391 ICE is the 1973 Philippino *V. cholerae* O1 that contained pJY1 [22]. To date, no report of SXT/R391 ICEs found in more ancient *V. cholerae* strains has ever been published.

Another reason might be the disconcerting low rate of conjugative transfer of SXT^{MO10} from *V. cholerae* [18]. Low transfer rate might be limiting the occurrence of successful mobilization of chromosomal DNA between *V. cholerae* strains, and therefore the spread of TCP. Yet nothing is known about the rate of intraspecific transfer of SXT^{MO10} in the environment, i.e., whether it transfers more efficiently or at even lower frequencies. However, the clear segregation of S and R exclusion group within the O139 and O1 serogroups of *V. cholerae* as discussed above rules in favor of the second possibility. *Vibrio cholerae* may in fact be a dead end from the ICEs of the SXT/R391 family.

Mobilization in an Hfr manner of the pathogenicity island VPI by SXT^{MO10} or a related ICE has not been demonstrated or even investigated to date. Since Hfr transfer does not generally allow maintenance of the mobilizing element, finding direct evidence of such event in populations of TCP$^+$ clinical or environmental *V. cholerae* might be impracticable.

9.9 Concluding Remarks

Today, the contribution of SXT/R391 ICE members to the dissemination of antibiotic resistance genes among environmental and clinical isolates of *V. cholerae* makes no doubt. However, antibiotic resistance genes represent only a fraction of the numerous genes transferred by these mobile elements. Many genes that are not involved in ICE's mobility, maintenance or regulation and for which no known function can currently be attributed are carried by these ICEs. Some of these genes are strictly conserved among the ICEs of the SXT/R391 family, whereas others appear to be ICE specific. In fact, more than half of the genes of SXT^{MO10} codes for proteins of unknown function. Most of these genes are specific to SXT^{MO10} as gene clusters inserted in hotspots. Similar situations are observed for the SXT/R391 ICEs that have been well characterized, i.e., R391, ICE*Spu*O1, ICE*Pda*Spa1, and ICE*Vch*Mex1. Clearly, ICEs of the SXT/R391 family provide an adaptive framework to pathogens such as *V. cholerae* that facilitate further gene acquisition and exchange. It is likely that at least some of these ICE-specific genes influence the fitness of their bacterial host in the environment or in the host that they colonize, or both. The presence of ICEs in *V. cholerae* isolates that do not confer any antibiotic resistance favors this hypothesis. The absence of antibiotic resistance gene from ICE*Vch*HKo1 indicates that selective pressure to become and remain resistant to antibiotics is not the sole explanation for the dissemination and persistence of SXT/R391 ICEs in Asian *V. cholerae*. The availability of the complete genome sequences of an increasing number of SXT/R391 family members from diverse γ-proteobacteria is offering a unique opportunity to look at the horizontally transferred gene pool and its impact on the behavior, fitness, and evolution of a large variety of pathogens.

The emergence of the SXT^{MO10}-containing *V. cholerae* O139 Bengal is relatively recent. It seems as if, while the existence of SXT/R391 family is ancient, the presence of members of this family in the species *V. cholerae* is an event that occurred in modern times. If this is true, we may find ourselves at the dawn of very disturbing times during which multiply resistant *V. cholerae* strains with epidemic or pandemic potential could spread and persist over extended periods. While it seems that plasmids conferring multiply drug resistance are not very stable in *V. cholerae* over long periods, ICE-borne antibiotic resistance genes might be stabilized into the chromosome of *V. cholerae* being inherited for countless generations.

In the near future it might be worth looking for bacteria in the environment that could be latent or hidden high-frequency donors of SXT/R391 ICEs. Comparative genomics using metagenomes could give clues about such possible primary hosts of SXT/R391 ICEs. For instance, Blast analysis of the conserved gene *int* gives a hit with a sequence matching of 96% in the marine metagenome 822,858 (accession number AACY022104488), which is the environmental genome of the Sargasso Sea. The primary hosts of SXT/R391 ICEs could be unculturable bacteria that could have provided *V. cholerae* with the progenitor of SXT^{MO10}, SXT^{MO10} itself or related ICEs a multiple number of times. Such hypothesis could explain the sudden bursts of emerging multiply antibiotic-resistant isolates of *V. cholerae* O1 and O139 in different parts of the world. These transfers would have been greatly facilitated by the treatment of cholera cases with DNA-damaging antibiotics such as ciprofloxacin that are known to stimulate the SOS response in bacteria and have been shown to promote the activation of transfer of SXT^{MO10} and other related ICEs. Knowing that systematic use of antibiotics promotes not only selection of drug-resistant bacteria but also dissemination of antibiotic resistance genes, one can wonder if on the long term and at a large scale, the remedy is not making the disease worse.

Acknowledgements V.B. holds a Canada research chair on molecular biology, impact and evolution of bacterial mobile elements.

References

1. Sack DA, Sack RB, Nair GB, Siddique AK. Cholera. Lancet. 2004;363:223–33.
2. Mwansa JC, Mwaba J, Lukwesa C, Bhuiyan NA, Ansaruzzaman M, Ramamurthy T, Alam M, Balakrish Nair G. Multiply antibiotic-resistant *Vibrio cholerae* O1 biotype El Tor strains emerge during cholera outbreaks in Zambia. Epidemiol Infect. 2007;135:847–53.
3. Thungapathra M, Amita, Sinha KK, Chaudhuri SR, Garg P, Ramamurthy T, Nair GB, Ghosh A. Occurrence of antibiotic resistance gene cassettes *aac(6′)-Ib, dfrA5, dfrA12*, and *ereA2* in

class I integrons in non-O1, non-O139 *Vibrio cholerae* strains in India. Antimicrob Agents Chemother. 2002;46:2948–55.
4. Ceccarelli D, Salvia AM, Sami J, Cappuccinelli P, Colombo MM. New cluster of plasmid-located class 1 integrons in *Vibrio cholerae* O1 and a *dfrA15* cassette-containing integron in *Vibrio parahaemolyticus* isolated in Angola. Antimicrob Agents Chemother. 2006;50:2493–9.
5. Dalsgaard A, Forslund A, Sandvang D, Arntzen L, Keddy K. *Vibrio cholerae* O1 outbreak isolates in Mozambique and South Africa in 1998 are multiple-drug resistant, contain the SXT element and the *aadA2* gene located on class 1 integrons. J Antimicrob Chemother. 2001;48:827–38.
6. Burrus V, Pavlovic G, Decaris B, Guedon G. Conjugative transposons: the tip of the iceberg. Mol Microbiol. 2002;46:601–10.
7. Burrus V, Waldor MK. Shaping bacterial genomes with integrative and conjugative elements. Res Microbiol. 2004;155:376–86.
8. Nair GB, Ramamurthy T, Bhattacharya SK, et al. Spread of *Vibrio cholerae* O139 Bengal in India. J Infect Dis. 1994;169:1029–34.
9. Cholera Working Group. Large epidemic of cholera-like disease in Bangladesh caused by *Vibrio cholerae* O139 synonym Bengal. Lancet. 1993;342:387–90.
10. Waldor MK, Mekalanos JJ. ToxR regulates virulence gene expression in non-O1 strains of *Vibrio cholerae* that cause epidemic cholera. Infect Immun. 1994;62:72–8.
11. Popovic T, Fields PI, Olsvik O, et al. Molecular subtyping of toxigenic *Vibrio cholerae* O139 causing epidemic cholera in India and Bangladesh, 1992–1993. J Infect Dis. 1995;171:122–7.
12. Waldor MK, Mekalanos JJ. Emergence of a new cholera pandemic: molecular analysis of virulence determinants in *Vibrio cholerae* O139 and development of a live vaccine prototype. J Infect Dis. 1994;170:278–83.
13. Calia KE, Murtagh M, Ferraro MJ, Calderwood SB. Comparison of *Vibrio cholerae* O139 with *V. cholerae* O1 classical and El Tor biotypes. Infect Immun. 1994;62:1504–6.
14. Rhine JA, Taylor RK. TcpA pilin sequences and colonization requirements for O1 and O139 *Vibrio cholerae*. Mol Microbiol. 1994;13:1013–20.
15. Iredell JR, Manning PA. Biotype-specific *tcpA* genes in *Vibrio cholerae*. FEMS Microbiol Lett. 1994;121:47–54.
16. Karaolis DK, Lan R, Reeves PR. The sixth and seventh cholera pandemics are due to independent clones separately derived from environmental, nontoxigenic, non-O1 *Vibrio cholerae*. J Bacteriol. 1995;177:3191–8.
17. Nair GB, Shimada T, Kurazono H, et al. Characterization of phenotypic, serological, and toxigenic traits of *Vibrio cholerae* O139 Bengal. J Clin Microbiol. 1994;32:2775–9.
18. Waldor MK, Tschape H, Mekalanos JJ. A new type of conjugative transposon encodes resistance to sulfamethoxazole, trimethoprim, and streptomycin in *Vibrio cholerae* O139. J Bacteriol. 1996;178:4157–65.
19. Hochhut B, Lotfi Y, Mazel D, Faruque SM, Woodgate R, Waldor MK. Molecular analysis of antibiotic resistance gene clusters in *Vibrio cholerae* O139 and O1 SXT constins. Antimicrob Agents Chemother. 2001;45:2991–3000.
20. Amita, Chowdhury SR, Thungapathra M, Ramamurthy T, Nair GB, Ghosh A. Class I integrons and SXT elements in El Tor strains isolated before and after 1992 *Vibrio cholerae* O139 outbreak, Calcutta, India. Emerg Infect Dis. 2003;9:500–2.
21. Yamamoto T, Nair GB, Albert MJ, Parodi CC, Takeda Y. Survey of in vitro susceptibilities of *Vibrio cholerae* O1 and O139 to antimicrobial agents. Antimicrob Agents Chemother. 1995;39:241–4.
22. Burrus V, Marrero J, Waldor MK. The current ICE age: biology and evolution of SXT-related integrating conjugative elements. Plasmid. 2006;55:173–83.
23. Marrero J, Waldor MK. The SXT/R391 family of integrative conjugative elements is composed of two exclusion groups. J Bacteriol. 2007;189:3302–5.
24. Alam M, Hasan NA, Sadique A, Bhuiyan NA, Ahmed KU, Nusrin S, Nair GB, Siddique AK, Sack RB, Sack DA, Huq A, Colwell RR. Seasonal cholera caused by *Vibrio cholerae*

serogroups O1 and O139 in the coastal aquatic environment of Bangladesh. Appl Environ Microbiol. 2006;72:4096–104.
25. Agrawal G, Jalgaonkar SV, Jagtap PM, Kamlakar UP, Deogade NG. Emergence and re-emergence of *Vibrio cholerae* O139: an epidemiological study during 1993–2002 at Nagpur, Central India. Indian J Med Sci. 2003;57:155–7.
26. Faruque SM, Chowdhury N, Kamruzzaman M, Ahmad QS, Faruque AS, Salam MA, Ramamurthy T, Nair GB, Weintraub A, Sack DA. Reemergence of epidemic *Vibrio cholerae* O139, Bangladesh. Emerg Infect Dis. 2003;9:1116–22.
27. Bani S, Mastromarino PN, Ceccarelli D, Le Van A, Salvia AM, Ngo Viet QT, Hai DH, Bacciu D, Cappuccinelli P, Colombo MM. Molecular characterization of ICEVchVie0 and its disappearance in *Vibrio cholerae* O1 strains isolated in 2003 in Vietnam. FEMS Microbiol Lett. 2007;266:42–8.
28. Ehara M, Nguyen BM, Nguyen DT, Toma C, Higa N, Iwanaga M. Drug susceptibility and its genetic basis in epidemic *Vibrio cholerae* O1 in Vietnam. Epidemiol Infect. 2004;132: 595–600.
29. Iwanaga M, Toma C, Miyazato T, Insisiengmay S, Nakasone N, Ehara M. Antibiotic resistance conferred by a class I integron and SXT constin in *Vibrio cholerae* O1 strains isolated in Laos. Antimicrob Agents Chemother. 2004;48:2364–9.
30. Mantri CK, Mohapatra SS, Ramamurthy T, Ghosh R, Colwell RR, Singh DV. Septaplex PCR assay for rapid identification of *Vibrio cholerae* including detection of virulence and int SXT genes. FEMS Microbiol Lett. 2006;265:208–14.
31. Ceccarelli D, Bani S, Cappuccinelli P, Colombo MM. Prevalence of *aadA1* and *dfrA15* class 1 integron cassettes and SXT circulation in *Vibrio cholerae* O1 isolates from Africa. J Antimicrob Chemother. 2006;58:1095–7.
32. Ahmed AM, Shinoda S, Shimamoto T. A variant type of *Vibrio cholerae* SXT element in a multidrug-resistant strain of *Vibrio fluvialis*. FEMS Microbiol Lett. 2005;242:241–7.
33. Burrus V, Quezada-Calvillo R, Marrero J, Waldor MK. SXT-related integrating conjugative element in New World *Vibrio cholerae*. Appl Environ Microbiol. 2006;72:3054–7.
34. Osorio CR, Marrero J, Wozniak RA, Lemos ML, Burrus V, Waldor MK. Genomic and functional analysis of ICEPdaSpa1, a fish-pathogen-derived SXT-related integrating conjugative element that can mobilize a virulence plasmid. J Bacteriol. 2008;190:3353–61.
35. Juiz-Rio S, Osorio CR, de Lorenzo V, Lemos ML. Subtractive hybridization reveals a high genetic diversity in the fish pathogen *Photobacterium damselae* subsp. *piscicida*: evidence of a SXT-like element. Microbiology. 2005;151:2659–69.
36. Korichi MN, Belhocine S, Rahal K. Plasmides IncJ identifiés pour la première fois dans *Vibrio cholerae* El Tor. Med Trop. 1997;57:249–52.
37. Hochhut B, Beaber JW, Woodgate R, Waldor MK. Formation of chromosomal tandem arrays of the SXT element and R391, two conjugative chromosomally integrating elements that share an attachment site. J Bacteriol. 2001;183:1124–32.
38. Beaber JW, Burrus V, Hochhut B, Waldor MK. Comparison of SXT and R391, two conjugative integrating elements: definition of a genetic backbone for the mobilization of resistance determinants. Cell Mol Life Sci. 2002;59:2065–70.
39. Boltner D, Osborn AM. Structural comparison of the integrative and conjugative elements R391, pMERPH, R997, and SXT. Plasmid. 2004;51:12–23.
40. Yokota T, Kuwahara S. Temperature-sensitive R plasmid obtained from naturally isolated drug-resistant *Vibrio cholerae* (biotype El Tor). Antimicrob Agents Chemother. 1977;11:13–20.
41. Beaber JW, Hochhut B, Waldor MK. Genomic and functional analyses of SXT, an integrating antibiotic resistance gene transfer element derived from *Vibrio cholerae*. J Bacteriol. 2002;184:4259–69.
42. Boltner D, MacMahon C, Pembroke JT, Strike P, Osborn AM. R391: a conjugative integrating mosaic comprised of phage, plasmid, and transposon elements. J Bacteriol. 2002;184: 5158–69.

43. Pembroke JT, Piterina AV. A novel ICE in the genome of *Shewanella putrefaciens* W3-18-1: comparison with the SXT/R391 ICE-like elements. FEMS Microbiol Lett. 2006;264: 80–8.
44. Mead S, Vaisman A, Valjavec-Gratian M, Karata K, Vandewiele D, Woodgate R. Characterization of polV_{R391}: a Y-family polymerase encoded by rumA'B from the IncJ conjugative transposon, R391. Mol Microbiol. 2007;63:797–810.
45. Kulaeva OI, Wootton JC, Levine AS, Woodgate R. Characterization of the *umu*-complementing operon from R391. J Bacteriol. 1995;177:2737–43.
46. Hochhut B, Waldor MK. Site-specific integration of the conjugal *Vibrio cholerae* SXT element into *prfC*. Mol Microbiol. 1999;32:99–110.
47. Burrus V, Waldor MK. Control of SXT integration and excision. J Bacteriol. 2003;185: 5045–54.
48. O'Halloran JA, McGrath BM, Pembroke JT. The orf4 gene of the enterobacterial ICE, R391, encodes a novel UV-inducible recombination directionality factor, Jef, involved in excision and transfer of the ICE. FEMS Microbiol Lett. 2007;272:99–105.
49. Lewis JA, Hatfull GF. Control of directionality in integrase-mediated recombination: examination of recombination directionality factors (RDFs) including Xis and Cox proteins. Nucleic Acids Res. 2001;29:2205–16.
50. McLeod SM, Burrus V, Waldor MK. Requirement for *Vibrio cholerae* integration host factor in conjugative DNA transfer. J Bacteriol. 2006;188:5704–11.
51. Maeda K, Nojiri H, Shintani M, Yoshida T, Habe H, Omori T. Complete nucleotide sequence of carbazole/dioxin-degrading plasmid pCAR1 in *Pseudomonas resinovorans* strain CA10 indicates its mosaicity and the presence of large catabolic transposon Tn*4676*. J Mol Biol. 2003;326:21–33.
52. Murata T, Ohnishi M, Ara T, Kaneko J, Han CG, Li YF, Takashima K, Nojima H, Nakayama K, Kaji A, Kamio Y, Miki T, Mori H, Ohtsubo E, Terawaki Y, Hayashi T. Complete nucleotide sequence of plasmid Rts1: implications for evolution of large plasmid genomes. J Bacteriol. 2002;184:3194–202.
53. Gilmour MW, Thomson NR, Sanders M, Parkhill J, Taylor DE. The complete nucleotide sequence of the resistance plasmid R478: defining the backbone components of incompatibility group H conjugative plasmids through comparative genomics. Plasmid. 2004;52:182–202.
54. Hochhut B, Marrero J, Waldor MK. Mobilization of plasmids and chromosomal DNA mediated by the SXT element, a constin found in *Vibrio cholerae* O139. J Bacteriol. 2000;182:2043–7.
55. Marrero J, Waldor MK. Interactions between inner membrane proteins in donor and recipient cells limit conjugal DNA transfer. Dev Cell. 2005;8:963–70.
56. Marrero J, Waldor MK. Determinants of entry exclusion within Eex and TraG are cytoplasmic. J Bacteriol. 2007;189:6469–73.
57. Beaber JW, Waldor MK. Identification of operators and promoters that control SXT conjugative transfer. J Bacteriol. 2004;186:5945–9.
58. Beaber JW, Hochhut B, Waldor MK. SOS response promotes horizontal dissemination of antibiotic resistance genes. Nature. 2004;427:72–4.
59. McGrath BM, O'Halloran JA, Pembroke JT. Pre-exposure to UV irradiation increases the transfer frequency of the IncJ conjugative transposon-like elements R391, R392, R705, R706, R997 and pMERPH and is recA+ dependent. FEMS Microbiol Lett. 2005;243:461–5.
60. Murphy DB, Pembroke JT. Transfer of the IncJ plasmid R391 to recombination deficient *Escherichia coli* K12: evidence that R391 behaves as a conjugal transposon. FEMS Microbiol Lett. 1995;134:153–8.
61. Heidelberg JF, Eisen JA, Nelson WC Clayton RA, Gwinn ML, Dodson RJ, Haft DH, Hickey EK, Peterson JD, Umayam L, Gill SR, Nelson KE, Read TD, Tettelin H, Richardson D, Ermolaeva MD, Vamathevan J, Bass S, Qin H, Dragoi I, Sellers P, McDonald L, Utterback T, Fleishmann RD, Nierman WC, White O. DNA sequence of both chromosomes of the cholera pathogen *Vibrio cholerae*. Nature. 2000;406:477–83.

62. Novick RP, Subedi A. The SaPIs: mobile pathogenicity islands of *Staphylococcus*. Chem Immunol Allergy. 2007;93:42–57.
63. Lesic B, Carniel E. Horizontal transfer of the high-pathogenicity island of *Yersinia pseudotuberculosis*. J Bacteriol. 2005;187:3352–8.
64. Schubert S, Dufke S, Sorsa J, Heesemann J. A novel integrative and conjugative element (ICE) of *Escherichia coli*: the putative progenitor of the *Yersinia* high-pathogenicity island. Mol Microbiol. 2004;51:837–48.
65. Waldor MK, Mekalanos JJ. Lysogenic conversion by a filamentous phage encoding cholera toxin. Science. 1996;272:1910–4.
66. McLeod SM, Kimsey HH, Davis BM, Waldor MK. CTXphi and *Vibrio cholerae*: exploring a newly recognized type of phage–host cell relationship. Mol Microbiol. 2005;57:347–56.
67. Davis BM, Waldor MK. Filamentous phages linked to virulence of *Vibrio cholerae*. Curr Opin Microbiol. 2003;6:35–42.
68. Manning PA. The tcp gene cluster of *Vibrio cholerae*. Gene. 1997;192:63–70.
69. Herrington DA, Hall RH, Losonsky G, Mekalanos JJ, Taylor RK, Levine MM. Toxin, toxin-coregulated pili, and the toxR regulon are essential for *Vibrio cholerae* pathogenesis in humans. J Exp Med. 1988;168:1487–92.
70. Taylor RK, Miller VL, Furlong DB, Mekalanos JJ. Use of *phoA* gene fusions to identify a pilus colonization factor coordinately regulated with cholera toxin. Proc Natl Acad Sci USA. 1987;84:2833–7.
71. Karaolis DK, Somara S, Maneval DR, Jr., Johnson JA, Kaper JB. A bacteriophage encoding a pathogenicity island, a type-IV pilus and a phage receptor in cholera bacteria. Nature. 1999;399:375–9.
72. Fullner KJ, Mekalanos JJ. Genetic characterization of a new type IV-A pilus gene cluster found in both classical and El Tor biotypes of *Vibrio cholerae*. Infect Immun. 1999;67:1393–404.
73. Blokesch M, Schoolnik GK. Serogroup conversion of *Vibrio cholerae* in aquatic reservoirs. PLoS Pathog. 2007;3:e81.
74. Karaolis DK, Kaper JB. *Vibrio cholerae* TCP: a trifunctional virulence factor? Response. Trends Microbiol. 1999;7:393.
75. Lee CA. *Vibrio cholerae* TCP: a trifunctional virulence factor? Trends Microbiol. 1999;7: 391–2;discussion 393.
76. Faruque SM, Asadulghani, Saha MN, Alim AR, Albert MJ, Islam KM, Mekalanos JJ. Analysis of clinical and environmental strains of nontoxigenic *Vibrio cholerae* for susceptibility to CTXPhi: molecular basis for origination of new strains with epidemic potential. Infect Immun. 1998;66:5819–25.
77. Said B, Smith HR, Scotland SM, Rowe B. Detection and differentiation of the gene for toxin co-regulated pili (*tcpA*) in *Vibrio cholerae* non-O1 using the polymerase chain reaction. FEMS Microbiol Lett. 1995;125:205–9.
78. Toma C, Nakasone N, Song T, Iwanaga M. *Vibrio cholerae* SXT element, Laos. Emerg Infect Dis. 2005;11:346–7.
79. Matthew M, Hedges RW, Smith JT. Types of beta-lactamase determined by plasmids in gram-negative bacteria. J Bacteriol. 1979;138:657–62.
80. Hedges RW. R factors from *Proteus mirabilis* and *P. vulgaris*. J Gen Microbiol. 1975;87: 301–11.
81. Coetzee JN, Datta N, Hedges RW. R factors from *Proteus rettgeri*. J Gen Microbiol. 1972;72:543–52.
82. Hedges RW. R factors from Providence. J Gen Microbiol. 1974;81:171–81.
83. Peters SE, Hobman JL, Strike P, Ritchie DA. Novel mercury resistance determinants carried by IncJ plasmids pMERPH and R391. Mol Gen Genet. 1991;228:294–9.

Chapter 10
Small Molecule Signaling Systems in *Vibrio cholerae*

Rupak K. Bhadra, Sangita Shah, and Bhabatosh Das

Abstract *Vibrio cholerae*, the causative agent of Asiatic cholera, still remains a major public health problem in most of the developing countries. Despite tremendous effort given in developing immunotherapeutics, availability of the whole genome sequence, and transcriptional profiling data, still a safe, effective, and long-lasting cholera vaccine is not available. One probable reason could be that our knowledge about stress adaptive mechanisms of *V. cholerae* with special reference to its survival both under in vitro and in vivo conditions is still incomplete. Indeed, a rapidly growing number of reports indicate that various intra- and extracellular small signaling molecules play critical roles in overcoming environmental onslaughts and allowing the pathogen to survive, grow, and cause the disease. *Vibrio cholerae* has two most important small molecule-mediated signaling pathways, which are extracellular quorum-sensing and intracellular 3′,5′-cyclic diguanylic acid (c-di-GMP) signaling system. Apart from these two systems, the intracellular (p)ppGpp molecule, called alarmone, also helps the organism to survive under nutritional scarcity and other stresses. The universal second messenger cAMP, another important small molecule, and its receptor protein, called CRP, have been implicated in virulence gene expression in *V. cholerae*, which may be due to its well-known overall impact on bacterial metabolism. Since these small molecules are crucial for the bacterial survival, the organism has evolved with multiple backup systems. If we consider that virulence gene expression of *V. cholerae* in human host is an outcome of stress-combating mechanism, then convergence of different small molecule-mediated signaling pathways is highly expected, but knowing how it is achieved is a real challenging job. This review focuses on how much we have learnt about this fascinating aspect of signaling systems of *V. cholerae* during post-genome sequencing era.

R.K. Bhadra (✉)
Infectious Diseases and Immunology Division, Indian Institute of Chemical Biology, Jadavpur, Kolkata 700 032, India
e-mail: rupakbhadra@iicb.res.in

T. Ramamurthy, S.K. Bhattacharya (eds.), *Epidemiological and Molecular Aspects on Cholera*, Infectious Disease, DOI 10.1007/978-1-60327-265-0_10,
© Springer Science+Business Media, LLC 2011

10.1 Introduction

Bacteria are unicellular self-replicating organisms and unlike multicellular higher eukaryotes, they are devoid of any sense organ. In spite of this, bacteria can respond quite efficiently to various environmental onslaughts including chemical, physical, and nutritional stresses they face relentlessly during their survival and growth under host-associated state or in their natural environmental habitats. They can sense environmental signals, quickly process it intracellularly through signal transduction mechanisms, and combat the situation by modulating expression of specific set of genes for a particular stimulus. Being a single cell, bacteria have got several advantages, namely it can make large number of progeny within a very short period of time, can directly receive the signals from environment, and can form spores to protect it from adverse situations. Probably for these reasons, most of the bacteria including pathogens are still prevalent on the Earth even after their evolution thousand and thousand of years ago, indicating that bacteria are fittest to survive in any sort of adverse environmental situation. This extraordinary feat is achieved by expressing right genes in right time and in a right place against a particular environmental cue. To do this, microorganisms synthesize and utilize several small molecules for sensing and responding various extra- and intracellular signals [1]. They are able to scan a mixture of small molecules to access information about both their extracellular and intracellular physiological status, and based on this information, they continuously interpret their circumstances and can react rapidly to any environmental changes. The cholera pathogen *Vibrio cholerae*, a gram-negative, non-invasive diarrheal pathogen, is a typical example regarding bacterial adaptation in human intestine during pathogenic state as well as when they reside in natural aquatic environment as a free-living organism [2]. In this chapter we discuss about recent progress in our understanding about extra- and intracellular small molecule signaling systems in *V. cholerae*, which help the pathogen in management of stress, pathogenesis, biofilm formation, and cell density-dependent quorum sensing.

10.2 Intracellular Small Molecule Signaling Systems

Orchestrated expression of genes in response to a variety of extracellular signals enables unicellular microorganism to survive and grow in a particular niche. Two cyclic ribonucleotides, namely cyclic adenosine 3′,5′-monophosphate or cAMP (Fig. 10.1) and cyclic guanosine 3′,5′-monophosphate or cGMP, are universally used as second messengers in intracellular signal transduction pathways in different organisms. The intracellular levels of cAMP and cGMP are tightly controlled by their rate of synthesis catalyzed by the enzymes adenylyl cyclase and guanylyl cyclase, respectively, and by phosphodiesterases, which hydrolyze the molecules [3, 4]. While cAMP-mediated signaling is common in both prokaryotes and eukaryotes, cGMP does not seem to be used by bacterial cells [1]. cAMP together with its receptor protein, called cAMP receptor protein (CRP), plays a pivotal role in bacterial metabolism in enteric bacteria by precisely regulating the utilization of carbon

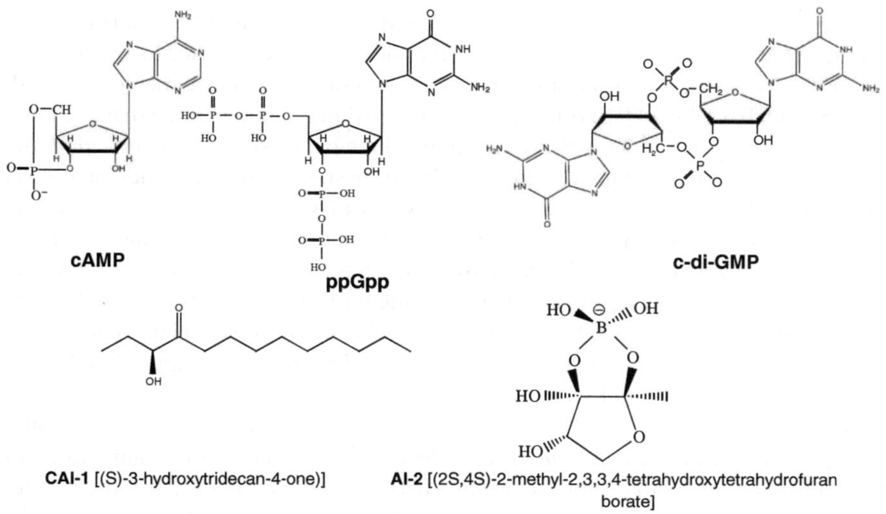

Fig. 10.1 Major small molecules involved in extra- and intracellular signaling in *V. cholerae*. Extracellular autoinducers CAI-1 and AI-2 are used for cell-to-cell communications and for intracellular signaling; cAMP, ppGpp, and c-di-GMP are the key players

and energy sources as shown under laboratory growth conditions [5, 3]. High intracellular cAMP leads the nucleotide to bind with CRP and changes its conformation followed by binding with DNA in a sequence-specific manner. In enteric bacteria, CRP functions as both positive and negative regulators of gene expression [3] and thus influences various cellular processes.

10.2.1 Intracellular cAMP-Mediated Regulation in Vibrio cholerae

Like the enteric bacteria *Escherichia coli* and *Salmonella enterica* serovar Typhimurium, *V. cholerae* genome also carries *cya* and *crp* genes which code for adenylate cyclase and CRP, respectively [6, 7]. Although at present very little information is available regarding the role of cAMP in the management of stress in *V. cholerae*, it has been reported that cAMP may be involved in virulence gene expression in this organism. It is now a well-established fact that the control of virulence gene expression in *V. cholerae* is strongly influenced by environmental conditions. Again when it infects human host, it faces plethora of stress-related stimuli and it is most likely that these multiple stimuli are the major determinants of expression of virulence factors. In *V. cholerae*, the two critical virulence factors are cholera toxin (CT) and toxin-coregulated pilus (TCP) and expression of both is tightly regulated by host-generated stimuli [8], although at present very little is known about the mechanism. Under in vitro conditions, it has been shown that the composition of growth media, its pH, salt concentration, and incubation temperature have profound influence in the expression of the major virulence factors CT and TCP by *V. cholerae* cells [6, 9]. Expression of these virulence factors is again under the coordinate control of three master regulators of virulence, namely TcpPH, ToxRS, and ToxT [9–11]. Skorupski and Taylor [6] first demonstrated that mutation of *cya* and *crp* genes of *V. cholerae* derepressed the expression of CT in *V. cholerae* at the non-permissive culture conditions. The study suggested that intracellular level of cAMP–CRP negatively regulates expressions of CT and TCP in *V. cholerae.* It was also shown by animal studies that mutation in the *crp* gene may influence pathogenesis of *V. cholerae* by regulating other processes necessary for optimal growth under in vivo conditions. The involvement of intracellular cAMP–CRP system in regulation of CT and TCP expression in *V. cholerae* suggests that carbon and energy sources in the environment may influence virulence gene expression, but the mechanism is currently not known.

Recently, Liang et al. [12] have reported that CRP is needed for fine regulation of quorum sensing in *V. cholerae.* In fact, a Δ*crp V. cholerae* strain expresses very low amount of the quorum-sensing master regulator HapR, leading to elevated expressions of *Vibrio* exopolysaccharide synthesis (*vps*)-related genes like *vpsA*, *vpsL,* and *vpsT*, leading to increase in biofilm formation. Since the activity of CRP is determined by intracellular cAMP (synthesized by the product of *cya* gene) levels, a comparative molecular analyses of phenotypes of the mutants like Δ*hapR,* Δ*crp*, and Δ*hapR* Δ*cya* indicated that cAMP–CRP system is involved in exopolysaccharide biosynthesis and rugose colony morphology by modulating HapR expression [13]. The rugose variant of *V. cholerae* has been shown to be more resistant to chlorinated water, osmotic, and oxidative stresses compared to the smooth morphotype [14–16]. Again, cAMP may positively influence expression of *V. cholerae*'s *rpoS*-encoded stationary phase σ^S factor, called RpoS, the master regulator of bacterial general stress response [17]. The cAMP–CRP complex activates the transcription of *rpoS* and *hapR*, the products of which in turn enhance hemagglutinin/protease, a putative virulence factor, which is needed during late phase of infection to detach *V. cholerae* cells from intestinal cell surface and thus allowing the pathogen to escape from host to environment. Although further research is needed to understand how intracellular cAMP level is influenced by environmental cues, which in turn intricately regulates gene functions related to quorum sensing, exopolysaccharide biosynthesis, smooth/rugose colonial morphology, and nutrient stress at stationary phase in *V. cholerae*, it is becoming quite apparent that cAMP is an important intracellular small molecule involved in signal transduction network.

10.2.2 Guanosine 3′-Diphosphate 5′-Triphosphate and Guanosine 3′,5′-Bis(diphosphate) [(p)ppGpp] as Cellular Alarmones in **Vibrio cholerae**

To respond to changing environmental conditions, particularly availability of nutrients, bacteria have evolved a multitude of cellular regulatory mechanisms. The global adaptive response to nutritional stress leads to rapid and complex cellular metabolic adjustments in bacterial cells known as stringent response and is characterized by the inhibition of rRNA synthesis, readjustment of metabolic pathways according to physiological requirements, and the induction of stationary phase genes needed for survival. The abrupt global changes associated with the stringent response are triggered by intracellular accumulation of two small molecules guanosine tetraphosphate (ppGpp, Fig. 10.1) and guanosine pentaphosphate (pppGpp), collectively called (p)ppGpp [18], which function as chemical messengers that permit bacteria to readjust their metabolism towards survival mode by inhibiting most of the growth-related pathways.

In *E. coli*, the stringent response has been well studied and primarily two genes, *relA* and *spoT*, have been identified as the genetic determinants responsible for the phenomenon. The *relA* gene codes for the enzyme (p)ppGpp synthetase and the *spoT* codes for a bifunctional enzyme with strong (p)ppGpp hydrolase and weak synthetase activities [18, 19]. Interestingly, RelA and SpoT respond differently to environmental conditions by some mechanisms, which are still not clearly understood. While RelA is activated during amino acid starvation, the synthetase activity of SpoT is activated upon carbon or energy starvation [18]. In vivo deletion mapping in *E. coli* concluded that the SpoT catalytic domains responsible for (p)ppGpp degradation and synthesis are localized within the N-terminal half of the protein on distinct but overlapping regions; amino acid residues 1–203 contain the domain for (p)ppGpp degradation, while residues 85–375 contain the domain for (p)ppGpp synthesis [20].

Like *E. coli*, *V. cholerae* genome also contains the *relA* (VC2451) and *spoT* (VC2710) homologs [7] and are involved in synthesis and hydrolysis of (p)ppGpp, respectively [21–24]. Recently, Das and Bhadra [21] provided evidences that there are fundamental differences in the metabolism of (p)ppGpp in *V. cholerae* compared with other bacterial systems, particularly the model gram-negative organism *E. coli*. Analyses of several Δ*relA* Δ*spoT* mutant strains of *V. cholerae* indicated the presence of RelA- and SpoT-independent (p)ppGpp synthesis pathway in *V. cholerae* [21]. The existence of a novel (p)ppGpp synthetase pathway in *V. cholerae* apart from the canonical genes *relA* and *spoT* is exciting and a completely new knowledge in this field of research. This finding also indicates that (p)ppGpp metabolism and its cellular homeostasis are really complex in *V. cholerae*. Like *V. cholerae*, the gram-positive organisms, namely *Streptococcus mutans* and *Bacillus subtilis* genomes, also contain novel genes coding for (p)ppGpp synthetases apart from canonical *rel* genes [25, 26].

10.2.2.1 Role of (p)ppGpp in *Vibrio cholerae*

RelA-dependent cellular accumulation of (p)ppGpp in *V. cholerae* upon amino acid starvation was first reported by Haralalka et al. [22]. During infection, *V. cholerae* cells pass through gastric acid barrier and encounter different types of stress that may be overcome by the wild-type cells due to fine regulation of stringent response effector molecules (p)ppGpp. Recently, Silva and Benitez [24] also studied the stringent response in *V. cholerae* and reported that the expression of virulence factors is not much affected in the *relA* mutant of *V. cholerae* cells. Since complete metabolic cycle of *V. cholerae* is not yet worked out, further studies are needed to find out the role of intracellular (p)ppGpp in regulation of nutritional and other stresses.

10.2.3 Bis-(3′,5′)-cyclic-di-guanosine Monophosphate (c-di-GMP)

The c-di-GMP (Fig. 10.1) has recently been recognized as an intracellular signaling molecule nearly universally used by bacteria [27, 28]. More than 20 years ago while studying regulation of cellulose synthesis in *Gluconacetobacter xylinus*, Ross et al. [29] discovered c-di-GMP as an intracellular signaling molecule. Importance of this cellular messenger in bacteria was further realized when whole genome sequence data of numerous microorganisms were analyzed [28]. Interestingly, genes coding for enzymes involved in synthesis and degradation of c-di-GMP are present unusually in high copy numbers in a particular organism and thus it is indeed a unique feature compared to other multicopy gene families. The c-di-GMP molecule is synthesized from two GTP molecules by the enzyme diguanylate cyclases (DGC) and it is degraded by c-di-GMP-specific phosphodiesterases (PDE), both of which are present in soluble or membrane-bound forms [30]. DGC activity is associated with Gly-Gly-Asp-Glu-Phe or Gly-Gly-Glu-Glu-Phe (GGDEF or GGEEF) domain-containing proteins, whereas c-di-GMP PDE activity has been shown for Glu-Ala-Lys (EAL) domain proteins [31, 32]. These domains, which were named for their most conserved amino acid sequence motifs, can occur either alone or in combination in a single polypeptide, and their activities seem to be controlled by various N-terminal sensory domains [28, 30]. The DGC and PDE domains containing proteins are found in most bacteria but they are absent in *Archaea* and eukaryotes [30]. GGDEF and EAL domains belong to the largest domain families with poorly understood molecular and physiological function in the protein databases [28, 33] and they often contain other known regulatory motifs like GAF domain responsible for binding with cyclic nucleotides and heme or PAS domain, which can detect light, redox potential, etc. It is believed that membrane-associated DGC and PDE proteins can relay external signals into the cytoplasm followed by activation/inactivation of intracellular c-di-GMP-mediated gene regulation and ultimately specific response to the external stimuli. Apart from EAL domain-containing proteins, the HD-GYP domain proteins have recently been shown to degrade c-di-GMP [34]. Altogether it appears that the metabolism of c-di-GMP in bacteria is complex.

10.2.3.1 c-di-GMP-Mediated Signaling in *Vibrio cholerae*

Analysis of the whole genome sequence of *V. cholerae* revealed that the organism possesses 31 genes encoding only for GGDEF domain-containing proteins, 10 genes encoding proteins with both GGDEF and EAL domains, and 12 genes encoding only EAL domain-containing proteins [35]. *Vibrio cholerae* genome also contains nine genes encoding proteins with HD-GYP domains [36]. Thus, it appears that a total of 62 genes encoding proteins with GGDEF, EAL, and HD-GYP domains are present in *V. cholerae* genome, which is astonishingly quite high compared to other bacteria. Therefore, it is a tremendous future challenge to establish the functions of each of these proteins and their exact role in *V. cholerae*. Toward this goal, recent research on c-di-GMP-mediated signal transduction pathways in *V. cholerae* indicates that this small intracellular molecule regulates the transcription of *vps* genes, the products of which are involved in the regulation of exopolysaccharide biosynthesis and biofilm formation [37]. *Vibrio cholerae* cells within biofilms are resistant to environmental stresses and biofilm-mediated attachment to abiotic surfaces may be critical for survival of *V. cholerae* in the environment [38, 39]. The quorum-sensing master regulator HapR, the product of the *hapR* gene, represses biofilm formation in *V. cholerae* [40–42] and thus the two processes are likely to be linked. In fact very recently, Waters et al. [43] have reported that HapR regulates the transcription of 14 genes encoding DGC and PDE proteins and the resultant effect is a reduction of intracellular c-di-GMP concentration at high cell density, leading to decrease in biofilm formation, indicating that c-di-GMP positively regulates biofilm formation in *V. cholerae*. They have also shown that HapR can directly bind to the promoter of *vpsT*, an important regulator of VPS production and biofilm formation. Again, it has been shown that c-di-GMP negatively regulates virulence gene expression in *V. cholerae* [44]. The response regulator VieA, an EAL domain-containing PDE protein, is needed to maintain a low cellular concentration of c-di-GMP in *V. cholerae* classical strain O395, leading to transcriptional activation of the principal virulence regulator ToxT followed by enhanced CT production. Thus, it is becoming increasingly clear that biofilm formation, quorum sensing, and virulence gene expression by *V. cholerae* cells are most likely the outcome of precise modulation of the intracellular levels of c-di-GMP through signal-response cascade, which also appears to be critical for the pathogen's ability to alternate between environment to the human host and again going back to the environment. Although at present it is not clear whether other intracellular signaling molecules like cAMP and/or (p)ppGpp are involved in c-di-GMP signal pathways, this is much likely since these three intracellular molecules are global modulators of gene expression under various environmental conditions and thus at some point the three signaling networks may converge.

The ability of *V. cholerae* to cause cholera epidemics is certainly dependent on its survival in aquatic environmental stress conditions. Under laboratory culture conditions, it has clearly been shown that the organism is capable of switching between two distinct morphological states, called rugose and smooth phases as discussed above. Lim et al. [37] have shown that intracellular level of c-di-GMP is quite high in rugose variant compared with the smooth variant. Furthermore, they also identified and characterized five differentially regulated genes, called *cdgA-E* for cyclic-di-guanylate, encoding proteins with GGDEF and/or EAL domains, which play critical roles in controlling rugosity, VPS production, motility, and biofilm formation. At present, however, it is not known how c-di-GMP promotes or represses the target genes. Amikam and Galperin [45], based on bioinformatics analysis, predicted that PilZ domain present in C-terminals of certain proteins could be the binding site of c-di-GMP. Recently, Ryjenkov et al. [46] and Pratt et al. [47] have tested this hypothesis in their laboratories using *E. coli* and *V. cholerae* as model organisms, respectively, and provided experimental evidences that the PilZ domain protein is indeed the intracellular receptor for c-di-GMP. Ryjenkov et al. [46] suggested that c-di-GMP binding to PilZ domain of the *E. coli* protein YcgR leads to conformational changes in that protein and c-di-GMP–YcgR complex is the active form, which initiates the downstream signal transduction cascade. They also provided evidence that the conserved motifs and amino acid residues are present in some PilZ domain proteins including YcgR, which serves as receptor for c-di-GMP. These motifs present in YcgR are ^{114}R*xxx*R^{118} and ^{145}D/N*z*S*x*GG^{151}, where *x* is any residue and *z* is a small hydrophobic residue. Moreover, site-directed mutational analysis indicated that the arginine residue present at 118 position of YcgR is essential for binding of c-di-GMP. Pratt et al. [47] reported that there are five PilZ-coding sequences in *V. cholerae* genome (VC0697, VC1885, VC2334, VCA0042, and VCA0735) and designated these genes as *plzA*, *plzB*, *plzC*, *plzD*, and *plzE*, respectively. They have shown that mutation of the PilZ genes *plzB*, *plzC*, and *plzD* leads to alterations in cellular processes known to be regulated by c-di-GMP, which include motility, biofilm formation, and virulence. However, they were able to show that only two *V. cholerae* PilZ proteins, PlzC and PlzD, bind specifically to c-di-GMP. Ryjenkov et al. [46] reported that like the *E. coli* YcgR, PlzD (VCA0042) protein of *V. cholerae* also contains the conserved motifs for c-di-GMP binding. However, at present the roles of other Plz proteins, namely PlzA, PlzB, and PlzE with respect to c-di-GMP binding, are not clear. Mutational analyses indicated that PlzA and PlzE coding genes are essential in *V. cholerae,* suggesting that certain PilZ domain-containing proteins may have some vital cellular functions. Although it has been discovered that c-di-GMP binds with specific motifs of a PilZ domain of Plz protein, how this complex influences gene expression is not known; even we do not know post-signal transduction fate of the complex, i.e., how the complex dissociates followed by degradation of c-di-GMP when a defined job of molecule is over.

10.2.3.2 c-di-GMP and In Vivo Gene Expression in *Vibrio cholerae*

Several lines of evidence indicate that c-di-GMP helps *V. cholerae* cells to adapt in host environment upon entry from environmental reservoirs. Upon entry into a human host, *V. cholerae* undergoes a quick shift from global environmental gene expression mode to virulence gene expression regime, particularly coordinated expression of CT and TCP and downregulation of *vps* genes required for biofilm formation. As discussed above, under in vitro growth conditions, high cellular level of c-di-GMP facilitates expression of *vps* genes and represses CT expression [37, 43, 48] (Fig. 10.2). Since *V. cholerae* has an intricate biofilm formation genetic network for its survival in diverse aquatic environmental niche and c-di-GMP is a critical intracellular determinant under such situation, it thus appears that host environment-generated cues most probably activate the c-di-GMP degradation pathway, leading to rapid decrease in intracellular level of the molecule and derepression of c-di-GMP-mediated various genetic circuits including virulence gene expression.

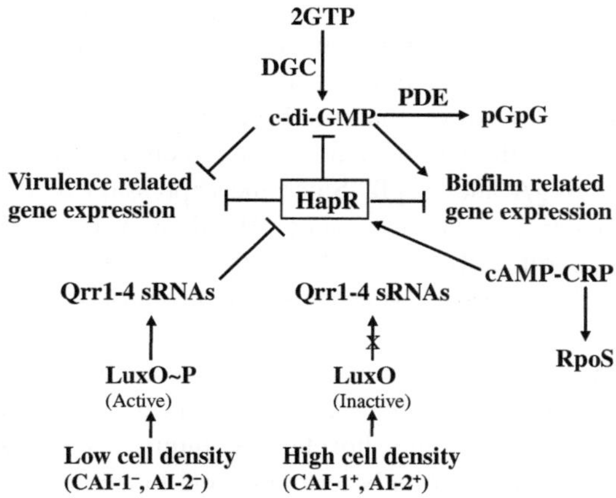

Fig. 10.2 Simplified diagram showing extra- and intracellular small molecule-mediated signaling circuits of *V. cholerae*. *Arrows* indicate positive regulation, while *T-bars* denote negative/inhibitory effects. Environmental cues activate DGC (diguanylate cyclase, GGDEF/GGEEF domain-containing proteins) or PDE (phosphodiesterase, EAL domain-containing protein) enzyme, which synthesizes or degrades c-di-GMP, respectively. c-di-GMP in turn positively and negatively regulates the genes involved in biofilm formation and virulence, respectively. Intracellular cAMP–CRP complex positively regulates the transcription of *hapR* and *rpoS* genes. When concentrations of AIs are low (CAI-1^-, AI-2^- conditions), the response regulator LuxO becomes activated by phosphorylation (LuxO~P) and positively influences expression of four small RNA (sRNA) genes *qrr1–4*, which encode Qrr1—4 RNA molecules. Now expressed Qrr1–4, sRNAs with the help of RNA-binding protein Hfq (not shown in the diagram) repress translation of the master transcriptional regulator HapR. Again, when concentrations of AIs are high (CAI-1^+, AI-2^+ conditions), LuxO is inactivated through dephosphorylation leading to drastic reduction of Qrr1–4 expression as denoted by "X" sign. In the absence of Qrr1–4 sRNAs, translational inhibition of *hapR* mRNA is no longer there and thus HapR is produced. Increase in cellular levels of HapR leads to repression of both biofilm formation and virulence factor production. From this diagram it appears that HapR plays a pivotal role in signaling cascade of *V. cholerae*

At present nothing is known about how the host-generated signals influence downregulation of c-di-GMP concentration. It may be mentioned here that passage through the gastric acid barrier is an essential step for *V. cholerae* cells to cause a productive infection and an adaptive response, called the acid tolerance response, allows the pathogen to survive acidic stress and facilitate colonization under in vivo conditions [49]. Again, Zhu and Mekalanaos [42] have shown that *V. cholerae* cells present in biofilms are much more resistant to killing by acid stress than are planktonic cells. Since the pathogen exists in the natural aquatic environment as biofilms [38] and the biofilm-associated *V. cholerae* cells have recently been shown to contain high concentration of c-di-GMP [37, 43], it is expected that acid stress confronted by these cells in initial phase of infection is an important signal leading to rapid reprogramming of gene expression, which may be achieved either through downregulation of c-di-GMP production or upregulation of EAL domain-containing PDE synthesis, which remains to be determined. Kovacikova et al. [50] have shown that

V. cholerae acgA and *acgB* genes code for proteins containing EAL and GGDEF domains, respectively, and overexpression of *acgA* caused decreased biofilm formation and increased motility, whereas overexpression of *acgB* had the opposite effect. Deletion of either gene, however, had no discernible effect on motility or biofilm formation. Interestingly, the virulence activator protein AphA negatively regulates the expressions of AcgA and AcgB. Since the expression of AphA is under the control of quorum-sensing master regulator HapR, it is at present not clear how c-di-GMP level is coordinately regulated with the overall scheme; still these results reveal a potential link between c-di-GMP-mediated control of biofilm formation and virulence gene expression in *V. cholerae*.

The *vieSAB* locus was identified during genetic screenings of *V. cholerae* genes, which were induced in the host environment, i.e., under in vivo situations [51, 52]. VieS is a polydomain sensor of the phosphorelay family that contains two periplasmic binding protein domains at its N terminus, and histidine kinase, receiver, and histidine phosphotransferase domains at its cytoplasmic C terminus. VieA is a protein that contains an N-terminal receiver domain, an EAL domain, a partial receiver domain, and a putative C-terminal helix-turn-helix (HTH) DNA-binding domain. VieB is a protein that contains both receiver and tetra tricopeptide repeat domains at its N terminus and a region with no similarity to known protein domains at its C terminus. Tischler and Camilli [48] found that expression of genes within the *vpsA-Q* locus was significantly decreased when VieA was overexpressed in classical biotype *V. cholerae*. Decreased *vpsA-Q* expression was a result of decreased transcription of *vpsR*, which encodes a positive regulator of the *vpsA-Q* genes, and the EAL domain of VieA was found to be necessary to prevent *vpsR* transcription. Mutational studies demonstrated that this effect, as well as the ability of VieSAB to induce *ctxAB* transcription [44], requires the PDE activity of the EAL domain, which this group also demonstrated biochemically [53]. Although the precise functions of the various domains of these proteins are yet to be determined, a model that fits the available data is one in which phosphorylation of the VieA receiver domain activates DNA binding, resulting in increased transcription of *vieA* (VieA appears to function as a positive autoregulator), and the resulting increased production of VieA decreases c-di-GMP levels through the PDE activity of its EAL domain, leading to decrease in biofilm formation. This may be further supported by the fact that overexpression of a DGC protein (VCA0956) was found to increase the VPS production and biofilm formation [48]. Furthermore, under in vivo condition, VieA PDE activity is also necessary for virulence in the infant mouse model [48] and expression of *vieSAB* operon occurs at an early stage of infection [52], strongly suggesting that c-di-GMP levels must be diminished for a productive infection as discussed above. It has been speculated that VieS may phosphorylate VieA in response to some as-yet unidentified signal present within the mammalian intestine. However, it is to be noted that unlike in classical biotype, the PDE protein VieA is not involved in the regulation of biofilm formation, motility, or virulence in the present pandemic strain of *V. cholerae* O1 El Tor biotype [54]; rather another in vivo-induced EAL protein (VC0130), called CdpA, has been identified [55]. Taken together, c-di-GMP-mediated gene expressions in *V. cholerae* cells under in vitro and in vivo conditions appear to be crucial and highly complex and future challenges are to determine how c-di-GMP concentration is precisely maintained in *V. cholerae* under a particular situation despite having so many copies of GGDEF/EAL/HD-GYP domain-coding genes and also to elucidate the molecular mechanisms of c-di-GMP-mediated gene expression in *V. cholerae*.

10.3 Extracellular Quorum Sensing in *Vibrio cholerae*

Quorum sensing is a process of cell-to-cell communication that bacteria use to assess their surrounding population density in order to coordinate the gene expression of the community [56]. Quorum sensing requires production, release, and detection of various extracellular signal molecules called autoinducers (AIs). In simplest scenario, accumulation of a threshold autoinducer concentration in surrounding environment, which is correlated with the increasing bacterial cell density, initiates a cascade of intracellular signal transductions that culminate in a population-wide adjustment of gene expression. Recent progress in quorum-sensing research on different organisms including pathogens indicates that quorum sensing modulates various bacterial cellular processes like bioluminescence, virulence, biofilm formation, antibiotic production, sporulation, and stress-related gene expression. In this article we will briefly discuss on quorum-sensing systems of *V. cholerae* so far known and also describe how various extracellular quorum-sensing signaling molecules intricately coordinate the gene expressions involved in stress management, virulence, biofilm formation, etc.

10.3.1 Quorum-Sensing Molecules and Pathways in Vibrio cholerae

Our present understanding on quorum sensing in *V. cholerae* is largely due to the elegant work carried out by Bassler's laboratory [41, 43, 57–60]. *Vibrio cholerae* has two major quorum-sensing systems (QSS): in system-1, cholerae quorum-sensing (Cqs) receptor present in the cell surface, called CqsS, interacts with the cholerae autoinducer-1 molecule or CAI-1 (Fig. 10.1), which is synthesized by an intracellular enzyme CqsA (CqsS/A system), and in system-2, the autoinducer molecule 2 (AI-2, Fig. 10.1), synthesized by the cytosolic enzyme LuxS, binds with the two-component LuxPQ cell surface receptors (LuxS/PQ system) for signal transduction. Apart from these confirmed QSS, experimental evidences indicated that two more QSS pathways (system-3 and system-4) might exist in *V. cholerae* [57]. However, currently the details of these two new QSS are not available. Presence of multiple QSS and their complex regulations most likely allow *V. cholerae* cells to survive and control its growth in various environmental conditions. It has been shown elegantly that in *V. cholerae,* autoinducer-mediated signals are channeled in parallel through both the CqsS/A and LuxS/PQ sensory systems via intracellular LuxU protein to the transcriptional activator protein LuxO, which then controls the cellular levels of the HapR protein, the master regulator of quorum sensing [41, 57, 58].

Although the intracellular signaling cascade is common for both the systems, the autoinducers, CAI-1 and AI-2, of *V. cholerae* are quite distinct from each other and function synergistically. While varieties of interconverting family of AI-2 molecules are produced from different species of bacteria, in all cases so far examined there is a common precursor molecule, called (*S*)-4,5-dihydroxypentane-2,3-dione (DPD), which is synthesized by the LuxS enzyme [57, 60]. Thus, AI-2 family of autoinducers are needed for both intra- and interspecies communications and giving rise to the idea that AI-2 represents a universal language fostering bacterial communication. AI-2 molecule synthesized by *V. cholerae* is the furanosylborate diester (2*S*, 4*S*)-2-methyl-2,3,3,4-tetrahydroxytetrahydrofuran borate [57, 61] (Fig. 10.1). It is to be noted that the cell surface-associated sensors LuxPQ and CqsS resemble proteins belonging to two-component signaling systems [56], each containing a conserved histidine kinase and response regulator domain but no DNA-binding motif.

10.3.2 Quorum Sensing-Regulated Virulence Gene Expression and Biofilm Formation in Vibrio cholerae

At low cell density condition, when AIs (CAI-1 and AI-2) are low, the AI receptors/sensors (CqsS and LuxPQ) remain active and by their kinase activities they funnel phosphate moiety to the response regulator LuxO protein converting it to LuxO~P. Now LuxO~P together with the alternative sigma factor σ^{54} activates expression of four different small RNAs (sRNAs) in *V. cholerae*, called *qrr1–4*, which along with the RNA-binding protein Hfq destabilize *hapR* mRNA leading to low cellular level of HapR [41]. In the absence of HapR, the virulence regulator AphA is expressed [62], which ultimately upregulates TcpP and ToxT expressions; ToxT in turn directly influences production of CT and TCP [42, 57, 63], the major virulence factors of *V. cholerae*. Low HapR condition also upregulates expression of *vps* genes (Fig. 10.2), the products of which regulate VPS synthesis needed for biofilm formation [42, 58, 63]. At high cell density, when extracellular AIs concentrations are increased, they bind to the sensors and the AI–sensor complexes behave as phosphatases leading to dephosphorylation of LuxO~P (Fig. 10.2). Since dephosphorylated LuxO is inactive, it is no longer able to activate the transcription of sRNAs *qrr1−4*; as a result, *hapR* mRNAs become stable and thus the cellular level of HapR is increased [41]. High cellular level of HapR represses the expression of AphA and finally downregulation of CT and TCP expression due to repression of *toxT* gene expression [62] (Fig. 10.2). Similarly, high cellular level of HapR also represses biofilm formation by inhibiting expression of VpsT [59, 43], the product of the gene *vpsT*, which regulates VPS synthesis needed for biofilm formation [64].

In *V. cholerae*, QSS acting through HapR enhances viability under certain stress conditions via upregulating expression of the stationary phase sigma factor *rpoS* [65]. RpoS can also stimulate *hapR* expression [66], suggesting the possible importance of a HapR/RpoS autoregulation loop in the face of various environmental stress factors. Although more information is needed in this aspect of gene regulation, it appears that interaction of quorum sensing and the stress response may play a survival role in biofilm-associated cells of *V. cholerae* in natural environmental conditions.

10.3.3 Relationship Between Quorum Sensing and c-di-GMP in Vibrio cholerae

From the above discussion it is understandable that two chemical signaling systems, extracellular quorum sensing through AIs and intracellular c-di-GMP signaling, intricately control both biofilm formation and virulence of *V. cholerae* under various environmental cues. Thus, there is a high possibility that the two signaling systems are linked somewhere to execute the above cellular processes [1]. In this respect the DNA-binding transcription factor HapR most likely plays a central role for connecting the two chemical signaling systems (Fig. 10.2). Again, it has been shown that for biofilm formation in *V. cholerae*, a high cellular level of c-di-GMP is needed (see above). Waters et al. [43] have shown very recently that at high cell density, HapR controls transcription of several genes encoding DGC and PDE proteins and the net effect was reduction in intracellular c-di-GMP level leading to decrease in biofilm formation (Fig. 10.2).

10.4 Concluding Remarks

Bacterial cells display a remarkable ability to respond to any physico-chemical changes in their surrounding environment. The task is even harder in the case of a pathogen like *V. cholerae*, which has a dual lifestyle, one in host-associated state and another in natural aquatic habitats. Thus, the survival of *V. cholerae* in these contrasting environmental conditions is critically dependent on orchestrated functioning of signal transduction systems. Availability of the whole genome sequence of *V. cholerae* [7] has helped tremendously to the researchers to unravel several unknown signaling circuits, but still much work remains to be done to understand clearly about how the pathogen switches its gene functions from environmental mode to virulence mode or *vice versa*. Recent research in this direction indicates that small molecules, like AIs, c-di-GMP, (p)ppGpp, and cAMP, play a pivotal role in extra- and intracellular signaling in *V. cholerae*. Although these molecules are involved in regulation of biofilm formation, stress survival, virulence gene expression as well as other related processes, still our knowledge is in a rudimentary stage regarding mechanisms of action of these small molecules. Again, it is conceivable that in natural situation, bacterial cells simultaneously face complex stimuli like temperature, salt concentration, nutrients, and oxygen tension, and they still manage in integrating all these external signals with the intracellular signaling circuits followed by expression of appropriate sets of genes to complete the stimulus-response cycle, which ultimately helps bacteria in adapting to that environment. Thus, there exists a high possibility that extra- and intracellular signaling relay systems are connected as shown in *V. cholerae* (Fig. 10.2) and that quorum sensing, virulence gene expression, biofilm formation, and intracellular c-diGMP signaling system are linked [42, 43, 55, 62]. In this regard, *V. cholerae* could be used as an excellent model system to study how other two important intracellular small molecules, cAMP and (p)ppGpp, modulate the functions of genes or proteins involved in quorum sensing and c-di-GMP-mediated signaling. It is also important to know how the other global regulators like RpoS impinge the regulatory circuits of quorum sensing and c-di-GMP signaling systems. At present very little is known about the exact functions of *V. cholerae*'s DGC and PDE proteins, how the genes encoding these proteins are regulated, and most importantly, what are the cellular targets of c-di-GMP and how this molecule regulates gene expression. The picture emerges from recent research that c-di-GMP is a linchpin intracellular molecule controlling multiple functions in *V. cholerae* as well as in other organisms. Since c-di-GMP is not found in eukaryotes, it may be used as a target for the development of antibacterial drugs, although at present it is not known what is the effect of total depletion of c-di-GMP from a bacterial cell.

Acknowledgments The work reported in this article was partially supported by the research grants SMM003 and MLP110 from Council of Scientific and Industrial Research (CSIR), Government of India.

References

1. Camilli A, Bassler BL. Bacterial small-molecule signaling pathways. Science. 2006;311: 1113–6.
2. Colwell RR. Global climate and infectious disease: the cholera paradigm. Science. 1996;274: 2025–31.
3. Harman JG. Allosteric regulation of the cAMP receptor protein. Biochim Biophys Acta. 2001; 1547:1–17.
4. Salter EA, Wierzbicki A. The mechanism of cyclic nucleotide hydrolysis in the phosphodiesterase catalytic site. J Phys Chem. 2007;111:4547–52.
5. Botsford JL, Harman JG. Cyclic AMP in prokaryotes. Microbiol Rev. 1992;56:100–22.
6. Skorupski K, Taylor RK. Control of the ToxR virulence regulon in *Vibrio cholerae* by environmental stimuli. Mol Microbiol. 1997;25:1003–9.
7. Heidelberg JF, Eisen JA, Nelson WC, Clayton RA, Gwinn ML, Dodson RJ, Haft DH, Hickey EK, Peterson JD, Umayam L, Gill SR, Nelson KE, Read TD, Tettelin H, Richardson D, Ermolaeva MD, Vamathevan J, Bass S, Qin H, Dragoi I, Sellers P, McDonald L, Utterback T, Fleishmann RD, Nierman WC, White O, Salzberg SL, Smith HO, Colwell RR, Mekalanos JJ, Venter JC, Fraser CM. DNA sequence of both chromosomes of the cholera pathogen *Vibrio cholerae*. Nature. 2000;406:477–83.
8. Lee SH, Hava DL, Waldor MK, Camilli A. Regulation and temporal expression patterns of *Vibrio cholerae* virulence genes during infection. Cell. 1999;99:625–34.
9. Miller VL, Mekalanos JJ. A novel suicide vector and its use in construction of insertion mutations: osmoregulation of outer membrane proteins and virulence determinants in *Vibrio cholerae* requires *toxR*. J Bacteriol. 1988;170:2575–83.
10. DiRita VJ, Parsot C, Jander G, Mekalanos JJ. Regulatory cascade controls virulence in *Vibrio cholerae*. Proc Natl Acad Sci USA. 1991;88:5403–7.
11. Hase CC, Mekalanos JJ. TcpP protein is a positive regulator of virulence gene expression in *Vibrio cholerae*. Proc Natl Acad Sci USA. 1998;95:730–4.
12. Liang W, Pascual-Montano A, Silva AJ, Jorge A. Benitez JA. The cyclic AMP receptor protein modulates quorum sensing, motility and multiple genes that affect intestinal colonization in *Vibrio cholerae*. Microbiology. 2007;153:2964–75.

13. Liang W, Silva AJ, Benitez JA. The cyclic AMP receptor protein modulates colonial morphology in *Vibrio cholerae*. Appl Environ Microbiol. 2007;73:7482–7.
14. Mori K, Aiba H. Evidence for negative control of *cya* transcription by cAMP and cAMP receptor protein in intact *Escherichia coli* cells. J Biol Chem. 1985;260:14838–43.
15. Wai SN, Mizunoe Y, Takade A, Kawabata SI, Yoshida SI. *Vibrio cholerae* O1 strain TSI–4 produces the exopolysaccharide materials that determine colony morphology, stress resistance, and biofilm formation. Appl Environ Microbiol. 1998;64:3648–55.
16. Yildiz FH, Schoolnik GK. *Vibrio cholerae* O1 El Tor: identification of a gene cluster required for the rugose colony type, exopolysaccharide production, chlorine resistance, and biofilm formation. Proc Natl Acad Sci USA. 1999;96:4028–33.
17. Silva AJ, Benitez JA. Transcriptional regulation of *Vibrio cholerae* hemagglutinin/protease by the cyclic AMP receptor protein and RpoS. J Bacteriol. 2004;186:6374–82.
18. Cashel M, Gentry DR, Hernandes VJ, Vinella D. The stringent response. In: Neidhardt FC, editor. *Escherichia coli* and *Salmonella typhimurium*: cellular and molecular biology. Washington, DC: American Society for Microbiology; 1996. pp. 1458–96.
19. Hogg T, Mechold U, Malke H, Cashel M, Hilgenfeld R. Conformational antagonism between opposing active sites in a bifunctional RelA/SpoT homolog modulates (p)ppGpp metabolism during the stringent response. Cell. 2004;117:57–68.
20. Gentry DR, Cashel M. Mutational analysis of the *Escherichia coli spoT* gene identifies distinct but overlapping regions involved in ppGpp synthesis and degradation. Mol Microbiol. 1996;19:1373–84.
21. Das B, Bhadra RK. Molecular characterization of *Vibrio cholerae* Δ*relA* Δ*spoT* double mutants. Arch Microbiol. 2008;189:227–38.
22. Haralalka S, Nandi S, Bhadra RK. Mutation in the *relA* gene of *Vibrio cholerae* affects in vitro and in vivo expression of virulence factors. J Bacteriol. 2003;185:4672–82.
23. Raskin MD, Judson N, Mekalanos JJ. Regulation of stringent response is the essential function of the conserved bacterial G protein CgtA in *Vibrio cholerae*. Proc Natl Acad Sci USA. 2007;104:4636–41.
24. Silva AJ, Benitez BA. *Vibrio cholerae* relaxed (*relA*) mutant expresses major virulence factors, exhibits biofilm formation and motility, and colonizes the suckling mouse intestine. J Bacteriol. 2006;188:794–800.
25. Lemos JA, Lin VK, Nascimento MM, Abranches J, Burne RA. Three gene products govern (p)ppGpp production by *Streptococcus mutans*. Mol Microbiol. 2007;65: 1568–81.
26. Nanamiya H, Kasai K, Nozawa A, Yun CS, Narisawa T, Murakami K, Natori Y, Kawamura F, Tozawa Y. Identification and functional analysis of novel (p)ppGpp synthetase genes in *Bacillus subtilis*. Mol Microbiol. 2007;67:291–304.
27. Jenal U. Cyclic di-guanosine-monophosphate comes of age: a novel secondary messenger involved in modulating cell surface structures in bacteria? Curr Opin Microbiol. 2004;7: 185–91.
28. Römling U, Gomelsky M, Galperin MY. C-di-GMP: the dawning of a novel bacterial signaling system. Mol Microbiol. 2005;57:629–39.
29. Ross P, Mayer R, Weinhouse H, Amikam D, Huggirat Y, Benziman M, de Vroom E, Fidder A, de Paus P, Sliedregt LA, van der Marel GA, van Boom JH. The cyclic diguanylic acid regulatory system of cellulose synthesis in *Acetobacter xylinum*. Chemical synthesis and biological activity of cyclic nucleotide dimer, trimer, and phosphothioate derivatives. J Biol Chem. 1990;265:18933–43.
30. Jenal U, Malone J. Mechanisms of cyclic-di-GMP signaling in bacteria. Annu Rev Genet. 2006;40:385–407.
31. Chang AL, Tuckerman JR, Gonzalez G, Mayer R, Weinhouse H, Volman G, Amikam D, Benziman M, Gilles-Gonzalez MA. Phosphodiesterase A1, a regulator of cellulose synthesis in *Acetobacter xylinum*, is a heme-based sensor. Biochemistry. 2001;40: 3420–6.

32. Christen M, Christen B, Folcher M, Schauerte A, Jenal U. Identification and characterization of a cyclic di-GMP-specific phosphodiesterase and its allosteric control by GTP. J Biol Chem. 2005;280:30829–37.
33. Galperin MY, Nikolskaya AN, Koonin EV. Novel domains of the prokaryotic two-component signal transduction systems. FEMS Microbiol Lett. 2001;203:11–21.
34. Ryan RP, Fouhy Y, Lucey JF, Crossman LC, Spiro S, He YW, Zhang LH, Heeb S, Cámara M, Williams P, Dow JM. Cell–cell signaling in *Xanthomonas campestris* involves an HD-GYP domain protein that functions in cyclic di-GMP turnover. Proc Natl Acad Sci USA. 2006;103:6712–7.
35. Galperin MY. Bacterial signal transduction network in genomic perspective. Environ Microbiol. 2004;6:552–67.
36. Ryan RP, Fouhy Y, Lucey JF, Dow JM. Cyclic di-GMP signaling in bacteria: recent advances and new puzzles. J Bacteriol. 2006;188:8327–34.
37. Lim B, Beyhan S, Meir J, Yildiz FH. Cyclic-diGMP signal transduction systems in *Vibrio cholerae*: modulation of rugosity and biofilm formation. Mol Microbiol. 2006;60:331–48.
38. Watnick PI, Kolter R. Steps in the development of a *Vibrio cholerae* El Tor biofilm. Mol Microbiol. 1999;34:586–95.
39. Faruque SM, Biswas K, Udden SM, Ahmad QS, Sack DA, Nair GB, Mekalanos JJ. Transmissibility of cholera: in vivo-formed biofilms and their relationship to infectivity and persistence in the environment. Proc Natl Sci Acad USA. 2006;103:6350–5.
40. Jobling MG, Holmes RK. Characterization of *hapR*, a positive regulator of the *Vibrio cholerae* HA/protease gene *hap*, and its identification as a functional homologue o the *Vibrio harveyi luxR* gene. Mol Microbiol. 1997;26:1023–34.
41. Lenz DH, Mok KC, Lilley BN, Kulkarni RV, Wingreen NS, Bassler BL. The small RNA chaperone Hfq and multiple small RNAs control quorum sensing in *Vibrio harveyi* and *Vibrio cholerae*. Cell. 2004;118:69–82.
42. Zhu J, Mekalanos JJ. Quorum sensing-dependent biofilms enhance colonization in *Vibrio cholerae*. Dev Cell. 2003;5:647–56.
43. Waters CM, Lu W, Rabinowitz JD, Bassler BL. Quorum sensing controls biofilm formation in *Vibrio cholerae* through modulation of cyclic di-GMP levels and repression of *vpsT*. J Bacteriol. 2008;190:2527–36.
44. Tischler AD, Camilli A. Cyclic diguanylate regulates *Vibrio cholerae* virulence gene expression. Infect Immun. 2005;73:5873–82.
45. Amikam D, Galperin MY. PilZ domain is part of the bacterial c-di-GMP binding protein. Bioinformatics. 2006;22:3–6.
46. Ryjenkov DA, Simm R, Römling U, Gomelsky M. The PilZ domain is a receptor for the second messenger c-di-GMP: the PilZ domain protein YcgR controls motility in enterobacteria. J Biol Chem. 2006;281:30310–4.
47. Pratt JT, Tamayo R, Tischler AD, Camilli A. PilZ domain proteins bind cyclic diguanylate and regulate diverse processes in *Vibrio cholerae*. J Biol Chem. 2007;282:12860–70.
48. Tischler AD, Camilli A. Cyclic diguanylate (c-di-GMP) regulates *Vibrio cholerae* biofilm formation. Mol Microbiol. 2004;53:857–69.
49. Merrell DS, Camilli A. Regulation of *Vibrio cholerae* genes required for acid tolerance by a member of the "ToxR-like" family of transcriptional regulators. J Bacteriol. 2000;182: 5342–50.
50. Kovacikova G, Lin W, Skorupski K. Dual regulation of genes involved in acetoin biosynthesis and motility/biofilm formation by the virulence activator AphA and the acetate-responsive LysR-type regulator AlsR in *Vibrio cholerae*. Mol Microbiol. 2005;57:420–33.
51. Camilli A, Mekalanos JJ. Use of recombinase gene fusions to identify *Vibrio cholerae* genes induced during infection. Mol Microbiol. 1995;18:671–83.
52. Lee SH, Angelichio MJ, Mekalanos JJ, Camilli A. Nucleotide sequence and spatiotemporal expression of the *Vibrio cholerae vieSAB* genes during infection. J Bacteriol. 1998;180: 2298–305.

53. Tamayo R, Tischler AD, Camilli A. The EAL domain protein VieA is a cyclic diguanylate phosphodiesterase. J Biol Chem. 2005;280:33324–30.
54. Beyhan S, Tischler AD, Camilli A, Yildiz FH. Transcriptome and phenotypic responses of *Vibrio cholerae* to increased cyclic di-GMP level. J Bacteriol. 2006;188:3600–13.
55. Tamayo R, Schild S, Pratt JT, Camilli A. Role of cyclic di-GMP during El Tor biotype *Vibrio cholerae* infection: characterization of the in vivo-induced cyclic di-GMP phosphodiesterase CdpA. Infect Immun. 2008;76:1617–27.
56. Miller MB, Bassler BL. Quorum sensing in bacteria. Annu Rev Microbiol. 2001;55:165–99.
57. Miller MB, Skorupski K, Lenz DH, Taylor RK, Bassler BL. Parallel quorum sensing systems converge to regulate virulence in *Vibrio cholerae*. Cell. 2002;110:303–14.
58. Hammer BK, Bassler BL. Quorum sensing controls biofilm formation in *Vibrio cholerae*. Mol Microbiol. 2003;50:101–4.
59. Lenz DH, Miller MB, Zhu J, Kulkarni RV, Bassler BL. CsrA and three redundant small RNAs regulate quorum sensing in *Vibrio cholerae*. Mol Microbiol;2005;58:1186–202.
60. Higgins DA, Pomianek ME, Kraml CM, Taylor RK, Semmelhack MF, Bassler BL. The major *Vibrio cholerae* autoinducer and its role in virulence factor production. Nature. 2007;450: 883–6.
61. Chen X, Schauder S, Potier N, Von Dorsselaer A, Pelczer I, Bassler BL, Hugson FM. Structural identification of a bacterial quorum sensing signal containing boron. Nature. 2002;415:545–9.
62. Kovacikova G, Skorupski K. Regulation of virulence gene expression in *Vibrio cholerae* by quorum sensing: HapR functions at the *aphA* promoter. Mol Microbiol. 2002;46:1135–47.
63. Zhu J, Miller MB, Vance RE, Dziejman M, Bassler BL, Mekalanos JJ. Quorum-sensing regulators control virulence gene expression in *Vibrio cholerae*. Proc Natl Acad Sci USA. 2002;99:3129–34.
64. Casper-Lindley C, Yildiz FH. VpsT is a transcriptional regulator required for expression of *vps* biosynthesis genes and the development of rugose colonial morphology in *Vibrio cholerae* O1 El Tor. J Bacteriol. 2004;186:1574–8.
65. Joelsson A, Kan B, Zhu J. Quorum sensing enhances the stress response in *Vibrio cholerae*. Appl Environ Microbiol. 2007;73:3742–6.
66. Yildiz FH, Liu XS, Heydorn A, Schoolnik GK. Molecular analysis of rugosity in a *Vibrio cholerae* O1 El Tor phase variant. Mol Microbiol. 2004;53:497–515.

Chapter 11
Vibrio cholerae Flagellar Synthesis and Virulence

Khalid Ali Syed and Karl E. Klose

Abstract Flagellar-mediated motility has been demonstrated to contribute to the pathogenesis of *Vibrio cholerae*. Nonmotile mutants of live attenuated cholera vaccine strains are significantly less reactogenic in human volunteers, but the exact contribution of motility to virulence appears to be multi-factorial. The flagellum of *V. cholerae* is a complex structure made up of multiple structural subunits (>40 proteins). Expression of flagellar genes proceeds via a transcription hierarchy. This flagellar regulatory cascade controls not only flagellar gene expression, but also influences the expression of additional (non-flagellar) genes with proven or implicated roles in virulence. Flagellar-mediated chemotaxis appears to be linked to transmission of *V. cholerae* from host to host, and thus epidemic spread of cholera. Motility also contributes to biofilm formation, which facilitates environmental persistence. Thus, the flagellum plays an integral part in the lifecycle of *V. cholerae*, both in the host as well as in the environment.

11.1 Introduction

Vibrio cholerae is a highly motile bacterium, due to the presence of a single polar sheathed flagellum (Fig. 11.1). Flagellar-mediated motility has been demonstrated to contribute to the pathogenesis of a number of bacteria, and motility has been implicated as a virulence determinant of *V. cholerae* [1]. Nonmotile mutants of *V. cholerae* cause less fluid accumulation in the rabbit ileal loop [2, 3] and also show defects in adhesion to isolated rabbit brush border cells [4]. Studies in human volunteers showed that nonmotile mutants of live attenuated cholera vaccine strains are significantly less reactogenic while still being able to colonize the gut [5, 6]. In addition to virulence, flagellar motility contributes to biofilm formation, a behavior believed to enhance environmental persistence [7, 8]. Thus, evidence has suggested

K.E. Klose (✉)
Department of Biology, South Texas Center for Emerging Infectious Diseases, University of Texas San Antonio, San Antonio, TX 78249-1644, USA
e-mail: karl.klose@utsa.edu

T. Ramamurthy, S.K. Bhattacharya (eds.), *Epidemiological and Molecular Aspects on Cholera*, Infectious Disease, DOI 10.1007/978-1-60327-265-0_11,
© Springer Science+Business Media, LLC 2011

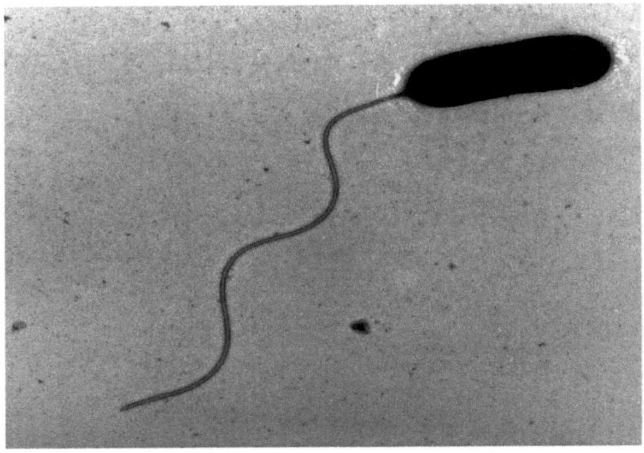

Fig. 11.1 Electron micrograph of *V. cholerae* showing the sheathed polar flagellum

that the flagellum plays an integral part in the lifecycle of *V. cholerae*, both in the host and in the environment. However, we are only now beginning to elucidate the exact connection between flagellar-mediated motility and *V. cholerae* virulence and biofilm formation.

11.2 The *Vibrio cholerae* Flagellar Transcription Hierarchy

Flagellar genes in *V. cholerae* encoding the structural components of the flagellum share high homology with the known flagellar components of other bacteria to ascribe putative functions to these ORFs. The genome sequence of *V. cholerae* revealed that the majority of flagellar genes are arranged in three loci on the large chromosome [9].

The flagellum is a complex structure made up of multiple structural subunits, and the assembly of this structure is exquisitely coordinated in a stepwise fashion, initiating inside the cell and building outward toward the flagellar tip [10]. The proximal portion of the flagellum embedded within the cytoplasmic membrane, encompassing the MS ring, is assembled first. This structure is associated with the export components of a type III secretion system, and the portions of the flagellum external to the cytoplasm are secreted through this structure in the order they are assembled. This structure is also associated with "switch" components, which communicate with the chemotaxis machinery to alter the rotation of the flagellum in response to external chemoattractants/repellents. The portion of the flagellum that spans the space between the inner and outer membranes comprises the majority of the "basal body," and it is assembled next, followed by the structure external to the cell, which consists of a flexible linker referred to as the hook, attached to the flagellar filament, which makes up the bulk of the flagellum [10, 11].

The filament in many bacteria is composed of a single repeating flagellin subunit, but interestingly *V. cholerae* appears to have five distinct flagellins within the filament [12], similar to other *Vibrio* spp. Another interesting aspect of the *Vibrio* flagellum that distinguishes it from most other bacterial flagella is the presence of a flagellar sheath. The sheath contains LPS [13] and thus appears to be an extension of the outer membrane, but nothing is known about how the sheath is formed or whether it functions to shield the highly immunogenic flagellins from the immune system, as has been hypothesized.

Given the large number of structural components that make up the flagellum (>40 proteins) and the requirement for an ordered assembly of these subunits, the expression of flagellar genes is also tightly regulated. In general, flagellar genes are expressed in the order in which they are needed for flagellar assembly, i.e., MS ring–switch–export apparatus genes are expressed first, followed by basal body–hook genes, and finally the flagellin genes are expressed last. To control the ordered expression of flagellar genes, bacteria typically express these genes within a flagellar transcription hierarchy, within which the expression of a class of genes is dependent on the expression and assembly of the previous class of genes. *Vibrio cholerae* has a four-tiered flagellar transcription hierarchy (Fig. 11.2) [14]. This transcription hierarchy is very similar to that found in *Pseudomonas aeruginosa* [15], which also has a single polar flagellum, and differs primarily in regulatory aspects from the well-characterized, three-tiered transcriptional hierarchy of the peritrichously flagellated organisms like *Salmonella typhimurium* [16].

There is a single class I gene, and it encodes the master regulator of the flagellar transcriptional hierarchy, FlrA. FlrA is a σ^{54}-dependent transcriptional activator that activates the expression of class II genes, which encode primarily the MS ring and export apparatus components, as well as chemotaxis and regulatory factors [14]. Transcription of the *flrA* homologue, *fleQ*, in *P. aeruginosa* is dependent on σ^{70} and repressed by the CRP homologue Vfr [17], but it is not known if the same is true in *V. cholerae*. A class II regulatory factor, FlhG, appears to repress *flrA* transcription in *V. cholerae* [18]; interestingly the *P. aeruginosa* homologue FleN represses the transcriptional activity of the FlrA homologue, but not its transcription.

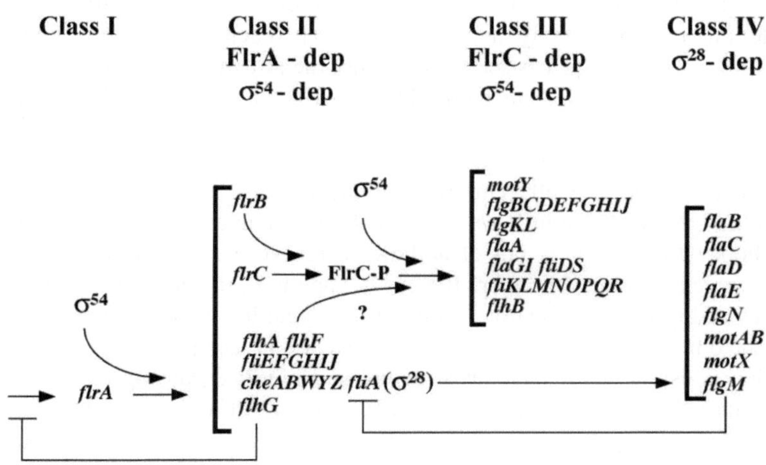

Fig. 11.2 The *V. cholerae* flagellar transcription hierarchy (described in text)

However, the phenotype resulting from inactivation of *flhG* (or *fleN*) is the same in both systems; the bacteria express multiple polar flagella [18]. This indicates that the presence of a single polar flagellum is due to tight control over the synthesis and/or the activity of FlrA.

FlrA activates σ^{54}-dependent transcription of class II genes, which, in addition to the MS ring–export apparatus components, also encode a number of regulatory factors: FlrB, FlrC, FlhF, FlhG, and FliA (σ^{28}), as well as the chemotaxis machinery [14]. FlrC is a σ^{54}-dependent transcriptional activator that activates class III flagellar genes. FlrB is an autokinase that transfers a phosphate to a conserved aspartate residue in FlrC, and phospho-FlrC is the transcriptionally active form; presumably phosphorylation stimulates the ATPase activity necessary for σ^{54}-dependent transcription [19]. It is possible that phosphorylation of FlrC only occurs upon completion of class II gene transcription and MS ring–export assembly; a similar checkpoint regulating the switch from class II to class III flagellar gene transcription involving phosphorylation of a σ^{54}-dependent activator exists in *Campylobacter jejuni*. In *C. jejuni*, the formation of a functional secretory apparatus is linked to the activation of class III genes, suggesting that a negative regulatory factor may be secreted to allow phosphorylation of the σ^{54}-dependent activator FlgR [20]. In *V. cholerae*, no negative regulator of FlrC-dependent transcription has been identified yet, but a positive regulatory protein, FlhF, contributes to class III gene expression, and thus it may stimulate phosphorylation of FlrC following completion of the export apparatus [18]. It is not clear if a functional export apparatus exists prior to the switch from class II to class III gene expression in *V. cholerae*, since some of the predicted components of the type III secretion (FlhB, FliP, FliQ, FliR) appear to be expressed as class III genes rather than class II genes (F. Yildiz and K. Klose, unpublished data).

Class III FlrC-dependent genes encode the basal body and hook, as well as some of the switch and export apparatus components. In *S. typhimurium*, a well-characterized checkpoint controls transcription of the flagellin genes upon completion of the basal body–hook structure [21]. When this structure is complete, an anti-sigma factor FlgM is secreted through the structure, allowing the alternate sigma factor σ^{28} to associate with RNAP and activate transcription of flagellin genes. *Vibrio cholerae* also secretes the anti-sigma factor FlgM, and this allows σ^{28} to associate with RNAP and activate class IV genes, which encode the additional flagellins FlaBCDE, as well as motor components [22].

While FlgM secretion in *V. cholerae* likely occurs upon completion of the basal body–hook structure, it is not clear how secretion occurs through the flagellar sheath, which presumably covers the entire flagellum, including the flagellar cap. Another oddity is the regulation of one of the flagellins, FlaA, as a class III gene, while the other four flagellins are regulated as class IV genes. FlaA is the only flagellin that is essential for flagellar structure and motility, causing us to name it a "core" flagellin, and its status as a class III gene would cause it to be temporally expressed prior to the other flagellin genes. Yet the class IV flagellins are expressed at high levels in a *flaA* mutant, indicating that the class III–class IV checkpoint occurs prior to incorporation of FlaA into the flagellum [12]. Since FlaBCDE are not essential for motility, it is not clear why the flagellum contains these alternate flagellins. One possibility is that differences in the flagellin composition may impart different functional characteristics under varying environmental conditions.

11.3 Motility and Virulence

Motility has been linked to virulence in *V. cholerae*. Studies have demonstrated that nonmotile mutants are defective for fluid accumulation and adherence in the rabbit ileal loop model [2, 3] and adherence to isolated rabbit brush borders [23]; nonmotile mutants are also defective for virulence in the rabbit RITARD model [2]. Nonmotile mutants of O1 El Tor biotype strains show colonization defects in the infant mouse model [24], while nonmotile classical mutants generally colonize similar to the motile wild-type strain [25]. Nonmotile mutants of live *V. cholerae* vaccine strains show reduced reactogenicity in human volunteers while still being able to colonize the intestine [5, 6]. Though a number of studies have implicated motility as being important for *V. cholerae* virulence, the exact connection between flagellar synthesis and cholera pathogenesis is still not clear.

In a study involving spontaneous mutants, Gardel and Mekalanos [25] found that nonmotile mutants showed increased expression of virulence factors (CT and TCP), while hyper-motile mutants produced less CT and TCP as compared to the wild-type *V. cholerae*. The authors proposed a model where virulence and motility are inversely related: motility is downregulated and virulence factor expression simultaneously upregulated when the organisms are colonized on the intestinal cell surface, and when not colonized, the organisms upregulate motility and downregulate virulence factor expression. Recently Silva et al. [3] provided evidence for this model when they found that transcription of *toxT*, *ctxA*, and *tcpA* is upregulated in a *V. cholerae* nonmotile (*motY*) strain. Further evidence for this model comes from Ghosh et al. [26], who found that the histone-like nucleoid structuring protein H-NS stimulates motility by stimulating *flrA* expression while repressing *ctxAB* and *tcpA* transcription.

Our laboratory has likewise found evidence that motility and virulence are inversely regulated in *V. cholerae.* Transcription profiling of flagellar regulatory mutants utilizing whole genome microarrays has demonstrated increased expression of genes encoding proven and putative virulence factors (F. Yildiz and K. Klose, unpublished data). One potential mechanism whereby the flagellar regulatory cascade controls diverse virulence genes appears to involve the modulation of cyclic diguanylate (c-dGMP) levels. c-dGMP levels in a cell can be modulated by regulating the expression of diguanylate cyclases, which contain a "GGDEF" domain, and our laboratory has found that several GGDEF proteins are regulated by the flagellar transcription hierarchy. Increased c-dGMP levels have been shown to suppress CT production and promote biofilm formation in *V. cholerae* [27–29]. The apparent inverse relationship between motility and virulence seems like a logical means by which *V. cholerae* could achieve the most efficient and beneficial expression patterns in the various niches it occupies. One imagines that motility would be beneficial in the early stages of infection, allowing the bacterium to reach the site of infection and colonize. However, once colonized, flagellar-based motility may be deleterious to *V. cholerae*, since flagellar synthesis expends a relatively large amount of resources and can take the bacterium away from the site of infection. Moreover, the flagellins are highly immunogenic, so downregulation of flagellin synthesis when inside the human host can be advantageous to the bacterium. Examples of inverse relationships between motility and virulence are found in other pathogenic bacterial species. For example, in *Bordetella* the virulence regulator BvgA represses expression of the master regulator of flagellar synthesis while activating virulence gene expression [30]. In *Salmonella*, the BarA/SirA regulatory proteins simultaneously stimulate virulence gene expression while repressing motility [31]. In *Pseudomonas syringae,* quorum sensing through *N*-acyl homoserine lactones represses motility and stimulates alginate production and hydrogen peroxide resistance [32] and in *P. aeruginosa*, a CRP homologue represses expression of the *flrA* homologue *fleQ* while stimulating expression of the quorum sensing regulator *lasR*, which stimulates expression of a number of genes that contribute to virulence [33].

11.4 Chemotaxis and Virulence

The chemotaxis components interact with the polar flagellum to alter its rotation and allow the bacterium to swim toward chemoattractants and away from chemorepellents [34]. Methyl-accepting chemoreceptors (MCP) in the membrane interact with these attractants or repellants, and the information is relayed through the chemotaxis proteins to the switch components associated with the flagellum, resulting in alteration of flagellar rotation from counterclockwise (CCW) to clockwise (CW). The *V. cholerae* genome contains a large number of MCPs (at least 43), indicating the importance of chemotaxis to the lifestyle of this organism [9]. Moreover, there are multiple paralogs of the chemotaxis components (e.g., four CheYs, three CheAs, and three CheWs) in the genome; however only those components transcribed within the class II *flhA* operon (*cheY-3, cheA-2, cheW-1*) are essential for chemotaxis [35–38].

Non-chemotactic mutants have been shown to outcompete chemotactic *V. cholerae* within the infant mouse small intestine [23, 36], and the enhanced colonization of the non-chemotactic strains has been shown to be due to these organisms' ability to colonize along the entire length of the intestine, rather than just the distal end, as the wild-type strains do [35]. In an elegant set of experiments, Camilli and colleagues demonstrated that when flagellar rotation is biased in CCW rotation (as in a chemotactic mutant), the bacteria swim in smooth straight runs, only changing direction every 33 s, and are increased for infectivity, whereas when flagellar rotation is biased in the CW rotation, the bacteria swim with little net movement, changing direction every 0.2 s, and are attenuated for colonization [35]. The implication of these results is that suppression of chemotaxis leads to increased infectivity.

Interestingly, *V. cholerae* emerging from humans within stool were found to be in a transient "hyperinfectious" state, and transcription profiling of the stool bacteria revealed a downregulation of chemotaxis genes [39]. This transient hyperinfectious state of human stool *V. cholerae* was shown to correlate specifically with suppression of *cheW-1*, which causes an inability to chemotax and enhances infectivity [36]. Thus, the rotation bias of the flagellum appears to be intimately linked to enhanced transmission of *V. cholerae* from host to host and may contribute to the epidemic spread of cholera.

11.5 Motility and Biofilm Formation

The ability of *V. cholerae* to persist in the environment is critical for its ability to cause pandemic disease. One mode of environmental persistence employed by a wide variety of microorganisms is to form large multicellular structures on surfaces known as biofilms. Biofilms provide protection to the bacteria from such stresses as harsh chemicals or antibiotics allowing for survival and persistence [40]. *Vibrio cholerae* biofilm formation is dependent on the expression of an exopolysaccharide (VPS), and expression of the VPS during growth on solid media results in a rugose colony phenotype [41–43].

Biofilm formation proceeds by free swimming cells first attaching to a surface, then forming a monolayer on the surface, and finally developing into large three-dimensional structures known as a mature biofilm with characteristic pillars and water channels [44]. It has been shown that motility is important for the initial stage of approaching and attaching to the surface [7], but once the cells have formed a monolayer, flagellar synthesis is downregulated [45]. Likewise, chemotaxis is important for the formation of a monolayer, but not subsequent steps in biofilm development [46]. Progression from monolayer to mature biofilm requires expression of VPS, and analysis of rugose (i.e., VPS-producing) *V. cholerae* revealed that flagellar synthesis is likewise downregulated in VPS-expressing cells [47].

The flagellum negatively regulates VPS exopolysaccharide expression in a number of *V. cholerae* strains [42, 48]. Specifically, nonflagellated mutants produce copious amounts of VPS and exhibit a rugose phenotype [42, 48] The sodium-driven flagellar motor is necessary for flagellar-dependent VPS expression, suggesting that the motor acts as a mechanosensor, signaling for the induction of VPS expression upon loss of the flagellum [48]. Thus, a model emerges where flagellar-dependent motility facilitates approach and attachment to a surface, and chemotaxis facilitates the formation of a monolayer on this surface. Upon formation of the monolayer, flagellar synthesis is downregulated, which in turn causes the upregulation of VPS production and facilitates the formation of the three-dimensional mature biofilm structures.

11.6 Summary

Studies have revealed that flagellar synthesis, motility, and chemotaxis contribute to both the virulence and environmental persistence of *V. cholerae*. Continued investigation is likely to reveal new complexities about how the regulatory pathways connect swimming behavior to the lifestyle of *V. cholerae* within the host as well as within the aquatic environment.

Acknowledgments K.E. Klose is funded by NIH AI43486 and AI51333.

References

1. Moens S, Vanderleyden J. Functions of bacterial flagella. Crit Rev Microbiol. 1996;22: 67–100.
2. Richardson K. Roles of motility and flagellar structure in pathogenicity of *Vibrio cholerae*: analysis of motility mutants in three animal models. Infect Immun. 1991;59:2727–36.
3. Silva AJ, Leitch GJ, Camilli A, Benitez JA. Contribution of hemagglutinin/protease and motility to the pathogenesis of El Tor biotype cholera. Infect Immun. 2006;74:2072–9.
4. Attridge SR, Rowley D. The role of the flagellum in the adherence of *Vibrio cholerae*. J Infect Dis. 1983;147:864–72.
5. Coster, TS, Killeen KP, Waldor MK, Beattie D, Spriggs D, Kenner JR, Trofa A, Sadoff J, Mekalanos JJ, Taylor DN. Safety, immunogenicity and efficacy of a live attenuated *Vibrio cholerae* O139 vaccine prototype, Bengal-15. Lancet. 1995;345:949–52.
6. Kenner JR, Coster TS, Taylor DN, Trofa AF, Barrera-Oro M, Hyman T, Adams JM, Beattie DT, Killeen, KP, Spriggs DR, Mekalanos JJ, Sadoff JC. Peru-15, an improved live attenuated vaccine candidate for *Vibrio cholerae* O1. J Infect Dis. 1995;172:1126–9.
7. Watnick PI, Kolter R. Steps in the development of a *Vibrio cholerae* biofilm. Mol Microbiol. 1999;34:586–95.
8. Watnick P, Kolter R. Biofilm, city of microbes. J Bacteriol. 2000;182:2675–9.
9. Heidelberg JF, Eisen JA, Nelson WC, Clayton RA, Gwinn ML, Dodson RJ, Haft DH, Hickey EK, Peterson JD, Umayam L, Gill, SR, Nelson KE, Read TD, Tettelin H, Richardson D, Ermolaeva MD, Vamathevan J, Bass S, Qin H, Dragoi I, Sellers P, McDonald L, Utterback T, Fleishmann RD, Nierman WC, White O, Salzberg SL, Smith HO, Colwell RR, Mekalanos JJ, Venter, JC., Fraser CM. DNA sequence of both chromosomes of the cholera pathogen *Vibrio cholerae*. Nature. 2000;406:477–83.
10. Macnab RM. How bacteria assemble flagella. Annu Rev Microbiol. 2003;57:77–100.
11. Macnab RM. Type III flagellar protein export and flagellar assembly. Biochim Biophys Acta. 2004;1694:207–17.
12. Klose KE, Mekalanos JJ. Differential regulation of multiple flagellins in *Vibrio cholerae*. J Bacteriol. 1998;180:303–16.
13. Fuerst JA, Perry JW. Demonstration of lipopolysaccharide on sheathed flagella of *Vibrio cholerae* O1 by protein A-gold immunoelectron microscopy. J Bacteriol. 1988;170:1488–94.
14. Prouty MG, Correa NE, Klose KE. The novel s^{54}- and s^{28}-dependent flagellar gene transcription hierarchy of *Vibrio cholerae*. Mol Microbiol. 2001;39:1595–609.
15. Dasgupta N, Wolfgang MC, Goodman AL, Arora SK, Jyot J, Lory S, Ramphal, R. A four-tiered transcriptional regulatory circuit controls flagellar biogenesis in *Pseudomonas aeruginosa*. Mol Microbiol. 2003;50:809–24.
16. Macnab RM. Flagella and motility. In: Neidhardt FC, Curtiss, RI, Ingraham, JL, Lin ECC, Low, KB, Magasanik B, Reznikoff WS, Riley M, Schaechter M, Umbarger HE, editors. *Escherichia coli and Salmonella*: cellular and molecular biology. Washington, DC: ASM Press; 1996. pp. 123–45.
17. Dasgupta N, Ferrell EP, Kanack KJ, West SE, Ramphal R. *fleQ*, the gene encoding the major flagellar regulator of *Pseudomonas aeruginosa*, is sigma70 dependent and is down-regulated by Vfr, a homolog of *Escherichia coli* cyclic AMP receptor protein. J Bacteriol. 2002;184:5240–50.
18. Correa NE, Peng F, Klose KE. Roles of the regulatory proteins FlhF and FlhG in the *Vibrio cholerae* flagellar transcription hierarchy. J Bacteriol. 2005;187:6324–32.

19. Correa NE, Lauriano CM, McGee R, Klose KE. Phosphorylation of the flagellar regulatory protein FlrC is necessary for *Vibrio cholerae* motility and enhanced colonization. Mol Microbiol. 2000;35:743–55.
20. Hendrixson DR, DiRita VJ. Transcription of sigma54-dependent but not sigma28-dependent flagellar genes in *Campylobacter jejuni* is associated with formation of the flagellar secretory apparatus. Mol Microbiol. 2003;50:687–702.
21. Hughes KT, Gillen KL, Semon MJ, Karlinsey JE. Sensing structural intermediates in bacterial flagellar assembly by export of a negative regulator. Science. 1993;262:1277–80.
22. Correa NE, Barker JR, Klose KE. The *Vibrio cholerae* FlgM homologue is an anti-sigma28 factor that is secreted through the polar sheathed flagellum. J Bacteriol. 2004;186: 4613–9.
23. Freter R, O'Brien PCM, Macsai MS. Role of chemotaxis in the association of motile bacteria with intestinal mucosa: *in vivo* studies. Infect Immun. 1981;34:234–40.
24. Lee SH, Butler SM, Camilli A. Selection for *in vivo* regulators of bacterial virulence. Proc Natl Acad Sci USA. 2001;98:6889–94.
25. Gardel CL Mekalanos JJ. Alterations in *Vibrio cholerae* motility phenotypes correlate with changes in virulence factor expression. Infect Immun. 1996;64:2246–55.
26. Ghosh A, Paul K, Chowdhury R. Role of the histone-like nucleoid structuring protein in colonization, motility, and bile-dependent repression of virulence gene expression in *Vibrio cholerae*. Infect Immun. 2006;74:3060–4.
27. Lim B, Beyhan S, Meir J, Yildiz FH. Cyclic-diGMP signal transduction systems in *Vibrio cholerae*: modulation of rugosity and biofilm formation. Mol Microbiol. 2006;60:331–48.
28. Tischler AD, Camilli A. Cyclic diguanylate (c-di-GMP) regulates *Vibrio cholerae* biofilm formation. Mol Microbiol. 2004;53:857–69.
29. Tischler, AD, Camilli A. Cyclic diguanylate regulates *Vibrio cholerae* virulence gene expression. Infect Immun. 2005;73:5873–82.
30. Akerley BJ, Cotter PA, Miller JF. Ectopic expression of the flagellar regulon alters development of the *Bordetella*–host interaction. Cell. 1995;80:611–20.
31. Goodier RI, Ahmer, BM. SirA orthologs affect both motility and virulence. J Bacteriol. 2001;183:2249–58.
32. Quinones B, Dulla G, Lindow SE. Quorum sensing regulates exopolysaccharide production, motility, and virulence in *Pseudomonas syringae*. Mol Plant Microbe Interact. 2005;18: 682–93.
33. Dasgupta, N, Ferrell EP, Kanack KJ, West SE, Ramphal R. *fleQ*, the gene encoding the major flagellar regulator of *Pseudomonas aeruginosa*, is sigma70 dependent and is down-regulated by Vfr, a homolog of *Escherichia coli* cyclic AMP receptor protein. J Bacteriol. 2002;184:5240–50.
34. Bren A, Eisenbach M. How signals are heard during bacterial chemotaxis: protein–protein interactions in sensory signal propagation. J Bacteriol. 2000;182:6865–73.
35. Butler SM, Camilli A. Both chemotaxis and net motility greatly influence the infectivity of *Vibrio cholerae*. Proc Natl Acad Sci USA. 2004;101:5018–23.
36. Butler SM, Nelson EJ, Chowdhury N, Faruque SM, Calderwood SB, Camilli A. Cholera stool bacteria repress chemotaxis to increase infectivity. Mol Microbiol. 2006;60:417–26.
37. Gosink KK, Kobayashi R, Kawagishi I, Hase CC. Analyses of the roles of the three *cheA* homologs in chemotaxis of *Vibrio cholerae*. J Bacteriol. 2002;184:1767–71.
38. Hyakutake A, Homma M, Austin MJ, Boin MA, Hase CC, Kawagishi I. Only one of the five CheY homologs in *Vibrio cholerae* directly switches flagellar rotation. J Bacteriol. 2005;187:8403–10.
39. Merrell DS, Butler SM, Qadri F, Dolganov NA, Alam A, Cohen MB, Calderwood SB, Schoolnik GK, Camilli A. Host-induced epidemic spread of the cholera bacterium. Nature. 2002;417:642–5.
40. Davey ME, O'Toole GA. Microbial biofilms: from ecology to molecular genetics. Microbiol Mol Biol Rev. 2000;64:847–67.

41. Wai SN, Mizunoe Y, Takade A, Kawabata SI, Yoshida SI. *Vibrio cholerae* O1 strain TSI-4 produces the exopolysaccharide materials that determine colony morphology, stress resistance, and biofilm formation. Appl Environ Microbiol. 1998;64:3648–55.
42. Watnick PI, Lauriano CM, Klose KE, Croal L, Kolter R. The absence of a flagellum leads to altered colony morphology, biofilm development and virulence in *Vibrio cholerae* O139. Mol Microbiol. 2001;39:223–35.
43. Yildiz FH, Schoolnik GK. *Vibrio cholerae* O1 El Tor: identification of a gene cluster required for the rugose colony type, exopolysaccharide production, chlorine resistance, and biofilm formation. Proc Natl Acad Sci USA. 1999;96:4028–33.
44. Costerton JW, Lewandowski Z, Caldwell DE, Korber DR, Lappin-Scott HM. Microbial biofilms. Annu Rev Microbiol. 1995;49:711–45.
45. Moorthy S, Watnick PI. Genetic evidence that the *Vibrio cholerae* monolayer is a distinct stage in biofilm development. Mol Microbiol. 2004;52:573–87.
46. Moorthy S, Watnick, PI. Identification of novel stage-specific genetic requirements through whole genome transcription profiling of *Vibrio cholerae* biofilm development. Mol Microbiol. 2005;57:1623–35.
47. Yildiz FH, Liu, XS, Heydorn A, Schoolnik GK. Molecular analysis of rugosity in a *Vibrio cholerae* O1 El Tor phase variant. Mol Microbiol. 2004;53:497–515.
48. Lauriano CM, Ghosh C, Correa NE, Klose, KE. The sodium-driven flagellar motor controls exopolysaccharide expression in *Vibrio cholerae*. J Bacteriol. 2004;186:4864–74.

Chapter 12
Filamentous Phages of *Vibrio cholerae* O1 and O139

Masahiko Ehara and M. John Albert

Abstract Bacteriophages are abundant biological entities and they play an enormous role in the adaptive evolution of bacteria. *Vibrio cholerae* O1 and O139 strains produce several types of spheroidal phages and filamentous phages, which are known to have genes encoding several virulence factors. Filamentous phages of *V. cholerae* O1 and O139 were classified into two types: fs1 and fs2. The fs1 has several subtypes and presumably play a great role in several epidemics. The receptor for filamentous phages fs1 and fs2 was shown to be the type IV fimbriae. Production of fimbriae was repressed even in the fimbriate strains due to infection with filamentous phages and this was confirmed using animal models. This chapter also describes background information regarding the development of effective cholera vaccine using phase variation in *V. cholerae*.

12.1 Introduction

Vibrio cholerae O1 and O139 strains produce several types of spheroidal phages and filamentous phages. However, their biological functions are not well understood. Bacteriophages are the most abundant biological entities on the planet and they play a profound role in driving the adaptive evolution of their hosts, the bacteria. Recent findings of filamentous phages of *V. cholerae* could lead us to better understand the cholera pathogenesis. VGJϕ, a novel filamentous phage of *V. cholerae* O139, was shown to integrate into the same chromosomal site as CTXϕ [1]. The filamentous phage fs2 of *V. cholerae* O139 also had the same *attP* with CTXϕ, suggesting that the fs2 also integrates into the bacterial *dif* locus. Furthermore, fs2 encoded *rstC*. RstC was shown to produce insoluble aggregates with RstR in vitro, causing inactivation of RstR [1]. Eventually, fs2 encoded a transcriptional antirepressor gene. Antirepression by RstC induces expression of the cholera toxin

M. Ehara (✉)
Department of Bacteriology, Institute of Tropical Medicine, Nagasaki University, Nagasaki 852-8523, Japan
e-mail: ehara@nagasaki-u.ac.jp

T. Ramamurthy, S.K. Bhattacharya (eds.), *Epidemiological and Molecular Aspects on Cholera*, Infectious Disease, DOI 10.1007/978-1-60327-265-0_12,
© Springer Science+Business Media, LLC 2011

(CT) genes and thus may contribute to the virulence of *V. cholerae*. Filamentous phages fs1 and fs2 use type IV fimbriae (mannose-sensitive hemagglutinin) as their receptors. Filamentous phages form turbid plaques without killing the host cells. Filamentous phages contain a single-stranded circular DNA genome and produce a double-stranded replicative form DNA (RF DNA) in infected cells, which can be recovered as a plasmid [2].

12.2 Isolation of Filamentous Phages from Stool Samples

Before the appearance of *V. cholerae* O139, little attention was paid to the roles of filamentous phages in the pathogenesis of cholera. Filamentous phages were isolated from all the diarrheal stool samples, independent of the duration of clinical symptoms. Stool samples were collected from patients who suffered due to *V. cholerae* O1 in 1997. The animal experiments conducted so far by using wild strains without infection of any filamentous phages will not reflect in vivo phenomenon. Cholera is not a simple infectious disease, as filamentous phages also play some roles in the pathogenesis of *V. cholerae* [3].

12.3 Typing and Subtyping of Filamentous Phages

Filamentous phages of *V. cholerae* O1 and O139 were classified into two types: fs1 and fs2 [4]. Since CTXϕ does not form a virion (virus structure), we do not consider it as a filamentous phage. CTXϕ may be a satellite phage, like RS1, which requires a helper phage to form a virion. Five types were classified in fs1 with *V. cholerae* strains isolated in Bangladesh, India, and Thailand when digested with *Hae*II or *Hin*dIII. These fs1 subtypes may increase in future outbreaks of cholera. There was no polymorphism in fs2 filamentous phages when we did restriction digestion with *Hae*II. So far, we have found two types of filamentous phages in strains of *V. cholerae* O1 and O139. It is possible to isolate new types of filamentous phages in future.

12.4 Filamentous Phages as a Tool for Molecular Epidemiology of *V. cholerae*

Specific filamentous phages may be found during cholera outbreaks. The fs1 subtype was detected in 1998 *V. cholerae* O1 and O139 strains isolated in Bangladesh. Primers were designed for the conserved *Hae*II fragments of fs1 and fs2, which amplified 959 bp and 499 bp fragments, respectively. Supernatant of stool specimens and whole cells were used as templates in the polymerase chain reaction (PCR). With both templates, phage sequences could be amplified. Figure 12.1 shows amplification of fs2 phage sequence from two strains of *V. cholerae* O1.

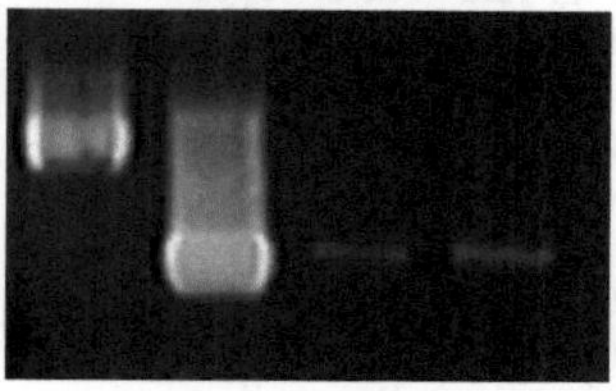

Fig. 12.1 Typing of filamentous phages produced by *V. cholerae* O1 strains 75 and 84 isolated in Vietnam in 2004. The phages were identical to fs2

Primers (5′–3′) used in these PCR assays were as follows:

fs1S TTCTCAGTGACTTCTGTTTC
fs1R CCGTTTTCTGTAATGCGTTT
fs2S AGCGCCAATCTAATGAGCTT
fs2R CGAAAAGCATTAGTAACCCC

12.5 Receptor for Filamentous Phages fs1 and fs2

The receptor for filamentous phages fs1 and fs2 was shown to be the type IV fimbriae (mannose-sensitive hemagglutinin) [4, 5]. This was confirmed by using two sets of *tcpA, mshA* deletion mutant strains. Two strains of *mshA* deletion mutant were unable to produce RF DNA for fs1 and fs2, suggesting that they are resistant to infection with fs1 and fs2 phages.

12.6 Role of Filamentous Phages in Pathogenesis of *V. cholerae*

When the fimbriate Bgd17 strain (*V. cholerae* O1, classical biotype, Inaba serotype) was infected with fs1 and fs2, the hemagglutination activity was decreased. Similar finding was observed when wild strain of *V. cholerae* O139 AI1854 was infected with fs1 and fs2. This suggests that the production of fimbriae was repressed even in the fimbriate Bgd17 strain due to infection with filamentous phages. This was confirmed in the rabbit ligated ileal loop test (Fig. 12.2). There was no change on the smooth surface of the O139 strain AI1854 infected with fs1 (Fig. 12.2a). However, there was a clear difference between the fimbriate Bgd17 strain and the same Bgd17 infected with fs2. The surface of the fimbriate Bgd17 strain infected with fs2 became smoother (Fig. 12.2b) when compared to the fimbriate Bgd17 strain (Fig. 12.2c). Filamentous structures of fimbriae change their shape to round particles during critical point drying during the preparation of samples for scanning electron microscopy. Thus, infection of *V. cholerae* with filamentous phages may play some role in the detachment of vibrios from epithelial cells of the small intestine.

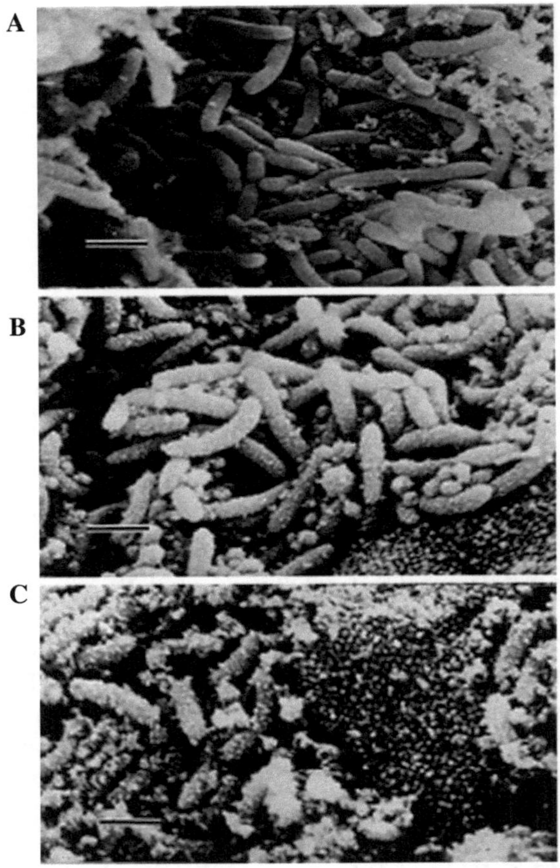

Fig. 12.2 The effect of a filamentous phage infection on fimbriation. (**a**) *V. cholerae* O139 strain AI1854 infected with fs1. (**b**) *V. cholerae* O1 classical, Inaba strain Bga17 (fimbriate) infected with fs2. (**c**) *V. cholerae* O1 Bgd17 (fimbriate). Bars indicate 1 μm

12.7 Genomic Organization of fs2

The sequence of the fs2 phage consisted of 8,651 nucleotides. The ORFs were designated according to the numbers of amino acids. Of the 11 ORFs identified in fs2, 9 were homologous to ORFs encoding previously reported peptides of M13 phage. The other two ORFs were found to be *rstC* and *rstB1* (truncated). These two ORFs are transcribed in the opposite direction of other ORFs of fs2.

RstC is known to form aggregates in vitro with RstR repressing the repressor that regulates transcription of the initiator replication protein, RstA (Davis et al. [1]). Thus, RstC enhances CT production by repressing RstR. Campos et al. [6] reported a novel type of specialized transduction consisting of site-specific cointegration of replicative form of filamentous phages, VGJϕ, and CTXϕ (or RS1) to produce a single hybrid molecule, which generates a single-stranded DNA hybrid genome that is packaged into hybrid viral particles. Their findings reinforced that TCP is not indispensable for the acquisition of CTXϕ.

12.8 *att* Site-Containing Region of fs2

The DNA sequence of fs2 revealed the presence of two sites homologous to *att* sequences known to function in integrative filamentous phages, such as VGJϕ and CTXϕ of *V. cholerae* (Fig. 12.3a). The *att*-like sites in fs2 were found to overlap in opposite directions (Fig. 12.3b). The *att* site in the viral plus strand was designated *att*-fs2-dir, and the opposite site was designated *att*-fs2-rev [7]; both sites mapped to the intergenic region between *orf500* and *rstC* (Fig. 12.3b). It is noteworthy that *att*-fs2-rev is completely identical to *att*-CTXϕ. This suggested that fs2 also cointegrates into the same site (*dif* site) with CTXϕ. This conclusion awaits experimental confirmation.

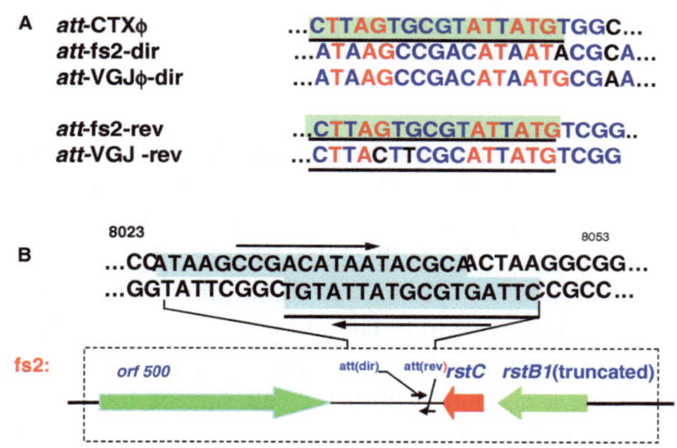

Fig. 12.3 Sequence alignment of the att regions of CTXϕ, fs2, and VGJϕ phages of *V. cholerae*

12.9 Strategies for Development of an Effective Cholera Vaccine Phase Variation of *V. cholerae*

Cholera is an acute diarrheal disease caused by *V. cholerae* O1 and O139. *V. cholerae* O1 and O 139 strains have a polar flagellum and type IV fimbriae as their cell surface appendages. To develop an effective vaccine against cholera, more attention should be paid to the phase variation of Gram-negative rods with flagella. In the case of *Salmonella* strains, the phase variation is easily observed by subculture in LB broth (Fig. 12.4). Flagellate phase bacteria are called as planktonic bacteria. To search for any colonization factor or adhesive factor, the planktonic bacteria are not

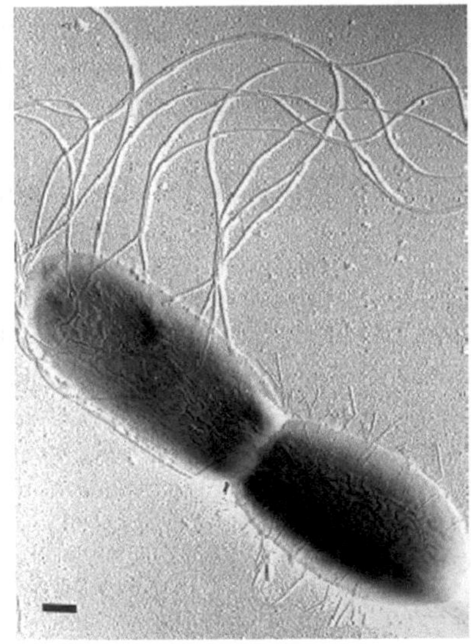

Fig. 12.4 The moment of phase variation of a strain of *Salmonella agona*. Bars indicate 100 nm

suitable. Fimbriate phase bacteria are suitable candidates for this purpose. Since the discovery of *V. cholerae* O1 by Robert Koch, little attention was paid to the fimbriate phase of *V. cholerae* O1. Most scientists working on the development of a cholera vaccine deal with planktonic vibrios (flagellate phase) only.

Fimbriate phase *V. cholerae* O1 strains exhibit properties different from flagellate phase organisms. Strains of fimbriate phase *V. cholerae* O1 have the property of autoagglutination in normal saline. They produce pellicle when cultured in liquid media. They have a unique motility for cell surface translocation (twitching motility) different from the motility imparted by flagella. A fimbriate phase strain of *V. cholerae* O1, strain Bgd17 was obtained by subculturing in TCG medium in the presence of chitin [8]. This medium was developed to increase the level of cyclic AMP in the cells due to the presence of thioproline. It is important not to include glucose in the medium. Glucose inhibits the production of cyclic AMP due to catabolic repression. Purified fimbriae from the fimbriate Bgd17 strain were highly hydrophobic and formed aggregates (Fig. 12.5).

Wild-type strains of *V. cholerae* O1 which exhibit rugose colonies and the fimbriate Bgd17 strain were transformed to ampicillin (Amp) resistance and the modes of resistance were compared between these two phases. The stationary phase of wild strains (non-fimbriate, resistant to Amp) exhibited the decreased OmpU protein production. On the other hand, the fimbriate Bgd17 strain (resistant to Amp) produced a new protein (CpxP).The dose–response of both mutant strains revealed differences between the wild-type (non-fimbriate) and fimbriate Bgd17. Under the stress conditions by Amp, a wild-type strain (V4) tries to survive by decreasing OmpU protein in a dose-dependent manner. The fimbriate Bgd17 strain tries to overcome the oxidative stress conditions by overproducing the stress combating protein (CpxP) also in

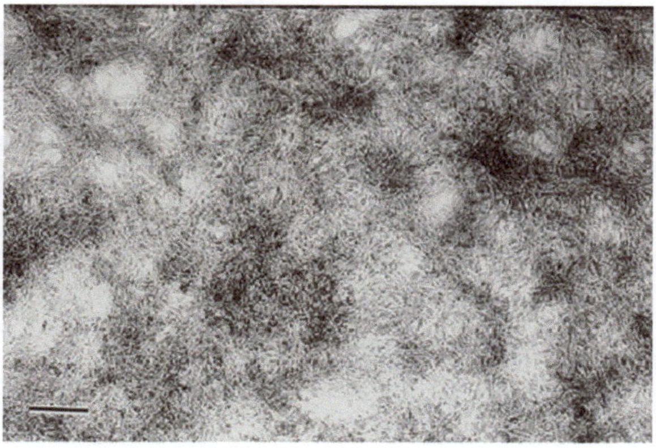

Fig. 12.5 Purified fimbriae from the fimbriate *V. cholerae* O1 Bgd17 strain. Bar indicates 100 nm

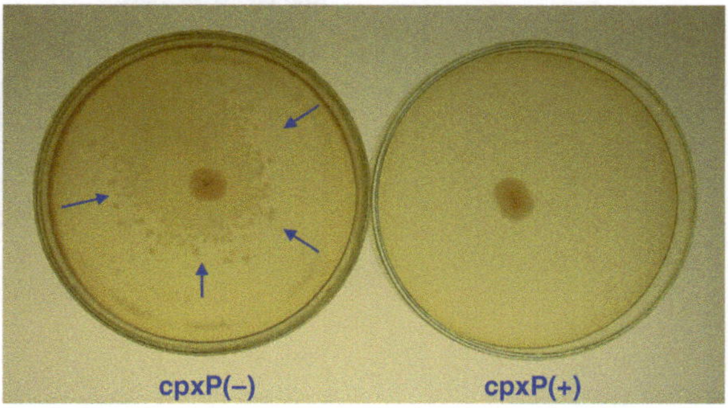

Fig. 12.6 Effect of cpxP on twitching motility. The presence of Cpx pathway is not known in *V. cholerae* O1, but its homologue may play some role in phase variation of *V. cholerae* O1 and O139. CpxP seems to repress the production of type IV fimbriae, a key factor for twitching motility

a dose-dependent manner. Interestingly, CpxP was produced in the absence of Amp once it transformed. Non-fimbriate wild strains of *V. cholerae* O1 V4 and V6 (isolated in Vietnam) and 98–42 (isolated in Laos) did not produce CpxP even in the presence of Amp. The twitching motility disappeared in the stressed Bgd17 strain that produces the CpxP protein (Fig. 12.6) [9]. The CpxP protein is known to be an inhibitor of the two-component signal transduction system, CpxA/R. The Cpx signaling pathway is known to monitor biogenesis of P-pili in *Escherichia coli* [10].

12.10 Development of Hyperfimbriate Strains of *V. cholerae* O1

V. cholerae O1 and O139 fimbrillin genes (*fimA* or *mshA*) were amplified by PCR and cloned into an *E. coli* pCR^{TM} vector. These clones were sequenced. The *fimA* sequences were found to be identical between *V. cholerae* O1 and O139. One of the clones was digested with *Eco*RI and inserted into the *Eco*RI site of pGEX-3X. The plasmid pVPP thus obtained was transferred into wild-type *V. cholerae* O1Bgd17 strain and its fimbriate strain by electroporation.

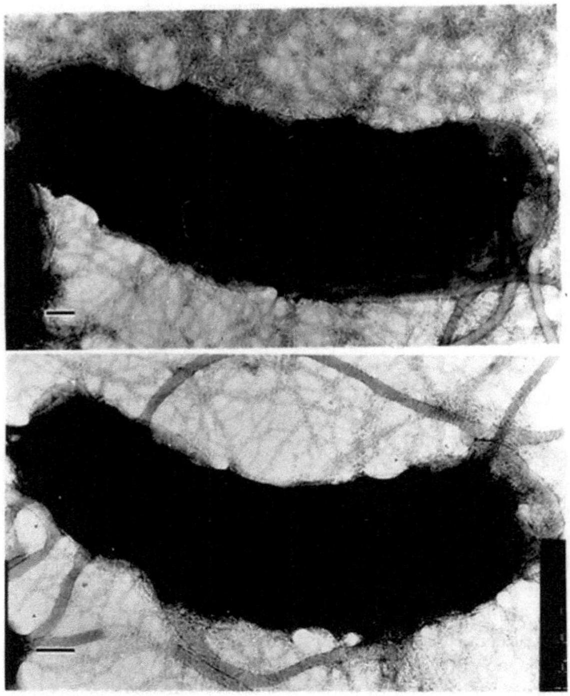

Fig. 12.7 Hyperfimbriate strains of *V. cholerae* O1, wild-type Bgd17 with pVPP (**a**), and the fimbriate Bgd17 with pVPP (**b**). Bars indicate 100 nm

The recombinant plasmid pVPP overexpressed mature fimbriae following induction of the *tac* promoter with isopropyl β-D-thiogalactopyranoside (Fig. 12.7) [11]. This candidate strain, potentially an effective cholera vaccine, is ready to be evaluated in cholera endemic areas.

References

1. Davis BM, Kimsley HH, Kane AV, Waldor MK. A satellite phage-encoded antirepressor induces repressor aggregation and cholera toxin gene transfer. EMBO J. 2002;21:4240–9.
2. Marvin DA, Hohn B. Filamentous bacterial viruses. Bacteriol Rev. 1969;33:172–209.
3. Nguyen DT, Nguyen BM, Tran HH, Ngo TC, Le TH, Nguyen HT, Albert MJ, Iwami M, Ehara M. Filamentous vibriophage fs2 encoding the *rstC* gene integrates into the same chromosomal region as the CTX phage. FEMS Microbiol Lett. 2008;284:225–30.
4. Ehara M, Shimodori S, Kojima E, Ichinose Y, Hirayama T, Albert MJ, Supawat K, Honma Y, Iwanaga M, Amako K. Characterization of filamentous phage of Vibrio cholerae O139 and O1. FEMS Microbiol Lett. 1997;154:293–301.
5. Shimodori S, Iida K, Kojima F, Takade A, Ehara M, Amako K. Morphological feature of a filamentous phage of *Vibrio cholerae* O130 Bengal. Microbiol Immunol. 1997;41:757–63.
6. Campos J, Martinez E, Marrero K, Silva Y, Rodriguez BL, Suzarte E, Ledon T, Fando R. Novel type of specialized transduction for CTXϕ or its satellite phage RS1 mediated by filamentous phage VGJϕ in *Vibrio cholerae.* J Bacteriol. 2003;185:7231–40.

7. Campos J, Martinez E, Suzarte E, Rodriguez BL, Marrero K, Silva Y, Ledon T. del Sol R, Fando R. VGJϕ, a novel filamentous phages of *Vibrio cholerae*, integrates into the same chromosomal site as CTXϕ. J Bacteriol. 2003;185:5685–96.
8. Ehara M, Iwami M, Ichinose Y, Shimodori S, Kangethe SK, Honma Y. Selective induction of fimbriate *Vibrio cholerae* O1. Trop Med 1991;33:93–107.
9. Nguyen DT, Ngo TC, Tran HH, Nguyen TPL, Nguyen BM, Morita K, et al. Two different mechanism of ampicillin resistance operating in strains of Vibrio cholerae O1 independent of resistance genes. FEMS Microbiol Lett. 2009;298:37–43.
10. Hung DL, Raivio TL, Jones CH,Silhavy TJ, Hultgren SJ. Cpx signaling pathway monitors biogenesis and affects assembly and expression of P pili. EMBO J. 2001;20:1508–18.
11. Ehara M, Iwami M, Ichinose Y, Hirayama T. Development of hyperfimbriated strains of *Vibrio cholerae* O1. Microbiol Immunol. 2000;44:439–46.

Chapter 13
Pathogenic Potential of Non-O1, Non-O139 *Vibrio cholerae*

Amit Sarkar, Ranjan K. Nandy, and Asoke C. Ghose

Abstract *Vibrio cholerae* organisms are known to exist in more than 200 different serogroups based on their "O" antigenic characters. Out of these, strains belonging to O1 or O139 serogroups have been implicated as the causative agent of cholera in the epidemic form. On the other hand, *V. cholerae* non-O1, non-O139 organisms, ubiquitously present in the aquatic environment, are usually nonpathogenic in nature. However, these strains can be isolated from sporadic cases or occasional outbreaks of gastroenteritis in man. Recently, there has been an upsurge in their isolation rate from diarrheal patients in the cholera endemic areas. Detailed characterization of these strains has demonstrated considerable diversity in their phenotypic as well as genotypic properties that may not be related to their pathogenic potential. While non-O1, non-O139 strains can cause gastroenteritis by mechanisms unrelated to those involved in epidemic cholera, some of these are shown to harbor gene clusters for cholera toxin (CTX) and toxin coregulated pilus (TCP) that are linked to the virulence potential of *V. cholerae* O1 or O139 strains. Data available so far suggest that the evolution of pathogenic strains (CTX^+ TCP^+) of non-O1, non-O139 *V. cholerae* may follow different mechanisms involving horizontal transfer of *Vibrio* pathogenicity island (VPI) and CTX gene clusters, exchange of "O" antigen biosynthesis genes, etc. However, mere acquisition of these gene clusters does not necessarily endow a non-O1, non-O139 strain with the ability to cause epidemic form of the disease which probably requires additional genetic traits that are unique to *V. cholerae* O1 or O139 strains.

13.1 Introduction

Vibrio cholerae organisms are enormously diverse in terms of serology as these have been shown to belong to more than 200 serogroups so far, based on their "O" antigenic characters [1, 2]. Initially, *V. cholerae* strains belonging to O1 serogroup

---

A.C. Ghose (✉)
National Institute of Cholera and Enteric Diseases, Beliaghata, Kolkata 700 010, India
e-mail: acghose@vsnl.net

T. Ramamurthy, S.K. Bhattacharya (eds.), *Epidemiological and Molecular Aspects on Cholera*, Infectious Disease, DOI 10.1007/978-1-60327-265-0_13,

© Springer Science+Business Media, LLC 2011

were only considered to be associated with epidemic or pandemic cholera [3–5]. All other *V. cholerae* strains that were biochemically similar to the epidemic-causing O1 strains, but not agglutinable with O1 antiserum, were referred to as "non-cholera vibrios" (NCV) or "non-agglutinable vibrios" (NAG) [3, 6]. Later on, these NCV or NAG vibrios were collectively referred to as "non-O group 1 *V. cholerae*" [7] or "non-O1 *V. cholerae*" [8]. After the emergence of *V. cholerae* O139 as the causative agent of the epidemic in 1991–1992 [9, 10], *V. cholerae* strains which were not agglutinable either by O1 or O139 antiserum were clubbed into a large heterogeneous group called "non-O1, non-O139" *V. cholerae* [1, 5]. Henceforth, the term "non-O1, non-O139 *V. cholerae*" will be used in this chapter, which will also include any reference to studies on NAG or NCV. A number of papers recently appeared in the literature dealing with different aspects of non-O1, non-O139 *V. cholerae* including their relatedness to O1 or O139 strains [11–28]. This chapter deals on the current status of non-cholera vibrios, with special reference to their pathogenic potential and their evolution.

13.2 Ecology and Epidemiology

Non-O1, non-O139 *V. cholerae* strains are autochthonous inhabitants of natural aquatic environments such as rivers, marshes, bay, and coastal areas [29–33]. These organisms were found to be widely distributed in the environment and isolated from surface and estuarine brackish waters, sewage, vegetables, seafood, chironomid egg masses etc. [34–39]. The isolation rate of these organisms from the environment normally exceeded that of *V. cholerae* O1 by several orders of magnitude even in the cholera-endemic areas [33]. The non-O1, non-O139 organisms were also isolated from stool samples of asymptomatic individuals [40] and domestic animals [41–44].

Although non-O1, non-O139 strains are largely nonpathogenic in nature, some of these were known to cause occasional outbreaks or sporadic cases of diarrhea or gastroenteritis [15–17, 45–52]. Clinical symptoms usually encountered were "cholera-like diarrhea" of mild-to-moderate severity, although bloody diarrhea was also encountered in some of the patients [33]. The presence of these organisms was reported in the stool samples of gastroenteritis patients in Asia [53–56], Africa [46], Europe [57], Australia [58], North America [7, 59, 60], and South America [61–63]. However, considerable variability was noted in the reported frequency of isolation of non-O1, non-O139 *V. cholerae* from stool samples of persons with diarrhea. Thus, the isolation rate was as high as 13% in patients with cholera-like mild watery diarrhea during cholera epidemics in Iraq [64], while in other parts of the world, e.g., Malaysia [65], Jordon [66], Thailand [15], Brazil [62], and India [67–69], these were isolated with variable frequencies. The isolation rate in Bangladesh was earlier reported to be ranging between 3 and 7% in diarrheal cases as compared to 0.01% isolation rate from stool samples of healthy village population [53, 70]. In another study, non-O1, non-O139 *V. cholerae* strains were isolated from 16% of 134 children and adults with diarrhea in Cancun, Mexico in 1983 [60]. Recently, these strains were shown to cause sporadic episodes or isolated outbreaks of diarrhea in

Khmers camps (Thailand) in 1993 [48], Lima (Peru) in 1994 [50], and east Delhi (India) in 1995 [51]. More recently, an upsurge in the isolation rate of non-O1, non-O139 strains was noted among the diarrheal patients in different parts of India [71, 72].

Apart from the sporadic cases or limited outbreaks, at least three explosive outbreaks have been reported so far from different parts of the world which could be attributed to non-O1, non-O139 *V. cholerae* [73]. A food-borne outbreak of cholera-like diarrhea due to non-O1, non-O139 *V. cholerae* (belonging to serogroups O5 and O37) was reported in Czechoslovakia in 1965 in a group of young men at an automobile training center affecting 56 persons with no fatalities [45]. The incubation period was estimated to be 20–30 h with symptoms like abdominal cramps and diarrhea that were resolved in majority of the cases by 24 h [45]. Three years later in 1968, a non-O1, non-O139 *V. cholerae* strain (O37) from a fecally contaminated well caused severe cholera-like outbreak in Khartoum (Sudan) [46, 74, 75]. About 400 cases were reported from this outbreak, out of which 125 people died [74]. It was probably the largest outbreak caused by any non-O1, non-O139 *V. cholerae* strain so far. In the year 1973, a severe cholera-like outbreak was caused by non-O1, non-O139 *V. cholerae* among aeroplane passengers arriving in Australia after stopovers in Bahrain and Singapore [58]. Food served in the flight was suspected as source of the outbreak.

The epidemiological data suggest that contaminated water and food (particularly seafood) are the important vehicles of transmission to cause sporadic cases or explosive outbreaks of non-O1, non-O139 *V. cholerae* disease, although transmission can occur through other routes as well [36, 76]. Possibility of person-to-person transmission of these infections and the "carrier status" in humans (which can last for 60 or more days) were also proposed earlier [34, 47]. Isolates of *V. cholerae* non-O1, non-O139 appear to have the capacity to survive and multiply in a wide range of foodstuff than do *V. cholerae* O1 strains, thereby highlighting the possibility of food-borne transmission of the former group [77].

Non-O1, non-O139 *V. cholerae* strains are also known to cause extraintestinal infection in man [36]. Thus, the organisms were reported to be isolated from blood, wound, ear, bile, sputum, and cerebrospinal fluid of human cases [34, 36, 78] that could be linked to their prior exposure to marine environment [52, 79, 80]. The non-O1, non-O139 *V. cholerae* was also associated with septicemia in human [56, 81–83] with high mortality rates. Some of the strains isolated from these patients were shown encapsulated and thus resisted serum bactericidal activity [33]. These records and the fact that some of these organisms can also cause septicemia, particularly in immunocompromised individuals [43, 84], suggest their invasive potential.

13.3 Strain Diversity

Non-O1, non-O139 *V. cholerae* strains isolated from clinical and environmental sources were shown to be quite diverse in terms of their serogroups [1], antibiotic resistance profiles [15, 52, 69, 85], pulse field gel electrophoresis (PFGE) patterns

[17, 19, 86], ribotypes [14, 26, 50, 87], multilocus enzyme electrophoresis analysis [21, 88], and other analytical methods [12, 18, 23, 27, 62, 89–91]. Analysis of a large number of O1, O139 and non-O1, non-O139 strains from Asia, Africa, and America showed that the same "O" serogroup may be present in genetically diverse strains exhibiting different zymovar patterns [21]. Conversely, the same zymovar may contain strains from different serogroups. These data collectively indicated that the majority of the non-O1, non-O139 strains were clonally unrelated to O1 or O139 strains. Environmental isolates of *V. cholerae* non-O1, non-O139 strains belonging to diverse serogroups or even within the same serogroup failed to show any clonal relatedness among them [15, 21]. On the other hand, predominance of certain serogroups such as O10, O12, O37, O53, and O141 was noted among the clinical isolates, particularly in those associated with localized outbreaks of gastroenteritis. For example, a localized diarrheal outbreak was caused in Lima, Peru, in 1995 by strains belonging to certain serogroups (O10, O12) of *V. cholerae* that exhibited identical ribotype patterns with homologous group [50]. In another study carried out in Calcutta, India, strains belonging to certain serogroups were isolated from 15 diarrheal cases [17]. Ribotyping and PFGE analyses revealed that the strains within the same serogroup (O11, O144) were clonally related. Some of the environmental *V. cholerae* non-O1, non-O139 strains harboring the heat-stable enterotoxin gene (*stn*) were predominantly identified as serogroup O14 [92]. Clonal relatedness between epidemiologically unrelated *V. cholerae* clinical O141 strains isolated from diverse geographical locations was documented by ribotyping and *ctx* genotyping analysis [16]. Incidentally, these strains harbored genes associated with CTX prophage and *Vibrio* pathogenicity island (VPI). Analysis of environmental strains of non-O1, non-O139 *V. cholerae* harboring VPI and CTX-like prophages showed genetic diversity amongst these strains with respect to their randomly amplified polymorphic DNA (RAPD) fingerprint types and virulence-associated gene clusters [18]. Recently, *Not*I-digested genomic DNA from several clinical and environmental *V. cholerae* strains belonging to different serogroups and harboring complete or incomplete cassettes of CTX and VPI-associated gene clusters were analyzed by PFGE [93]. The analysis showed genetic diversity amongst these strains except those belonging to the O37 serogroup, which appeared to be clonally related (Fig. 13.1). The strains CK2 (ATCC 25872) and CK4 (ATCC 25874) were associated with a large outbreak in Czechoslovakia in 1965, while the strain S7 was involved in another explosive outbreak in Sudan in 1968.

13.4 Toxins and Toxigenic Factors

Increasing numbers of literature from different parts of the world have shown that the non-O1, non-O139 *V. cholerae* strains are the causative agents of sporadic cases and outbreaks of diarrhea in man [17, 33, 45, 46, 48, 50, 51, 58]. Volunteer studies also confirmed that non-O1, non-O139 *V. cholerae* can cause human diarrhea [36]. However, pathophysiological mechanisms underlying the diarrhea caused by these organisms are not completely understood till date as no common virulence

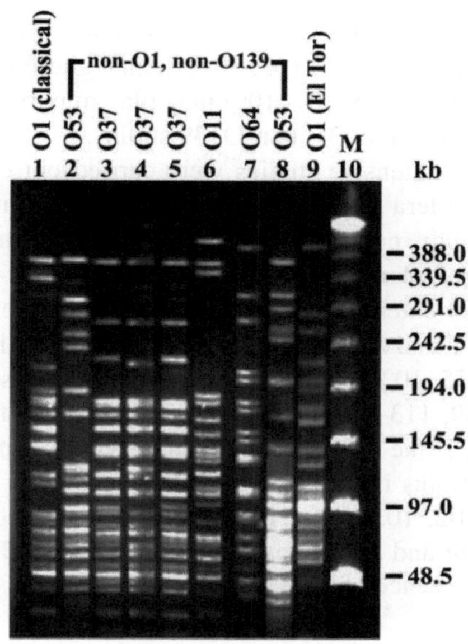

Fig. 13.1 PFGE analysis of *Not*I-digested genomic DNA of non-O1, non-O139 *V. cholerae* strains harboring incomplete or complete cassettes of VPI and CTX [93]. Similar data generated with *V. cholerae* O1 strains are also presented in the figure for a comparison. *Vibrio cholerae* strains used were O395 (lane 1), AS18 (2), S7 (3), CK2 (4), CK4 (5), AS16 (6), CG15 (7), 10259 (8), and Co840 (9). Positions of marker DNA (M) are indicated

factor(s) could be identified in strains isolated from the reported cases. As a matter of fact, a large number of non-O1, non-O139 *V. cholerae* strains isolated from localized outbreaks were found to be lacking all known virulence traits of toxigenic strains of *V. cholerae* O1 despite the fact that some of these isolates produced cytotoxic effect in their culture supernatant [17, 50]. On the other hand, some of the non-O1, non-O139 *V. cholerae* strains isolated from Osaka airport (Japan) quarantine station were shown to produce a heat-stable enterotoxin (later on designated as NAG-ST) that closely resembled the heat-stable toxin (ST) produced by enterotoxigenic *Escherichia coli* [94]. Other studies supported this observation through demonstration of the presence of NAG-ST gene (*stn/sto*) in 2.3–13.1% of clinical or environmental non-O1, non-O139 *V. cholerae* strains isolated from different geographical regions [95–97]. In fact, an outbreak of cholera-like illness in Thailand was caused by NAG-ST producing non-O1, non-O139 *V. cholerae* strains [48].

The presence of a gene cluster related to the repeat in toxin (RTX) family could be demonstrated in clinical and environmental strains of non-O1, non-O139 *V. cholerae* [16, 98]. The RTX toxin was shown to be responsible for the cytotoxic activity and acute inflammatory response [99] induced by *V. cholerae* strains. Recent genomic analysis of non-O1, non-O139 *V. cholerae* strains [28] predicted a protein with high degree of similarity to the thermostable-direct hemolysin (TDH) of *Vibrio parahaemolyticus*. Further, the TDH secretion could be facilitated by the type III secretion system (TTSS) shown to be present in certain pathogenic strains of non-O1, non-O139 *V. cholerae*. The presence of another type VI secretion system was recently documented in *V. cholerae*, which was proposed to be responsible for the virulence of certain non-O1, non-O139 strains [100]. Several other putative virulence factors like El Tor hemolysin [27, 101, 102], cell-associated hemagglutinins [50, 55, 103, 104], hemagglutinin–protease [105, 106], and other cytotoxins [19, 50, 55, 107–109] may play important role in the induction of diarrhea by some of the non-O1, non-O139 strains.

Extensive studies were carried out over the years to ascertain the presence of cholera toxin (CT) or CT-like toxin in non-O1, non-O139 *V. cholerae*. In an early study, non-O1, non-O139 *V. cholerae* strains isolated from a cholera-like epidemic in Sudan were shown to produce a heat-labile toxin almost identical to *V. cholerae* O1 enterotoxin [75]. Later studies, however, demonstrated that majority of the clinical and environmental isolates of non-O1, non-O139 *V. cholerae* did not produce CT [55, 103, 109–112] or even did not possess the structural genes for CT [19, 52, 68, 69, 113–115]. Nevertheless, presence of *ctxAB* genes and production of CT and/or CT-like toxin were documented in a number of non-O1, non-O139 *V. cholerae* strains isolated from clinical as well as environmental sources [16, 23, 39, 54, 97, 102, 103, 116–122]. A few non-O1, non-O139 *V. cholerae* strains also harbored *zot* and *ace* genes (in addition to *ctxAB*), thereby suggesting the possibility of the presence of CTX genetic element in these strains [13, 39, 102, 121].

13.5 CTX Prophage (Genetic Element) and VPI

The presence of genes related to CTX prophage could be demonstrated in a number of clinical and environmental strains of non-O1, non-O139 *V. cholerae* [13, 16, 18, 23, 26, 93, 123]. Restriction fragment length polymorphism (RFLP) analysis showed that the genetic organizations of the *ctx* region of these strains were quite heterogeneous and different from those of the epidemic-causing O1 strains. While majority of the strains carried a single copy of the CTX genetic element, a few strains also had two copies of the same [13, 23, 123]. Interestingly, some of these strains possessed only two copies (instead of three or more copies found in the epidemic-causing O1 or O139 strains) of the heptanucleotide (TTTTGAT) repeats in the intergenic region upstream of the *ctxAB* [24]. Nucleotide sequence analysis of *rstR* and *ctxAB* genes suggested the presence of new alleles of CTX prophage in some of the non-O1, non-O139 strains [23, 123]. These studies also identified the existence of pre-CTX prophages that probably served as progenitor(s) of new alleles.

It was believed earlier that non-O1, non-O139 *V. cholerae* was devoid of genes for another critical determinant of *V. cholerae* pathogenicity, i.e., the gene for the toxin-coregulated pilus protein or TcpA [124, 125]. However, the presence of *tcpA* gene was recently reported in a limited number of strains [13, 16, 122, 126–128]. Interestingly, a few of these strains possessed new alleles of *tcpA* gene that differed from *tcpA* of both classical and El Tor biotype strains [18, 20, 22, 23, 127, 128]. Further analysis revealed that the sequence divergence in these alleles was primarily located in the carboxyl terminal half of the 20-kDa TcpA protein which was likely to influence the reactivity of the pilus or pilus protein to antibodies and/or ligands or receptors of biological interest [129, 130]. However, the new TcpA variants were biologically active to facilitate bacterial colonization and to act as CTXϕ receptor [20, 24, 25].

Analysis of the VPI genes of a non-O1, non-O139 strain belonging to the serogroup O53 was carried out with respect to *tcp* gene cluster and certain other VPI-associated genes [24]. While the nucleotide sequence of *tcpA* of the strain differed by 26–28% from that of O1 classical and El Tor biotype strains, respectively, partial sequence analysis of certain other VPI-associated genes and intergenic regions of the strain showed only minor variations (0.4–4.8%) from corresponding sequences in O1 strains. The GC content of the partially sequenced region of VPI of the O53 strain was found to be 34.6%, which was quite comparable to the GC content (35.5%) of the VPIs of O1 strains [131], but differed from the overall GC content (47–48%) of the *V. cholerae* genome [132]. In an earlier study, Karaolis et al. [131] demonstrated that most of the divergence between the VPIs of sixth (classical) and seventh (El Tor) pandemic strains were located in or around the *tcpA* gene constituting the central region of VPI, but not its left or right segments. Thus, the generation of new variants of *tcpA* in the non-O1, non-O139 *V. cholerae* strain (O53) may arise as a result of horizontal transfer and recombinational events involving *tcpA* and adjacent genes located in the central segment of VPI [20, 131]. Divergence of certain other VPI-associated genes in the environmental strains of non-O1, non-O139 *V. cholerae* from the corresponding genes in pathogenic O1 strains was also recently documented [18, 25]. For example, this divergence ranged between 20 and 70% at the protein level for the TcpF and 36% for the ToxT of VPI, although the strains shared the same TcpA sequence of O1 classical type. The segment-wise divergence of VPI-associated genes between the clinical and certain environmental strains of *V. cholerae* probably indicates the VPIs to be composed of genes in the modular form and selective genetic exchange of these modules between different VPIs may give rise to further diversity in these genetic islands [18].

13.6 Pathogenic Potential

Isolation of non-O1, non-O139 *V. cholerae* strains from clinical cases around the world had documented the pathogenic potential of these strains. Human volunteer studies [36] also established that some of these strains can colonize the intestine and cause severe diarrhea through the production of toxins like NAG-ST. Diarrheagenic potential of these strains was also established in experimental animals using rabbit illeal loop [19, 26, 103, 133], removable intestinal tie adult rabbit diarrhea (RITARD) [25, 50, 133], and suckling mouse models [93, 103, 133]. In fact, enrichment of *V. cholerae* strains present in the environmental waters through passage in the rabbit intestine had helped in the selective isolation of strains capable of colonizing infant mice [26]. Interestingly, 56.8% of these strains were shown to possess genes for the TCP alone or both TCP and CT. However, this and other studies also showed that the non-O1, non-O139 strains can colonize the intestine and induce diarrhea even in the absence of TCP, CT, and NAG-ST [17, 26, 36]. Clearly, pathogenic potential of some of the non-O1, non-O139 *V. cholerae* strains is likely to be mediated by diarrheagenic factors unrelated to those associated with the epidemic-causing strains of *V. cholerae* O1 or O139 serogroups. Although non-O1, non-O139 strains were shown to release different cytotoxic factors, their relevance in the induction of diarrhea in vivo is far from clear. Similarly, the role of different types of pili, other than TCP [134–137] and capsules [138, 139] in the intestinal colonization process is yet to be established in humans.

Isolation of non-O1, non-O139 strains harboring CTX and VPI-associated genes had been reported from clinical as well as environmental sources. However, the documentation of the mere presence of virulence-associated genes may not provide sufficient evidences toward the pathogenic or the epidemic potential of these strains, which is likely to be dependent on the functionality of the entire virulence gene cluster and the complex signaling pathways that are needed to couple appropriate environmental signals to virulence gene expression at an adequate level. To address these questions, several clinical and environmental isolates of non-O1, non-O139 *V. cholerae* strains harboring *ctxAB* and *tcpA* gene clusters were subjected to evaluate their pathogenic potential through in vitro and in vivo experiments (Table 13.1). These included strains that were involved in the outbreaks in Sudan (S7) and Czechoslovakia (CK2 and CK4). Out of ten strains tested, only three (S7, CK2, and CK4) produced significantly higher amounts of CT in vitro as compared

Table 13.1 Cholera toxin (CT) production, colonization efficiency, and lethal toxicity (LD_{50}) values of *V. cholerae* strains[a] belonging to non-O1, non-O139 serogroups

Strain	Serogroup	Amount of CT[b] produced in AKI medium (ng/ml/unit of opacity)	Increase in CFU (folds)[c]	LD_{50} value[d]
V5	O6	3.7	n.d.[e]	$>1 \times 10^8$
NS70	O6	328	3.1	4.6×10^7
A16	O11	3.4	1.3	$> 1 \times 10^8$
S7	O37	2,475	150	9.3×10^6
CK2	O37	2,600	97	7.8×10^6
CK4	O37	2,200	107	6.9×10^6
10259	O53	1.6	12	2×10^7
AS18	O53	344	77	1.2×10^7
CG15	O64	4.2	1.7	$>1 \times 10^8$
Co366	O1 (El Tor)	4500	345	5.1×10^6

[a]Strains were isolated from clinical and environmental sources and shown to possess incomplete or complete cassettes of VPI and CTX [93]
[b]Assayed by the GM_1-ELISA method and expressed as nanograms of toxin per milliliter of culture supernatant per unit opacity of bacterial suspension measured at 540 nm
[c]Each mouse in the experimental group (at least four in each group) was challenged with the organism taken in 0.1 ml of saline. Control animals received normal saline (0.1 ml) only. Extent of intestinal colonization is expressed as colony forming units (CFU) of bacteria recovered from each mouse intestine
[d]LD_{50} values were determined by orally feeding groups of mice (at least six in each group) with different doses of bacteria and recording the number of surviving mice after 24 h of challenge
[e]n.d., Not determined

to those of the remaining strains. In fact, the amounts of CT produced by these three strains were quite comparable to O1 strains belonging to both the biotypes. These three strains also possessed the ability to colonize in vivo when grown in the mouse intestine and exhibited lethal toxicity (LD_{50}) that were comparable to those of *V. cholerae* O1 strains. The other non-O1, non-O139 strains exhibited poor colonization ability and lethal toxicity when assayed under comparable conditions. In a recent study, non-O1, non-O139 *V. cholerae* strains isolated from the environment were shown to colonize mouse intestine and induce fluid accumulation [140].

Coordinately regulated expression of CT and TcpA is an essential (although perhaps not sufficient) feature of epidemic-causing strains of *V. cholerae*, which differentiated these from non-epidemic strains. Expression of CT and TcpA is regulated at the transcriptional level that involves several regulatory proteins which include the transmembrane proteins ToxR, TcpP, and TcpH of the TcpPH operon and the cytosolic protein ToxT [141, 142]. Coordinated expression of CT and TcpA was demonstrated in the O37 serogroup strains S7, CK2, and CK4, which involved transcriptional activation of *toxR*, *tcpPH*, and *toxT* genes [24, 93]. In contrast, the other non-O1, non-O139 strains harboring CTX and VPI failed to exhibit similarly regulated expression of the corresponding genes, which probably explains their low-level expression of CT and TcpA and poor pathogenic potential. As a matter of fact, majority of the non-O1, non-O139 *V. cholerae* strains harboring CT genes were earlier shown to produce considerably smaller amounts of CT than those produced by their O1 or O139 counterparts [13, 24, 25, 103]. The possible mechanisms for this may include (i) the absence of *toxT* gene or presence of a *toxT* allele functionally deficient from the canonical *toxT*; (ii) poor or lack of synthesis of ToxT as a result of defects or deviations in the signaling pathway that couples appropriate environmental signals to *toxT* activation; and (iii) failure to activate *ctxAB* operon (despite the production of an adequate amount of functional ToxT) due to the lack of sufficient number of the heptanucleotide repeats in its promoter region.

13.7 Evolutionary Perspective

It is reasonable to argue that *V. cholerae* initially emerged as an organism capable of its survival in environmental niche. Its adaptation to human host necessitated the acquisition of additional genes, some of which coded for pathogenic factors. The generation of strains with epidemic or pandemic potential required the acquisition of additional traits that were needed to cause explosive outbreaks as well as to facilitate their regional or global dissemination. Out of different serogroups of *V. cholerae*, only O1 strains were able to acquire all these genetic traits to express their epidemic/pandemic potential. More recently, *V. cholerae* O139 strains, shown to be derived from an O1 El Tor strain, were found to be responsible for cholera-like epidemics [5]. Comparative genomic analysis of several *V. cholerae* O1 classical, El Tor (pre-pandemic and pandemic), and O139 strains isolated over the decades revealed high degree of genetic similarity among these strains [143]. On the other hand, similar analysis carried out on four diarrheagenic non-O1, non-O139 strains indicated these strains to be quite divergent from each other as well as from representative O1 or O139 strains [28]. All the four strains, which belonged to different serogroups and were devoid of *ctx* and *tcp*, apparently lacked ~6% (> 250) of the genes that were present in the seventh pandemic O1 strains [28]. Evolutionary divergence among non-O1, non-O139 strains was documented by O'Shea et al. [27] through comparative nucleotide sequence analysis of the housekeeping genes as well as by studying the distribution of 12 virulence-associated genetic regions. This and other studies [11, 21, 144] suggested that majority of the non-O1, non-O139 strains may be phylogenetically quite different from *V. cholerae* O1 strains. However, available data also indicate that, as in the case of O139 strains, certain non-O1, non-O139 strains probably arose from *V. cholerae* O1 progenitor strains through exchange of genes involved in O-antigen biosynthesis [22, 27, 145]. These include the strains belonging to O37 serogroup that were involved in Sudan and Czechoslovakia outbreaks. Similar conclusion was drawn earlier by comparative analyses of *IS1004* DNA fingerprinting [89] and multilocus enzyme electrophoresis [88] profiles of the O37 (Sudan) and O1 strains. Incidentally, the Czechoslovakia outbreak was caused by *V. cholerae* strains which showed striking similarity with the Sudan strain in PFGE analysis (Fig. 13.1) [93]. Similarly, a comparison of the *IS1004* fingerprint [89] and 16S rRNA–23S rRNA intergenic spacer region ribotype [87] profiles between the O1 classical and Czechoslovakia strains suggested that these strains were closely related. Thus, it is likely that *V. cholerae* O37 strains that were involved in Sudan and Czechoslovakia outbreaks arose from *V. cholerae* O1 classical strains through the replacement of O1 by O37 antigenic character. In fact, evolution of an O37 strain from an O1 classical strain was recently documented which involved recombination-based replacement of *wbe* gene cluster by *wb∗* of O37 [22]. An analysis of *wav* gene cluster type (involved in the biosynthesis of core oligosaccharide of *V. cholerae* lipopolysaccharide or LPS) showed that clinical isolates belonging to O1, O139, and O37 serogroups shared identical type 1 *wav* gene cluster, while the other non-O1, non-O139 strains contained one of the four other types (type 2 to type 5) of this cluster [144]. These results suggest the predominance of type 1 *wav* gene cluster in strains associated with clinical cholera (which include Sudan and Czechoslovakia strains of O37 serogroup), while the other non-O1, non-O139 strains are evolutionarily divergent in their core oligosaccharide (OS) biosynthesis gene cluster. The evolutionary relatedness of the O37 strains (S7, CK2, and CK4) to that of O1 classical strains was further suggested by the demonstration of VPI (with classical type of *tcpA* gene) and CTX genetic elements in these strains with three to four copies of heptanucleotide (TTTTGAT) repeats in the *zot–ctx* intergenic region [93].

Sequential acquisition of VPI and CTX or pre-CTX prophage by non-O1, non-O139 *V. cholerae* through horizontal gene transfer was already proposed earlier [18, 23]. These include clonally related clinical strains belonging to O141 serogroup that were isolated between 1984 and 1994 from sporadic cases occurring at diverse geographical location [16]. However, further analysis suggested that the O141 strains were evolutionarily distinct from the epidemic-causing strains of *V. cholerae* [27]. This is also exemplified by the observation that a large number of non-O1, non-O139 strains (which include these O141 clinical strains) possess genes related to "type III secretion system" or TTSS [28]. Incidentally, TTSS genes were shown to be present in *V. parahaemolyticus* and certain other species of *Vibrio* but not in epidemic-causing strains of *V. cholerae*.

Based on the above considerations, a model for the evolution of pathogenic strains non-O1, non-O139 *V. cholerae* (VPI^{+}, CTX^{+}) belonging to the serogroups O37 and O141 is proposed (Fig. 13.2). However, the fact remains that mere acquisition of VPI and CTX may not endow a *V. cholerae* strain the ability to cause an epidemic or a pandemic, which may very well depend on the genetic background of the host strain. So far, strains belonging to O1, and O139, serogroups were shown to cause cholera epidemics/pandemics around the world. Thus, a major question to address in this regard is the failure of the non-O1 strains (including those derived from O1 strains by exchange of "O" antigen biosynthesis genes) to cause persistent epidemics or pandemics like O1 strains. For example, the strains which caused explosive outbreaks of cholera-like diarrhea in Sudan and Czechoslovakia failed to reappear in the same or other parts of the world. Incidentally, the isolation rate of *V. cholerae* O139 strains (the emergence of which was earlier thought to herald the beginning of the eighth pandemic) gradually declined in the cholera endemic parts of India and Bangladesh [146, 147]. Thus, it is quite possible that certain "O" serogroups may endow the *V. cholerae* strains with the selective advantage to persist in the environment resulting in transmission within a population as well as to

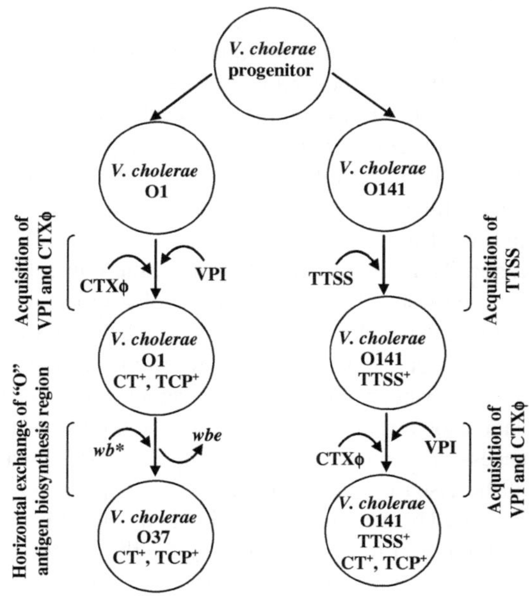

Fig. 13.2 A model showing the evolution of pathogenic *V. cholerae* strains belonging to serogroups O37 and O141

move from one geographical location to another. Interestingly, the clinical isolates of *V. cholerae* O141, though failed to cause explosive outbreaks or epidemics, were capable of regional or even global dissemination [16].

Epidemic-causing *V. cholerae* strains belonging to O1 serogroup adopt several strategies for their survival in the environment under adverse conditions during the interepidemic period. These include (i) exopolysaccharide production responsible for their conversion to rugose phenotype with high frequency [148, 149], (ii) transformation into a "viable but non-culturable" state [150–153], and (iii) association with planktonic and/or abiotic surfaces to form biofilms [154–156]. However, differences were noted between *V. cholerae* strains in regard to their capacity to adapt to these strategies for their prolonged survival in the environment. Further, the prevalence of serogroup-specific phages in the cholera-endemic areas may influence the environmental persistence of *V. cholerae* strains during the seasonal outbreak of cholera [157, 158]. Whether the apparent failure of some of the pathogenic non-O1, non-O139 *V. cholerae* strains to persist in the environment and disseminate can be attributed to these factors remains to be established.

13.8 Conclusion

Non-O1, non-O139 *V. cholerae* organisms, ubiquitously present in the aquatic environment, are usually nonpathogenic in nature. However, these organisms can be isolated from sporadic cases or occasional outbreaks of gastroenteritis. Recently, there has been an upsurge in their isolation rate from diarrheal patients in the Indian subcontinent as well as in certain other parts of the world. Phenotypic as well as genotypic analyses suggest that this group of organisms is extremely diverse, although certain pathogenic strains within the same serogroup (for example, O37 and O141) may have a clonal origin. Such diversity is also reflected in the failure to attribute their pathogenic properties to a common set of virulence factors. While some strains can induce diarrhea by mechanisms unrelated to those involved in O1 cholera, certain others are shown to harbor CTX and TCP gene clusters that are related to those found in epidemic-causing strains of *V. cholerae* O1 and O139. Interestingly, the available data suggest that the evolution of pathogenic strains of non-O1, non-O139 *V. cholerae* (CTX^+ TCP^+) may follow different lineages. Thus, like the O139 strains, the clinical O37 strains (involved in outbreaks in Czechoslovakia and Sudan) probably evolved from O1 strains (CTX^+ TCP^+) through the exchange of "O" antigen biosynthesis gene clusters. On the other hand, the pathogenic strains belonging to O141 serogroup were likely to evolve from a non-O1 background through horizontal acquisition of VPI and CTX clusters. The data also suggest that mere acquisition of VPI and CTX genetic elements does not necessarily endow a non-O1 *V. cholerae* strain with the ability to cause severe form of the disease which depends on the expression of adequate levels of TCP and CT regulated by a complex signaling pathway. However, the ability to persist in the environment and to disseminate regionally or globally perhaps depends on additional factors that may include specific "O" serogroup characters. Evidently, *V. cholerae*

O1 (and to some extent O139) strains have been able to fulfill both the requirements as these strains were implicated in the epidemics so far recorded in the history of cholera.

Acknowledgments ACG is the recipient of an Emeritus Scientist award from the Indian Council of Medical Research (ICMR), New Delhi, India.

References

1. Shimada T, Arakawa E, Itoh K, Okitsu T, Matsushima A, Asai Y, Yami S, Nakazato T, Nair GB, Albert MJ, Takeda Y. Extended serotyping scheme for *Vibrio cholerae*. Curr Microbiol. 1994;28:175–8.
2. Yamai S, Okitsu T, Shimada T, Katsube Y. Distribution of serogroups of *Vibrio cholerae* non-O1 non-O139 with specific reference to their ability to produce cholera toxin, and addition of novel serogroups. Kansenshogaku Zasshi. 1997;71:1037–45.
3. Gardner AD, Venkatraman VK. The antigens of the cholera group of vibrios. J Hyg. 1935;35:262–82.
4. Pollitzer R. Cholera. Monograph no. 43. WHO Geneva; 1959.
5. Kaper JB, Morris JG Jr., Levine MM. Cholera. Clin Microbiol Rev. 1995;8:48–86.
6. Sakazaki R. Bacteriology of *Vibrio* and related organisms. In: Barua D, Greenough WB III, editors. Cholera. New York: Plenum Medical Book Co.; 1992. pp. 37–55.
7. Morris JG Jr., Wilson R, Davis BR, Wachsmuth IK, Riddle CF, Wathen HG, Pollard RA, Blake PA. Non-O group 1 *Vibrio cholerae* gastroenteritis in the United States: clinical, epidemiologic, and laboratory characteristics of sporadic cases. Ann Intern Med. 1981;94:656–8.
8. Yamamoto K, Takeda Y, Miwatani T, Craig JP. Evidence that a non-O1 *Vibrio cholerae* produces enterotoxin that is similar but not identical to cholera enterotoxin. Infect Immun. 1983;41:896–901.
9. Albert MJ, Siddique AK, Islam MS, Faruque ASG, Ansaruzzaman M, Faruque SM, Sack RB. Large outbreak of clinical cholera due to *Vibrio cholerae* non-O1 in Bangladesh. Lancet. 1993;341:704.
10. Ramamurthy T, Garg S, Sharma R, Bhattacharya SK, Nair GB, Shimada T, Takeda T, Karasawa T, Kurazano H, Pal A, Takeda Y. Emergence of novel strain of *Vibrio cholerae* with epidemic potential in southern and eastern India. Lancet. 1993;341:703–4
11. Karaolis DK, Lan R, Reeves PR. The sixth and seventh cholera pandemics are due to independent clones separately derived from environmental, nontoxigenic, non-O1 *Vibrio cholerae*. J Bacteriol. 1995;177:3191–8.
12. Rivera IG, Chowdhury MA, Huq A, Jacobs D, Martins MT, Colwell RR. Enterobacterial repetitive intergenic consensus sequences and the PCR to generate fingerprints of genomic DNAs from *Vibrio cholerae* O1, O139, and non-O1 strains. Appl Environ Microbiol. 1995;61:2898–904.
13. Ghosh C, Nandy RK, Dasgupta SK, Nair GB, Hall RH, Ghose AC. A search for cholera toxin (CT), toxin coregulated pilus (TCP), the regulatory element ToxR and other virulence factors in non-O1/non-O139 *Vibrio cholerae*. Microb Pathog. 1997;22:199–208.
14. Dalsgaard A, Forslund A, Mortensen HF, Shimada T. Ribotypes of clinical *Vibrio cholerae* non-O1 non-O139 strains in relation to O-serotypes. Epidemiol Infect. 1998;121:535–45.
15. Dalsgaard A, Forslund A, Bodhidatta L, Serichantalergs O, Pitarangsi C, Pang L, Shimada T, Echeverria P. A high proportion of *Vibrio cholerae* strains isolated from children with diarrhoea in Bangkok, Thailand are multiple antibiotic resistant and belong to heterogenous non-O1, non-O139 O-serotypes. Epidemiol Infect. 1999;122:217–26.
16. Dalsgaard A, Serichantalergs O, Forslund A, Lin W, Mekalanos J, Mintz E, Shimada T, Wells JG. Clinical and environmental isolates of *Vibrio cholerae* serogroup O141 carry the

CTX phage and the genes encoding the toxin-coregulated pili. J Clin Microbiol. 2001;39:4086–92.

17. Sharma C, Thungapathra M, Ghosh A, Mukhopadhyay AK, Basu A, Mitra R, Basu I, Bhattacharya SK, Shimada T, Ramamurthy T, Takeda T, Yamasaki S, Takeda Y, Nair GB. Molecular analysis of non-O1, non-O139 *Vibrio cholerae* associated with an unusual upsurge in the incidence of cholera-like disease in Calcutta, India. J Clin Microbiol. 1998;36:756–63.
18. Mukhopadhyay AK, Chakraborty S, Takeda Y, Nair GB, Berg DE. Characterization of VPI pathogenicity island and CTXphi prophage in environmental strains of *Vibrio cholerae*. J Bacteriol. 2001;183:4737–46.
19. Singh DV, Matte MH, Matte GR, Jiang S, Sabeena F, Shukla BN, Sanyal SC, Huq A, Colwell RR. Molecular analysis of *Vibrio cholerae* O1, O139, non-O1, and non-O139 strains: clonal relationships between clinical and environmental isolates. Appl Environ Microbiol. 2001;67:910–21.
20. Boyd EF, Waldor MK. Evolutionary and functional analyses of variants of the toxin-coregulated pilus protein TcpA from toxigenic *Vibrio cholerae* non-O1/non-O139 serogroup isolates. Microbiology. 2002;148:1655–66.
21. Freitas FS, Momen H, Salles CA. The zymovars of *Vibrio cholerae*: multilocus enzyme electrophoresis of *Vibrio cholerae*. Mem Inst Oswaldo Cruz. 2002;97:511–6.
22. Li M, Shimada T, Morris JG Jr., Sulakvelidze A, Sozhamannan S. Evidence for the emergence of non-O1 and non-O139 *Vibrio cholerae* strains with pathogenic potential by exchange of O-antigen biosynthesis regions. Infect Immun. 2002;70:2441–53.
23. Li M, Kotetishvili M, Chen Y, Sozhamannan S. Comparative genomic analyses of the *Vibrio* pathogenicity island and cholera toxin prophage regions in nonepidemic serogroup strains of *Vibrio cholerae*. Appl Environ Microbiol. 2003;69:1728–38.
24. Sarkar A, Nandy RK, Nair GB, Ghose AC. *Vibrio* pathogenicity island and cholera toxin genetic element-associated virulence genes and their expression in non-O1 non-O139 strains of *Vibrio cholerae*. Infect Immun. 2002;70:4735–42.
25. Faruque SM, Kamruzzaman M, Meraj IM, Chowdhury N, Nair GB, Sack RB, Colwell RR, Sack DA. Pathogenic potential of environmental *Vibrio cholerae* strains carrying genetic variants of the toxin-coregulated pilus pathogenicity island. Infect Immun. 2003;71:1020–5.
26. Faruque SM, Chowdhury N, Kamruzzaman M, Dziejman M, Rahman MH, Sack DA, Nair GB, Mekalanos JJ. Genetic diversity and virulence potential of environmental *Vibrio cholerae* population in a cholera-endemic area. Proc Natl Acad Sci USA. 2004;101:2123–8.
27. O'Shea YA, Reen FJ, Quirke AM, Boyd EF. Evolutionary genetic analysis of the emergence of epidemic *Vibrio cholerae* isolates on the basis of comparative nucleotide sequence analysis and multilocus virulence gene profiles. J Clin Microbiol. 2004;42:4657–71.
28. Dziejman M, Serruto D, Tam VC, Sturtevant D, Diraphat P, Faruque SM, Rahman MH, Heidelberg JF, Decker J, Li L, Montgomery KT, Grills G, Kucherlapati R, Mekalanos JJ. Genomic characterization of non-O1, non-O139 *Vibrio cholerae* reveals genes for a type III secretion system. Proc Natl Acad Sci USA. 2005;102:3465–70.
29. Kaper J, Lockman H, Colwell RR, Joseph SW. Ecology, serology, and enterotoxin production of *Vibrio cholerae* in Chesapeake Bay. Appl Environ Microbiol. 1979;37:91–103.
30. Colwell RR, Shimada T, West PA, Maneval D, Remmers EF, Illiot EL, Carlson NE. Ecology of pathogenic vibrios in Chesapeake Bay. In: Colwell RR, editor. Vibrios in the environment. New York: Wiley; 1984. pp. 367–87.
31. Kaysner CA, Abeyta C Jr., Wekell MM, DePaola A Jr., Stott RF, Leitch JM. Incidence of *Vibrio cholerae* from estuaries of the United States West Coast. Appl Environ Microbiol. 1987;53:1344–8.
32. Nair GB, Sarkar BL, De SP, Chakraborti MK, Bhadra RK, Pal SC. Ecology of *Vibrio cholerae* in the freshwater environs of Calcutta, India. Microb Ecol. 1988;8:203–16.

33. Morris JG Jr. Non-O group 1 *Vibrio cholerae* strains not associated with epidemic disease. In: Wachsmuth IK, Blake PA, Olsvik O, editors. *Vibrio cholerae* and cholera: molecular to global perspectives. Washington, DC: ASM Press; 1994. pp. 103–15.
34. Blake PA, Allegra DT, Snyder JD, Barrett TJ, McFarland L, Caraway CT, Feeley JC, Craig JP, Lee JV, Puhr ND, Feldman RA. Cholera possible endemic focus in the United States. N Engl J Med. 1980;302:305–9.
35. Hood MA, Ness GE. Survival of *Vibrio cholerae* and *Escherichia coli* in estuarine waters and sediments. Appl Environ Microbiol. 1982;43:578–84.
36. Morris JG Jr. Non-O group 1 *Vibrio cholerae*: a look at the epidemiology of an occasional pathogen. Epidemiol Rev. 1990;12:179–91.
37. Broza M, Gancz H, Halpern M, Kashi Y. Adult non-biting midges: possible windborne carriers of *Vibrio cholerae* non-O1 non-O139. Environ Microbiol. 2005;7:576–85.
38. Isaac-Marquez AP, Lezama-Davila CM, Eslava-Campos C, Navarro-Ocana A, Cravioto-Quintana A. Serotypes of *Vibrio cholerae* non-O1 isolated from water supplies for human consumption in Campeche, Mexico and their antibiotic susceptibility pattern. Mem Inst Oswaldo Cruz. 1998;93:17–22.
39. Jiang S, Chu W, Fu W. Prevalence of cholera toxin genes (*ctxA* and *zot*) among non-O1/O139 *Vibrio cholerae* strains from Newport Bay, California. Appl Environ Microbiol. 2003;69:7541–4.
40. Lowry PW, McFarland LM, Peltier BH, Roberts NC, Bradford HB, Herndon JL, Stroup DF, Mathison JB, Blake PA, Gunn RA. *Vibrio* gastroenteritis in Louisiana: a prospective study among attendees of a scientific congress in New Orleans. J Infect Dis. 1989;160:978–84.
41. Sack RB. A search for canine carriers of *Vibrio*. J Infect Dis. 1973;127:709–12.
42. Rhodes JB, Schweitzer D, Ogg JE. Isolation of non-O1 *Vibrio cholerae* associated with enteric disease of herbivores in western Colorado. J Clin Microbiol. 1985;22:572–5.
43. Sanyal SC. Epidemiology and pathogenicity of non-O1 *Vibrio* species and related organisms. In: Barua D, Greenough WB III, editors. Cholera. New York: Plenum Medical Book Company; 1992. pp. 57–67.
44. Visser IJ, Vellema P, van Dokkum H, Shimada T. Isolation of *Vibrio cholerae* from diseased farm animals and surface water in The Netherlands. Vet Rec. 1999;144:451–2.
45. Aldova E, Laznickova K, Stepankova E, Lietava J. Isolation of nonagglutinable vibrios from an enteritis outbreak in Czechoslovakia. J Infect Dis. 1968;118:25–31.
46. Kamal AM. Outbreak of gastroenteritis by non-agglutinable (NAG) vibrios in the republic of the Sudan. J Egypt Public Health Assoc. 1971;46:125–73.
47. Morris JG Jr., Takeda T, Tall BD, Losonsky GA, Bhattacharya SK, Forrest BD, Kay BA, Nishibuchi M. Experimental non-O group 1 *Vibrio cholerae* gastroenteritis in humans. J Clin Invest. 1990;85:697–705.
48. Bagchi K, Echeverria P, Arthur JD, Sethabutr O, Serichantalergs O, Hoge CW. Epidemic of diarrhea caused by *Vibrio cholerae* non-O1 that produced heat-stable toxin among Khmers in a camp in Thailand. J Clin Microbiol. 1993;31:1315–7.
49. World Health Organization. Cholera in 1992. Wkly Epidemiol Rec. 1993;68:149–56.
50. Dalsgaard A, Albert MJ, Taylor DN, Shimada T, Meza R, Serichantalergs O, Echeverria P. Characterization of *Vibrio cholerae* non-O1 serogroups obtained from an outbreak of diarrhea in Lima, Peru. J Clin Microbiol. 1995;33:2715–22.
51. Rudra S, Mahajan R, Mathur M, Kathuria K, Talwar V. Cluster of cases of clinical cholera due to *Vibrio cholerae* O10 in east Delhi. Indian J Med Res. 1996;103:71–3.
52. Dalsgaard A, Forslund A, Hesselbjerg A, Bruun B. Clinical manifestations and characterization of extra-intestinal *Vibrio cholerae* non-O1, non-O139 infections in Denmark. Clin Microbiol Infect. 2000;6:625–7.
53. McIntyre OR, Feeley JC, Greenough III WB, Benenson AS, Hassan SI, Saad A. Diarrhea caused by non-cholera vibrios. Am J Trop Med Hyg. 1965;14:412–8.
54. Spira WM, Daniel RR, Ahmed QS, Huq A, Yusuf A, Sack DA. Clinical features and pathogenicity of O group 1 non-agglutinating *Vibrio cholerae* and other vibrios isolated from

cases of diarrhea in Dacca, Bangladesh. In: Zinnaka Y, Takeya K, editors. Proceedings of the 14th joint conference US–Japan Cooperative Medical Science Program; 1978. pp. 137–53.

55. Ramamurthy T, Bag PK, Pal A, Bhattacharya SK, Bhattacharya MK, Shimada T, Takeda T, Karasawa T, Kurazono H, Takeda Y, Nair GB. Virulence patterns of *Vibrio cholerae* non-O1 strains isolated from hospitalised patients with acute diarrhoea in Calcutta, India. J Med Microbiol. 1993;39:310–7.
56. Ko WC, Chuang YC, Huang GC, Hsu SY. Infections due to non-O1 *Vibrio cholerae* in southern Taiwan: predominance in cirrhotic patients. Clin Infect Dis. 1998;27:774–80.
57. Back E, Ljunggren A, Smith Jr. H. Non-cholera vibrios in Sweden. Lancet. 1974;303:723–4.
58. Dakin WP, Howell DJ, Sutton RG, O'Keefe MF, Thomas P. Gastroenteritis due to non-agglutinable (non-cholera) vibrios. Med J Aust. 1974;2:487–90.
59. Morris JG, Jr., Black RE. Cholera and other vibrioses in the United States. N Engl J Med. 1985;312:343–50.
60. Finch MJ, Valdespino JL, Wells JG, Perez-Perez G, Arjona F, Sepulveda A, Bessudo D, Blake PA. Non-O1 *Vibrio cholerae* infections in Cancun, Mexico. Am J Trop Med Hyg. 1987;36:393–7.
61. Kay BA, Sack RB, Spira WM, Guerra HE, Guerrero CE, Chaparro E, Yi AE, Salazar-Lindo E, Chea E, Wachsmuth IK. *Vibrio cholerae* non-O1 isolated from five people with diarrhoea in Lima. Lancet. 1984;1:218.
62. Thompson FL, Thompson CC, Vicente AC, Theophilo GN, Hofer E, Swings J. Genomic diversity of clinical and environmental *Vibrio cholerae* strains isolated in Brazil between 1991 and 2001 as revealed by fluorescent amplified fragment length polymorphism analysis. J Clin Microbiol. 2003;41:1946–50.
63. Bidinost C, Saka HA, Aliendro O, Sola C, Panzetta-Duttari G, Carranza P, Echenique J, Patrito E, Bocco JL. Virulence factors of non-O1, non-O139 *Vibrio cholerae* isolated in Cordoba, Argentina. Rev Argent Microbiol. 2004;36:158–63.
64. El-Shawi N, Thewaini AJ. Non-agglutinable vibrios isolated in the 1966 epidemic of cholera in Iraq. Bull WHO. 1969;40:163–6.
65. Dutt AK, Alwi S, Velauthan T. A shellfish-borne cholera outbreak in Malaysia. Trans R Soc Trop Med Hyg. 1971;65:815–8.
66. Shehabi AA, Abu Rajab AB, Shaker AA. Observations on the emergence of non-cholera vibrios during an outbreak of cholera. Jordan Med J. 1980;14:125–7.
67. Mukhopadhyay AK, Garg S, Nair GB, Kar S, Ghosh RK, Pajni S, Ghosh A, Shimada T, Takeda T, Takeda Y. Biotype traits and antibiotic susceptibility of *Vibrio cholerae* serogroup O1 before, during and after the emergence of the O139 serogroup. Epidemiol Infect. 1995;115:427–34.
68. Bhattacharya MK, Dutta D, Bhattacharya SK, Deb A, Mukhopadhyay AK, Nair GB, Shimada T, Takeda Y, Chowdhury A, Mahalanabis D. Association of a disease approximating cholera caused by *Vibrio cholerae* of serogroups other than O1 and O139. Epidemiol Infect. 1998;120:1–5.
69. Chakraborty S, Garg P, Ramamurthy T, Thungapathra M, Gautam JK, Kumar C, Maiti, S, Yamasaki S, Shimada T, Takeda Y, Ghosh A, Nair GB. Comparison of antibiogram, virulence genes, ribotypes and DNA fingerprints of *Vibrio cholerae* of matching serogroups isolated from hospitalised diarrhoea cases and from the environment during 1997–1998 in Calcutta, India. J Med Microbiol. 2001;50:879–88.
70. Black RE, Merson MH, Brown KH. Epidemiological aspects of diarrhea associated with known enteropathogens in rural Bangladesh. In: Chen LC, Scrimshaw NS, editors. Diarrhea and malnutrition. New York: Plenum Publishing Corporation; 1983. pp. 73–86.
71. Chatterjee S, Ghosh K, Raychoudhuri A, Chowdhury G, Bhattacharya MK, Mukhopadhyay AK, Ramamurthy T, Bhattacharya SK, Klose KE, Nandy RK. Incidence, virulence factors, and clonality among clinical strains of non-O1, non-O139 *Vibrio cholerae* isolates from hospitalized diarrheal patients in Kolkata, India. J Clin Microbiol. 2009;47: 1087–95.

72. Chandrasekhar MR, Krishna BV, Patil AB. Changing characteristics of *Vibrio cholerae*: emergence of multidrug resistance and non-O1, non-O139 serogroups. Southeast Asian J Trop Med Public Health. 2008;39:1092–7.
73. Cheasty T, Said B, Threlfall EJ. *V. cholerae* non-O1: implications for man? Lancet. 1999;354: 89–90.
74. World Health Organization. Outbreak of gastroenteritis by non-agglutinable (NAG) vibrios. Wkly Epidemiol Rec. 1969;44:1–28.
75. Zinnaka Y, Carpenter CCJ, Jr. An enterotoxin produced by non-cholera vibrios. Johns Hopkins Med J. 1972;131:403–11.
76. Namdari H, Klaips CR, Hughes JL. A cytotoxin-producing strain of *Vibrio cholerae* non-O1, non-O139 as a cause of cholera and bacteremia after consumption of raw clams. J Clin Microbiol. 2000;38:3518–9.
77. Roberts D, Gilbert RJ. Survival and growth of non-cholera vibrios in various foods. J Hyg (Lond). 1979;82:123–31.
78. Hughes JM, Hollis DG, Gangarosa EJ, Weaver RE. Non-cholera *Vibrio* infections in the United States. Clinical, epidemiologic, and laboratory features. Ann Intern Med. 1978;88:602–6.
79. Bonner JR, Coker AS, Berryman CR, Pollock HM. Spectrum of *Vibrio* infections in a Gulf Coast community. Ann Intern Med. 1983;99:464–9.
80. Hoge CW, Watsky D, Peeler RN, Libonati JP, Israel E, Morris JG Jr. Epidemiology and spectrum of *Vibrio* infections in a Chesapeake Bay community. J Infect Dis. 1989;160: 985–93.
81. Chong Y, Kwon OH, Lee SY, Kim BS, Min JS. Non-O group 1 *Vibrio cholerae* septicemia and peritonitis—report of two cases. Yonsei Med J. 1985;26:82–4.
82. Safrin S, Morris JG Jr., Adams M, Pons V, Jacobs R, Conte JE Jr. Non-O:1 *Vibrio cholerae* bacteremia: case report and review. Rev Infect Dis. 1988;10:1012–7.
83. Wistrom J. A case of non-O:1 *Vibrio cholerae* bacteremia from northern Europe. J Infect Dis. 1989;160:732.
84. Anderson AM, Varkey JB, Petti CA, Liddle RA, Frothingham R, Woods CW. Non-O1 *Vibrio cholerae* septicemia: case report, discussion of literature, and relevance to bioterrorism. Diagn Microbiol Infect Dis. 2004;49:295–7.
85. Garg P, Chakraborty S, Basu I, Datta S, Rajendran K, Bhattacharya T, Yamasaki S, Bhattacharya SK, Takeda Y, Nair GB, Ramamurthy T. Expanding multiple antibiotic resistance among clinical strains of *Vibrio cholerae* isolated from 1992–7 in Calcutta, India. Epidemiol Infect. 2000;124:393–9.
86. Kotetishvili M, Stine OC, Chen Y, Kreger A, Sulakvelidze A, Sozhamannan S, Morris JG Jr. Multilocus sequence typing has better discriminatory ability for typing *Vibrio cholerae* than does pulsed-field gel electrophoresis and provides a measure of phylogenetic relatedness. J Clin Microbiol. 2003;41:2191–6.
87. Chun J, Huq A, Colwell RR. Analysis of 16S–23S rRNA intergenic spacer regions of *Vibrio cholerae* and *Vibrio mimicus*. Appl Environ Microbiol. 1999;65: 2202–8.
88. Beltran P, Delgado G, Navarro A, Trujillo F, Selander RK, Cravioto A. Genetic diversity and population structure of *Vibrio cholerae*. J Clin Microbiol. 1999;37:581–90.
89. Bik EM, Gouw RD, Mooi FR. DNA fingerprinting of *Vibrio cholerae* strains with a novel insertion sequence element: a tool to identify epidemic strains. J Clin Microbiol. 1996;34:1453–61.
90. Jiang SC, Matte M, Matte G, Huq A, Colwell RR. Genetic diversity of clinical and environmental isolates of *Vibrio cholerae* determined by amplified fragment length polymorphism fingerprinting. Appl Environ Microbiol. 2000;66:148–53.
91. Theophilo GN, Rodrigues DP, Leal NC, Hofer E. Distribution of virulence markers in clinical and environmental *Vibrio cholerae* non-O1/non-O139 strains isolated in Brazil from 1991 to 2000. Rev Inst Med Trop Sao Paulo. 2006;48:65–70.

92. Sarkar B., Bhattacharya T, Ramamurthy T, Shimada T, Takeda Y, Nair GB. Preferential association of the heat stable enterotoxin gene (*stn*) with environmental strains of *Vibrio cholerae* belonging to the O14 serogroup. Epidemiol Infect. 2002;129:245–51.
93. Sarkar A. Molecular analysis of cholera toxin genetic element and toxin coregulated pilus related gene clusters demonstrable in certain strains of non-O1/ non-O139 *Vibrio cholerae*. Ph.D. thesis, Jadavpur University, Kolkata; 2005.
94. Arita M, Takeda T, Honda T, Miwatani T. Purification and characterization of *Vibrio cholerae* non-O1 heat-stable enterotoxin. Infect Immun. 1986;52:45–9.
95. Hoge CW, Sethabutr O, Bodhidatta L, Echeverria P, Robertson DC, Morris JG Jr. Use of a synthetic oligonucleotide probe to detect strains of non-serovar O1 *Vibrio cholerae* carrying the gene for heat-stable enterotoxin (NAG-ST). J Clin Microbiol. 1990;28:1473–6.
96. Pal A, Ramamurthy T, Bhadra RK, Takeda T, Shimada T, Takeda Y, Nair GB, Pal SC, Chakrabarti S. Reassessment of the prevalence of heat-stable enterotoxin (NAG-ST) among environmental *Vibrio cholerae* non-O1 strains isolated from Calcutta, India, by using a NAG-ST DNA probe. Appl Environ Microbiol. 1992;58:2485–9.
97. Chen AZ, Cai DS, Zhu SY, Huang ZX, Liao HL, Yan GY, Zhu W. Study on biology and epidemiological characteristics of *Vibrio cholerae* non-O1 strains isolated from environmental waters in Foshan city. Zhonghua Yu Fang Yi Xue Za Zhi. 2004;38:47–9.
98. Chow KH, Ng TK, Yuen KY, Yam WC. Detection of RTX toxin gene in *Vibrio cholerae* by PCR. J Clin Microbiol. 2001;39:2594–7.
99. Fullner KJ, Boucher JC, Hanes MA, Haines III GK, Meehan BM, Walchle C, Sansonetti PJ, Mekalanos JJ. The contribution of accessory toxins of *Vibrio cholerae* O1 El Tor to the proinflammatory response in a murine pulmonary cholera model. J Exp Med. 2002;195:1455–62.
100. Pukatzki S, Ma AT, Sturtevant D, Krastins B, Sarracino D, Nelson WC, Heidelberg JF, Mekalanos JJ. Identification of a conserved bacterial protein secretion system in *Vibrio cholerae* using the *Dictyostelium* host model system. Proc Natl Acad Sci USA. 2006;103:1528–33.
101. Ichinose Y, Yamamoto K, Nakasone N, Tanabe MJ, Takeda T, Miwatani T, Iwanaga M. Enterotoxicity of El Tor-like hemolysin of non-O1 *Vibrio cholerae*. Infect Immun. 1987;55:1090–3.
102. Kurazono H, Pal A, Bag PK, Nair GB, Karasawa T, Mihara T, Takeda Y. Distribution of genes encoding cholera toxin, zonula occludens toxin, accessory cholera toxin, and El Tor hemolysin in *Vibrio cholerae* of diverse origins. Microb Pathog. 1995;18:231–5.
103. Datta-Roy K, Banerjee K, De SP, Ghose AC. Comparative study of expression of hemagglutinins, hemolysins, and enterotoxins by clinical and environmental isolates of non-O1 *Vibrio cholerae* in relation to their enteropathogenicity. Appl Environ Microbiol. 1986;52: 875–9.
104. Datta-Roy K, Dasgupta C, Ghose AC. Hemagglutination and intestinal adherence properties of clinical and environmental isolates of non-O1 *Vibrio cholerae*. Appl Environ Microbiol. 1989;55:2403–6.
105. Booth BA, Finkelstein RA. Presence of hemagglutinin/protease and other potential virulence factors in O1 and non-O1 *Vibrio cholerae*. J Infect Dis. 1986;154:183–6.
106. Ghosh A, Saha DR, Hoque KM, Asakura M, Yamasaki S, Koley H, Das SS, Chakrabarti MK, Pal A. Enterotoxigenicity of mature 45-kilodalton and processed 35-kilodalton forms of hemagglutinin protease purified from a cholera toxin gene-negative *Vibrio cholerae* non-O1, non-O139 strain. Infect Immun. 2006;75:2937–46.
107. Saha PK, Koley H, Nair GB. Purification and characterization of an extracellular secretogenic non-membrane-damaging cytotoxin produced by clinical strains of *Vibrio cholerae* non-O1. Infect Immun. 1996;64:3101–8.
108. Begum K, Ahsan CR, Ansaruzzaman M, Dutta DK, Ahmad QS, Talukder KA. Toxin(s), other than cholera toxin, produced by environmental non-O1, non-O139 *Vibrio cholerae*. Cell Mol Immunol. 2006;3:115–21.

109. Draskovicova M, Karolcek J, Winkler. Experimental toxigenicity of NAG vibrios. Zentralbl Bakteriol. 1977;237:65–71.
110. Madden JM, McCardell BA, Read RB. *Vibrio cholerae* in shellfish from US coastal waters. Food Technol. 1982;36:93–6.
111. Gyobu Y, Kodama H, Uetake H, Katsuda S. Studies on the enteropathogenic mechanism of non-O1 *Vibrio cholerae* isolated from the environment and fish in Toyama Prefecture. Microbiol Immunol. 1984;28:735–45.
112. Shehabi AA, Drexler H, Richardson SH. Virulence mechanisms associated with clinical isolates of non-O1 *Vibrio cholerae*. Zentralbl Bakteriol Mikrobiol Hyg. 1986; 261:232–9.
113. Kaper JB, Nataro JP, Roberts NC, Siebeling RJ, Bradford HB. Molecular epidemiology of non-O1 *Vibrio cholerae* and *Vibrio mimicus* in the US Gulf Coast region. J Clin Microbiol. 1986;23:652–4.
114. Nair GB, Oku Y, Takeda Y, Ghosh A, Ghosh RK, Chattopadhyay S, Pal SC, Kaper JB,Takeda T. Toxin profiles of *Vibrio cholerae* non-O1 from environmental sources in Calcutta, India. Appl Environ Microbiol. 1988;54:3180–2.
115. Guhathakurta B, Sasmal D, Pal S, Chakraborty S, Nair GB, Datta A. Comparative analysis of cytotoxin, hemolysin, hemagglutinin and exocellular enzymes among clinical and environmental isolates of *Vibrio cholerae* O139 and non-O1, non-O139. FEMS Microbiol Lett. 1999;179:401–7.
116. Roberts NC, Seibeling RJ, Kaper JB, Bradford HB Jr. Vibrios in the Louisiana Gulf Coast environment. Microb Ecol. 1982;8:299–312.
117. Nishibuchi M, Seidler RJ. Medium-dependent production of extracellular enterotoxins by non-O-1 *Vibrio cholerae*, *Vibrio mimicus*, and *Vibrio fluvialis*. Appl Environ Microbiol. 1983;45:228–31.
118. Spira WM, Fedorka-Cray PJ, Pettebone P. Colonization of the rabbit small intestine by clinical and environmental isolates of non-O1 *Vibrio cholerae* and *Vibrio mimicus*. Infect Immun. 1983;41:1175–83.
119. Yamamoto K, Takeda Y, Miwatani T, Craig JP. Purification and some properties of a non-O1 *Vibrio cholerae* enterotoxin that is identical to cholera enterotoxin. Infect Immun. 1983;39:1128–35.
120. Hanchalay S, Seriwatana J, Echeverria P, Holmgren J, Tirapat C, Moseley SL, Taylor DN. Non-O1 *Vibrio cholerae* in Thailand: homology with cloned cholera toxin genes. J Clin Microbiol. 1985;21:288–9.
121. Karasawa T, Mihara T, Kurazono H, Nair GB, Garg S, Ramamurthy T, Takeda Y. Distribution of the *zot* (zonula occludens toxin) gene among strains of *Vibrio cholerae* O1 and non-O1. FEMS Microbiol Lett. 1993;106:143–5.
122. Chakraborty S, Mukhopadhyay AK, Bhadra RK, Ghosh AN, Mitra R, Shimada T, Yamasaki S, Faruque SM, Takeda Y, Colwell RR, Nair GB. Virulence genes in environmental strains of *Vibrio cholerae.* Appl Environ Microbiol. 2000;66:4022–8.
123. Bhattacharya T, Chatterjee S, Maiti D, Bhadra RK, Takeda Y, Nair GB, Nandy RK. Molecular analysis of the *rstR* and *orfU* genes of the CTX prophages integrated in the small chromosomes of environmental *Vibrio cholerae* non-O1, non-O139 strains. Environ Microbiol. 2006;8:526–634.
124. Taylor R, Shaw C, Peterson K, Spears P, Mekalanos J. Safe, live *Vibrio cholerae* vaccines? Vaccine. 1988;6:151–4.
125. Waldor MK, Mekalanos JJ. Emergence of a new cholera pandemic: molecular analysis of virulence determinants in *Vibrio cholerae* O139 and development of a live vaccine prototype. J Infect Dis. 1994;170:278–83.
126. Said B, Smith HR, Scotland SM, Rowe B. Detection and differentiation of the gene for toxin co-regulated pili (*tcpA*) in *Vibrio cholerae* non-O1 using the polymerase chain reaction. FEMS Microbiol Lett. 1995;125:205–9.

127. Novais RC, Coelho A, Salles CA, Vicente AC. Toxin-co-regulated pilus cluster in non-O1, non-toxigenic *Vibrio cholerae*: evidence of a third allele of pilin gene. FEMS Microbiol Lett. 1999;171:49–55.
128. Nandi B, Nandy RK, Vicente AC, Ghose AC. Molecular characterization of a new variant of toxin-coregulated pilus protein (TcpA) in a toxigenic non-O1/Non-O139 strain of *Vibrio cholerae*. Infect Immun. 2000;68:948–52.
129. Chattopadhyaya R, Ghose AC. Model of *Vibrio cholerae* toxin coregulated pilin capable of filament formation. Protein Eng. 2002;15:297–304.
130. Craig L, Taylor RK, Pique ME, Adair BD, Arvai AS, Singh M, Lloyd SJ, Shin DS, Getzoff ED, Yeager M, Forest KT, Tainer JA. Type IV pilin structure and assembly: X-ray and EM analyses of *Vibrio cholerae* toxin-coregulated pilus and *Pseudomonas aeruginosa* PAK pilin. Mol Cell. 2003;11:1139–50.
131. Karaolis DK, Lan R, Kaper JB, Reeves PR. Comparison of *Vibrio cholerae* pathogenicity islands in sixth and seventh pandemic strains. Infect Immun. 2001;69:1947–52.
132. Heidelberg JF, Eisen JA, Nelson WC Clayton RA, Gwinn ML, Dodson RJ, Haft DH, Hickey EK, Peterson JD, Umayam L, Gill SR, Nelson KE, Read TD, Tettelin H, Richardson D, Ermolaeva MD, Vamathevan J, Bass S, Qin H, Dragoi I, Sellers P, McDonald L, Utterback T, Fleishmann RD, Nierman WC, White O. DNA sequence of both chromosomes of the cholera pathogen *Vibrio cholerae*. Nature. 2000;406:477–83.
133. Spira WM, Fedorka-Cray PJ. Production of cholera toxin-like toxin by *Vibrio mimicus* and non-O1 *Vibrio cholerae*: batch culture conditions for optimum yields and isolation of hypertoxigenic lincomycin-resistant mutants. Infect Immun. 1983;42:501–9.
134. Nakasone N, Iwanaga M. Pili of *Vibrio cholerae* non-O1. Infect Immun. 1990;58:1640–6.
135. Yamashiro T, Iwanaga M. Purification and characterization of a pilus of a *Vibrio cholerae* strain: a possible colonization factor. Infect Immun. 1996;64:5233–8.
136. Kuroki H, Toma C, Nakasone N, Yamashiro T, Iwanaga M. Gene analysis of *Vibrio cholerae* NAGV14 pilus and its distribution. Microbiol Immunol. 2001;45:417–24.
137. Sengupta TK, Sengupta DK, Ghose AC. A 20-kDa pilus protein with haemagglutination and intestinal adherence properties expressed by a clinical isolate of non-O1 *Vibrio cholerae*. FEMS Microbiol Lett. 1993;112:237–42.
138. Chen Y, Stine OC, Morris JG, Johnson JA. Genetic variation of capsule/LPS biogenesis in two serogroup O31 *Vibrio cholerae* isolates. FEMS Microbiol Lett. 2007;273:133–9.
139. Chen Y, Bystricky P, Adeyeye J, Panigrahi P, Ali A, Johnson JA, Bush CA, Morris JG Jr., Stine OC. The capsule polysaccharide structure and biogenesis for non-O1 *Vibrio cholerae* NRT36S: genes are embedded in the LPS region. BMC Microbiol. 2007;7:20.
140. Bag PK, Bhowmik P, Hajra TK, Ramamurthy T, Sarkar P, Majumder M, Chowdhury G, Das SC. Putative virulence traits and pathogenicity of *Vibrio cholerae* non-O1,non-O139 isolates from surface waters in Kolkata, India. Appl Environ Microbiol. 2008;74:5635–44.
141. Skorupski K, Taylor RK. A new level in the *Vibrio cholerae* ToxR virulence cascade: AphA is required for transcriptional activation of the *tcpPH* operon. Mol Microbiol. 1999;31: 763–71.
142. Yu RR, DiRita VJ. Regulation of gene expression in *Vibrio cholerae* by ToxT involves both antirepression and RNA polymerase stimulation. Mol Microbiol. 2002;43:119–34.
143. Dziejman M, Balon E, Boyd D, Fraser CM, Heidelberg JF, Mekalanos JJ. Comparative genomic analysis of *Vibrio cholerae*: genes that correlate with cholera endemic and pandemic disease. Proc Natl Acad Sci USA. 2002;99:1556–61.
144. Nesper J, Kraiss A, Schild S, Blass J, Klose KE, Bockemuhl J, Reidl J. Comparative and genetic analyses of the putative *Vibrio cholerae* lipopolysaccharide core oligosaccharide biosynthesis (*wav*) gene cluster. Infect Immun. 2002;70:2419–33.
145. Blokesch M, Schoolnik GK. Serogroup conversion of *Vibrio cholerae* in aquatic reservoirs. PLoS Pathog. 2007;3:e81.
146. Basu A, Garg P, Datta S, Chakraborty S, Bhattacharya T, Khan A, Ramamurthy T, Bhattacharya SK, Yamasaki S, Takeda Y, Nair GB. *Vibrio cholerae* O139 in

Calcutta, 1992–1998: incidence, antibiograms, and genotypes. Emerg Infect Dis. 2000;6: 139–47.

147. Albert MJ, Nair GB. *Vibrio cholerae* O139 Bengal-10 years on. Rev Med Microbiol. 2005; 16:135–43.
148. Ali A, Rashid MH, Karaolis DK. High-frequency rugose exopolysaccharide production by *Vibrio cholerae*. Appl Environ Microbiol. 2002;68:5773–8.
149. Matz C, McDougald D, Moreno AM, Yung PY, Yildiz FH, Kjelleberg S. Biofilm formation and phenotypic variation enhance predation-driven persistence of *Vibrio cholerae*. Proc Natl Acad Sci USA. 2005;102:16819–24.
150. Colwell RR, Brayton PR, Grimes DJ, Roszak DR, Huq SA, Palmer LM. Viable but nonculturable *Vibrio cholerae* and related pathogens in the environment: implication for release of genetically engineered microorganisms. Biotechnology. 1985:817–20.
151. Reidl J, Klose KE. *Vibrio cholerae* and cholera: out of the water and into the host. FEMS Microbiol Rev. 2002;26:125–39.
152. Binsztein N, Costagliola MC, Pichel M, Jurquiza V, Ramirez FC, Akselman R, Vacchino M, Huq A, Colwell R. Viable but nonculturable *Vibrio cholerae* O1 in the aquatic environment of Argentina. Appl Environ Microbiol. 2004;70:7481–6.
153. Alam M, Sultana M, Nair GB, Sack RB, Sack DA, Siddique AK, Ali A, Huq A, Colwell RR. Toxigenic *Vibrio cholerae* in the aquatic environment of Mathbaria, Bangladesh. Appl Environ Microbiol. 2006;72:2849–55.
154. Yildiz FH, Schoolnik GK. *Vibrio cholerae* O1 El Tor: identification of a gene cluster required for the rugose colony type, exopolysaccharide production, chlorine resistance, and biofilm formation. Proc Natl Acad Sci USA. 1999;96:4028–33.
155. Zhu J, Mekalanos JJ. Quorum sensing-dependent biofilms enhance colonization in *Vibrio cholerae*. Dev Cell. 2003;5:647–56.
156. Faruque SM, Biswas K, Udden SM, Ahmad QS, Sack DA, Nair GB, Mekalanos JJ. Transmissibility of cholera: in vivo-formed biofilms and their relationship to infectivity and persistence in the environment. Proc Natl Acad Sci USA. 2006;103:6350–5.
157. Faruque SM, Islam MJ, Ahmad QS, Faruque AS, Sack DA, Nair GB, Mekalanos JJ. Self-limiting nature of seasonal cholera epidemics: Role of host-mediated amplification of phage. Proc Natl Acad Sci USA. 2005;102:6119–24.
158. Weitz JS, Hartman H. Phage in the time of cholera. Lancet Infect Dis. 2006;6:257–8.

Chapter 14
Proteases Produced by *Vibrio cholerae* and Other Pathogenic Vibrios: Pathogenic Roles and Expression

Sumio Shinoda

Abstract Pathogenic vibrios produce various pathogenic factors such as enterotoxin, hemolysin, cytotoxin, protease, siderophore, adhesive factor, and hemagglutinin. Direct toxic factors such as enterotoxin, hemolysin, and cytotoxin are related to the symptoms, whereas siderophore and adhesive factors may cause indirect factors, which play roles in the establishment of the infection. The proteases produced by pathogenic vibrios are long recognized to play pathogenic roles in the infection. Zn metalloprotease produced by *Vibrio cholerae* is thought to play indirect pathogenic roles such as processing other protein toxin(s) or supporting bacterial growth in the digestive tract. Hemagglutinin/protease (HA/P) produced by *V. cholerae* was the one first noticed which has 609 amino acid residues and a molecular size of 69.3 kDa. Further characterization of this protein revealed that the C-terminal peptide mediates hemagglutination or binding to the cell surface in many vibrios. HA/P is secreted via the type II secretion pathway at the cell pole and thought to play a role by promoting mucin gel penetration, detachment, and the spread of infection along the gastrointestinal tract. Special interaction of HA/P on chironomid egg masses was shown to be an important factor for the survival of *V. cholerae* in aquatic environments. Vibrios have a complex quorum-sensing system containing three kinds of autoinducers. In some, HA/P plays a role in controlling the quorum-sensing machinery. A cysteine protease domain was found in repeat in toxin (RTX) of *V. cholerae*. RTX is a large multifunctional toxin that causes actin cross-linking and is processed by proteolytic action during translocation into host cells. Other proteases such as *Vibrio vulnificus* protease (VVP) is well known for cell damage and association with major virulence. Thus, proteases produced by pathogenic vibrios play a variety of pathological roles.

S. Shinoda (✉)
Faculty of Science, Okayama University of Science, Okayama 700-0005, Japan
e-mail: shinoda@dls.ous.ac.jp

T. Ramamurthy, S.K. Bhattacharya (eds.), *Epidemiological and Molecular Aspects on Cholera*, Infectious Disease, DOI 10.1007/978-1-60327-265-0_14,
© Springer Science+Business Media, LLC 2011

14.1 Introduction

Proteases play many physiological roles in the life cycle of an organism, but the enzymes produced by pathogenic microorganisms occasionally act as toxic factors. Many of the bacterial proteases are metalloproteases which have zinc(II) ion in the catalytic site, although other types of proteases such as serine protease are also produced by pathogenic bacteria. For example, the light chain of botulinum toxin is a metalloprotease having HEXXH motif as the zinc-binding site [1]. Tetanus toxin [2], *Bacteroides fragilis* enterotoxin [3], and *Bacillus anthracis* lethal toxin [4] are known to be Zn metalloproteases. Pathogenic roles of bacterial protease are summarized in Table 14.1.

Table 14.1 Pathogenic roles of bacterial proteases

Growth stimulation by supplying peptides/amino acids
Production of bradykinin by activation of kallikrein–kinin cascade
Histamine release from mast cells
Inactivation of plasma protease inhibitor
Inactivation of immunoglobulins or complement system
Disturbance of coagulation system
Stimulation of bacterial invasion by collagenolytic activity
Stimulation of septicemia
Processing of protein toxins
Stimulation of toxin distribution
Stimulation of viral infection

Pathogenic vibrios produce various pathogenic factors including enterotoxin, hemolysin, cytotoxin, protease, siderophore, adhesive factor, hemagglutinin, and so on. Of these, enterotoxin, hemolysin, and cytotoxin are direct toxic factors causing the symptoms, whereas siderophore and adhesive factor are indirect factors, which play roles in the establishment of the infection. The proteases produced by pathogenic vibrios are also recognized to play pathogenic roles in the infection, directly and indirectly. Many of vibrios produce Zn metalloproteases which are correlated to each other, immunologically and genetically.

As shown in Table 14.2, the main *Vibrio* proteases are zinc metalloproteases or serine proteases. It is noted that many of the metalloproteases belong to the thermolysin family. Zn metalloprotease produced by *Vibrio cholerae* is thought to play indirect pathogenic roles such as processing other protein toxin(s) or supporting bacterial growth in the digestive tract, whereas that produced by *Vibrio vulnificus* is thought to be a direct toxic factor causing skin damage. This paper reviews *V. cholerae* protease comparing it with proteases of related vibrios.

14.2 *Vibrio cholerae* Protease

14.2.1 Hemagglutinin/Protease (HA/P)

Of the *Vibrio* proteases, hemagglutinin/protease (HA/P) produced by *V. cholerae* was the one first noticed. HA/P was initially reported to be a hemagglutinin and

Table 14.2 Proteases produced by vibrios

Species	Protease	Reference
Human pathogens		
Intestinal infections		
V. cholerae	Zinc metalloprotease (T[a])	[19]
V. fluvialis	Zinc metalloprotease (T[a])	[76]
V. mimicus	Zinc metalloprotease (T[a])	[75]
	Zinc metalloprotease (X[b])	[72]
V. parahaemolyticus	Serine protease	[65, 66]
	Zinc metalloprotease (X[b])	[70, 71]
Extra-intestinal infections		
V. alginolyticus	Serine protease	[68]
	Zinc metalloprotease (X[b])	[74]
V. metschnikovii	Serine protease	[67]
V. vulnificus	Zinc metalloprotease (T[a])	[9, 62]
Non-human pathogens and fish pathogens		
V. anguillarum[c]	Zinc metalloprotease (T[a])	[77, 78]
V. carchariae[c]	Serine protease	[69]
V. harveyi[c]	Zinc metalloprotease (T[a])	[80]
V. pelagius[c]	Zinc metalloprotease (T[a])	[79]
V. penaeicida[c]	Cysteine protease	[87]
V. proteolyticus	Zinc metalloprotease (T[a])	[82, 83]
V. tubiashii[c]	Zinc metalloprotease (T[a])	[81]

[a] Thermolysin family
[b] Metalloprotease other than thermolysin–family protease
[c] Fish pathogen

termed cholera lectin by Finkelstein et al. [5], and its protease activity was recognized later [6, 7]. They carried out cloning of the *hapA* gene encoding the HA/P precursor, which has 609 amino acid residues and a molecular size of 69.3 kDa [8]. The open reading frame for *hapA* reveals the presence of a 24-amino acid signal peptide followed by a 171-amino acid N-terminal propeptide. The precursor is processed to a 32-kDa mature protease by autodigestion or other proteolytic activity.

Mature HA/P has hemagglutinating activity in addition to proteolytic activity. In *V. vulnificus*, as shown later, the mature Zn metalloprotease liberates a 10-kDa C-terminal peptide by an autoproteolytic reaction, and the N-terminal moiety still possesses proteolytic activity but no hemagglutinating activity, suggesting that the C-terminal peptide mediates hemagglutination or binding to the cell surface [9, 10]. Honda et al. [11] showed the existence of a monoclonal antibody against HA/P that neutralized the proteolytic but not the hemagglutinating activity, suggesting that the conformation of active domains of HA/P is similar to that of *V. vulnificus* protease: the proteolytic (catalytic) domain on the N-terminal region and the hemagglutinating (binding) domain on the C-terminal.

The deduced amino acid sequence revealed that the mature HA/P had a similarity of 61.5% with *Pseudomonas aeruginosa* elastase and a zinc-binding motif peculiar to the thermolysin family protease. Zn metalloproteases are presently divided into four superfamilies, and the zincin superfamily is characterized by having the

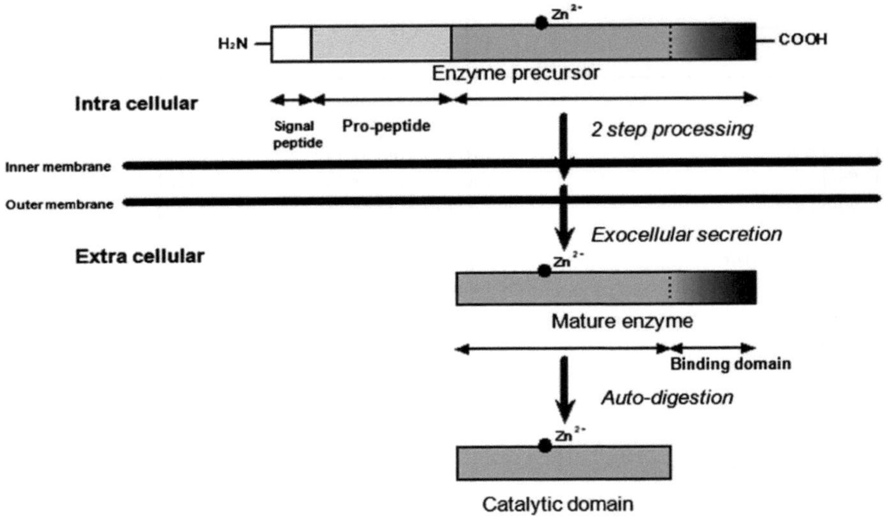

Fig. 14.1 Processing of Zn metalloprotease produced by vibrios

HEXXH motif as the zinc-binding site [12]. The thermolysin family is a member of the zincin superfamily and characterized by E at the 25th position from the first H of the above motif. As shown in Table 14.2, most *Vibrio* metalloproteases belong to the thermolysin family. Precursors of these Zn metalloproteases of vibrios are processed to mature enzymes after liberating N-terminal signal peptide and propeptide and secreted exocellularly (Fig. 14.1). HA/P is secreted via the type II secretion pathway at the cell pole [13, 14].

Although a toxic effect of HA/P on the cultured cells has been reported [15], it is thought to act as an indirect pathogenic factor supporting growth and translocation of the vibrios in the digestive tract. Thickness of the human intestinal mucus layer was estimated to be around 150 μm, so infecting diarrheagenic bacteria must overcome a physical barrier 50 times their length to reach the epithelial surface. HA/P is thought to play a role by promoting mucin gel penetration, detachment, and the spread of infection along the gastrointestinal tract [16–19]. Benitez et al. [17] showed that inactivation of HA/P increased adherence of *V. cholerae* to cultured human intestinal cell, but they also postulated that HA/P mucinase activity mediates with a similar mechanism both penetration of the vibrios into the mucus barrier in vivo and detachment from the mucin meshwork [18]. Silva et al. [19] showed that purified mucin enhanced HA/P production in carbon source starvation by a HapR-independent mechanism and that expression of *hapA* was required for translocation through a mucin-containing gel. Silva et al. [20] examined the contribution of HA/P and motility to the pathogenesis of El Tor vibrios and concluded that both motility and HA/P were necessary for full expression of enterotoxicity.

Furthermore, HA/P can process the protein toxins to the active forms by nicking the A subunit of cholera toxin [21] or liberating the 15-kDa N-terminal region from 79-kDa immature El Tor hemolysin [22]. Although HA/P is important for full expression of enterotoxicity of *V. cholerae* in its role of detaching and spreading in the intestinal tract as shown above, it suppresses adherence of the vibrios to epithelial cells. This property was used in the development of vaccine. Inactivation of *hap* gene is expected to decrease excretion of the resulting vaccine vector strain by increasing the duration of adherence and thus increase its immunogenicity [23]. A live cholera vaccine strain *V. cholerae* 638 attenuated by deletion of the CTXΦ phage from *V. cholerae* C7258 (O1, El Tor Ogawa) and by insertion of the *Clostridium thermocellum* endoglucanase A gene into the HA/P coding sequence was constructed [23]. The vaccine strain 638 was tested for safety and immunogenicity [24] and its effectiveness was confirmed in healthy volunteers in Cuba [25].

Vibrio cholerae is an inhabitant of aquatic environments. The vibrios live in water in a free-living or biofilm form and occasionally associates with various organisms such as zoo- or phytoplanktons [26, 27]. Broza and Halpren [28] suggested that egg masses of the non-biting midge *Chironomus* sp. (Dipter) harbor and serve as a nutritive source for *V. cholerae*, thereby providing a natural reservoir for the vibrios. They purified the extracellular factor for the degradation of chironomid egg masses from *V. cholerae* O1 and O139 strains and identified it as HA/P [29]. The substrate in the egg mass was characterized as a glycoprotein. Chironomids are the most widely distributed insect in freshwater. Females deposit egg masses at the water's edge, and each egg mass contains eggs embedded in a gelatinous matrix. HA/P is thought to digest the egg masses and prevent them from hatching. Halpern et al. [30] showed that most of the *V. cholerae* inhabiting the egg mass are in a viable but non-culturable (VBNC) state. Various bacteria other than *V. cholerae* are also found in a chironomid egg mass. All isolates of *V. cholerae* were capable of degrading the egg mass, whereas almost none of the other bacteria isolated from the egg masses possessed this ability. These findings suggest the special interaction of HA/P and the egg masses and its important roles for the survival of *V. cholerae* in aquatic environments.

14.2.2 Quorum-Sensing Regulation of HA/P Production

Quorum sensing is a bacterial regulation system for coordinated gene expression in response to a change in cell-population density. All bacteria possessing this system synthesize and secrete chemical signal molecules called autoinducers that increase in concentration as a function of cell density. In a low cell density, since the concentration is at a low level, the autoinducer has no effect on the target system. However, once the concentration reaches the threshold level, the bacterium senses the autoinducer and signals generated are transmitted into the bacterial cells, which leads to an alteration in the gene expression. Although the quorum-sensing system was first discovered in light production by *Vibrio fischeri* [31, 32], its contribution has been expanded to various gene expression systems including protease production, symbiosis, virulence, competence, conjugation, antibiotic production, motility, sporulation, and biofilm formation of gram-positive and gram-negative bacteria [33, 34].

The LuxI/LuxR type quorum-sensing system that was first identified in *V. fischeri* is a general system in gram-negative bacteria. LuxI is the autoinducer synthetase that produces the acylhomoserine lactone (AHL) autoinducer, whereas LuxR is a transcriptional activator promoting transcription of the target gene. On the other hand, gram-positive bacteria generate oligopeptide autoinducers. The third signal molecule termed AI-2, of which the chemical structure is furanosyl borate diester, was identified in another light-producing bacterium *Vibrio harveyi* [35]. This species has a complex quorum-sensing system [36] containing three kinds of autoinducers, AHL autoinducer (AI-1), AI-2, and CAI-1, the latter having been recently found in *V. cholerae* [34]. Among these autoinducers, AI-2 is detected by the sensor proteins LuxPQ. Subsequently, LuxQ dephosphorylates LuxO, a response regulator protein, which results in derepression of LuxR, a transcriptional activator of luciferase genes.

LuxO negatively regulates luminescence expression by activating a putative downstream repressor of the luciferase operon (*luxCDABE*) [37]. The LuxO homologues have been found in pathogenic vibrios and those of *V. cholerae* can regulate various functions involved in HA/P production [37, 38]. HapR, a regulator protein of HA/P production, is closely homologous to LuxR [39]. Zhu et al. [40] revealed that the ToxR regulon was repressed in the *luxO* mutant and that this effect was mediated by another negative regulator HapR. They showed that LuxO repressed *hapR* expression early in log phase growth and that constitutive expression of *hapR* blocked ToxR regulon expression. They also suggested regulation of various cellular processes including motility, protease production, biofilm formation, and virulence gene expression. Vance et al. [41] isolated a mutant-deficient HA/P due to a point mutation in the *luxO* quorum-sensing regulator. With this mutant they demonstrated the role of LuxO as a central switch that coordinately regulated virulence-related phenotypes such as protease production and biofilm formation. Liu et al. [42] identified another transcriptional regulator VqmA that activates *hapR* expression at low cell density. VqmA stimulated quorum-sensing regulation at lower cell densities and this stimulation bypassed the known LuxO–RNA regulatory circuit.

HapR also downregulates the expression of *tcpPH* which regulates ToxT together with ToxR/S. HapR decreases the *tcpPH* transcription indirectly by repressing the transcription of *aphA* encoding AphA, an upstream activator of TcpP/H [43]. It is known that the global response of bacteria such as *Escherichia coli* to starvation and stress requires *rpoS*, which is induced to express an alternative sigma factor *s*. Reduction of production and secretion of HA/P by *rpoS* mutation was demonstrated by Yildiz and Schoolnik [44]. Silva and Benitez [45] showed that *rpoS* was required in the transcription of *hapA*. In the same study, they demonstrated that CRP, cyclic AMP receptor protein, enhanced the transcription of *rpoS* and *hapR*.

14.2.3 Other Proteases of Vibrio cholerae

Possibility of the role of another Zn metalloprotease of *V. cholerae* in aquatic ecology was also suggested. Vaitkevicius et al. [46] studied growth and survival of the vibrio in *Caenorhabditis elegans* and found that protease(s) regulated by the LuxO–HapR pathway was involved. However, HA/P, the major protease regulated by the pathway, had no role, and a protease PrtV was identified as the factor. PrtV is a Zn metalloprotease which was identified by cloning downstream of the *hly* region encoding El Tor hemolysin by Ogierman et al. [47], but this has not yet been well characterized.

A cysteine protease domain was found in RTX toxin of *V. cholerae* [48]. RTX is a large multifunctional toxin that causes actin cross-linking and is processed by proteolytic action during translocation into host cells. Sheahan et al. [48] carried out in vivo transfection study and in vitro characterization of purified recombinant protein and revealed that the processing was due to autoproteolytic action of the cysteine protease domain of RTX toxin.

14.3 Proteases Produced by Other Vibrios

Many *Vibrio* species produce Zn metalloprotease similar to HA/P, but some vibrios produce other proteases, mainly serine protease. These proteases also play various roles in the pathogenic mechanism.

14.3.1 Vibrio vulnificus *Protease (VVP)*

It is thought that *V. vulnificus* protease (VVP) acts as a direct toxic factor in contrast to HA/P, which acts as an indirect pathogenic factor. The first isolation of *V. vulnificus* was from a leg ulcer [49]. However, it was reported as a *Vibrio parahaemolyticus* infection at that time because of similar bacteriological characteristics of the isolate to *V. parahaemolyticus* [50]. These two species were subsequently recognized as separate species due to the difference in their lactose fermentation [51]. The bacterium causes two types of illness, the primary septicemia and the wound infection [52, 53]. The former is remarkable for its high fatality rate. *Vibrio vulnificus* is now recognized as being among the most rapidly fatal of human pathogens. In the majority of cases, primary septicemia is associated with the consumption of raw seafood, especially shellfish such as oysters, contaminated with the vibrios. In the United States, 95% of all seafood-related deaths are due to *V. vulnificus*, most commonly from the consumption of raw oysters [53]. Primary septicemia due to *V. vulnificus* is an opportunistic infection, that is, most patients with septicemia have an underlying disease(s) of liver dysfunction, alcoholic cirrhosis, or hemochromatosis, which leads to an increased plasma iron level and decreased host defense system. In two-thirds of patients, edematous or hemorrhagic secondary skin lesions appear on the extremities and the trunk. Symptoms of the digestive tract such as diarrhea or vomiting are very rare. Wound infection is characterized by the development of edema, erythema, or necrosis around a new wound exposed to seawater. This type of infection can occur in healthy persons as well as in compromised hosts, and may occasionally progress to septicemia. Differing from *V. cholerae*, a protease (VVP) produced by *V. vulnificus* is thought to be the major pathogenic factor of the vibrios. VVP is a 45-kDa zinc metalloprotease of the thermolysin family and may be an important virulence factor for skin lesion [54]. We have documented that the most drastic pathological action of VVP is its vascular permeability-enhancing action through the release of histamine from mast cells and activation of the factor XII–plasma kallikrein–kinin cascade [55–57]. Bradykinin generated by activation of the kallikrein–kinin cascade is a well-known mediator of inflammation. Miyoshi et al. [58] suggested that VVP facilitates the development of a systemic infection by disturbing the plasma proteinase–proteinase inhibitor system, and Chang et al. [59] documented further analysis of the role of VVP in this system recently.

In addition to enhancing vascular permeability, VVP also induces a hemorrhagic reaction, which is one of the typical skin lesions in *V. vulnificus* infection, by digesting type IV collagen of blood vessel basement membrane gel [60]. Contribution to the utilization of heme by *V. vulnificus* is also an important pathogenic role of VVP [61].

We carried out cloning of the *vvp* gene encoding a precursor of VVP and showed similarity of the sequence to other metalloproteases of the thermolysin family produced by vibrios [62]. Thermolysin family proteases are commonly produced as a precursor form having a signal peptide and an N-terminal propeptide, $\sim$25 and 170 amino acid residues, respectively, and processed to a mature form by two step reactions. As shown in Fig. 14.1, mature VVP exocellularly excreted consists of two functional domains: the N-terminal 35-kDa domain (VVP-N) catalyzing the proteolytic reaction and the C-terminal 10-kDa domain (VVP-C) mediating the efficient association with protein substrate or cell membrane [9]. VVP-N is sufficient for catalytic reaction, but VVP-C is necessary for full activity of intact VVP such as digestion of large protein molecules, hemagglutination, histamine release from mast cells, and hemorrhage.

Production of VVP is also regulated by a quorum-sensing system. Existence of the AI-2 system, but not the AI-1 system, in *V. vulnificus* and its contribution to the VVP production has been suggested. For instance, Kim et al. [63] reported the decrease of VVP production by disruption of the *luxS* gene encoding the synthetase of the AI-2 precursor. We also examined the regulation system for VVP production in the culture medium and demonstrated a close relation between expression of the *vvp* gene and *luxS* gene, suggesting control of VVP production by the AI-2-dependent quorum-sensing system [64]. However, this system might function more effectively at 26°C than at 37°C. In human serum, VVP production and the *vvp* expression increased at 37°C in proportion to the concentration of ferric ion without an increase in the expression of *luxS*. Thus, VVP production in host serum may be regulated by a system other than the AI-2-dependent quorum-sensing system.

14.3.2 Vibrio parahaemolyticus *and Others*

Although many pathogenic vibrios including *V. cholerae* and *V. vulnificus* produce similar metalloprotease, as described above, *V. parahaemolyticus,* another important pathogenic vibrio, produces a serine protease as its major protease. Ishihara et al. [65] purified a major protease (VPP1) of *V. parahaemolyticus* and demonstrated it to be a serine protease having a molecular weight of 43 kDa. Lee et al. [66] also reported purification of a similar protease and demonstrated the cytotoxicity against CHO, HeLa or Vero cells, and the mouse lethal toxicity, but the pathogenic role has remained to be examined. VPP1 purified showed the immunological cross-reactivity with both *Vibrio metschnikovii* VapT protease and *Vibrio alginolyticus* protease and revealed similarity of the amino acid sequence to that of VapT protease [67]. A serine protease as a pathogenic factor against prawn produced by *V. alginolyticus* was purified by Chen et al. [68]. *Vibrio carchariae,* a fish pathogen, also produced a serine protease that is thought to be a pathogenic factor causing gastroenteritis [69].

Yu and Lee [70] and Kim et al. [71] reported cloning of the genes encoding metalloproteases/collagenases which were designated as PrtV and VppC, respectively, from *V. parahaemolyticus.* Those are also zincin superfamily metalloproteases but differing from the thermolysin family. Although PrtV and VppC are produced by the *E. coli* transformant harboring those genes cloned, production from *V. parahaemolyticus* itself has not been observed; therefore, their pathogenic significance is unclear. Existence of the protease genes similar to *prtV* was shown in *Vibrio mimicus* [72, 73] and *V. alginolyticus* [74].

Chowdhury et al. [75] purified a thermolysin-like metalloprotease from a pathogenic strain of *V. mimicus*, a species closely related to *V. cholerae*, and the metalloprotease purified showed strong similarity to HA/P. On the other hand, Lee et al. [72] reported the existence of a protease in another strain of *V. mimicus*, which is similar to PrtV protease of *V. parahaemolyticus.* This protease is also a metalloprotease, but it is different from the thermolysin family.

Production of thermolysin family proteases similar to HA/P or VVP has been documented in various vibrios, not only in human pathogens [76] but also in fish pathogens such as *Vibrio anguillarum* [77, 78], *Vibrio pelagius* [79], *V. harveyi* [80], or *Vibrio tubiashii* [81]. Furthermore, a thermolysin family protease designated as vibriolysin is produced by a non-pathogenic species *Vibrio proteolyticus* [82, 83]. Of these vibrio metalloproteases of the thermolysin family produced by a non-human pathogen, that of *V. anguillarum* has been the most extensively examined [79, 80, 84–86]. Several groups have reported the quorum-sensing control system and pathophysiological effects on the digestive tract of host fishes. A cysteine protease-like exotoxin as a virulence factor for white shrimp was isolated from *Vibrio penaeicida* [87].

14.4 Conclusions

The proteases produced by pathogenic vibrios play a variety of pathological roles: direct roles by digesting many kinds of host proteins or indirect roles by processing other pathogenic protein factors. Especially VVP from *V. vulnificus* is thought to be a major pathogenic factor; however, some contradictions of the pathogenic roles were also reported [88]. The precise pathogenic roles of many *Vibrio* proteases remain to be identified.

Although HA/P is not a direct toxic factor of *V. cholerae*, its significance is an undeniable fact. Production of HA/P is regulated together with major pathogenic factors such as CT (cholera toxin) or TCP (toxin-co-regulated pilus) by a quorum-sensing system. HA/P is necessary for full expression of pathogenicity of the vibrios by supporting growth and translocation in the digestive tract. Processing of protein toxins such as CT or El Tor hemolysin is also an important pathogenic role. Non-O1 or non-O139 *V. cholerae* rarely causes extra-intestinal tract infection in a compromised host. In that case, HA/P is thought to have a direct toxic effect similar to VVP of *V. vulnificus* such as causing skin edema, necrosis, or hemorrhage. Roles for supply of peptide/amino acid nutrients or adhesion to aquatic organisms are also important.

Organisms produce a great number of proteases as the essential physiological factor for their life. Although the major protease produced by *V. cholerae* is HA/P, existence of another Zn metalloprotease PrtV and cysteine protease domain in RTX toxin was also demonstrated. Additional minor proteases are thought to function to maintain the bacterial physiology. Further investigation of various proteases is necessary for understanding pathogenicity and ecology of the vibrios.

References

1. Schiavo G, Poulain B, Benfenati F, DasGupta BR, Montecucco C. Novel targets and catalytic activities of bacterial protein toxins. Trends Microbiol. 1993;1:170–4.
2. Bryant AE, Stevens DL. Clostridial toxins in the pathogenesis of gas gangrene. In: Alouf JF, Popoff MR, editors. Comprehensive sourcebook of bacterial protein toxins, 3rd ed. New York: Academic Press; 2006. pp. 919–29.
3. D'Abusco AS, Sanfilippo L, Menozzi MG, Pantosti A. Activity and role of BFT, an enterotoxin produced by *Bacteroides fragilis*. J Natural Toxins. 2000;9:267–80.
4. Leppla SH. *Bacillus anthracis* toxins. In: Alouf JF, Popoff MR, editors. Comprehensive sourcebook of bacterial protein toxins, 3rd ed. New York: Academic Press; 2006. pp. 323–47.
5. Finkelstein RA, Arita M, Clements JD, Nelson ET. Isolation and purification of an adhesive factor ("cholera lectin") from *Vibrio cholerae*. In: Proceedings of the 13th joint conference on cholera. US–Japan Cooperative Medical Science Program, National Institute of Health, Bethesda, MD; 1978. pp. 137–51.
6. Finkelstein RA, Hanne LF. Purification and characterization of the hemagglutinin (cholera lectin) produced by *Vibrio cholerae*. Infect Immun. 1982;36:1199–208.
7. Hanne LF, Finkelstein RA. Characterization and distribution of the hemagglutinins produced by *Vibrio cholerae*. Infect Immun. 1982;36:209–14.
8. Hase CC, Finkelstein RA. Cloning and nucleotide sequence of the *Vibrio cholerae* hemagglutinin/protease (HA/protease) gene and construction of an HA/protease-negative strain. J Bacteriol. 1991;173:3311–7.
9. Miyoshi S, Wakae H, Tomochika K, Shinoda S. Functional domains of a zinc metalloprotease from *Vibrio vulnificus*. J Bacteriol. 1997;179:7606–9.
10. Miyoshi S, Kawata K, Hosokawa M, Tomochika K, Shinoda S. Histamine-releasing reaction induced by the N-terminal domain of *Vibrio vulnificus* metalloprotease. Life Sci. 2003;72:2235–42.
11. Honda T, Hata-Naka A, Lertpocasombat K, Miwatani T. Production of monoclonal antibodies against a hemagglutinin/protease of *Vibrio cholerae* non-O1. FEMS Microbiol Lett. 1991;62:227–30.

12. Hooper NM. Families of zinc metalloproteases. FEBS Lett. 1994;354:1–6.
13. Sandkvist M, Michel LO, Hough LP, Morales VM, Bagdasarian M, Koomkey M, DiRita VJ Bagdasarian M. General secretion pathway (*eps*) genes required for toxin secretion and outer membrane biogenesis in *Vibrio cholerae.* J Bacteriol. 1997;179:6994–7003.
14. Scott ME, Dossani ZY, Sandkvist M. Directed polar secretion of protease from single cells of *Vibrio cholerae* via the type II secretion pathway. Proc Natl Acad Sci USA. 2001;98: 13978–83.
15. Wu Z, Nybom P, Sundqvist T, Magnusson KE. Endogenous nitric oxide in MDCK-I cells modulates the *Vibrio cholerae* haemagglutinin/protease (HA/P)-mediated cytotoxicity. Microb Pathog. 1998;24:321–6.
16. Finkelstein RA, Boesman-Finkelstein M, Chang Y, Hase CC. *Vibrio cholerae* hemagglutinin/protease, colonial variation, virulence, and detachment. Infect Immun. 1992;60: 472–8.
17. Benitez JA, Spelbrink RG, Phillips TE, Stanley CM, Boesman-Finkelstein M, Finkelstein RA. Adherence of *Vibrio cholerae* to cultured differentiated human intestinal cells: an in vitro colonization model. Infect Immun. 1997;65:3473–7.
18. Benitez JA, Silva AJ, Finkelstein RA Environmental signals controlling production of hemagglutinin/protease in *Vibrio cholerae.* Infect Immun. 2001;69:6549–53.
19. Silva A, Pham K, Benitez JA Haemagglutinin/protease expression and mucin gel penetration in El Tor biotype *Vibrio cholerae.* Microbiology. 2003;149:1883–91.
20. Silva AJ, Leitch GJ, Camilli A, Benitez J. Contribution of hemagglutinin/protease and motility to the pathogenesis of El Tor biotype cholera. Infect Immun. 2006;74:2072–9.
21. Booth BA, Boesman-Finkelstein M, Finkelstein RA. *Vibrio cholerae* hemagglutinin/protease nicks cholera enterotoxin. Infect Immun. 1984;45:558–60.
22. Nagamune K, Yamamoto K, Naka A, Matsuyama J, Miwatani T, Honda T. In vitro proteolytic processing and activation of the recombinant precursor of El Tor cytolysin/hemolysin (pro-HlyA) of *Vibrio cholerae* by soluble hemagglutinin/protease of *V. cholerae*, trypsin, and other proteases. Infect Immun. 1996;64:4655–8.
23. Robert A, Silva A, Benitez JA, Rodriguez BL, Fando R, Campos J, Sengupta DK, Boesman-Finkelstein M, Finkelstein RA. Tagging a *Vibrio cholerae* El Tor candidate vaccine strain by disruption of its hemagglutinin/protease gene using a novel reporter enzyme: *Clostridium thermocellum* endoglucanase A. Vaccine. 1996;14:1517–22.
24. Benítez JA, García L, Silva A, García H, Fando R, Cedré B, Pérez A, Campos J, Rodríguez BL, Pérez JL, Valmaseda T, Pérez O, Pérez A, Ramírez M, Ledón T, Jidy MD, Lastre M, Bravo L, Sierra G. Preliminary assessment of the safety and immunogenicity of a new CTXPhi-negative, hemagglutinin/protease-defective El Tor strain as a cholera vaccine candidate. Infect Immun. 1999;67:539–45.
25. Garcia L, Jidy MD, Garcia H, Rodriguez BL, Fernandez R, Ano G, Cedre B, Valmaseda T, Suzarte E, Ramirez M, Pino Y, Campos J, Menendez J, Valera R, Gonzalez D, Gonzalez I, Perez O, Serrano T, Lastre M, Miralles F, Del Campo J, Maestre JL, Perez JL, Talavera A, Perez A, Marrero K, Ledon T, Fando R. The vaccine candidate *Vibrio cholerae* 638 is protective against cholera in healthy volunteers. Infect Immun. 2005;73:3018–24.
26. Huq A, Colwell RR, Rahman R, Ali A, Chowdhury MA, Parveen S, Sack DA, Russek-Cohen E. Detection of *Vibrio cholerae* O1 in the aquatic environment by fluorescent-monoclonal antibody and culture methods. Appl Environ Microbiol. 1990;56:2370–3.
27. Huq A, Sack RB, Nizam A, Longini IM, Nair GB, Ali A, Morris JG Jr., Khan MN, Siddique AK, Yunus M, Albert MJ, Sack DA, Colwell RR. Critical factors influencing the occurrence of *Vibrio cholerae* in the environment of Bangladesh. Appl Environ Microbiol. 2005;71: 4645–54.
28. Broza M, Halpern M. Pathogen reservoirs. Chironomid egg masses and *Vibrio cholerae.* Nature. 2001;412:40.
29. Halpern M, Gancz H, Broza M, Kashi Y. *Vibrio cholerae* hemagglutinin/protease degrades chironomid egg masses. Appl Environ Microbiol. 2003;69:4200–4.

30. Halpern M, Landsberg O, Raats, D, Rosenberg E. Culturable and VBNC *Vibrio cholerae:* interactions with chironomid egg masses and their bacterial population. Microb Ecol. 2007;53:285–93.
31. Nealson KH, Platt T, Hastings JW. Cellular control of the synthesis and activity of the bacterial luminescent system. J Bacteriol. 1970;104:313–22.
32. Eberhard A. Inhibition and activation of bacterial luciferase synthesis. J Bacteriol. 1972;109:101–5.
33. Whitehead NA, Barnard AM, Slater H, Simpson NJ, Salmond GP. Quorum-sensing in Gram-negative bacteria. FEMS Microbiol Rev. 2001;25:365–404.
34. Miller MB, Skorupski K, Lenz DH, Taylor RK, Bassler BL. Parallel quorum sensing systems converge to regulate virulence in *Vibrio cholerae*. Cell. 2002;110:303–14.
35. Mok KC, Wingreen NS, Bassler BL. *Vibrio harveyi* quorum sensing: a coincidence detector for two autoinducers controls gene expression. Embo J. 2003;22:870–81.
36. Henke JM, Bassler BL. Three parallel quorum sensing systems regulate gene expression in *Vibrio harveyi.* J Bacteriol. 2004;186:6902–14.
37. Lilley BN, Bassler BL. Regulation of quorum sensing in *Vibrio harveyi* by LuxO and sigma-54. Mol Microbiol. 2000;36:940–54.
38. Raychaudhuri S, Jain V, Dongre M. Identification of a constitutively active variant of LuxO that affects production of HA/protease and biofilm development in a non-O1, non-O139 *Vibrio cholerae* O110. Gene. 2006;369:126–33.
39. Jobling MG, Holme RK. Characterization of *hapR*, a positive regulator of the *Vibrio cholerae* HA/protease gene *hap*, and its identification as a functional homologue of the *Vibrio harveyi luxR* gene. Mol Microbiol. 1997;26:1023–34.
40. Zhu J, Miller MB, Vance RE, Dziejman M, Bassler BL, Mekalanos JJ. Quorum-sensing regulators control virulence gene expression in *Vibrio cholerae*. Proc Natl Acad Sci USA. 2002;99:3129–34.
41. Vance RE, Zhu J, Mekalanos JJ. A constitutively active variant of the quorum-sensing regulator LuxO affects protease production and biofilm formation in *Vibrio cholerae.* Infect Immun. 2003;71:2571–6.
42. Liu Z, Hsiao A, Joelsson A, Zhu J. The transcriptional regulator VqmA increases expression of the quorum-sensing activator HapR in *Vibrio cholerae*. J Bacteriol. 2006;188: 2446–53.
43. Kovacikova G, Skorupski K. Regulation of virulence gene expression in *Vibrio cholerae* by quorum sensing: HapR functions at the *aphA* promoter. Mol Microbiol. 2002;46:1135–47.
44. Yildiz FH, Schoolnik GK. Role of *rpoS* in stress survival and virulence of *Vibrio cholerae*. J Bacteriol. 1998;180:773–84.
45. Silva AJ, Benitez JA. Transcriptional regulation of *Vibrio cholerae* hemagglutinin/protease by the cyclic AMP receptor protein and RpoS. J Bacteriol. 2004;186:6374–82.
46. Vaitkevicius K, Lindmark B, Ou G, Song T, Tom C, Iwanag M, Zhu J, Andersson A, Hammarstrom ML, Tuck S, Wai SN. A *Vibrio cholerae* protease needed for killing of *Caenorhabditis elegans* has a role in protection from natural predator grazing. Proc Natl Acad Sci USA. 2006;103:9280–5.
47. Ogierman MA, Fallarino A, Riess T, Williams SG, Attridge SR, Manning PA Characterization of the *Vibrio cholerae* El Tor lipase operon *lipAB* and a protease gene downstream of the *hly* region. J Bacteriol. 1997;179:7072–80.
48. Sheahan KL, Cordero CL, Satchell KJ. Autoprocessing of the *Vibrio cholerae* RTX toxin by the cysteine protease domain. EMBO J. 2007;26:2552–61.
49. Roland FP. Leg gangrene and endotoxin shock due to *Vibrio parahaemolyticus*. An infection acquired in New England coastal water. New Engl J Med. 1970;282:1306.
50. Hollis DG, Weaver RE, Baker CN, Thronsberry C. Halophilic *Vibrio* species isolated from blood cultures. J Clin Microbiol. 1976;3:425–31.
51. Farmer JJ. *Vibrio* (*Beneckea*) *vulnificus*: the bacterium associated with sepsis, septicemia, and the sea. Lancet. 1979;ii:903.

52. Blake PA, Weaver RE, Hollis DG. Disease of humans other than cholera caused by vibrios. Ann Rev Microbiol. 1980;34:341–67.
53. Klontz KC, Lieb S, Schreiber M, Janowski HT, Baldy L, Gunn RA. Syndromes of *Vibrio vulnificus* infections: clinical and epidemiological features in Florida cases, 1981–1987. Ann Intern Med. 1988;109:318–23.
54. Miyoshi S, Shinoda S. Exocellular toxic factors produced by *Vibrio vulnificus*. J. Toxicol Toxin Rev. 1993;12:253–88.
55. Miyoshi S, Sugiyama K, Suzuki Y, Furuta H, Miyoshi N, Shinoda S. Enhancement of vascular permeability due to histamine-releasing effect of *Vibrio vulnificus* protease. FEMS Microbiol Lett. 1987;40:95–8.
56. Miyoshi N, Miyoshi S, Sugiyama K, Suzuki Y, Furuta H, Shinoda S. Activation of the plasma kallikrein–kinin system by *Vibrio vulnificus* protease. Infect Immun. 1987;55:1936–9.
57. Miyoshi S, Watanabe H, Kawase T, Yamada H, Shinoda S. Generation of active fragments from human zymogens in the bradykinin-generating cascade by extracellular proteases from *Vibrio vulnificus* and *V. parahaemolyticus*. Toxicon. 2004;44:887–93.
58. Miyoshi S, Narukawa H, Tomochika K, Shinoda S. Actions of *Vibrio vulnificus* metalloprotease on human plasma proteinase–proteinase inhibitor systems: a comparative study of native protease with its derivative modified by polyethylene glycol. Microbiol Immunol. 1995;39:959–66.
59. Chang AK, Ki HY, Park JE, Acharya P, Park IS, Yoon SM, You HJ, Hahm KS, Park JK, Lee JS. *Vibrio vulnificus* secretes a broad-specificity metalloprotease capable of interfering with blood homeostasis through prothrombin activation and fibrinolysis. J Bacteriol. 2005;187:6909–16.
60. Miyoshi S, Nakazawa H, Kawata K, Tomochika K, Tobe K, Shinoda S. Characterization of the hemorrhagic reaction caused by *Vibrio vulnificus* metalloprotease, a member of the thermolysin family. Infect Immun. 1998;66:4851–5.
61. Nishina Y, Miyoshi S, Nagase A, Shinoda S. Significant role of an exocellular protease in utilization of heme by *Vibrio vulnificus*. Infect Immun. 1992;60:2128–32.
62. Shinoda S, Miyoshi S, Wakae H Rahman M, Tomochika K. Bacterial proteases as pathogenic factors, with special emphasis on vibrio proteases. J Toxicol Toxin Rev. 1996;15:327–39.
63. Kim SY, Lee SE, Kim YR, Kim CM, Ryu PY, Choy HE, Chung SS, Rhee JH. Regulation of *Vibrio vulnificus* virulence by the LuxS quorum-sensing system. Mol Microbiol. 2003;8:1647–64.
64. Kawase T, Miyoshi S, Sultan Z, Shinoda S. Regulation system for protease production in *Vibrio vulnificus*. FEMS Microbiol Lett. 2004;240:55–9.
65. Ishihara M, Kawanishi A, Watanabe H, Tomochika K, Miyoshi S, Shinoda S. Purification of a serine protease of *Vibrio parahaemolyticus* and its characterization. Microbiol Immunol. 2002;46:298–303.
66. Lee CY, Cheng MF, Yu MS, Pan MJ. Purification and characterization of a putative virulence factor, serine protease, from *Vibrio parahaemolyticus*. FEMS Microbiol Lett. 2002;209:31–7.
67. Kwon YT, Kim JO, Moon SY, Yoo YD, Rho HM. Cloning and characterization of the gene encoding an extracellular alkaline serine protease from *Vibrio metschnikovii* strain RH530. Gene. 1995;152:59–63.
68. Chen FR, Liu PC, Lee KK. Purification and partial characterization of a toxic serine protease produced by pathogenic *Vibrio alginolyticus*. Microbios. 1999;98:95–111.
69. Lee KK, Yii KC, Yang TI, Hong HI, Liu PC. Protease and virulence of the extracellular products produced by *Vibrio carchariae* after growth on various media. Z Naturforsch. 1999;54:383–6.
70. Yu MS, Lee CY. Expression and characterization of the prtV gene encoding a collagenase from *Vibrio parahaemolyticus* in *Escherichia coli*. Microbiology. 1999;145:143–50.
71. Kim SK, Yang JY, Cha J. Cloning and sequence analysis of a novel metalloprotease gene from *Vibrio parahaemolyticus* 04. Gene. 2002;283:277–86.
72. Lee JH, Kim GT, Lee JY, Jun HK, Yu JH, Kong IS. Isolation and sequence analysis of metalloprotease gene from *Vibrio mimicus*. Biochim Biophys Acta. 1998;1384:1–6.

73. Lee JH, Ahn SH, Lee EM, Kim YO, Lee SJ, Kong IS. Characterization of the enzyme activity of an extracellular metalloprotease (VMC) from *Vibrio mimicus* and its C-terminal deletions. FEMS Microbiol Lett. 2003;223:293–300.
74. Takeuchi H, Shibano Y, Morihara K, Fukushima J, Inami S, Keil B, Gilles AM, Kawamoto S, Okuda K. Structural gene and complete amino acid sequence of *Vibrio alginolyticus* collagenase. Biochem J. 1992;281:703–8.
75. Chowdhury MA, Miyoshi S, Shinoda S. Purification and characterization of a protease produced by *Vibrio mimicus*. Infect Immun. 1990;58:4159–62.
76. Miyoshi S, Sonoda Y, Wakiyama H, Rahman MM, Tomochika K, Shinoda S, Yamamoto S, Tobe K. An exocellular thermolysin-like metalloprotease produced by *Vibrio fluvialis*: purification, characterization, and gene cloning. Microb Pathog. 2002: 33:127–34.
77. Norqvist A, Norrman B, Wolf-Watz H. Identification and characterization of a zinc metalloprotease associated with invasion by the fish pathogen *Vibrio anguillarum*. Infect Immun. 1990;58:3731–6.
78. Milton DL, Norqvist A, Wolf-Watz H. Cloning of a metalloprotease gene involved in the virulence mechanism of *Vibrio anguillarum*. J Bacteriol. 1992;174:7235–44.
79. Farto R, Perez MJ, Fernandez-Briera A, Nieto TP. Purification and partial characterisation of a fish lethal extracellular protease from *Vibrio pelagius*. Vet Microbiol. 2002;89:181–94.
80. Teo JW, Zhang LH, Poh CL. Cloning and characterization of a metalloprotease from *Vibrio harveyi* strain AP6. Gene. 2002;303:147–56.
81. Delston RB, Kothary MH, Shangraw KA, Tallm BD. Isolation and characterization of a zinc-containing metalloprotease expressed by *Vibrio tubiashii*. Can J Microbiol. 2003;49:25–9.
82. David VA, Deutch AH, Sloma A, Pawlyk D, Ally A, Durham DR. Cloning, sequencing and expression of the gene encoding the extracellular neutral protease, vibriolysin, of *Vibrio proteolyticus*. Gene. 1992;112:107–12.
83. Shinoda S, Yamamoto S, Miyoshi S. Vibriolysin. In: Handbook of proteolytic enzymes. 2nd ed. Amsterdam: Elsevier; 2004. pp. 399–401.
84. Croxatto A, Chalker VJ, Lauritz J, Jass J, Hardman A, Williams P, Camara M, Milton DL. VanT, a homologue of *Vibrio harveyi* LuxR, regulates serine, metalloprotease, pigment, and biofilm production in *Vibrio anguillarum*. J Bacteriol. 2002;184:1617–29.
85. Denkin SM, Nelson DR. Regulation of *Vibrio anguillarum empA* metalloprotease expression and its role in virulence. Appl Environ Microbiol. 2004;70:4193–204.
86. Staroscik AM, Denkin SM, Nelson DR. Regulation of the *Vibrio anguillarum* metalloprotease EmpA by posttranslational modification. J Bacteriol. 2005;187:257–60.
87. Aguirre-Guzman G, Ascencio F, Saulnier D. Pathogenicity of *Vibrio penaeicida* for white shrimp *Litopenaeus vannamei*: a cysteine protease-like exotoxin as a virulence factor. Dis Aquat Organ. 2005;67:201–7.
88. Shao CP, Hor L. Metalloprotease is not essential for *Vibrio vulnificus* virulence in mice. Infect Immun. 2000;68:3569–73.

Chapter 15
Toxins of *Vibrio cholerae* and Their Role in Inflammation, Pathogenesis, and Immunomodulation

Kamini Walia and Nirmal Kumar Ganguly

Abstract Disease manifestations in cholera are primarily attributed to the secretion of cholera toxin (CT). However, the discovery of additional secretory virulence factors in *Vibrio cholerae* has invoked interest in their potential role related to pathogenesis, inflammation, and immune modulation in its disease attributes. Cholera was earlier thought to be a non-inflammatory diarrhea, but several reports provide evidence for an inflammatory response in cholera disease. These findings have come from studies with *Vibrio cholerae* strains devoid of CT gene *ctx* but induced inflammatory response. Studies have shown that while the CT essentially elicits an anti-inflammatory cytokine response, the accessory CTs have been associated with development of a proinflammatory cytokine response in gut. It is postulated that the increase in the inflammatory response to the *V. cholerae* (Δctx) infection could be due to the absence of the immunomodulatory activity of the B subunit of CT that blocks the secretion of proinflammatory cytokines by macrophages, dendritic cells, and epithelial cells. Further studies are needed to understand the potential role of each of these toxins in the inflammation and immunomodulation in cholera disease.

15.1 Introduction

Until recently, it was thought that clinical manifestations of cholera result primarily from interaction between cholera toxin (CT) and intact intestinal epithelial cells. However, the occurrence of mild-to-moderate residual diarrhea in human volunteers fed with live oral cholera vaccine strains [1] or genetically engineered mutants [2, 3] incapable of producing whole biologically active CT [4] prompted investigators to search for additional toxins that could contribute to the pathogenesis of cholera. This led to the discovery of new cholera toxin (NCT) [5], zonula occludens

N.K. Ganguly (✉)
DBT Cell for UNESCO Regional Center for Biotechnology and Translational Health Science and Technology Institute, National Institute of Immunology, Aruna Asaf Ali Marg, New Delhi, India
e-mail: nkganguly@nii.res.in

T. Ramamurthy, S.K. Bhattacharya (eds.), *Epidemiological and Molecular Aspects on Cholera*, Infectious Disease, DOI 10.1007/978-1-60327-265-0_15,
© Springer Science+Business Media, LLC 2011

Table 15.1 New toxins secreted by different serogroups of *V. cholerae*

Name of toxin	Mode of action	References
Accessory cholera enterotoxin	Increases membrane permeability by creating ion-permeable pore	[7]
Cytotonic protein	CHO elongation	[77]
Heat-stable toxin (ST)	Similar to cholera toxin	[75]
Hemolysin cytolysin	Cytotoxic for erythrocytes and mammalian cells, villus degeneration, accumulation of cellular exudate	[1, 64]
New cholera toxin	Not known	[5]
Non-membrane damaging cytotoxin	Cell rounding of CHO cells and HeLa cells without membrane damage	[72]
Shiga-like toxin	Cytotoxicity in HeLa cells	[73]
WO7 toxin	Electrolyte imbalance	[61]
Zonula occludens toxin	Increases membrane permeability by affecting tight junctions	[6, 57, 58]

toxin (zot) [6], accessory cholera enterotoxin (ace) [7], and many more (Table 15.1). The role of the toxins other than CT in the pathogenesis of disease due to *Vibrio cholerae* is largely unknown. These toxins clearly cannot cause cholera gravis; however, toxins other than CT may contribute to diarrhea and may also be responsible for certain symptoms associated with CT-negative strains. Additionally, such toxins may serve as secondary secretogenic molecules when conditions for producing CT are not optimal. After ingestion of pathogenic *V. cholerae*, bacteria colonize in the small intestine and produce a number of virulence factors, notably the CT. This toxin targets and activates the adenylate cyclase within host epithelial cells, provoking the net secretion of chloride ions and water into the intestinal lumen, resulting in the extensive diarrhea that is characteristic of this disease [8]. As such, the pathology of cholera has traditionally been considered non-inflammatory [9]. However, several reports provide evidence for an inflammatory response in cholera disease. Lymphocytes and mononuclear cells have been observed in the intestinal lamina propria in biopsy specimens obtained from the gut of cholera patients [10, 11], and increased levels of lactoferrin, myeloperoxidase, and prostaglandins have been measured in stool samples from infected humans [12, 13]. There have been reports wherein leukocytes and erythrocytes were detected in the stools from cholera patients [14], and elevated concentrations of nitric oxide metabolites were observed in urine and serum samples [15, 16]. *V. cholerae* vaccine strains have also been found to cause symptoms consistent with inflammation in human volunteers [17]. Owing to the discovery of the additional toxins and the recent findings on the inflammatory aspect of cholera disease, scientists have been working toward finding a plausible evidence for the involvement of other virulence factors in the cholera disease. This chapter reviews and discusses the importance of CT as well as other toxins in the pathogenesis, inflammation, and immunomodulation in cholera disease, which in turn might influence development of immunity to the disease.

15.2 Cholera Enterotoxin

In 1884, Robert Koch isolated *V. cholerae* and proposed that the agent responsible for cholera produces "a special poison" which acts on epithelium and symptoms of cholera could be "regulated essentially as a poisoning" [17]. The significant works of De in Calcutta during 1950–1960 breached several qualms pertaining to the enteric toxin produced by *V. cholerae*. Three of his works, viz. ligated intestinal loop method for studying cholera in rabbit model [18], ileal loop model to demonstrate the association of some strains of *Escherichia coli* with diarrhea [19], and most importantly his discovery of toxigenic effect of *V. cholerae* in 1959 in the cell-free culture filtrate of *V. cholerae* that stimulated a specific cellular response [20]. Almost during the same time, Dutta and colleagues [21] working in Bombay demonstrated that ingestion of sterile lysates of *V. cholerae* 569B produced fatal diarrhea in infant rabbits. This toxin was later purified to homogeneity by Finkelstein and LoSpalluto [22]. The ultimate proof of the role of CT associated with human disease was an experiment reported by Levine et al. [23] where ingestion of 25 μg of CT by the volunteers resulted in 20 l of rice-water stool. Like *E. coli* heat-labile enterotoxin (LTI), Shiga/verotoxin (responsible for hemolytic uremic syndrome), and pertussis toxin, CT belongs to the AB5 subunit family. In CT, five identical peptides (11 kDa) assemble into a highly stable pentameric ring termed the B subunit (55 kDa). The B subunit exhibits specific and high-affinity binding to the oligosaccharide domain of ganglioside GM1 and functions to tether the toxin to the plasma membrane of host cells. The specificity and the stability of toxin binding to GM1 dictate toxin function, likely by affecting toxin trafficking into the cell. CT contains a well-defined, ER-targeting KDEL motif that increases the efficiency of toxin function, presumably by facilitating toxin entry into the ER [24].

CT enters polarized epithelial cells through a complex pathway involving apical endocytosis and retrograde membrane traffic through Golgi cisternae to ER [24–26]. It is currently believed that CT must enter the ER for the A1 peptide to unfold and translocate into the cytosol. After membrane translocation, the A1 peptide may move to the adenylyl cyclase complex on the cytoplasmic surface of the basolateral membrane by diffusion through the cytosol. The B subunit, unlike the A1 peptide, does not translocate across cell membranes. Rather, the B subunit remains membrane associated (presumably bound to GM1) and eventually moves back out of the secretory pathway by vesicular traffic to the cell surface [27]. B subunit (and a small fraction of holotoxin) can move from its original site of binding on the apical (or mucosal) cell surface to the basolateral (or serosal) cell surface by first moving through Golgi cisternae and possibly ER. This process is also called "indirect" transcytosis [27]. This ability of CT to breach the epithelial barrier by crossing through epithelial cells within transport vesicles may contribute to the potent effects of orally delivered CT on mucosal and systemic immune responses. CT affects one of the most important regulatory factors of eukaryotic cell, namely adenylate cyclase, which results in characteristic rice-water stools. The A1 peptide (22 kDa) is the enzymatically active subunit that activates adenylyl cyclase inside the cell by catalyzing the ADP ribosylation of the heterotrimeric GTPase Gsa. After CT binds to intact cells, there is a lag of 15–60 min before adenylate cyclase is activated [28]. Normally, adenylate cyclase is activated or inactivated in response to a variety of stimuli. Regulation of adenylate cyclase is mediated via GTP-binding proteins or G proteins. The specific G protein involved in the activation of adenylate cyclase is Gs protein. CT acts by ADP ribosylation of alpha subunit of GS1 protein. Upon ADP ribosylation, the intrinsic GTPase activity of Gs protein is inhibited, thereby resulting in constitutive activation of the adenylate cyclase. Potential modes of activation include (i) A1 translocation through the membrane with diffusion through the cytoplasm to the adenylate cyclase, (ii) A1 modification of brush border membrane Gs protein, which then diffuses to basolateral membrane to activate adenylate cyclase, and (iii) endocytosis of CT with delivery of active A1 in endosomal membrane to the basolateral adenylate cyclase [29, 30].

After initial increase in cAMP, there is slow onset of inhibition of Na^+ absorption and stimulation of $C1^-$ secretion as observed in rabbit ileum in vivo [31]. CT may stimulate intestinal secretions, which serve to augment cyclic AMP-mediated intestinal secretion. In vitro and in vivo data strongly implicate the role of prostaglandins of E-series (PGE1 and PGE2) [32, 33] and platelet-activating factor (PAF) [34] in the pathogenesis of intestinal secretion stimulated by CT. The role of leukotrienes and other metabolites of arachidonic acid in causing intestinal secretions has also been suggested [35]. A model has been suggested in which cAMP levels increased by CT serve not only to activate protein kinase A but also to regulate transcription of phospholipase or phospholipase-activating protein. The activated phospholipase could act on membrane phospholipids to produce arachidonic acid, a precursor of prostaglandins and leukotrienes [36].

CT has also been shown to alter the activity of enteric nervous system possibly by stimulating release of serotonin, 5-hytroxytryptamine [5-HT], and vasoactive intestinal peptide (VIP) in human small bowel [37]. Both of these peptide hormones can directly stimulate intestinal epithelial cell secretion, suggesting a possible role for these hormones in CT-mediated diarrhea.

15.2.1 Immune Modulation by Cholera Toxin

CT is a powerful in vivo immunomodulator that is being widely used in experimental animals, both as a mucosal adjuvant and as a carrier for covalently linked antigens [38–40]. However, interest in its use as adjuvant became much more widely established as they were able to subvert the normal processes of mucosal tolerance, potentiating immunity to co-administered antigens after oral delivery [41, 42]. In an attempt to understand the mechanisms underlying the immunological properties of CT, several studies have focused on the effects of this protein on specific immune cell populations. Because the main receptor for the toxins is the ubiquitous GM1 ganglioside, CT is able to interact polyclonally with B and T cells, irrespective of their antigen specificity, as well as with antigen-presenting cells (APCs) such as macrophages and dendritic cells (DCs).

15.2.1.1 B and T Cells

Both recombinant CT (rCT) and rCT-B have been shown to induce isotype differentiation of B cells [43, 44]. In this regard, rCT acts synergistically with IL-4 to drive isotype switching to the IgG1 antibody subclass [43]. rCT was shown to be 1,000-fold more efficient at driving isotype differentiation than was rCTB, while this effect could be reproduced using a combination of rCTB and dibutyryl cAMP. Both the enzymatic activity and the receptor binding contribute to the stimulatory effects of the holotoxin [44]. In addition, rCTB induces B-cell activation, as shown by the upregulation of major histocompatibility complex (MHC) class II and costimulatory molecules on B cells [45–47]. One consequence of receptor binding by the toxin and the B subunit is the lowering of the threshold required for the differentiation of B cells into antibody-secreting cells. It is also possible that B cells acquire an enhanced ability to present antigen to T cells, further promoting the immune response. In contrast to the stimulatory effects on B cells and professional APCs, rCTB [48] also induces programmed cell death (apoptosis) of CD8+ T cells as depletion of CD8+ T cells by rCTB has also been shown to occur in vivo. By contrast, the CD4+ T-cell subset is relatively insensitive to rCTB-induced apoptosis [49].

15.2.1.2 Monocytes and Macrophages

CT also enhances expression of the costimulatory molecule B7.2 on macrophages. Injection of mice with an anti-B7.2 antibody inhibited both the mucosal immunogenicity and adjuvant properties of CT in vivo [50]. CT treatment of endotoxin-stimulated macrophages has been shown to increase production of IL-10, IL-6, and IL-β, and decrease production of IL-12, TNF-α, and nitric oxide [51]. These data partially explain the predilection of CT to selectively drive Th2-type responses when used as an adjuvant in vivo. The B subunit of CT bound to an APC can also directly co-stimulate cytokine production from T cells. Whole CT but not rCTB treatment also inhibited secretion of IL-12 by monocytes [52]. CT also enhances the APC function of macrophages in vitro [53, 54]. CT treatment of macrophages results in enhanced MHC class II-restricted presentation of a peptide from hen egg lysozyme to a T-cell hybridoma and a corresponding increase in T-cell proliferation [54].

15.2.1.3 Dendritic Cells

CT treatment of human dendritic cells (DCs) induces their maturation as shown by upregulation of expression of MHC class II and costimulatory molecules B7.1 and B7.2 [55]. CT-treated DCs are then able to prime naive T cells in vitro, driving their polarization toward the Th2 phenotype. CT enhances the costimulatory potential of DC via upregulation of CD80 and to a lesser extent CD-86 and reduced surface expression of CCR5. CT also modulates DC to preferentially activate IL-10-secreting cells but not Th1 cells. The presence of low doses of LPS however seems to be crucial for induction of IL-10 by CT. CT inhibits the expression of CD40 and ICAM-1[56]. Activation of DCs might play an important role in the adjuvant properties of CT in vivo.

15.3 The Other Toxins of *Vibrio cholerae*

A number of additional toxins have been identified in *V. cholerae*, which was earlier known to secrete only the CT. These additional toxins play a significant role in the pathogenesis and immune modulation. This hypothesis is currently being explored, especially on their role in the pathogenesis.

15.3.1 Zona Occludens Toxin (Zot)

In 1991, Fasano et al. [6] reported that *V. cholerae* produces a toxin that increased the permeability of the small intestinal mucosa by affecting the structure of intracellular tight junctions or zonula occludens. The *zot* gene encodes a 44.8-kDa peptide [57] whose native form has not yet been purified. Crude Zot diminishes the resistance of rabbit ileal tissue in Ussing chamber without causing detectable changes in the potential difference [6]. The onset of action of crude Zot is immediate and activity is reversible. Zot acts on zonula occludens or tight junctions possibly through a rearrangement of F-actin. Zot possibly contributes to production of diarrhea in cholera by altering the permeability of intestinal tissue [58]. Several signal transduction mechanisms (e.g., calcium, PKC, tyrosine kinase, cyclic AMP) have been shown to regulate tight junctions in cells. Zot effect on intestinal cells is not mediated by cAMP but through protein kinase C (PKC) [58].

15.3.2 Accessory Cholera Enterotoxin (Ace)

The third toxin identified in *V. cholerae* O1 is Ace (for accessory cholera enterotoxin), a 111.3-kDa protein encoded by the *ace* gene [7]. Ace is predicted to be an amphipathic molecule and forms multimers, which insert into the eukaryotic cell membrane, creating an ion-permeable pore. In rabbit ileal tissue mounted in Ussing chambers, crude Ace stimulates a delayed Isc (current needed to nullify potential difference) and PD (potential difference). Inoculation of ligated rabbit ileal segments with a CT- and Zot-negative *V. cholerae* strain containing the cloned *ace* gene results in fluid accumulation. The protein sequence of this toxin shares homology with eukaryotic ion-transporting proteins including gastric fibrosis transmembrane regulator human plasma membrane calcium pump and calcium-transporting ATPase from rat brain (CFTR) [59]. In addition, Ace also shows sequence similarity to a virulence protein of *Salmonella* Dublin, *spv B*, which is essential for virulence in mice [60]. The C-terminal region of Ace shows 47% amino acid similarity with residues 2 to 20 of the toxin from *Staphylococcus aureus* [7].

15.3.3 WO7 Toxin

V. cholerae WO7 (serogroup O1) strain isolated from a patient with diarrhea produces an extracellular toxin despite the absence of *ctx*, *zot*, and *ace* genes from its genome. The toxin elongates CHO cells, produces fluid accumulation in ligated rabbit ileal loops, and agglutinates freshly isolated rabbit erythrocytes. The toxin is heat-labile and sensitive to proteases and has a subunit structure consisting of two subunits with molecular masses of about 58 and 40 kDa as estimated by sodium dodecyl sulfate–polyacrylamide gel electrophoresis. GM1 was found to be the physiologic receptor for WO7 toxin on the enterocytes [61]. The toxin showed maximum binding to GM1 and interacted with a 20-kDa glycoprotein present on the cell membrane of mice enterocytes in a GM1-specific manner. The analysis of biochemical parameters in enterocytes triggered with this toxin revealed a significant increase in intracellular calcium concentration and a massive secretion of Cl^-. However, no absorption of Na^+ was observed under the same condition. This toxin also elevated the level of cAMP as well as PKA [62]. Thus, the novel toxin, although distinct from CT, showed some functional homology to it and may be one of the key factors in inducing electrolyte imbalance within intestinal cells, causing a cholera-like symptom. WO7 toxin was also found to be structurally and functionally distinct from other toxins including CT and the enterotoxic activities expressed by WO7 toxin appeared to contribute to the pathogenesis of disease associated with *V. cholerae* O1 strains [61]. Further studies are needed to understand the role of WO7 toxin in cholera-like diarrhea.

The elucidation of immune responses evoked by *V. cholerae* WO7 in the gut of mice revealed striking differences as compared to those elicited by *V. cholerae* 569B infection. PP and SP lymphoid cells from *V. cholerae* WO7-infected animals elaborated significant amounts of IL-2, IFN-γ, and IL-12 by 7 days p.i., suggesting a Th1 type of response. However, by 15 days p.i., the PP and SP lymphoid cells secreted only IL-6 and IL-10 with traces of IFN-γ. On the other hand, infection by *V. cholerae* 569B yielded mainly Th2-type responses at Peyer's patches as well as the splenic level. Infection with both *V. cholerae* WO7 and 569B strains induced toxin-specific IgA-secreting cells at the gut and spleen level along with IgG1-secreting cells, indicating that both *V. cholerae* WO7 and 569B strains evoke an antigen-specific Th2 type of response in the gut as well as spleen. The persistence of IgA along with Th1-type cytokines indicates an alternate induction mechanism since mucosal IgA responses are usually associated with Th2-type responses [63]. The WO7 toxin also induced programmed cell death of PP lymphoid cells and enterocytes in culture but not the splenic lymphoid cells (unpublished data).

15.3.4 Hemolysin–Cytolysin

V. cholerae O1 and O139 strains also produce a cytotoxic protein, which produces fluid accumulation in animal models. This toxin is known as El Tor hemolysin or hemolysin cytolysin. It has been found to be cytotoxic for a variety of erythrocytes and mammalian cells in culture and was found to be rapidly lethal for mice [1]. The hemolysin addition has been shown to cause considerable villus degeneration, accumulation of cellular exudates and mucin, and a collection of inflammatory cells in lamina propria [64]. The cytotoxin is initially made as an 82-kDa protein and processed in two steps to a 65-kDa active protein [65]. When added to planar lipid bilayers, this forms anion-selective channels [66]. The purified cytotoxin produced fluid accumulation in both rabbits and mice [67]. In classical *V. cholerae* strain 568B, an 11-bp deletion leads to truncated *hlyA* gene product of 27 kDa [68]. Alm et al. [69] suggested that the ability to lyze sheep erythrocytes lies in C-terminal end of hemolysin cytolysin and enterotoxic activity lies in N-terminal and within truncated 27-kDa product termed "HlyA." Ikigai et al. [70] showed that El Tor hemolysin (ETH) spontaneously assembles as oligomeric aggregates on the membranes of rabbit erythrocyte ghosts and liposomes. They proposed that ETH damages the membranes of both erythrocytes and liposomes by intercalating water-soluble ETH into target membrane and by forming toxin oligomers with large water-filled pore. Another protein the hemolysin-coregulated protein (HP) has been shown to be regulated by HlyA [71].

15.3.5 Non-membrane Damaging Cytotoxin (NMDCY)

Saha et al. [72] reported that some clinical strains of non-O1 *V. cholerae* produce an extracellular factor that evokes a rapid and dramatic cytotoxic response, which manifests as cell rounding of CHO and HeLa cells, without accompanying membrane damage. NMDCY was not inhibited by antitoxins against CT, Shiga-like toxin, heat-labile toxin of enterotoxigenic *E. coli*, and El Tor hemolysin. This indicated that NMDCY does not bear any immunologic relationship with known toxins and hemolysins. This toxin has been shown to have a molecular weight of 35 kDa, is heat-labile, and sensitive to trypsin. They also demonstrated that NMDCY could be one of the factors contributing to the pathogenesis of the disease associated with *V. cholerae* non-O1 strains.

15.3.6 Shiga-Like Toxin

In 1984, O'Brien et al. reported that O1 *V. cholerae* produced a Shiga-like toxin [73]. This toxin was identified on the basis of cytotoxicity in HeLa cells, which was neutralized by antibody raised against Shiga toxin purified from *Shigella dysenteriae* type 1 strain [73]. It was hypothesized that this toxin was responsible for diarrhea in volunteers who ingested genetically engineered Δ*ctx* strains. Consistent with this hypothesis, CVD 103-HgR strain, which does not produce detectable Shiga-like activity, caused little or no reactogenicity in human volunteers [74]. However, genes encoding Shiga-like toxin activity have not been identified from any *V. cholerae* strains.

15.3.7 Heat-Stable Enterotoxin (ST)

Some strains of non-O1 *V. cholerae* produce a toxin that shares 50% protein sequence homology with the heat-stable enterotoxin (ST) of *E. coli* [75]. Production of this toxin has been associated with diarrhea in volunteers [76].

15.3.8 New Cholera Toxin (NCT)

In 1983, Sanyal and colleagues reported that environmental strains of *V. cholerae* O1 that lack genes encoding CT could cause fluid accumulation in ligated rabbit ileal loops [5]. Fluid accumulation and rabbit skin permeability reactions were observed when either whole cell or culture filtrates were used. This toxin was termed "new cholera toxin" (NCT) and proposed as the cause of diarrhea in volunteers fed with genetically engineered CT^- *V. cholerae* strains [5]. Toxin activity of NCT was neutralized by the crude homologous antiserum, but not the one raised against CT. No additional biochemical or genetic characterization of this toxin has been reported.

15.3.9 Secreted CHO Cell-Elongating Protein (S-CEP)

A novel cytotonic factor named secreted CHO cell-elongating protein (S-CEP) was purified from culture supernatants of CVD 103-HgR [4, 77]. Five fractionation steps yielded electrophoretically pure S-CEP with an M_r of 79,000. A partially purified preparation caused fluid accumulation in the sealed infant mouse model. The amino terminus bore a unique sequence with strong homology to a cytotonic toxin of El Tor *V. cholerae*. No homology was found with the recently described novel *V. cholerae* cytotoxin by Walia et al. [61]. S-CEP as a toxin was found to be active as a single major polypeptide chain of 79 kDa.

15.3.10 Repeat in Toxin (RTX)

A novel toxin in *V. cholerae* that belongs to the RTX (repeat in toxin) family of toxins, which are generally produced by several pathogenic gram-negative bacteria [78], was recently discovered. The RTX toxin gene cluster in *V. cholerae* encodes the presumptive cytotoxin (*rtxA*), an acyltransferase (*rtxC*), and an associated ATP-binding cassette transporter system (RtxB and RtxD, two proteins for toxin transportation). It is physically linked to the core element in the *V. cholerae* genome, although its activity is independent of the core element [78]. Phenotypically, these genes are proven to be associated with cytotoxicity in HEp-2 cells. Studies have demonstrated that the RTX toxins of *V. cholerae* caused actin depolymerization and cross-linking in HEp-2 cells [79]. Similar actin rearrangement or condensation was observed in HEp-2 cells and was caused by a protein encoded by the *eaeA* gene of enteropathogenic and hemorrhagic *E. coli*, leading to the effacement of microvilli, with subsequent hemorrhagic colitis and bloody diarrhea [80]. Chow et al. [81] demonstrated the occurrence of RTX toxins in strains of *V. cholerae* except those exhibiting the classical biotype. Deletion of *rtxA* in *ctxAB* mutant led to less severe pathology and decreased levels of proinflammatory molecules IL-6 and macrophage inflammatory protein (MIP)-2 [82]. These data suggest that RTX contributes to severity of acute inflammatory response.

15.4 Role of Additional Toxins in Inflammation and Immunomodulation in Cholera Disease

A key goal in the development of an effective live oral cholera vaccine is the characterization of the bacterial mechanism(s) that provokes a reactogenic response in humans. When the first-generation recombinant live oral cholera vaccines were originally being planned and constructed, it was assumed that simple deletion of the genes encoding CT (*ctx*) would be sufficient to render these strains safe and totally avirulent [83–85]. However, although these initial Δ*ctx* strains no longer caused the severe diarrhea characteristic of cholera, they still produced a variety of other symptoms in several human volunteers, including mild residual diarrhea, nausea, vomiting, fever, and abdominal cramps [4, 8, 86]. This led to the discovery of a number of additional toxins in *V. cholerae* as described above. Some studies have been carried out lately to elucidate the role of these additional toxins in the pathogenesis and inflammation in the cholera disease. These studies indicate that, although earlier considered as a non-inflammatory diarrhea, cholera does trigger inflammatory events in the gut. A study carried out by Qadri et al. [12] in adults and children infected with *V. cholerae* O1 Ogawa El Tor and O139 strains showed elevated levels of non-specific immune mediators of innate defense system, i.e., prostaglandins E2(PGE2), leukotriene B4(LTB4), lactoferrin (Lf), myeloperoxidase (MPO), superoxide dismutase (SOD), and nitric oxide (NO) metabolites. These findings suggest that following *V. cholerae* infection, inflammatory processes as well as innate immune system are activated. Inflammation and innate components are believed to be a prerequisite for the host to mount an appropriate adaptive immune response to an infection [87]. It is possible that in cholera, different mediators of the innate system are activated leading to better protection. However, the inflammation observed is of a low-grade nature that might clear the bacteria or be helpful in mounting an appropriate adaptive immune response. Consistent with these are the findings from studies done with *V. cholerae* WO7 strain. Analysis of biochemical parameters in enterocytes triggered by this toxin revealed a significant increase in intracellular calcium concentration and a massive secretion of Cl^-. However, no absorption of Na^+ was observed under the same condition. This toxin also elevated the level of cAMP as well as PKA [62]. WO7 toxin, which was more hemorrhagic than *V. cholerae* 569B, also induced a Th1 response, followed by a Th2 kind of response leading to production of IgA and IgG1 antibodies. Although WO7 did not carry any of the *ctx, zot,* or *ace* genes, the evoked Th2 responses were similar to those of CT-producing *V. cholerae* strains.

Excessive cell infiltration and inflammation was seen in the patients infected with Δ*ctx V cholerae* strains. The development of inflammatory diarrhea in the absence of CT may be due to loss of immunomodulatory signaling by CT [88]. CT has been shown in vitro to block production of proinflammatory cytokines TNF-α, IL-1, and IL-12 by LPS-stimulated monocytes and dendritic cells [51, 55]. Thus, in addition to stimulating fluid secretion through activation of adenylate cyclase activity [89], CT suppresses induction of inflammation during *V. cholerae* infection. The importance of these factors in the reactogenicity of cholera vaccines in

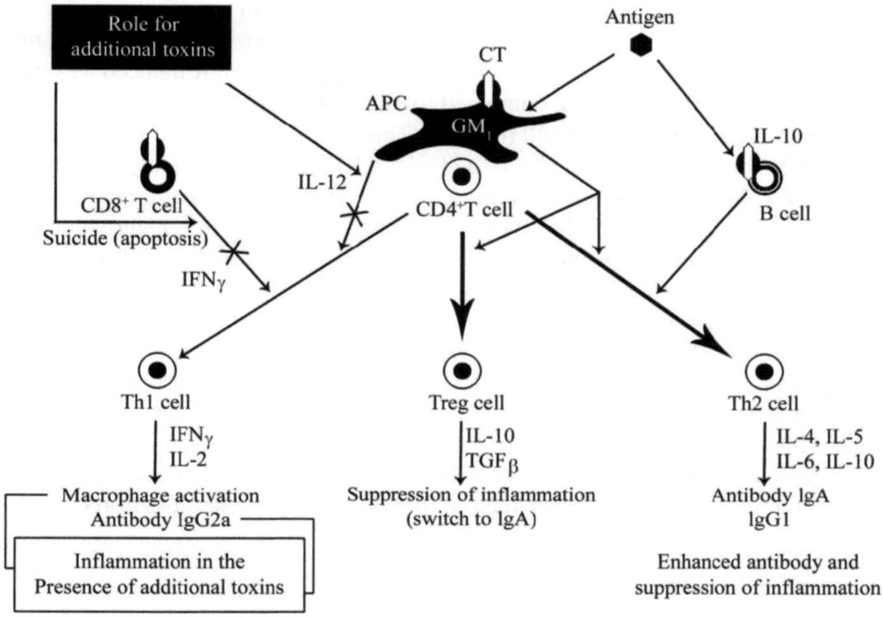

Fig. 15.1 Interaction of the CT with its receptor ganglioside GM1 modulates the activation of antigen-presenting cells (APCs), T cells, and B cells within sites local to delivery. Cytokine production by APCs is modulated, leading to enhanced production of interleukin 10 (IL-10), with a concomitant suppression of IL-12 synthesis. In this way, the major cytokine that drives T helper 1 (Th1) differentiation is blocked in favor of the production of a factor able to promote T regulatory (Treg)-cell differentiation. It is possible that accessory cholera toxins might induce proinflammatory cytokines in the absence of CT immunomodulation, in vaccine strains having *ctx* deletions. Some local CD8+ T cells also undergo apoptosis in the presence f CT and are thus unavailable as a potential source of interferon-γ (IFN-γ) that might otherwise favor Th1 differentiation. B cells are polyclonally activated, a process that might involve them further in antigen presentation and thus favor the induction of Th2 immunity

human volunteers has not been conclusively established yet. This increase in the inflammatory nature of *V. cholerae* infection could be due to the absence of the immunomodulatory activity of the B subunit of CT that blocks the secretion of proinflammatory cytokines by macrophages, dendritic cells, and epithelial cells in response to bacterial lipopolysaccharide (LPS) [90] due to downregulation of mitogen-activated protein kinase pathways [91] (Fig.15.1).

Many additional toxins were reported initiate a proinflammatory response. However, the exact role of each of these in causing inflammation and immunomodulation at the cellular level is not clear. In one of the studies carried out by Stokes et al. [92], they found that the expression of chemokine genes was not uniformly activated in the host cells infected with different *V. cholerae* vaccine strains. Rather, these components were differentially affected by the different strains. Strain CVD101 induced greater expression of IL-8 and MIP-3, which is more reactogenic than

CVD103-HgR that subsequently translated into changes in protein levels for IL-8 and is also well tolerated. This observation is consistent with the findings of a separate study, which demonstrated that the strain CVD103-HgR induced less transcription and synthesis of IL-8 compared to other attenuated strains of *V. cholerae* in another intestinal epithelial cell line [93]. This is further supported by the finding that the El Tor vaccine strains JBK70 and CVD110, which were highly reactogenic in human volunteers [74, 94]. In addition, they induced greater expression of these genes than does strain CVD112, which was less reactogenic in human volunteers [95]. It is noteworthy that the well-tolerated strain CVD103-HgR did not induce any anti-inflammatory cytokines (e.g., IL-4 and IL-10). The induction of proinflammatory cytokines appears to be an important component of the cellular response to *V. cholerae*. Beyond an extensive understanding of the mechanism of CT, its role in pathogenesis and immunomodulation, the role of the additional toxins and the knowledge of the responses of host cells to infection with *V. cholerae* are limited. Although significant strides have been made in understanding the genetics of toxin production, similar achievements for elucidating the role of all these additional toxins in immunity to disease is a missing link in the vaccine design efforts.

References

1. Honda T, Finkelstein RA. Purification and characterization of hemolysin produced by *Vibrio cholerae* biotype El Tor: another toxic substance produced by cholera vibrios. Infect. Immunol. 1979;26:1020–7.
2. Kaper JB, Lockman H, Baldini MM, Levine MM. A recombinant live oral cholera vaccine. Biol Technol. 1984;2:345–9.
3. Kaper JB, Lockman H, Baldini MM, Levine MM. Recombinant nontoxinogenic *Vibrio cholerae* strains as attenuated cholera vaccine candidates. Nature. 1984;308:655–8.
4. Levine MM, Kaper JB, Herrington D, Ketley J, Morris JG, Clemens ML, Black RE, Tall B, Hall R. Volunteer studies of deletion mutants of *Vibrio cholerae* O1 prepared by recombinant techniques. Infect Immunol. 1988;56:161–7.
5. Saha S, Sanyal SC. Immunobiological relationships of enterotoxins produced by cholera toxin gene positive (CT^+) and negative (CT^-) strains of *Vibrio cholerae* O1. J Med Microbiol. 1990;32:33–7.
6. Fasano A, Baudry B, Pumpkin DW, Wasserman SS, Tale BD, Ketley JM, Kaper JB. *Vibrio cholerae* produces a second enterotoxin, which affects intestinal tight junctions. Proc Natl Acad Sci USA. 1991;88:5242–6.
7. Truckiss M, Galan JE, Michalski J, Fasano A, Kaper JB (1993). Accessory cholera enterotoxin (Ace), the third toxin of *Vibrio cholerae* virulence cassette. Proc Natl Acad Sci USA. 1993;90:5267–71.
8. Kaper JB, Morris JG, Levine MM. Cholera. Clin Microbiol Rev. 1995;8:48–86.
9. Bennish ML. Cholera: pathophysiology, clinical features, and treatment. In: Wachsmuth IK, Blake PA, Olsvik O, editors. *Vibrio cholerae* and cholera: molecular to global perspectives. Washington, DC: ASM Press; 1994. pp. 229–55.
10. Gangarosa EJ, Beisel WR, Benyajati CM, Sprinz H, Piyaratn P. The nature of the gastrointestinal lesion in Asiatic cholera and its relation to pathogenesis: a biopsy study. J Trop Med Hyg. 1960;9:125–35.
11. Pastore GG, Schiraldi GF, Sforza E, Schiraldi O. A bioptic study of gastrointestinal mucosa in cholera patients during an epidemic in southern Italy. Am J Dig Dis. 1976;21:613–7.

12. Qadri F, Raqib R, Ahmed F, Rahman T, Wenneras C, Das SK, Alam NH, Mathan M, Svennerholm AM. Increased levels of inflammatory mediators in children and adults infected with *Vibrio cholerae* O1 and O139. Clin Diagn Lab Immunol. 2002;9:221–9.
13. Silva TM, Schleupner MA, Tacket CO, Steiner TS, Kaper JB, Edelman R, Guerrant R. New evidence for an inflammatory component in diarrhea caused by selected new, live attenuated cholera vaccines and by El Tor and O139 *Vibrio cholerae*. Infect Immunol. 1996;64:2362–4.
14. Saha DR, Niyogi SK, Nair GB, Manna B, Bhattacharya SK. Detection of faecal leucocytes and erythrocytes from stools of cholera patients suggesting an evidence of an inflammatory response in cholera. Indian J Med Res. 2000;112:5–8.
15. Janoff EN, Hayakawa H, Taylor DN, Fasching CE, Kenner JR, Jaimes E, Raij L. Nitric oxide production during *Vibrio cholerae* infection. Am J Physiol. 1997;273:1160–7.
16. Rabbani GH, Islam S, Chowdhury AK, Mitra AK, Miller MJ, Fuchs G. Increased nitrite and nitrate concentrations in sera and urine of patients with cholera or shigellosis. Am J Gastroenterol. 2001;96:467–72.
17. Koch R. An address on cholera and its bacillus. Br Med J. 1884;2:403–7.
18. De SN, Chatterje DN. An experimental study of the mechanism of action of *Vibrio cholerae* on the intestinal mucous membrane. J Pathol Bacteriol. 1953;66:559–62.
19. De SN, Bhattacharya K., Sarkar JK. A study of the pathogenicity of strains of *Bacterium coli* from acute and chronic enteritis. J Pathol Bacteriol. 1956;71:201–9.
20. De SN. Enterotoxicity of bacteria-free culture-filtrate of *Vibrio cholerae*. Nature. 1959;83:1533–4.
21. Dutta NK, Pawe MW, Kulkarni DR. Role of cholera toxin in experimental cholera. J Bacteriol. 1959;78:594–5.
22. Finkelstein RA, LoSpalluto JJ. Pathogenesis of experimental cholera: preparation and isolation of choleragen and choleragenoid. J Exp Med. 1969;130:185–202.
23. Levine MM, Kaper JB, Black RE, Clamens ML. New knowledge on pathogenesis of bacterial enteric infections as applied to vaccine development. Microbiol Rev. 1983;47: 510–50.
24. Lencer WI, Constable C, Moe S, Jobling M, Webb HM, Ruston S, Madara JL, Hirst T, Holmes R. Targeting of cholera toxin and *Escherichia coli* heat labile toxin in polarized epithelia: role of C-terminal KDEL. J Cell Biol. 1995;131:951–62.
25. Majoul IV, Bastiaens PIH, Söling H-D (1996). Transport of an external Lys-Asp-Glu-Leu (KDEL) protein from the plasma membrane to the endoplasmic reticulum: studies with cholera toxin in Vero cells. J Cell Biol. 1996;133:777–89.
26. Orlandi PA. Protein-disulfide isomerase-mediated reduction of the A subunit of cholera toxin in a human intestinal cell line. J Biol Chem. 1997;272:4591–9.
27. Lencer WI, Moe S, Rufo PA, Madara JL. Transcytosis of cholera toxin subunits across model human intestinal epithelia. Proc Natl Acad Sci USA. 1995;92:10094–8.
28. Gill DM, King CA. The mechanism of action of cholera toxin in pigeon erythrocytes lysates. J Biol Chem. 1975;250:6424–32.
29. Lencer WI, Depp C, Neutra MR, Madara JL. Mechanism of cholera toxin action of a polarized human intestinal epithelial cell line: role of vesicular traffic. J Cell Biol. 1992;117: 1197–209.
30. Lencer WI, Almedia JB, Moe S, Slow JL, Ausiello DA, Madara JL. Entry of cholera toxin into polarized human epithelial cells: identification of an early brefeldin A sensitive event required for A1 peptide generation. J Clin Invest. 1993;92:2941–51.
31. Field M. Secretion of electrolytes and water by mammalian small intestine. In: Johnson LR editor. Physiology of gastrointestinal tract. New York: Raven Press; 1981. pp. 963–82.
32. Peterson J, Molina NC, Houston WC, Fuder RC. Elevated camp in intestinal epithelial cells during experimental cholera and salmonellosis. Toxicon 1983;21:761–75.
33. Triadafilopoulos G, Pothoulakis C, Wiss R, Giampaolo C, LaMont JT. Comparative study of *Clostridium difficile* toxin A and cholera toxin in rabbit ileum. Gastroenterology. 1989;97:1186–92.

34. Guerrant RL, Fang RD, Thielman NM, Fonletes MC. Role of platelet activating factor (PAF) in the intestinal epithelial secretory and Chinese hamster ovary (CHO) cell cytoskeletal responses to cholera toxin. Proc Natl Acad Sci USA. 1994;91:9655–8.
35. Gaginella TS. Eicosanoid-mediated intestinal secretion. In: Lebenthal E, Duffed ME, editors. Textbook of secretory diarrhoea. New York: Raven Press; 1990. pp. 15–30.
36. Peterson JW, Reitmeyer JC, Jackson CA, Ansari GAS. Protein synthesis is required for cholera toxin induced stimulation of arachidonic acid metabolism. Biochem Biophys Acta. 1991;1092:79–84.
37. Cooke HJ. Hormones and neurotransmitters regulating intestinal ion transport. In: Field M editor. Diarrhoeal diseases. New York: Elsevier; 1991. pp. 23–48.
38. Lycke N Holmgren J. Strong adjuvant properties of cholera toxin on gut mucosal immune responses to orally presented antigens. Immunology. 1986;59:301.
39. Chen K-S, Strober W. Cholera holotoxin and its B subunit enhance Peyer's patch B cell responses induced by orally administered influenza virus: disproportionate cholera toxin enhancement of the IgA B cell response. Eur J Immunol. 1990;20:433.
40. Clarke CJ, Wilson AD, Williams NA, Strokes CR. Mucosal priming of T-lymphocyte responses to fed protein antigens using cholera toxin as an adjuvant. Immunology. 1991;72:323.
41. Elson CO, Ealding W. Generalized systemic and mucosal immunity in mice after mucosal stimulation with cholera toxin. J Immunol. 1984;132:2736–41.
42. Clements JD, Hartzog NM, Lyon FL. Adjuvant activity of *Escherichia coli* heat-labile enterotoxin and effect on the induction of oral tolerance in mice to unrelated protein antigens. Vaccine. 1988;6:269–77.
43. Lycke N, Severinson E, Strober W. Cholera toxin acts synergistically with IL-4 to promote IgG1 switch differentiation. J Immunol. 1990;145:3316–24.
44. Lycke NY. Cholera toxin promotes B cell isotype switching by two different mechanisms. cAMP induction augments germ-line Ig H-chain RNA transcripts whereas membrane ganglioside GM1-receptor binding enhances later events in differentiation. J Immunol. 1993;150:4810–21.
45. Francis ML, Ryan J, Jobling MG, Holmes RK, Moss J, Mond JJ. Cyclic AMP-independent effects of cholera toxin on B cell activation. II. Binding of ganglioside GM1 induces B cell activation. J Immunol. 1992;148:1999–2005.
46. Nashar TO, Webb HM, Eaglestone S, Williams NA, Hirst TR. Potent immunogenicity of the B subunits of *Escherichia coli* heat-labile enterotoxin: receptor binding is essential and induces differential modulation of lymphocyte subsets. Proc Natl Acad Sci USA. 1996;93:226–30.
47. Nashar TO, Hirst TR, Williams NA. Modulation of B-cell activation by the B subunit of *Escherichia coli* enterotoxin: receptor interaction up-regulates MHC class II, B7, CD40, CD25 and ICAM-1. Immunology. 1997;91:572–8.
48. Yankelevich B, Soldatenkov VA, Hodgson J, Polotsky AJ, Creswell K, Mazumder A. Differential induction of programmed cell death in CD8+ and CD4+ T cells by the B subunit of cholera toxin. Cell Immunol. 1996;168:229–34.
49. Elson CO, Holland SP, Dertzbaugh MT, Cuff CF, Anderson AO. Morphologic and functional alterations of mucosal T cells by cholera toxin and its B subunit. J Immunol. 1995;154:1032–40.
50. Cong Y, Weaver CT, Elson CO. The mucosal adjuvanticity of cholera toxin involves enhancement of costimulatory activity by selective up-regulation of B7.2 expression. J Immunol. 1997;159:5301–8.
51. Cong Y, Oliver AO, Elson CO. Effects of cholera toxin on macrophage production of co-stimulatory cytokines. Eur J Immunol. 2001;31:64–71.
52. Ta-Kai L, Barbara SF. Cholera toxin B subunit binding to an antigen presenting cell directly co-stimulates cytokine production from T cell clone. Int Immunol. 1996;8:1849–56.
53. Bromander A, Holmgren J, Lycke N. Cholera toxin stimulates IL-1 production and enhances antigen presentation by macrophages in vitro. J Immunol. 1991;146:2908–14.

54. Matousek MP, Nedrud JG, Harding CV. Distinct effects of recombinant cholera toxin B subunit and holotoxin on different stages of class II MHC antigen processing and presentation by macrophages. J Immunol. 1996;156:4137–45.
55. Gagliardi MC, Sallusto F, Marinaro M, Langenkamp A, Lanzavecchia A, De Magistris MT. Cholera toxin induces maturation of human dendritic cells and licences them for Th2 priming. Eur J Immunol. 2000;30:2394–403.
56. Lavelle EC, McNeela E, Armstrong ME, Leavy O, Higgins SC, Mills KHG. Cholera toxin promotes induction of regulatory T cells specific for bystander antigens by modulating dendritic cell activation. J Immunol. 2000;171:2384–92.
57. Baudry B, Fasano A, Ketley J, Kaper JB. Cloning of gene (*zot*) encoding a new toxin produced by *Vibrio cholerae*. Infect Immunol. 1992;60:428–34.
58. Fasano A, Florentino C, Donelli G, Uzzau S, Kaper JB, Margaratten K, Ding X, Guandaliv D, Comstock L, Goldblum SE. Zonula occludens toxin modulates tight junctions, through protein kinase C dependent actin reorganization in vitro. J Clin Invest. 1995;96:710–20.
59. Bear CE, Li C, Kartner N, Bridges RJ, Jensen TJ, Ramjeesingh M, Riordan JR. Purification and functional reconstitution of cystic fibrosis transmembrane conductance regulator (CFTR). Cell. 1992;68:809–18.
60. Krause M, Roudier C, Fierer J, Harwood J, Guiney D. Molecular analysis of virulence locus of *Salmonella* Dublin plasmid psD12. Mol Microbiol. 1991;5:307–16.
61. Walia K, Ghosh S, Singh H, Nair GB, Ghosh A, Sahni G, Vohra H, Ganguly NK. Purification and characterization of novel toxin produced by *Vibrio cholerae* O1. Infect Immunol. 1999;67:5215–22.
62. Bhattacharyya S, Gosh S, Shant J, Ganguly NK, Majumdar S. Effect of WO7-toxin on gut physiological response in mice. Microb Pathog. 2004;37:1–9.
63. Walia K, Vohra H, Ganguly NK. Spectrum of gut immunologic reactions: selective induction of distinct responses to *Vibrio cholerae* WO7 and its toxin. Microbiol Immunol. 2000;44: 931–40.
64. McCardell B, Madden JM, Shah DB. Isolation and characterization of cytolysin produced by *Vibrio cholerae* serogroup non-O1. Can J Microbiol. 1985;31:711–20.
65. Yamamoto K, Ichinose Y, Shinagawa H, Makino K, Nakata A, Iwanaga M, Honda T, Miwatani. Two step processing for activation of cytolysin/hemolysin of *Vibrio cholerae* O1 biotype El Tor: nucleotide sequence of the structural gene (*hlyA*) and characterization of processed products. Infect Immunol. 1990;58:4106–16.
66. Krasilnikov OV, Murathkhodjair JN, Voronov AE, Yezepchuk YV. The ionic channels formed by cholera toxin planar bilayer lipid membranes are entirely attributable to its B subunit. Biochem Biophys Acta. 1991;1067:166–70.
67. Ichinose Y, Yamamoto K, Nakasone N, Tanabe MJ, Takeda T, Miwatani T, Iwanaga M. Enterotoxicity of El Tor- like hemolysin of non-O1 *Vibrio cholerae*. Infect Immunol. 1987;55:1090–3.
68. Alm RA, Strocher UH, Manning PA. Extracellular proteins of *Vibrio cholerae*: nucleotide sequence of structural gene (*hlyA*) for the hemolysin of hemolytic El Tor strain 017 and characterization of *hlyA* mutation in the non-hemolytic classical strain 569B. Mol Microbial. 1988;2:481–8.
69. Alm RA, Mayrhofer G, Kotlarski I, Manning PA. Amino-terminal domain of the El Tor hemolysin of *Vibrio cholerae* O1 is expressed in classical strains and is cytotoxic. Vaccine. 1991;9:588–94.
70. Ikigai H, Akatsuka A, Tsujiyama H, Nakae T, Shimamura T. Mechanism of membrane damage by El Tor hemolysin of *Vibrio cholerae* O1. Infect Immunol. 1996;64:2968–73.
71. Willaims SG, Varcoe LT, Attridge SR, Manning PA. *Vibrio cholerae* Hcp, a secreted protein coregulated with HlyA. Infect Immunol. 1996;64:283–9.
72. Saha PK, Koley H, Nair GB. Purification and characterization of an extracellular secretogenic non-membrane damaging cytotoxin produced by clinical strains of *Vibrio cholerae* non-O1. Infect Immunol. 1996;64:3101–8.

73. O'Brien Ad, Chen ME, Holmes RK, Kaper J, Levine MM. Environmental strains and human isolates of *Vibrio cholerae* and *V. parahemolyticus* produce a *Shigella dysenteriae* (Shiga) like cytotoxin. Lancet. 1984;I:77–8.
74. Levine MM, Kaper JB, Herrington D, Ketley J, Losonsky G, Tacket CO, Tale B, Cryz S. Safety, immunogenicity and efficacy of recombinant live oral cholera vaccines CVD103 and CVD 103 HgR. Lancet. 1988;ii:467–70.
75. Ogawa A, Kato JI, Watanabe H, Nair GB, Takeda T. Cloning and nucleotide sequence of heat-stable enterotoxin gene from *Vibrio cholerae* non O1 isolated from patients with traveller's diarrhea. Infect Immunol. 1990;58:3325–9.
76. Morris JG, Jr. Non-O group 1 *Vibrio cholerae* a look at the epidemiology of occasional pathogen. Epidemiol Rev. 1990;12:179–91.
77. Sathyamoorthy V, Hall RH, Mccardell BA, Kothary MH, Aahn SJ, Ratnayake S. Purification and characterization of a cytotonic protein expressed in vitro by the live cholera vaccine candidate CVD 103-HgR. Infect Immunol. 2000;68:6062–5.
78. Lin W, Fullner KJ, Clayton R, Sexton JA, Rogers MB, Calia KE, Calderwood SB, Fraser C, Mekalanos JJ. Identification of a *Vibrio cholerae* RTX toxin gene cluster that is tightly linked to the cholera toxin prophage. Proc Natl Acad Sci USA. 1999;96:1071–6.
79. Fullner KJ, Mekalanos JJ. In vivo covalent cross-linking of cellular actin by the *Vibrio cholerae* RTX toxin. EMBO. 2000;19:5315–23.
80. Knutton S, Baldwin T, Williams PH, McNeish AS. Actin accumulation at sites of bacterial adhesion to tissue culture cells: basis of a new diagnostic test for enteropathogenic and enterohemorrhagic *Escherichia coli*. Infect Immunol. 1989;57:1290–1298.
81. Chow KH, NG TK, Yuen KY, Yami WC. Detection of RTX toxin gene in *Vibrio cholerae* by PCR. J Clin Microbiol. 2001;39:2594–7.
82. Fullner KJ, Boucher JC, Hanes MA, Kenneth Haines G III, Meehan BM, Walchle C, Sansonetti PJ, Mekalanos JJ. The Contribution of accessory toxins of *Vibrio cholerae* O1 El Tor to the proinflammatory response in a murine pulmonary cholera model. J Exp Med. 2002;195:1455–62.
83. Kaper JB, Lockman H, Baldini MM, Levine MM. A recombinant live oral cholera vaccine. Bio Technol. 1984;2:345–59.
84. Kaper JB, Lockman H, Baldini MM, Levine, MM. Recombinant non-toxinogenic *Vibrio cholerae* strains as attenuated cholera vaccine candidates. Nature. 1984;308:655–658.
85. Mekalanos JJ, Swartz DJ, Pearson GD, Harford N, Groyne F, de Wilde M. Cholera toxin genes: nucleotide sequence, deletion analysis and vaccine development. Nature. 1983;306:551–7.
86. Herrington DA, Hall RH, Losonsky G, Mekalanos JJ, Taylor RK, Levine MM. Toxin, toxin-coregulated pili, and the toxR regulon are essential for *Vibrio cholerae* pathogenesis in humans. J Exp Med. 1988;168:1487–92.
87. Marcinkiewicz J. Neutrophil chloramines: missing links between innate and acquired immunity. Immunol Today. 1997;18:577–80.
88. Williams NA, Hirst TR, Nashar TO. Immune modulation by the cholera-like enterotoxins: from adjuvant to therapeutic. Immunol Today. 1999;20:95–101.
89. Lencer WI. Microbes and microbial toxins: paradigms for microbial—mucosal interactions: *Vibrio cholerae*: invasion of the intestinal epithelial barrier by a stably folded protein toxin. Am J Physiol Gastrointest Liver Physiol. 2001;280:781–6.
90. Satchell KJF. Activation and suppression of proinflammatory immune response by *Vibrio cholerae* toxins. Microbes Infect. 2003;5:1241–7.
91. Chen P, Li J, Barnes J, Kokkonen GC, Lee JC, Liu Y. Restraint of pro-inflammatory cytokine biosynthesis by mitogen activated protein kinase phosphatase-1 in lipopolysaccharide-stimulated macrophages. J Immunol. 2002;169:6408–16.
92. Stokes NR, Zhou X, Meltzer SJ, Kaper JB. Transcriptional responses of intestinal epithelial cells to infection with *Vibrio cholerae*. Infect Immunol. 2004;72:4240–8.
93. Rodriguez BL, Rojas A, Campos J, Ledon T, Valle E, Toledo W, Fando R. Differential interleukin-8 response of intestinal epithelial cell line to reactogenic and nonreactogenic candidate vaccine strains of *Vibrio cholerae*. Infect Immunol. 2001;69:613–6.

94. Tacket CO, Losonsky G, Nataro JP, Cryz SJ, Edelman R, Fasano A, Michalski J, Kaper JB, Levine MM. Safety and immunogenicity of live oral cholera vaccine candidate CVD 110, a Δ*ctx*A Δzot Δace derivative of El Tor Ogawa *Vibrio cholerae*. J Infect Dis. 1993;168: 1536–40.
95. Tacket CO, Losonsky G, Nataro JP, Comstock L, Michalski J, Edelman R, Kaper JB, Levine MM. Initial clinical studies of CVD 112 *Vibrio cholerae* O139 live oral vaccine: safety and efficacy against experimental challenge. J Infect Dis. 1995;172:883–6.

Chapter 16
Vibrio cholerae Hemolysin: An Enigmatic Pore-Forming Toxin

Kalyan K. Banerjee and Budhaditya Mazumdar

Abstract *Vibrio cholerae* cytolysin/hemolysin (VCC) is an extracellular membrane-damaging toxin expressed by the majority of *V. cholerae* El Tor and non-O1 strains. The mature toxin has a native monomeric molecular weight of 65,000 and induces colloid osmotic lysis in a wide spectrum of eukaryotic cells including rabbit erythrocytes and human enterocytes by forming transmembrane, β-barrel heptameric pores with an internal diameter of ~1.5 nm. The purified toxin is enterotoxigenic in ligated rabbit ileal loop, causes death of adult mice by oral inoculation, induces vacuolation and death of cells in culture, and is apoptogenic to murine intestinal B-1a cells. However, the role of VCC in human cholera is still being debated. The three-dimensional structure of the VCC pro-toxin has recently been solved and has revealed similarity in pore formation mechanism with the small pore-forming toxin of *Staphylococcus aureus* α-hemolysin. However, VCC has unique functional characteristics not shared by the α-hemolysin or other pore-forming toxins. We present a summary of the recent studies on VCC that shed light on how a water-soluble protein transforms into an integral membrane protein by exploiting its intrinsic thermodynamic properties as well as its affinity for specific protein and lipid components of the target membrane and point out the missing links that need to be bridged before we comprehend its in vivo and in vitro mode of action in molecular details.

16.1 Introduction

Vibrio cholerae hemolysin/cytolysin (VCC) is a membrane-damaging pore-forming toxin [PFT; 1, 2] excreted in a water-soluble form in the growth medium by the majority of El Tor O1 and non-O1 strains [3, 4]. The mature form of the fully active toxin, which has a native monomeric molecular weight of 65,000, kills a wide

K.K. Banerjee (✉)
Division of Biochemistry, National Institute of Cholera and Enteric Diseases, Kolkata 700 010, India
e-mail: banerjeekalyan2003@yahoo.co.in

T. Ramamurthy, S.K. Bhattacharya (eds.), *Epidemiological and Molecular Aspects on Cholera*, Infectious Disease, DOI 10.1007/978-1-60327-265-0_16,
© Springer Science+Business Media, LLC 2011

spectrum of eukaryotic cells including rabbit and human erythrocytes, enterocytes, and lymphocytes by destroying selective permeability of the plasma membrane by implanting themselves as transmembrane heptameric β-barrel diffusion channels in the membrane lipid bilayer [5, 6]. Historically, the hemolytic activity of El Tor isolates toward sheep erythrocytes was used to distinguish the El Tor from the non-hemolytic classical biotype of *V. cholerae* that produced a more severe disease and was prevalent till the sixth pandemic in the early twentieth century [7]. Since 1960, however, an increasing number of El Tor isolates were reported to have poor or no hemolytic activity [8] and some classical strains produced hemolysis, mediated apparently by a biochemically distinct, but as yet uncharacterized cytolysin [9, 10], questioning the reliability of hemolysis as the sole phenotypic marker to identify the biotype. During the last 50 years, VCC has evolved from a possible diagnostic tool for biotype classification to a prominent member of a large, heterogonous family of extremely potent membrane-damaging toxins with very intriguing molecular and functional properties. In the present review, we focus on the structure–function relationship of VCC and, in the process, try to identify the missing links in our knowledge that have made this toxin one of the most well-characterized proteins with the least understood role in pathogenesis of the human cholera.

16.2 Expression, Isolation, and Purification

VCC is synthesized as an 82 kDa preprocytolysin [11] encoded by the *hlyA* gene located in the smaller of the two chromosomes, chromosome 2 (1,072,314 bp) of the bacteria [12]. Interestingly, the small chromosome has been suggested to have a megaplasmid ancestry and does not include any of the genes encoding essential bacterial functions or the major virulence factors like cholera toxins or adhesins. It is reasonable to assume, therefore, that the *hlyA* gene is a relatively recent acquisition by the bacteria; nevertheless, the gene is conserved in El Tor and non-O1 strains of clinical as well as environmental origins [13], suggesting that VCC might play an as yet unclear role in bacterial physiology, survival in animal hosts or in the environment. Following cleavage of the 25-amino acid signal peptide involved in targeting the translated cytolysin to the bacterial inner membrane [11], the protein is exported to the culture medium as the hemolytically inactive 79 kDa procytolysin [14]. The pro-toxin is converted to the 65 kDa fully active, mature toxin by removal of the 14 kDa N-terminal Pro-region by bacterial proteases in the culture medium, e.g., the Zn-dependent hemagglutinin/protease (Hap). A number of exogenous proteases, e.g., trypsin, chymotrypsin, subtilisin, and thermolysin can also process the maturation of the pro-toxin at cleavage sites in the octapeptide region (Leu-146 to Ser-153;14). A 50 kDa variant of the cytolysin, VCC^{50}, is generated spontaneously by prolonged exposure of the toxin to bacterial proteases, e.g., under conditions of overnight culture by cleavage at the carboxy-terminus end [15]. The toxin has been purified to homogeneity in several laboratories, e.g., by $(NH_4)_2SO_4$ fractionation of the overnight culture supernatant and binding to Sephadex G-200 [16] and ethanol precipitation of the culture supernatant followed by isoelectric focusing in sucrose density gradient and hydroxyapatite chromatography [17]. We purify the toxin from the early stationary phase culture supernatant in brain–heart infusion broth by adsorption to phenyl-Sepharose CL-4B followed by anion-exchange chromatography on polybuffer exchanger (PBE)-94 (Pharmacia) and hydrophobic interaction chromatography on phenyl-Sepharose CL-4B using 0–80% ethylene glycol gradient elution [18, 19]. The purified toxin induces hemolysis of rabbit erythrocytes at a concentration of 100 pM.

16.3 Structure and Biophysical Characteristics

Before we discuss the three-dimensional structure and solution properties of VCC, it is worthwhile to take a glance at the features that distinguish PFTs from all other classes of proteins [1–2, 20, 21]. PFTs may exist in one of the two stable forms: a water-soluble native monomer, a form in which they are released from the bacteria and are cytolytically active, and a rigid, SDS-stable transmembrane β-barrel multimer serving as a diffusion channel. Apparently, the dimorphic existence of PFTs violates the dogma that the amino acid sequence of a protein uniquely defines its native conformation. Transition of the globular to the integral membrane protein conformation is irreversible and brought about in a specific sequence of biochemical events involving four conformationally and organizationally distinct states of the toxin, viz. monomer in water, membrane-bound monomer, self-assembly on the target cell surface to the pre-pore multimer, and finally the β-barrel integral membrane protein with the architecture and function of a diffusion channel (Fig. 16.1). In contrast to α-helical integral membrane proteins like glycophorin A or G protein-coupled receptors, PFTs resemble porins of the outer membrane of Gram-negative bacteria and mitochondria [22] in having no uninterrupted stretch of hydrophobic amino acids to span the hydrocarbon core of the membrane bilayer.

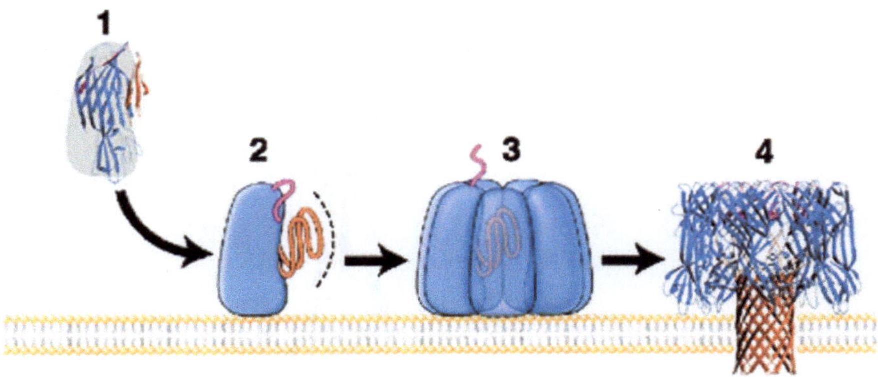

Fig. 16.1 Hypothetical model of pore formation by β-barrel membrane proteins. *1*. Water-soluble monomer. *2*. Membrane-bound monomer. *3*. Pre-pore multimeric assembly on the membrane surface. *4*. Membrane-inserted pore

Instead, each PFT monomer contributes a single β-polypeptide strand to the terminal β-barrel. PFTs, that include VCC [5, 6], aerolysin from *Aeromonas hydrophila* [20] and *Staphylococcus aureus* α-toxin [23], form small channels of internal diameter 1–2 nm comprising a relatively small number, usually seven, monomers. In contrast, the cholesterol-dependent cytolysins, a prominent group of PFTs that depend critically on membrane cholesterol for initiation of the chain of events leading to bilayer perforation [24], form channels of diameter ~25 nm by self-assembly of ~30–40 monomers. Previously, there was some confusion in the literature about whether self-assembly of PFTs is a pre- or post-insertion event [21, 25]. It is now accepted, on considerations of the energetic advantage of inserting a preformed oligomer with the amphipathic β-hairpin protruding outward in comparison to the monomer and also some experimental evidence, that the pre-pore first assembles to a pre-pore multimer on the target cell surface and then inserts into the bilayer [1, 21, 26]. Insertion is generally thought to be spontaneous unless interrupted, e.g., by mutation [27].

The structure of the 79 kDa VCC pro-toxin has recently been resolved by X-ray crystallography at 2.3 Å and that of the VCC heptamer at 3.5 Å resolution (Fig. 16.2; [6]). The toxin has a central cytolysin domain homologous to that of staphylococcal α-toxin [23] and is thought to be involved in oligomerization as well as in constituting the stem spanning the target lipid bilayer. The stem region is relatively enriched in charged and polar amino acids, suggesting that locating this domain into the nonpolar bilayer core would be thermodynamically less compatible than

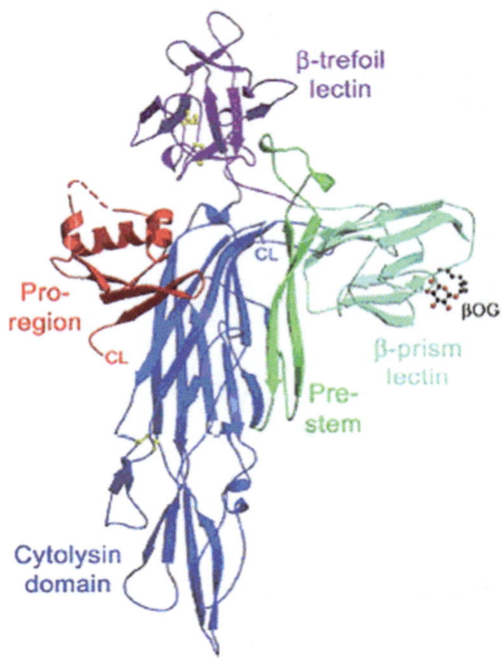

Fig. 16.2 Crystallographic structure of the VCC pro-toxin at 2.3 Å resolution [6]

staphylococcal α-toxin. Interestingly, the central cytolysin domain is followed by two lectin domains, designated, respectively, the β-trefoil domain that bears homology to the galactose-binding domain of the plant toxin ricin and the β-prism domain resembling the sugar-binding domain of the plant lectin jacalin. We may note that that the 50 kDa truncated VCC variant, VCC^{50}, lacks the C-terminus β-prism lectin domain.

Our knowledge of solution properties of PFTs in general is relatively meager. The 65 kDa VCC possesses features that seem to be somewhat unusual for a water-soluble globular protein. We reported that the apparently water-soluble, native VCC monomer resembles membrane proteins, perhaps in a limited sense, by partitioning quantitatively to the detergent phase in preference to the aqueous phase [19]. Furthermore, the intrinsic amphipathicity of the cytolysin drives the toxin from water to synthetic lipid vesicles or liposomes and biomembranes in a receptor-independent manner. Monitoring of the VCC–liposome interaction by tryptophan fluorescence emission spectrophotometry revealed movement of the aromatic amino acids from the core of the protein to the lipid–water interface of the target membrane (Fig. 16.3). It is possible that the partial unfolding of the toxin on the cell surface facilitates its subsequent reconstitution into the β-barrel pore (Fig. 16.4). Surfactant-like activity may account for nonspecific interaction with chromatographic matrices and also for nonstoichiometric aggregation of VCC in water.

It is a common consensus that self-assembly of a PFT monomer to the β-barrel multimer is triggered by interaction of the toxin with the target membrane. We found that VCC oligomerizes slowly in water to the SDS-stable heptamer in the absence of membrane components [19]. The propensity of the VCC monomer to oligomerize spontaneously explains why we isolated the toxin from the bacterial culture supernatant by ammonium sulfate precipitation as hemolytically inactive

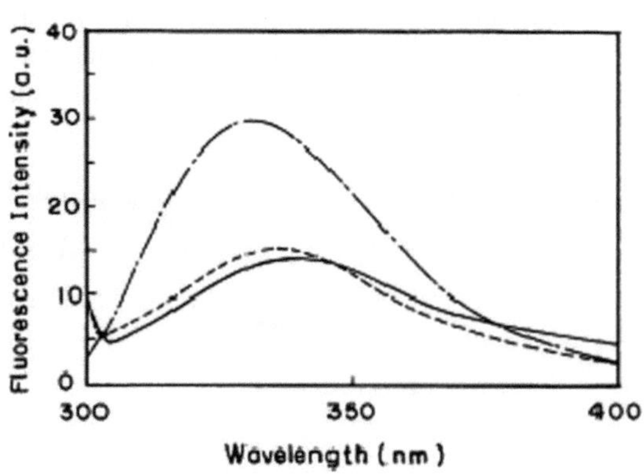

Fig. 16.3 Partial unfolding of the VCC monomer induced by partitioning to the lipid–water interface of the phosphatidylcholine–cholesterol (—) and phosphatidyletanolamine–cholesterol (---) vesicles. The tryptophan fluorescence emission spectrum of the VCC monomer is included for comparison (-.-.-). The VCC monomer was incubated with the lipid vesicles at a protein concentration of 50 μg/ml and at a lipid:protein ratio of 10:1 (19)

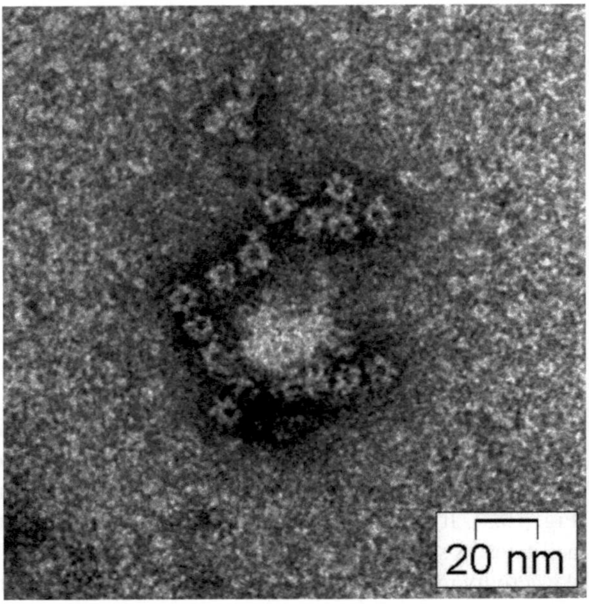

Fig. 16.4 VCC and VCC[50] assemble to β-barrel channels in liposomes. Transmission electron micrographs of VCC (A) and VCC[50] (B) incubated with PC-CL vesicles at a lipid:protein weight ratio of 100:1. The specimens were negatively stained with 2% uranyl acetate

multimeric lectin with strong hemagglutinating activity toward rabbit erythrocytes and not as the hemolytically active monomer [28, 29]. In a more recent study, we showed by combining urea gradient gel electrophoresis and fluorescence spectrofluorometry that a moderate perturbation of the native conformation of the VCC monomer by 2 M urea triggered quantitative conversion to the heptamer (Fig. 16.5; [30]). At higher concentrations of urea, we observed competition between unfolding of the monomer to a random coil configuration and reconstitution into the hemolytically inactive hemagglutinating oligomer. We interpreted these observations by postulating that the VCC monomer represents a quasi-stable conformation

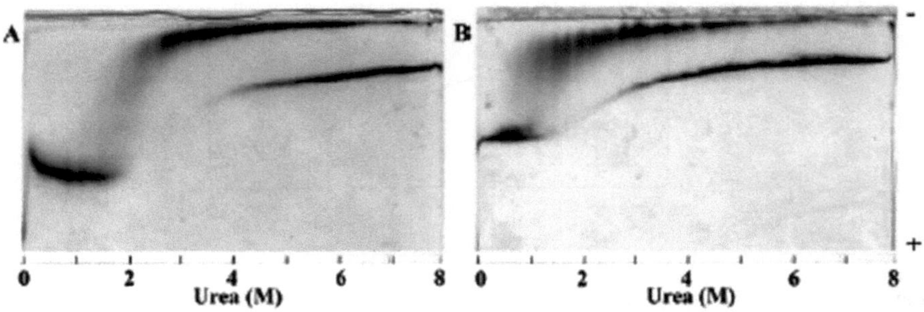

Fig. 16.5 Transverse urea gradient gel electrophoresis of VCC. The purified cytolysin was subjected to electrophoresis at pH 8.8 in 7.5% polyacrylamide gel containing transverse 0–8 M gradient of urea and visualized by staining in Coomassie Brilliant Blue (30)

corresponding to a local energy minimum and the heptameric channel a stable conformation corresponding to an absolute energy minimum. Because VCC relaxes to a more stable heptamer on removal of the conformational constraints by restricted unfolding, we may compare the monomer to a stretched string subject to physical constraints. Interestingly, pro-VCC renatures under similar conditions [30], presumably because the pro-VCC monomer with its oligomerization domain locked by the 14 kDa Pro-peptide [6] has no option to relax to the oligomer. Earlier, Nagamune et al. [31] postulated, on the basis of the homology of the Pro-region with the heat-shock protein, Hsp90, an intramolecular chaperone-like role for the Pro-peptide in renaturation of the denatured VCC monomer to the native toxin. Collectively, these observations suggest that in spite of apparent dimorphism, VCC is intrinsically committed to a single terminal structure that may correspond to the heptameric pore.

16.4 Receptor Specificity, Membrane Binding, and Bilayer Insertion

Any protein designed to act on specific target cells interacts with cell surface molecules. The VCC monomer binds to a host of plasma membrane constituents of eukaryotic cells though their relevance to the pore-forming activity of the toxin is not always clear. We were the first to report that VCC interacted with soluble as well as erythrocyte surface glycoconjugates with multiple β1-galactosyl moieties with an association constant of $9.4 \times 10^{-7}\ M^{-1}$ [18]. The presence of two contiguous lectin-like domains homologous to the sugar-binding domains of the plant lectins ricin and jacalin in VCC were later corroborated by the crystallographic study of Olson and Gouaux [6]. β1-Galactosyl-terminated glycoconjugates inhibited strongly the hemolytic activity of the cytolysin toward rabbit erythrocytes. Intriguingly, they also inhibited interaction of the toxin with liposomes devoid of incorporated glycoconjugates implying that soluble carbohydrate ligands did not compete with cell surface glycoconjugates. Furthermore, enzymatic destruction of erythrocyte surface galactose residues caused a marked increase in susceptibility of cells to VCC, questioning the functional significance of target cell surface carbohydrates in binding and subsequent events [18]. Zhang et al. [32] reported that the human erythrocyte glycophorin B might act as a receptor for VCC, apparently through a protein–protein or protein–carbohydrate type of interaction. Since VCC acts on cells lacking glycophorin, the role of this interaction in pore formation remains ambiguous.

Arguably, there is more definitive evidence, though confusing at some points, for a role of membrane cholesterol in toxin-targeting, self-assembly to the heptamer and in bilayer insertion [19, 33–37]. Cholesterol enhances the kinetics of self-assembly of the VCC monomer on the target cell surface by several orders of magnitude [33, 34] and may serve the cytolysin to distinguish eukaryotic target cells from the prokaryotes. Pore formation in liposomes is strongly promoted by the presence of sphingolipids, especially monohexosylceramide and free ceramide along with cholesterol [34]. By quantifying binding of VCC to six cholesterol analogs, Ikigai et al. [37] concluded that the interaction is stereospecific, with the β-OH group and the C–C double bond between C-5 and C-6 of the sterol making a critical contribution. In contrast, sphingolipids do not exhibit stereospecificity in their augmenting effects on pore formation by VCC [35]. To interpret the role of sphingolipids, Zitzer et al. [35] proposed that VCC does not interact directly with sphingolipids; rather ceramides promote interaction of the cytolysin with cholesterol by lowering the free energy of the sterol and thereby changing its configuration in the membrane bilayer by shielding its bulky hydrophobic group from exposure to water. As noted above, global amphipathicity of VCC drives sequestration of the toxin from the aqueous phase to the lipid bilayer surface [19]. So, it is unlikely that cholesterol is important in targeting the cytolysin to the cell surface. It is more likely that cholesterol serves as a concentration platform to promote oligomerization of the toxin monomer. We may recall that cholesterol–sphingolipid-rich lipid microdomains or rafts play a critical role in pore formation by cholesterol-dependent cytolysins, e.g., perfringolysin O [38]. Interestingly, the VCC heptamer does not bind cholesterol [33, 36] implying that cholesterol or lipid rafts are not involved in penetration of the lipid bilayer core by the cytolysin.

The mechanism of insertion of VCC or any PFT into the target membrane bilayer should be viewed in the context of innate complexity of translocation of proteins from water into the hydrocarbon core. In living cells, newly synthesized polypeptides emerging from ribosomes are translocated into or across the endoplasmic reticulum (ER) membrane by an extremely complex process that involves synchronization of folding and bilayer insertion and is coordinately regulated by the protein-based molecular machinery or the translocon complex [39, 40]. In contrast, perforation of the membrane bilayer by water-soluble PFTs is thought to be a simple, self-driven process. How PFTs simplify a very complex biological process is indeed baffling. However, some observations question the spontaneity of passage of the VCC heptamer from the cell surface or the lipid–water interface [41] to the bilayer core. Liposomes with a lipid composition most favorable for toxin binding and oligomerization are more than 1000-fold less susceptible to permeabilization by VCC than erythrocytes [34]. Similarly, we have found that pore formation by VCC could be essentially abrogated at the stage of self-assembly on the erythrocyte surface by nicking the C-terminus β-prism lectin domain. These observations hint at the possibility of some as yet unknown interaction between VCC and some biomembrane component that plays a critical role in enabling the β-barrel heptamer to span the bilayer. Valeva et al. [42] demonstrated by using environment-sensitive fluorophores that the staphylococcal α-toxin heptameric pore penetrates the lipid bilayer of the sensitive human lymphocyte plasma membrane but not of the resistant human fibroblasts. Vijayvargia et al. [43] presented evidence of significant enhancement of pore-forming activity of the α-toxin by its interaction with the protein Caveolin-1 of caveolae of mammalian cells. Because the α-toxin binds to and oligomerizes in contact with liposomes containing phosphatidylcholine, such interactions seem to positively regulate bilayer insertion of the toxin, either by catalyzing transition of the pre-pore to the pore state, e.g., by exposing the amphipathic β-hairpin to anchor to the hydrocarbon core, or by providing the energy for transporting a not too amphipathic β-barrel multimer into the hydrocarbon core. Collectively, these data suggest that the concept of spontaneous insertion of the PFT into the membrane bilayer is a gross oversimplification. Furthermore, they illustrate how elusive it is to identify the functional receptor of VCC or any PFT, because a membrane component can modulate the susceptibility of the bilayer by affecting any of the stages of membrane permeabilization, e.g., membrane binding, oligomerization, pre-pore-to-pore conversion, or bilayer insertion.

16.5 Interaction of VCC with Nonerythroid Cells: Relevance to Disease

An agent that destroys the selective permeability of the plasma membrane separating the cytosol and the environment would eventually kill the cell by colloid osmotic lysis. Even at sublytic concentrations of the toxin when there is not much swelling to cause rupturing of the membrane, alteration in ion permeability would have a profound effect on cell physiology. Ichinose et al. [44] provided the first indication that VCC might have a role in pathogenesis of cholera by demonstrating that the purified toxin induces secretion of hemorrhagic fluid in rabbit-ligated ileal loop and diarrhea in suckling mice. Although the mechanism of intestinal secretion in rabbit and mice is not clear, VCC induces death of human intestinal cells by forming small transmembrane pores that allow rapid efflux of K^+ but not influx of Ca^{2+} ions leading eventually to irreversible depletion of ATP [45]. Notably, VCC exerts its cytocidal effect at a concentration as low as 10 ng, suggesting that the cytolysin is potentially as important a virulence factor as cholera toxin. Recently, Olivier et al. [46] demonstrated a predominating role for VCC in causing death of 4- to 6-week-old mice with secondary contribution by the multifunctional autoprocessing RTX toxin. Collectively, these studies based on animal models and tissue culture cells suggest strongly that VCC might substitute for the role of cholera enterotoxin in diarrheagenic Δ*ctx* strains. However, two ΔhlyA and Δ*ctx* *V. cholerae* strains CVD104 and CVD105 constructed by recombinant DNA technology caused diarrhea in 33% of the subjects in a human volunteer study [47] implying that VCC is unlikely to be the sole or even a prominent cause of diarrhea in cholera toxin-negative strains. It seems that for some strange reason, VCC is not enterotoxic in human cholera in spite of being an extremely potent toxin toward human cells. It is tempting to speculate if the lability of the cytolytically active VCC monomer reflected in its transformation to the inactive oligomer [30] limits its survival in the human gut and thereby its relevance to pathogenesis of cholera.

The purified VCC induces extensive vacuolation of HeLa and Vero cells leading eventually to cell death [48, 49]. Moschioni et al. [50] demonstrated that vacuolation caused by VCC is mechanistically different from that induced by the VacA cytotoxin of *Helicobacter pylori* [51] and depends critically on the formation of anion-selective channel in the plasma membrane rather than on the inhibition of the specific vacuolar ATPase activity. The vacuoles are derived from intracellular organelles like late endosomes and the *trans*-Golgi network. VCC, in common with several other PFTs like listeriolysin O [52] and staphylococcal α-toxin [53], induces apoptosis of mouse peritoneal B-1a cells [54]. Notably, the VCC monomer causes upregulation of the Toll-like receptor-2 (TLR-2) identified as a death receptor for transduction of apoptotic signals. Incubation of the B-1a cells with anti-mouse TLR-2 antibody inhibits expression of the receptor without blocking apoptosis, indicating that the cytolysin activates caspases via intrinsic death pathway independent of death receptor signaling. Interestingly, the hemolytically inactive VCC heptamer is not apoptogenic but evokes secretion of IgA by the B-1a cells. It seems unlikely that apoptotic activity of VCC plays a direct role in pathogenesis of diarrhea. However, it may help the bacteria to evoke the immune surveillance of the host by killing the B-1a cells involved in IgA production. We may mention that the VCC monomer and the heptamer are distinguished by macrophage, monitored by the expression of co-stimulatory molecules, CD80 and CD86, presumably due to their difference in protease susceptibility [55].

16.6 Conclusion

We have presented a summary of our knowledge of the structure and function of VCC. However, the facts are yet to be woven into a self-consistent picture that allows one to interpret its biological activity and role in virulence in terms of its structure and biophysical properties. We do not know why evolution has segregated the *hlyA* gene in chromosome 2 that does not encode any of the major virulence factors of *V. cholerae*, nor do we know why the toxin folds into an amphipathic molecule when it is supposed to be active in water and into a more hydrophilic heptamer (unpublished observation, Mazumdar and Banerjee) when it is supposed to be functional as an integral membrane protein. We have identified the diverse membrane components that bind to VCC; however, we have not been able to identify without ambiguity the receptor involved in membrane permeabilization and cell death. We do not know why VCC has a limited, if any, role in virulence of the organism in spite of being an extremely potent toxin. In fact, we are not even sure if evolution has designed VCC to be a virulence factor in the host as the monomer or serve some unknown role either in bacterial physiology or in its survival in the environment as the more stable heptameric hemagglutinating lectin. Recent studies have added new insights to the mode of action of VCC; however, they have made this protein more enigmatic than what it was a decade earlier.

Acknowledgments We acknowledge with pleasure the contributions of N. Saha and K. Chattopadhyay of this laboratory in this study. I (KKB) express my gratitude to Prof A. C. Ghose for suggestion to work on this project and to Prof A.C. Ghose, Dr S. Chakraborty, Dr T. Biswas, Dr A. N. Ghosh, and Dr G. B. Nair for stimulating discussion.

References

1. Heuck AP, Tweten, RK, Johnson, AE. β-Barrel pore-forming toxins. Biochemistry. 2001;40:9065–73.
2. Tilley SJ, Saibil HR. The mechanism of pore formation by bacterial toxins. Curr Opin Struct Biol. 2006;16:230–6.

3. Kaper JB, Morris JG, Levine, MM. Cholera. Clin Microbiol Rev. 1995;8:48–86.
4. Yamamoto K, Ichinose Y, Nakasone N, Tanabe M, Nagahama M, Sakurai J, Iwanaga M. Identity of hemolysins produced by *Vibrio cholerae* non- O1 and *V. cholerae* O1, biotype El Tor. Infect Immun. 1986;51:927–31.
5. Zitzer A, Walev I, Palmer M, Bhakdi S. Characterization of *Vibrio cholerae* cytolysin as an oligomerizing pore-forming toxin. Med Microbiol Immunol. 1995;184:37–44.
6. Olson R, Gouaux E. Crystal structure of the *Vibrio cholerae* cytolysin (VCC) pro-toxin and its assembly into a heptameric transmembrane pore. J Mol Biol. 2005;350:997–1016.
7. Pollitzer R. Cholera. Geneva: World Health Organization; 1959.
8. Barret TJ, Blake PA. Epidemiological usefulness of changes in hemolytic activity of *Vibrio cholerae* biotype El Tor during the seventh pandemic. J Clin Microbiol. 1981;13:126–9.
9. Alm RA, Stroeher UH, Manning PA. Extracellular proteins of *Vibrio cholerae*: nucleotide sequence of the structural gene (*hlyA*) for the hemolysin of the hemolytic El Tor strain 017 and characterization of the *hlyA* mutation of the nonhemolytic classical strain 569B. Mol Microbiol. 1988;2:481–8.
10. Richardson K, Michalski J, Kaper JB. Hemolysin production and cloning of two hemolysin determinants from classical *Vibrio cholerae*. Infect Immun. 1986;54:415–20.
11. Yamamoto K, Ichinose Y, Shinagawa H, Makino K, Nakata A, Iwanaga M, Honda T, Miwatani T. Two-step processing for activation of the cytolysin/hemolysin *Vibrio cholerae* 01 biotype El Tor: nucleotide sequence of the structural gene (*hlyA*) and characterization of the processed products. Infect Immun. 1990;58:4106–16.
12. Heidelberg JF, Eisen JA, Nelson WC, Clayton RA, Gwinn ML, Dodson RJ, Haft DH, Hickey EK, Peterson JD, Umayam L, Gill SR, Nelson KE, Read TD, Tattelin H, Richardson D, Ermolaeva MD, Vamathevan J, Bass S, Qin H, Dragoi I, Sellers P, McDonald L, Utterback T, Fleishmann RD, Nierman WC, White O, Salzberg SL, Smith HO, Colwell RR, Mekalanos JJ, Venter JC, Fraser CM. DNA sequence of both chromosomes of the cholera pathogen *Vibrio cholerae*. Nature. 2000;406:477–83.
13. Brown MH, Manning PA. Hemolysin genes of *Vibrio cholerae*: presence of homologous DNA in non-hemolytic and hemolytic non-O1 strains. FEMS Microbiol Lett. 1985;30:197–201.
14. Nagamune K, Yamamoto K, Naka A, Matsuyama J, Miwatani T, Honda T. In-vitro proteolytic processing and activation of the recombinant precursor of El Tor cytolysin/hemolysin (Pro-HlyA) of *Vibrio cholerae* by soluble hemagglutinin/protease of *V. cholerae*, trypsin, and other proteases. Infect Immun. 1996;64:4655–8.
15. Zitzer A, Walev I, Palmer M, Bhakdi S. Characterization of *Vibrio cholerae* cytolysin as an oligomerizing pore-forming toxin. Med Microbiol Immunol. 1995;184:37–44.
16. Yamamoto K, Al-Omani M, Honda T, Takeda Y, Miwatani, T. Non-O1 *Vibrio cholerae* hemolysin: purification, partial characterization and immunological relatedness to El Tor hemolysin. Infect Immun. 1984;45:192–6.
17. Zitzer A, Palmer M, Weller U, Wassenaar T, Biermann C, Tranum-Jensen J, Bhakdi S. Mode of primary binding to target membranes and pore formation induced by *Vibrio cholerae* cytolysin (hemolysin). Eur J Biochem. 1997;247:209–16.
18. Saha N, Banerjee KK. Carbohydrate-mediated regulation of interaction of *Vibrio cholerae* hemolysin with erythrocyte and phospholipid vesicle. J Biol Chem. 1997;272:162–7.
19. Chattopadhyay K, Bhattacharyya D, Banerjee KK. *Vibrio cholerae* hemolysin: Implication of amphiphilicity and lipid-induced conformational change for its pore-forming activity. Eur J Biochem. 2002;269:4351–8.
20. Abrami L, Fivaz M, van der Goot FG. Adventures of a pore-forming toxin at the target cell surface. Trends Microbiol. 2000;8:168–72.
21. Bayley H, Jayasinghe L, Wallace M. Prepore for a breakthrough. Nature. 2005;12:385–6.
22. Tamm LK, Arora A, Kleinschmidt JH. Structure and assembly of β-barrel membrane proteins. J Biol Chem. 2001;276:32399–402.
23. Song L, Hobaugh MR, Shustak C, Cheley S, Bayley H, Gouaux JE. Structure of Staphylococcal α-hemolysin, a heptameric transmembrane pore. Science. 1996;274:1859–66.

24. Tweten RK. Cholesterol-dependent cytolysins, a family of versatile pore-forming toxins. Infect Immun. 2005;73:6199–209.
25. Ojcius DM, Young JDM. Cytolytic proteins and peptides: is there a common structural motif? Trends Biochem Sci. 1991;16:225–9.
26. Walker B, Braha O, Cheley S, Bayley H. An intermediate in the assembly of a pore-forming protein trapped with a genetically-engineered switch. Chem Biol. 1995;2:99–105.
27. Walker BJ, Krishnasastry M, Zorn L Bayley H. Assembly of the oligomeric membrane pore formed by Staphylococcal alpha-hemolysin examined by truncation mutagenesis. J Biol Chem. 1992;267:21782–6.
28. Banerjee KK, Ghosh AN, Dutta-Roy K, Pal SC, Ghose AC. Purification and characterization of a novel hemagglutinin from *Vibrio cholerae*. Infect Immun. 1990;58:3698–705.
29. Saha N, Banerjee KK. Carbohydrate-dependent binding of the cell-free hemagglutinin of *Vibrio cholerae* to glycoprotein and glycolipid. J Bacteriol. 1995;177:758–64.
30. Chattopadhyay K, Banerjee KK. Unfolding of *Vibrio cholerae* hemolysin induces oligomerization of the toxin monomer. J Biol Chem. 2003;278:38470–75.
31. Nagamune K, Yamamoto K, Honda T. Intramolecular chaperone-activity of the Pro-region of *Vibrio cholerae* El Tor cytolysin. J Biol Chem. 1997;272:1338–43.
32. Zhang D, Takahashi J, Seno T, Tani Y, Honda T. Analysis of receptor for *Vibrio cholerae* El Tor hemolysin with a monoclonal antibody that recognizes glycophorin B of human erythrocyte membrane. Infect Immun. 1999;67:5332–7.
33. Ikigai H, Akatsuka A, Tsujiyama H, Nakae T, Shimamura T. Mechanism of membrane damage by El Tor hemolysin of *Vibrio cholerae* O1. Infect Immun. 1996;64:2968–73.
34. Zitzer A, Zitzer O, Bhakdi S, Palmer M. Oligomerization of *Vibrio cholerae* cytolysin yields a pentameric pore and has a dual specificity for cholesterol and sphingolipids in the target membrane. J Biol Chem. 1999;274:1375–80.
35. Zitzer A, Bittman R, Verbicky CA, Erukulla RK, Bhakdi S, Weis S, Valeva A, Palmer M. Coupling of cholesterol and cone-shaped lipids in bilayers augments membrane permeabilization by the cholesterol-specific toxins streptolysin O and *Vibrio cholerae* cytolysin. J Biol Chem. 2001;276:14628–33.
36. Harris JR, Bhakdi S, Meissner U, Scheffler D, Bittman R, Li G, Zitzer A, Palmer M. Interaction of the *Vibrio cholerae* cytolysin (VCC) with cholesterol, some cholesterol esters, and cholesterol derivatives: a TEM study. J Struct Biol. 2002;139:122–35.
37. Ikigai H, Otsuru H, Yamamoto K, Shimamura T. Structural requirements of cholesterol for binding to *Vibrio cholerae* hemolysin. Microbiol Immunol. 2006;50:751–7.
38. Waheed AA, Shimada Y, Heijnen HF, Nakamura M, Inomata M, Hayashi M, Iwashita S, Slot JW, Ohno-Iwashita Y. Selective binding of perfringolysin O derivative to cholesterol-rich membrane microdomains (rafts). Proc Natl Acad Sci USA. 2001;98:4926–31.
39. Rapoport TA, Goder V, Heinrich SU, Matlack KE. Membrane protein integration and the role of the translocation channel. Trends Cell Biol. 2004;14:568–75.
40. Schnell DJ, Hebert DN. Protein translocons: multifunctional mediators of protein translocation across membranes. Cell. 2003;112:491–505.
41. White SH, Ladokhin AS, Jayasinghe S, Hristova K. How membranes shape protein structure. J Biol Chem. 2001;276:32395–8.
42. Valeva A, Walev I, Pinkernell M, Walker B, Bayley H, Bayley H, Palmer M, Bhakdi S. Transmembrane β-barrel of staphylococcal α-toxin forms in sensitive but not in resistant cells. Proc Natl Acad Sci USA. 1997;94:11607–11.
43. Vijayvargia R, Suresh CG, Krishnasastry MV. Functional form of Caveolin-1 is necessary for the assembly of α-hemolysin. Biochem Biophys Res Commun. 2004;324:1130–6.
44. Ichinose Y, Yamamoto K, Nakasone N, Tanabe MJ, Takeda T, Miwatani T, Iwanaga M. Enterotoxicity of El Tor-like hemolysin of non-O1 *Vibrio cholerae*. Infect Immun. 1987;55:1090–3.

45. Zitzer A, Wassenaar TM, Walev I, Bhakdi S. Potent membrane-permeabilizing and cytocidal action of *Vibrio cholerae* cytolysin on human intestinal cells. Infect Immun. 1997;65: 1293–8.
46. Olivier V, Haines GK3rd, Tan Y, Fullner SKJ. Hemolysin and the MARTXVc toxin are virulence factors during intestinal infection of mice with *Vibrio cholerae* El Tor O1 strains. Infect Immun. 2007;75:5035–42.
47. Levine MM, Kaper JB, Herrington D, Losonsky G, Morris JG, Clements ML, Black RE, Tall B, Hall R. Volunteer studies of deletion mutants of *Vibrio cholerae* O1 prepared by recombinant techniques. Infect Immun. 1988;56:161–7.
48. Coelho A, Andrade JR, Vicente AC, Dirita VJ. Cytotoxic cell vacuolating activity from *Vibrio cholerae* hemolysin. Infect Immun. 2000;68:1700–5.
49. Mitra R, Figueroa P, Mukhopadhyay AK, Shimada T, Takeda Y, Berg DE, Nair GB. Cell vacuolation, a manifestation of the El Tor hemolysin of *Vibrio cholerae*. Infect Immun. 2000;68:1928–33.
50. Moschioni M, Tombola F, Bernard MD, Coelho A, Zitzer A, Zoratti, M, Montecucco, C. The *Vibrio cholerae* hemolysin anion channel is required for cell vacuolation and death. Cell. Microbiol. 2002;4:397–409.
51. Cover TL, Blaser MJ. Purification and characterization of the vacuolating toxin from *Helicobacter pylori*. J Biol Chem. 1992;267:10570–5.
52. Carrero JA, Calderon B, Unanue ER. Listeriolysin from *Listeria monocytogenes* is a lymphocyte apoptogenic molecule. J Immunol. 2004;172:4866–74.
53. Bantel H, Sinha B, Domschke W, Peters G, Schultze-Osthoff K, Janicke, RU. α-Toxin is a mediator of *Staphylococcus aureus*–induced cell death and activates caspases via the intrinsic death pathway independently of death receptor signaling. J Cell Biol. 2001;155:637–47.
54. Mukherjee G, Biswas A, Banerjee KK, Biswas T. *Vibrio cholerae* hemolysin is apoptogenic to peritoneal B-1a cells but its oligomer shepherd the cells for IgA response. Mol Immunol. 2008;45:266–70.
55. Ray A, Chattopadhyay K, Banerjee KK, Biswas T. Macrophage distinguishes *Vibrio cholerae* hemolysin from its protease insensitive oligomer by time dependent and selective expression of CD80–CD86. Immunol Lett. 2003;89:143–7.

Chapter 17
Integron-Mediated Antimicrobial Resistance in *Vibrio cholerae*

Amit Ghosh and T. Ramamurthy

Abstract The disease cholera is the result of infection with the toxigenic strains of *Vibrio cholerae*. Even though the oral rehydration therapy is the main stay for the treatment of cholera patients, administration of antimicrobials is a common existing practice as it shortens the volume and duration of diarrhea, duration of hospitalization, and excretion of the causative pathogen. Due to excessive use of antibiotics, many groups of infectious bacteria have become resistant rendering these drugs ineffective. There are several mechanisms by which the bacteria become resistance to the antimicrobials and possession or acquirement of integrons is one of the prime recognized genetic factors for the resistance. Integrons are naturally occurring gene capture and expression systems by which bacteria can acquire genes for the resistance. Class 1 integrons are very common in *V. cholerae* and other vibrios. Several resistance gene cassettes have been indentified in vibrios and the distribution of these genes seems to be specific for certain regions where the antimicrobial drugs are extensively used. This chapter describes about the types of integrons in *V. cholerae*, mechanisms by which the organism uptakes the resistance genes, and their general distribution in different geographical locations.

17.1 Introduction

Vibrio cholerae is a major pathogen among diarrhea-causing organisms. Apart from oral rehydration therapy, oral administration of antimicrobial drugs for treatment is a common practice in the developing countries. Antimicrobial therapy shortens the volume and duration of diarrhea, thereby reducing the fluid requirements, duration of hospitalization, and excretion of the causative pathogen. In addition, several antibiotics are being used as a prophylactic measure during cholera outbreaks and by the travelers visiting cholera-endemic regions.

A. Ghosh (✉)
National Institute of Cholera and Enteric Diseases, Kolkata 700010, India
e-mail: amitghosh24@yahoo.com

T. Ramamurthy, S.K. Bhattacharya (eds.), *Epidemiological and Molecular Aspects on Cholera*, Infectious Disease, DOI 10.1007/978-1-60327-265-0_17,

© Springer Science+Business Media, LLC 2011

Combination therapy of two or more antimicrobials has helped to extend the effective treatment, but occurrences of cross-resistance and co-resistance have increased considerably. It is generally accepted that the main cause of antimicrobial resistance is the inappropriate use of these compounds in clinical medicine, aquaculture, poultry, animal husbandry, and veterinary practice [1]. Emergence of antimicrobial resistance in previously susceptible bacterial population is a very serious threat and now it is a worldwide problem. The determination of antimicrobial susceptibility is often crucial for the optimal therapy for cholera. In some cases, the presence of resistance gene is highly predictive for clinical outcome of antimicrobial therapy. However, an antimicrobial resistance gene, which is not expressed in vitro, may express in vivo.

For a long time, *V. cholerae* unlike many other pathogenic bacteria did not become drug resistant. As pointed out by Sack et al. [2], widespread resistance was not known prior to 1977, though there were occasional reports [3, 4]. In a worldwide survey conducted in 1976, 97% of all strains examined were found to be sensitive to all commonly used antibiotics [5]. However, the scenario changed rapidly mainly due to the indiscriminate use of the antibiotics and in a survey conducted in Bangladesh in 1980, around 18% of all isolates tested turned out to be drug resistant [6]. This trend is continuing unabated and more and more strains are being found to be resistant to a variety of drugs. As in other microorganisms, resistance of *V. cholerae* to antimicrobials could be due to a variety of reasons: lack of target sites, provision of bypass pathways, changes in the target sites, efflux, etc. [7]. Lack of target sites for a drug or the presence of chromosomal species-specific genes that mediate resistance to the drug in a bacterial species makes it intrinsically resistant to that drug. This resistance is species and genus specific and defines the range of activity of an antibiotic. While the lack of target sites or the presence of species-specific genes makes a microbe intrinsically resistant to specific antibiotics, other mechanisms like the ones listed earlier can be acquired by the bug through mutations in housekeeping genes or through the acquisition of mobile foreign genes from elsewhere. Transposons, after their discovery, were thought to be the main vehicle for gene transfer between DNA molecules. Subsequently it was discovered that a resistance determinant carried by a transposon was often located within a genetic element which could capture individual resistance genes. This element was given the name integron [8] and is now known to be a key member in the dissemination of drug resistance genes because their association with mobile genetic elements facilitates their movement across phytogenetic boundaries [9]. In this chapter we will provide a brief description of integrons and review the existing information on their role in the dissemination of antimicrobial resistance in *V. cholerae*. For a broader perspective, the reader is referred to an excellent review by Rowe-Magnus et al. [10].

17.2 General Characteristics and Classes of Integrons

Integrons are naturally occurring gene capture and expression systems by which bacteria can acquire "resistance" genes [11]. It has a specific integration and excision system for the capture, integration, and expression of gene cassettes that encode antimicrobial resistance. Integrons consist of a gene (*intI*) encoding a site-specific recombinase (integrase) belonging to the tyrosine-recombinase family [12], a recombination site, *attI*, where the captured gene or gene cassettes are located, and a promoter Pc located within *intI*, which drives the transcription of the captured gene. Capture of gene cassettes, which are discrete units of circularized DNA harboring (usually) resistance genes [13], is mediated by the *intI*-coded integrase through site-specific RecA independent recombination between the *attl* site of the integrons and the *attC* (originally called 59 be) of the gene cassettes. The *attC* sites are a diverse family of sequences that act as the recognition sites for the site-specific integrase. The length of *attC* varies from 57 to 141 bp and their sequence similarities are restricted to their boundary regions defined by inverse (RYYYAAC) and forward (G$^{\downarrow}$TTRRRY) core sequences [14, 15] (R, purine; Y, pyrimidine; ↓, point of recombination). Once a gene cassette is captured, the promoter Pc promotes its transcription, leading eventually to the synthesis of the encoded protein. Integrons are quite unique among integration/excision systems, because (i) they permit tandem insertion of multiple gene cassettes into a single *attl* site thereby facilitating multidrug resistance and (ii) the *intI* gene is located adjacent to the *attl* receptor site, and not in a gene cassette.

All integrons discovered so far can be broadly classified into two categories: the "mobile integrons," which are associated with mobile DNA elements and which play a major role in the dissemination of resistance genes, and the "superintegrons," which are sedentary and have been found on the chromosomes of g-proto bacteria. Parenthetically, it may be mentioned here that some reservations have been expressed on this terminology (please see [16, 17]).

17.3 Mobile Integrons (MIs)

Integrons are classified based on the sequence of the *intI* protein, which displays 40–60% identity. Based on this criterion, till date, five classes of integrons have been defined, classes 1, 2, 3, SXT-ICE, and an unnamed class found on a *Vibrio salmonicida* plasmid (GenBank accession no. A1277063) [18]. All these are linked with mobile DNA elements, which are known to be the vehicles for intra- and inter-species gene transfer [18]. Each class of integrons carries a specific *intI* gene and an *attI* site. Class 1 integrons represent the most widely found class among the clinical isolates of pathogenic bacteria. These integrons are characterized by the presence of two conserved segments (5′- and 3′-CS). The 5′-CS contains *intI, attI*, and the promoter Pc. The 3′-CS harbors the genes *sul1*, conferring resistance to sulfonamides, *qac* EΔI, conferring resistance to quaternary ammonium compounds, and an ORF5, whose function is unknown [13] (Fig. 17.1). Class 1 integrons are associated with transposons derived from Tn 402, which in turn can be a part of a larger transposon like Tn 21 [18, 19].

Class 1 integrons are most widespread among clinical isolates and have been found in 40–70% of the gram-negative pathogens [20, 21]. They have been detected

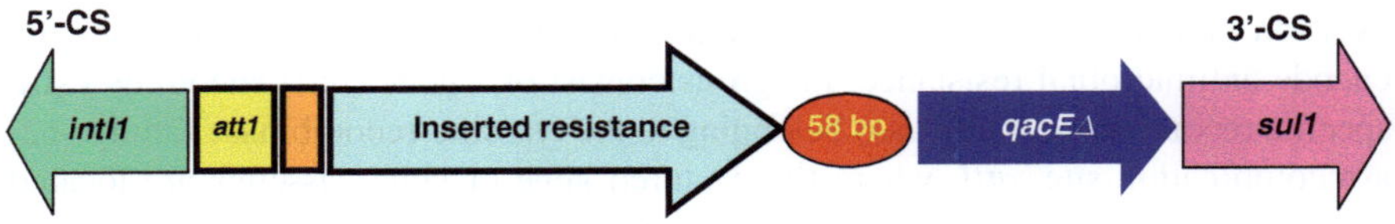

Fig. 17.1 Schematic representation of a class 1 integron

in gram-positive species as well [22–24]. Around 100 gene cassettes, which code for antibiotic resistance, have been detected within the integrons [13, 17]. Class 1 integrons have also been reported to be present widely in different serogroups of *V. cholerae*. Though distributed most extensively among the clinical isolates of gram-negative bacteria, class 1 integrons have also been found in environmental bacteria isolated from sediment samples [22]. In a recent report, Gillings et al. [25] have shown that class 1 integrons can be found on the chromosomes of non-pathogenic soil and freshwater *Betaproteobacteria* and based on the analysis of available data they have speculated that environmental *Betaproteobacteria* could be the original source of class 1 integrons.

Class 2 integrons are found on transposon Tn7 and related elements [26, 27]. The integron on Tn7 usually carries three integrated gene cassettes *dhfr*1–*sat*–*aad*A1 adjacent to a defective integrase gene, though other genes have also been found [28, 29]. Class 2 integrons lack 3′-CS but have genes which promote Tn7 transposition [13]. In contrast to class 1 integrons, only six different resistance cassettes have been found to be associated with class 2 integrons. These integrons have been detected in a number of organisms [22, 30–34]; however, except for one report, class 2 integrons have not been detected in *V. cholerae* [35].

Class 3 integrons are extremely rare and only four examples are known [36–38]. Two were detected in pathogenic microbes *Serratia marcescens* and *Klebsiella pneumoniae,* which harbored resistance gene cassettes, while the other two have recently been discovered in two environmental bacterial strains *Delftia acidovorans* and *Delftia tsuruhatensis.* Class 3 integrons in these strains carry three to four gene cassettes of unknown functions [38]. Class 3 integrons have essentially the same structural organization as classes 1 and 2, and are thought to be associated with a transposon. Class 3 integron integrase protein shows 61% homology with the integrases of class 1 integrons [36, 39]. As of now, class 3 integrons have not been discovered in *V. cholerae*,

The fourth class of integron is a part of the SXT^{ET} element of *V. cholerae* [40]. In 1996, Waldor et al. discovered a "new" integrative conjugative element (ICE) in *V. cholerae* O139 strain M010, originally isolated in India in 1992, which they named SXT. This element was found to code for resistance to sulfamethoxazole, trimethoprim, and streptomycin [41]. Later, a variant form of SXT was discovered in *V. cholerae* O1 El Tor strains isolated in Bangladesh and India in 1994, which was similar but not identical to the SXT originally discovered in *V. cholerae* MO10 [40]. To distinguish between the two, the "original" SXT was re-named SXT^{M010} and the one discovered later in *V. cholerae* O1 El Tor was called SXT^{ET} [40] (SXT was first considered a conjugative transposon [41], then a CONSTIN [42], and now an ICE [43]). Subsequent research showed that "conserved" backbone of SXT contains three "hotspots" for insertion of additional DNA sequences [44] and that SXT^{ET} contains in one of the hotspots a novel class of integron (class 4) that harbors the *dfr*A1 gene cassette. [40].

Besides these four, recently a fifth class of integron has been discovered on the plasmid pRVSI of a *Vibrio salmonicida* strain (GenBank accession no. AJ277063) [9]. To date it remains the only example known of this class of integrons.

17.4 Superintegrons (SIs)

This class of integrons was first discovered in *V. cholerae* as the result of investigations on the relationship between MI cassette arrays and a chromosomal repetitive sequence element (termed VCR for *V. cholerae* repeat) found on the chromosome 2 of *V. cholerae* [45–47]. It was found that the MIs and SIs shared an identical structural organization and the specific integrase VchInt1A carried by the *V. cholerae* was related to the integrases of MIs. Subsequently SIs have been detected in many bacteria and are now known to be an integral component of many g-proteobacterial genomes (see [10, 17] for an overview). In spite of their overall structural similarities, however, there are two characteristics which differentiate SIs from MIs: (i) in contrast to the *attC* sites in the MIs, which are variable in length and sequence, *attC* sites in an SI are highly homologous and (ii) SIs harbor a large number of cassettes (Fig. 17.2). Analysis of the gene sequence of *V. cholerae* El Tor N16961 revealed the presence of 216 ORFs in an array of 179 gene cassettes, most of which are of unknown functions and do not have any counterparts in the database. There are a few, however, which could be involved in virulence and metabolic function [48, 49]. Though no drug resistance genes could be detected; nevertheless a few cassettes displaying significant homology to a number of drug resistance genes

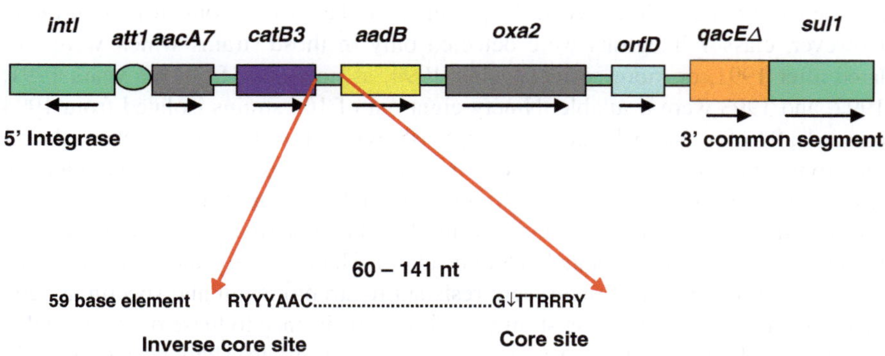

Fig. 17.2 Schematic representation of superintegron. The length of the *attC* (59-be) varies considerably and their sequence similarities are primarily restricted to their boundaries, which correspond to the inverse core site (RYYYAAC) and the core site (G↓TTRRRY). R is a purine, Y is a pyrimidine, and the insert arrow is a recombination point

could be identified [49], suggesting that under appropriate conditions, these could acquire the ability to express resistance phenotypes.

Antiquity of SI (this structure has been detected in vibrios from the nineteenth century, pre-dating antibiotic era [47]) and the structural identity of SIs and MIs led Rowe-Magnus et al. [50] to hypothesize that SIs could be the ancestors of MIs. Indeed, it has been demonstrated that an MI can directly recruit a cassette from an SI and confer a resistance phenotype [9]. By applying appropriate selection pressure, a novel chloramphenicol acetyltransferase gene cassette *Cat*B9 was discovered in a *V. cholerae* SI by Rowe-Magnus et al. [9]. Three other resistance cassettes CARB-7 and CARB-9, coding for novel carbenicillinase, and *qnr*, responsible for resistance to quinolones, have been identified in different strains of non-O1, non-O139 *V. cholerae*. The first two were discovered in two environmental strains of non-O1, non-O139 *V. cholerae* isolated in Argentina [51, 52] and the third one in two *V. cholerae* O1 clinical strains, one isolated in Brazil and the other one in Vietnam [53]. In the light of these findings it appears likely that more such cassettes will be discovered in different *V. cholerae* strains in days to come.

17.5 Epidemiology of *Vibrio cholerae* with Class 1 Integrons

Class 1 integrons have been detected in various strains of *vibrios* isolated in many parts of the world. Investigations have mostly been focused on their detection and the characterization of the gene cassettes present on them.

17.5.1 South-east Asia

The first report on the presence of class 1 integrons in *V. cholerae* came from Dalsgaard et al. [54]. An analysis of 134 *V. cholerae* O1 strains isolated in Vietnam between 1979 and 1996 revealed the presence of class 1 integrons in 73% of them. However, class 1 integrons were detected only in those strains which were isolated after 1991, or more correctly after 1994, as no strains from the years 1991, 1992, and 1993 were available. Ninety-eight out of 104 strains isolated from 1994 onwards, 94 of which belonged to a single ribotype RI, were found to carry class 1 integrons. Only one type of cassette, *ant* (3″) *1a* (same as *aad*A1 [55]) conferring resistance to streptomycin and spectinomycin, could be detected. It is interesting to note that though the majority of these strains had the same ribotype, RI, as the class 1 integron-bearing strains isolated in Samutsakorn, Thailand [56], and Calcutta, India, in 1992 and though all of them were resistant to streptomycin and spectinomycin, gene cassettes present in these strains conferring resistance to these drugs were different: *ant* (3″)*1a* in Vietnam and Calcutta and *aad*A2 in Thailand strains [57]. A follow-up study on strains isolated in 1995, 2000, and 2002 [58] showed that all strains isolated in 2000 and 2002 lacked class 1 integrons, but strains isolated in 1995 carried class 1 integrons, carrying *aad*A1 gene cassette. This observation was in conformity with that observed by Dalsgaard et al. [54]. It appears that sometime after 1995, class 1 integron-bearing strains were replaced by those lacking it, as the primary agent for causing cholera in Vietnam.

In a similar study a total of 176 clinical and environmental isolates from Thailand belonging to different serogroups, besides O1 and O139, were examined [57]. Out of these, 69 strains were from the Children Hospital, Bangkok, collected over a 2-year period from 1993 to 1995. These strains were all devoid of cholera toxin genes and belonged to 37 different serotypes. Nine of these strains contained class 1 integrons with the gene cassettes *aad*A2 (conferring resistance to streptomycin and spectinomycin), *dfr*A15 (resistant determinant for trimethoprim), and *bla*P1 (conferring resistance to ampicillin); five of these had *bla*P1 and *dfr*A15, one had *aad*A2, and three had *aad*A2 and *bla*P1. Class 1 integrons could be detected in a larger proportion (11/19) strains of non-O1, non-O139 *V. cholerae*, which were recovered from the stool samples of patients with diarrhea and from an epidemic of a cholera-like illness among Khmer refugees in 1990. Ten carried the gene cassettes *aad*A2 and *dfr*A1.*orf*C located on two separate integrons. The remaining one had only the gene cassette *dfr*A1.*orf*C borne on an integron. In contrast to these findings with the clinical isolates of non-O1, non-O139 *V. cholerae*, it was found that none of the 34 environmental strains of non-O1, non-O139 *V. cholerae*, isolated from water, sediment, and shrimp in a shrimp production area, harbored class 1 integrons. A class 1 integron could, however, be detected in 1 non-O1, non-O139 strain out of the 12 strains isolated from seafood samples. This integron-bearing strain had two class 1 integrons both of which carried the same two gene cassettes *aad*A2 and *bla*P1, which were also detected in some of the non-O1, non-O139 strains obtained from the Children Hospital, Bangkok.

When 31 *V. cholerae* O1 clinical isolates recovered from patients in Samutsakorn between 1982 and 1996 were screened, 20 strains isolated during, after, and within a 10-month period before 1993–1994 O139 outbreak in Thailand showed the presence of class 1 integron with *aad*A2 gene cassette in them. All these strains belonged to the same ribotype as the integron-bearing *V. cholerae* O1 strains isolated in Calcutta in 1992 and in Vietnam after 1990, suggesting that they could have originated from the same clone but acquired different integron-borne gene cassettes, as they came to be established as the main cholera-causing strain in these countries. No class 1 integron could be detected in the O139 strains isolated during the 1993–1994 O139 outbreak period. One interesting outcome of the Thailand study was the discovery of plasmid-borne class 1 integrons in *V. cholerae.* Three strains isolated in Samutsakorn in 1982 were found to harbor class 1 integrons with *aad*B gene cassette (encoding resistance to gentamicin, kanamycin, and tobramycin) on a 150-kb conjugative plasmid. No chromosomally located integron was found in these strains. The fact that this plasmid was not found later in strains isolated in Samutsakorn or elsewhere [56] suggested that it was subsequently lost.

Studies carried out with 50 *V. cholerae* O1 strains in Laos (a country neighboring Vietnam and Thailand) isolated between 1993 and 2000, a period which partially overlapped the periods covering the studies carried out both in Vietnam and in Thailand, showed that 48% of the strains, all of which were isolated before 1997, carried class 1 integrons harboring *aad*A1 gene cassette on their chromosomes [59].

Thus, gene compositions of O1 Laotian strains were essentially similar to those of the Vietnamese strains. None of the strains isolated after 1997 were found to contain class 1 integrons.

17.5.2 South Asia: India

In 1996 there was an unusual upsurge in the number of cholera cases (more correctly, cholera-like illness indistinguishable from clinical cholera [60]) due to *V. cholerae* strains belonging to serotypes other than O1 and O139. All these strains were found to be resistant to multiple antibiotics and in some cases to as many as 11 antibiotics. To examine the possible contribution of class 1 integron-borne gene cassettes in multidrug resistance, 94 non-O1, non-O139 strains (out of which 17 were environmental isolates) isolated in Calcutta, India, during 1997–1998 were screened for the presence of class 1 integrons [61]. Twenty-two out of the 94 strains, all of which were of clinical origin, were found to harbor class 1 integrons. Seven different types of gene cassettes, *dfr*A1, *dfr*A5, *dfr*A12, *dfr*A15 coding for trimethoprim resistance, *aac*(6′)-1b for amikacin and tobramycin resistance, *aad*A1, *aad*A2 for resistance to streptomycin and spectinomycin, and *ere*A2 for resistance to erythromycin, could be identified. Gene cassettes *dfr*A5, *dfr*A12, *aac*(6′)-Ib, and *ere*A2 were found in *V. cholerae* for the first time; presence of *ere*A2 at that time was known only in one other bacteria, *Providencia stuartii*. Cassettes *dfr15* and *dfr*A12 were subsequently detected among *Vibrio* strains in Africa, but *ereA2* and *dfr*A5, occurring in tandem within a single class 1 integron in a lone isolate in Calcutta, were not detected anywhere else so far. In 14 strains out of the 22 integron-bearing ones, class 1 integrons were found to be chromosomally located, whereas in another 5, these were detectable only on the plasmids ranging in size from 50 to 370 kb. Most interestingly, in the remaining three integron-bearing strains, class 1 integron was found both on the chromosome and on the plasmids. Though plasmid-borne integrons harboring a single or double gene cassette had been detected earlier in *V. cholerae* O1 strains isolated in Thailand and Guinea-Bissau [57, 62], their presence on plasmids of different molecular sizes and that too with five different types of cassettes, *dfr*A1, *dfrA12*, *dfrA15 aadA2*, and *aac*(6′)-Ib, in 8% of the total number of strains examined seems to suggest that large plasmids carrying integrons may not be very rare among the non-O1, non-O139 strains of *V. cholerae* in circulation in India.

When a similar search for class 1 integrons was extended to *V. cholerae* O1 strains isolated in Calcutta, India, during three different periods, March–December 1992, July–November 1993, and March 1994–June 1995, i.e., before, during, and after the first cholera outbreak caused by *V. cholerae* O139, majority of the strains isolated before the advent of O139 exhibited the presence of class 1 integrons [63]. In contrast, very few of the strains isolated later harbored integrons. All integron-bearing strains, irrespective of their period of isolation, harbored only the *aad*A1 gene cassette, determinant for streptomycin and spectinomycin resistance, even though the strains displayed resistance to multiple antibiotics besides streptomycin.

Moreover, none of the strains harbored any plasmid. This finding was quite in contrast to the results obtained with *V. cholerae* non-O1, non-O139 strains described above. It was observed that strains isolated in Calcutta during March–December 1992 had the same ribotype as those which appeared in Samutsakorn, Thailand, in 1993, after the *V. cholerae* O139 outbreak there. Unlike Samutsakorn strains, however, which harbored the *aad*A2 gene cassette, Calcutta strains carried *aad*A1. It was shown earlier [64] that post-O139, O1 strains isolated in Calcutta could have migrated to Guinea -Bissau in Africa and it was speculated that this strain could subsequently have acquired from an unknown source a 150-kb plasmid carrying a class 1 integron, bearing the [*ant* (3′)*1a*] gene cassette, [62]. In the light of this, it is tempting to speculate that both the Calcutta strain alluded to above and the Samutsakorn strain could have originated from the same clone. Work carried out by Amita et al. [65] showed that only about 10% of the post *V. cholerae* O139, O1 strains screened harbored class 1 integrons, a situation clearly different from what was found for the post O139, non-O1, non-O139 vibrios, where 30% of the strains carried class 1 integrons [61]. When this study was extended to *V. cholerae* O1 strains isolated between 1999 and 2004, both in Calcutta and in Chandigarh, a city 300 km north of Delhi, the same trend was observed; None of the 40 strains obtained in Calcutta and the 106 strains isolated in Chandigarh was found to carry class 1 integrons. In contrast, 9 out of 84 *V. cholerae* non-O1, non-O139 strains, obtained from the same locale during the same period, showed the presence of class 1 integrons [63]. Three strains harbored *dfrA1* in association with an unknown *orf, dfr*A1–*orf*, a combination which was also detected in some strains of non-O1, non-O139 *V. cholerae* isolated in Calcutta during 1996–1997 [63]. Remaining strains had class 1 integrons with single cassettes; three had *dfr15,* while the other two had *aadA1* and *bla*Pl. In every case, integrons appeared to be chromosomally located.

A comprehensive study covering a longer, almost the entire period (1992–2000) covered by the other studies mentioned above and also involving *V. cholerae* strains besides the O1 and O139 strains isolated from the same geographical location, Calcutta, was undertaken by Shi et al. [66]. Three to six strains each of O1, O139, and non-O1, non-O139 serogroups were randomly selected from each year's collection. A total of 44 O1, 45 O139, and 44 non-O1, non-O139 strains were examined. Seven O1, six non-O1, non-O139, and one O139 strains were found to carry class 1 integrons. All strains excepting one harbored a single copy of class 1 integron, and all of them carried a variant form of *aad*A1 gene (termed *aadA1C*), which differed from the other homologues, *aadA1*a and *aadA2b* (GenBank accession numbers X12870 and M95287, respectively), by 6–10 nucleotides. The remaining O1 strain appeared to have two class 1 integrons, each carrying four gene cassettes in tandem, *aar3* (resistance to rifampicin), *aacA4* (resistance to kanamycin and gentamicin), *dfrA1* (resistance to trimethoprim), and an open reading *orf*C. Incidentally, this is the largest number of gene cassette within a single integron in *V. cholerae* found so far. Class 1 integrons were also found in 6 out of the 44 non-O1, non-O139 strains, 3 of which were found to carry *dfrA1–orf*C. The other three carried two class 1 integrons, each with a different cassette, while one strain had *dfrA15* and *aadA2*, another had *dfrA15* and *blaP1*, and the third one had *aadA2* and *blaP1*. These authors also reported the presence of a class 1 integron in a *V. cholerae* O139 strain for the first time. This particular strain isolated in 1992 contained the cassette *dfr*A1–*orf*C, which, as stated above, was also detected in a few non-O1, non-O139 strains isolated in the same place.

17.5.3 Africa

Epidemiology of class 1 integron-bearing *V. cholerae* strains in the African continent has been investigated fairly extensively. The first such study was carried out by Dalsgaard et al. [2] with strains isolated in 1996–1997 during an outbreak in Guinea-Bissau, a country located in the west coast of Africa. Analysis of 10 strains of *V. cholerae* O1 revealed the presence of class 1 integrons in 7 of them. In each strain, the integron bearing the gene cassettes *ant*(3″)*Ia* and *dfrA12* was located on a 150-kb plasmid. It may be recalled that a 150-kb plasmid harboring class 1 integrons was also detected in three *V. cholerae* O1 strains isolated in the Thailand way back in 1982 [57] However, the gene cassette harbored there was *aad*A1. Also, it is not known if these two plasmids were the same or different.

In contrast to what was found in Guinea-Bissau, plasmid-borne integrons were not detected in the strains isolated from the Mozambiquian migrant workers in South Africa, during an outbreak in 1998 [67]. An analysis of 20 such strains revealed that though two of the strains did carry plasmids, they did not harbor any class 1 integron. Seventeen out of the 20 strains had chromosomally located class 1 integrons, all bearing the *aadA2* gene cassette. From this work it appeared that *aadA2* gene cassette was circulating in Mozambique, as most isolates were obtained from the migrant workers coming to South Africa from Mozambique. This inference, however, was not corroborated by the findings of Ciccarelli et al. [68], who examined 12 *V. cholerae* O1 strains isolated in Mozambique between August 1997 and June 1998, for the presence of class 1 integrons. Their work showed that while all the 12 strains examined contained *aadA1* genes, 3 also had *dfrA15* cassettes. Class 1 integron in all these strains was located on the chromosome. These cassettes were also found to be present in six other strains isolated at different times in four other African countries, namely Somalia, Rwanda, Swaziland, and Zimbabwe [68], while two isolates from Somalia were of 1985 vintage, 2 from Rwanda, 1 from Swaziland, and 2 from Zambia were of more recent origins being isolated in 1994, 1998, and 1998, respectively.

The gene cassettes *aad*A1 and *dfr*A15 were chromosomally located in the strains isolated in Rwanda, Swaziland, and Zimbabwe. In contrast, the strain isolated in Somalia harbored the class 1 integron bearing these cassettes on a plasmid. Since no investigation on their clonal relationships was carried out, it is not known whether these strains had a common origin. Class 1 integrons were also detected in 4 out of 10 environmental strains of *V. cholerae* isolated in the Maputo area of Mozambique between April 2002 and June 2003 [69]. However, unlike in the *V. cholerae* O1 clinical strains isolated in Mozambique in 1997–1998 or in South Africa in 1998 from the Mozambique migrant workers, the predominant gene cassette in these strains was found to be *blaP1*. This cassette was not detected in any *V. cholerae* O1 strain in the earlier studies.

Three other countries where investigations on the occurrence of class 1 integrons have been carried out are Algeria, Angola, and Zambia. In the Algerian study, eight *V. cholerae* O1 strains, out of which two were isolated way back in 1980, and ten non-O1 strains recovered from different cities between 1980 and 2003 were examined [70]. All *V. cholerae* O1 strains excepting two, one from 1980 and the other from 1997, were found to harbor class 1 integrons. While a single integron carrying *aadA1* was detected in a strain from 1994, two integrons harboring the drug cassettes *aacC, aadA1, dfr2, aadA6, dfr15,* and *aad*A1 could be found in the isolates from 1980, 1981, and 1986. As regards, the non-O1 strains, class 1 integrons could be detected only in four strains. Out of these, one strain isolated in 2001 possessed two class 1 integrons with one carrying the *aad*A7 gene and the other *dfr*1.*orf*C. Remaining three non-O1 isolates each had a single class 1 integron, with either one or two drug cassettes; while two of them had either *aadA1* or *dfr7,* the third one carried *aadA5.dfr17*. Since no attempt was made to establish the location of class 1 integrons in any of the strains, it is not known whether they were plasmid or chromosome borne.

Angolan studies involved examination of 13 *V. cholerae* strains isolated from different provinces [71]. Out of these, 11 were *V. cholerae* O1 of clinical origin. Of the remaining two, one was a *V. cholerae* O1 strain and the other a *V. cholerae* non-O1 strain, both isolated from a river. All *V. cholerae* O1 strains were found to carry three class 1 integrons on a 100-kb plasmid named p3iANG by the investigators. All three integrons were located within a 19-kb region of the plasmid with one integron separated by a 6.2-kb spacer region from the remaining two present in tandem. The first integron had the *blaP1* gene cassette. In the other two, occurring in tandem, the one closest to the first integron had *aadA15*, while the other had *aacH.aad8*. Interestingly, the 6.2 kb spacer region itself carried the resistant determinants *cat18, aph, sul2,* and *tetG* capable of conferring resistance to chloramphenicol, kanamycin, sulfonamide, and tetracycline, respectively. The plasmid p3iANG did not seem to bear any relationship with the 150-kb integron-bearing plasmids discovered in Guinea-Bissau and Thailand.

To understand the genetic basis of multidrug resistance of *V. cholerae* O1 strains isolated from eight outbreaks between February 1990 and December 2004 in Zambia, a country where cholera has been breaking out frequently since 1971, Mwansa et al. [72] screened 69 strains of *V. cholerae* O1 strains isolated between 1996 and 2004 for the presence of class 1 integrons. Their work revealed that 26 out of the 31 strains isolated during 1996–1997 carried class 1 integrons, as revealed by the presence of *intI*. But none of the strains isolated later carried an integron. No information about the gene cassettes harbored by these strains is available.

17.5.4 Europe

Even though there have been reports on the sporadic outbreaks of cholera in early 1990s in many countries of Europe [73–76], only the strains from Albania and Italy were examined for the presence of class 1 integrons. In September 1994, an epidemic of cholera broke out in Albania [76]. Soon the disease moved to the Italian coastal town of Bari, situated on the coast of Adriatic Sea opposite Albania. Analysis of 15 strains from Albania and 12 from Italy, all of which were later found to be clonally related [77] and showed resistance to streptomycin, spectinomycin, tetracycline, trimethoprim, sulfathiazole, and 0/129 (2,4-diamino-6,7-diisopropylpterdine), revealed the presence of a chromosomally located class 1 integron harboring *aad*A1 gene. All other resistance markers to the other antibiotics, including the *dfr*A1 gene, which confers resistance to trimethoprim, were found to be located on a 60-kb plasmid but were not a part of an integron. It was interesting to note that *dfr*A1 gene, which nearly always is found to be integron borne, was not a part of an integron in these strains.

17.5.5 South America

In January 1991, cholera re-appeared in Latin America after a gap of 100 years. It emerged first in Peru and then within a short time spread to seven other countries. Between 1991 and 1994, there were over 1 million cases of cholera and 10,000 deaths [78]. The strains, which first appeared in Peru, were initially sensitive to antibiotics tested [79]. Soon after this incidence, drug-resistant strains made their appearance and toward the later part of 1991, a percentage of isolates in Ecuador displayed multidrug resistance. Examination of strains isolated in Brazil between 1991 and 1999 showed that nearly all clinical *V. cholerae* O1 strains examined were resistant to one or more antibiotics. However, in spite of the fairly widespread occurrence of drug resistance in South America in 1990, studies on the detection and distribution of class 1 integrons have been rather limited. Campos et al. [80], examined 134 *V. cholerae* O1 strains of clinical origin and 14 non-O1 strains of environmental origin isolated in Brazil between 1991 and 1999 and found that none harbored a class 1 integron. This result was thus quite in contrast to the results obtained with *V. cholerae* strains isolated in other parts of the world during the same period.

17.6 Complex Class 1 and Class 2 Integrons in *Vibrio cholerae*

Complex class 1 integrons (also known as *orf*513 bearing class 1 integrons), which were first described in 1990 [81, 82], are a variant form of the conventional class 1 integron in which there is partial duplication of the 3′-CS. These thus contain a complete 5′-CS and two copies of a portion of the 3′-CS. The first copy contains 1355 bases of a typical 3′-CS ending at 24 nucleotides after the stop codon of the *sul1* gene [81, 82]. The structure of the region between the 5′-CS and the first 3′-CS is identical to the typical class 1 integrons with one or more gene cassettes located there. Following the first 3′-CS, there is a region designated "common region" (CR), which is present in all such integrons, and a region "unique" to each complex class 1 integrons. The "common region" (now called ISCR1) [83] accommodates a gene *orf*513, which could be coding for an atypical transposase. The "unique" region varies in contents in different complex class 1 integrons and could incorporate different gene cassettes [82–86].

After the re-emergence of cholera in Latin America in 1991, *V. cholerae* O1 became one of the most important pathogens in Argentina. Presence of extended-spectrum beta-lactamases could be detected in about 1.5% of the strains received at the National Reference Laboratory in that country between 1992 and 1998 [87]. On examination, an *orf*513 bearing complex class 1 integron *Inv*117 was detected in one of the strains, *V. cholerae* O1 El Tor M1516 located on a 150-kb conjugation plasmid pASI. Analysis of the plasmid DNA sequence revealed the presence of gene cassettes $bla_{\text{CTX-M2}}$ and *aac*(6′)-1b, coding for aminoglycoside acetyltransferase and an extended-spectrum beta-lactamase belonging to CTX-Mβ lactamase family [88]. Whether such complex class 1 integrons are present in other *V. cholerae* strains isolated elsewhere or the drug-resistant Argentinean strains harbored conventional class 1 integrons is not known.

Strains belonging to different serotypes including O1 and O139 have also been screened for the presence of class 2 integrons in two different laboratories [35, 66]. Class 2 integrons could be detected only in two strains, one belonging to serotype O27 isolated in India in October 2001 and the other belonging to serogroup O39 isolated in July 2003 in Bangladesh. The strain from India harbored several plasmids, while the one from Bangladesh had none. In both the cases, however, class 2 integrons appeared to be chromosomally located and carried the gene cassettes *dfr*A1, *sat*1, and *aad*A1, normally found on class 2 integron accounting partly for the multidrug-resistant phenotype displayed by these strains. None of the O1 or O139 strains examined harbored class 2 integrons.

17.7 Class 1 Integrons in Other Vibrios

Besides *V. cholerae*, only a few other vibrios have been examined for the presence of class 1 integrons. Among these, *Vibrio fluvialis* has been examined in somewhat greater detail [89, 90]; *V. fluvialis* has been responsible over the years for sporadic outbreaks of diarrhea in many parts of the world [91–93]. Nineteen strains of *V. fluvialis* isolated between 1998 and 2002 from patients admitted to the Infectious Diseases Hospital, Calcutta, with acute cholera-like diarrhea were examined. Eight strains were found to carry class 1 integrons. Five of these integron-bearing strains also harbored plasmids but whether the integrons were plasmid or chromosome borne was not investigated. Three of the eight integron-bearing strains did not carry any gene cassettes on their integrons. Two carried *dfr*A1–*orf*C and the remaining three had *aac*(3)-Id-*aad*A7, *aad*A7, and *dfr*A15. Since all integron-bearing strains excepting one, belonged to the same ribotype, it appears that all of them could have originated from a single clone. The cassette *aac*(3)-Id had been discovered earlier by Ahmed et al. [89] in a *V. fluvialis* strain isolated in India in 2002 from a 6-month-old infant suffering from diarrhea. Cloning and sequencing of the gene showed that *aac*(3)-Id represented a new type of aminoglycoside acyltransferase not known earlier. Subsequently this gene was also detected in *Salmonella enterica* serovar Newport [94]. One interesting finding by Srinivasan et al. [90] was the detection of "null" integrons in a sizeable proportion (37%) of the integron-bearing strains, even though the number of strains tested itself was small. This is quite in contrast to what has been found for *V. cholerae* strains examined in India or elsewhere in the world. Further, since two out of three such strains isolated in 1998 and 2000 had the same ribotype as the integron-bearing strains isolated in 2002, it is tempting to speculate that the latter strains could have evolved from the earlier ones by capturing gene cassettes from some unknown sources. Besides *V. fluvialis*, class 1 integrons have also been detected in two other vibrios, *Vibrio parahaemolyticus* and *Vibrio alginolyticus*. Class 1 integrons were detected in two strains of *V. parahaemolyticus* isolated in Angola in 1992 and 1993 [71]. Both the strains harbored the gene cassette *dfrA15* located on a class 1 integron and though both strains carried a 80-kb multiresistance plasmid, no evidence of its association with the integrons could be found. Taviani et al. [69] screened 10 *V. cholerae* O1, 6 *V. parahaemolyticus*, 2 *V. alginolyticus*, and 1 *V. fluvialis* strains isolated from the environmental sources in Mozambique in 2003. Besides *V. cholerae* O1 strains, only the *V. alginolyticus* strain was found to harbor a class 1 integron. However, unlike the O1 strains which contained *bla*P1 and *dfr*A15 gene cassettes, *V. alginolyticus* had *aad*A2.

17.8 Conclusion

Emergence and spread of antibiotic resistance in pathogenic microbes is now endemic world over and is posing a serious challenge to effective therapy [95]. Integrons, which capture and disseminate antibiotic resistance genes, play a vital role in the spread of antibiotic resistance among microbes. Though they were first identified in the late 1980s, it is now clear that they played a role even in the first outbreak of multidrug resistance in the 1950s [9]. Till date, more than 100 different gene cassettes, borne on integrons, have been discovered. But despite their multiplicity, very few resistance integrons as detailed in this article, have been detected in vibrios with *V. cholerae* being the only exception. But even there, only a few types of drug cassettes are seen frequently. In most cases it was found that the gene cassettes borne on class 1 integrons could not fully account for the multidrug-resistant phenotype displayed by the *V. cholerae* strain harboring the integron, suggesting that despite their importance in the diffusion of drug resistance, integrons play only a limited role in the development of drug resistance in vibrios.

Like most other genetic exchange processes in bacteria, capture and expulsion of gene cassettes by an integron is a random process. If the random "capture" of a particular resistance cassette in a particular milieu provides an advantage to a bacterium harboring an integron, then that bacterium achieves selective advantage over the other competing bugs in that milieu and "flourishes." In this context, it appears enigmatic why only the resistance integrons encoding resistance to trimethoprim and aminoglycosides are found more abundantly in *V. cholerae* O1 strains, even though a number of other drugs are also used routinely in the treatment of cholera. Similarly it also remains an enigma why among the *Vibrio* spp., integrons are found more frequently in *V. cholerae* and that too among the non-O1, non-O139 strains.

Acknowledgments Authors gratefully acknowledge the financial assistance to their laboratories from the Department of Biotechnology, Govt. of India, Indian Council of Medical Research and the Council of Scientific and Industrial Research, Govt. of India.

References[1]

1. Fluit AC, Schmitz F-J. Resistance integrons and super-integrons. Clin Microbiol Infect. 2004;10:272–88.
2. Sack DA, Lyke C, McLaughlin C, Suwanvanichkij V. Antimicrobial resistance in shigellosis, cholera and campylobacteriosis, WHO/CDS/CSR/DRS/2001.8. Geneva: World Health Organization; 2001.
3. Kuwahara S, Goto S, Kimura M. Drug-sensitivity of El Tor *Vibrio* strains isolated in the Philippines in 1964 and 1965. Bull World Health Org. 1667;37:763–71.
4. Kobari K, Takakura I. Nakatomi M, Soqame S, Uylanqco C. Antibiotic-resistant strains of El Tor *Vibrio* in the Philippines and the use of furalazine for chemotherapy. Bull World Health Org. 1970;43:365–71.
5. O'Grady F, Lewis MJ, Pearson NJ. Global surveillance of antibiotic sensitivity of *Vibrio cholerae*. Bull World Health Org. 1976;54:181–5.
6. Glass RI, Huq I, Alim AR, Yunus M. Emergence of multiply antibiotic-resistant *Vibrio cholerae* in Bangladesh. J Infect Dis. 1980;142:939–42.
7. Davies J. Microbes have the last word. EMBO. 2007;8:616–21.
8. Stokes HW, Hall RM. A novel family, of potentially mobile DNA elements, encoding site-specific gene-integration functions: integrons. Mol Microbiol. 1989;3:1669–83.
9. Rowe-Magnus DA, Guerout AM, Mazel D. Bacterial resistance evolution by recruitment of super-integron gene cassettes. Mol Microbiol. 2002;43:1657–69.
10. Rowe-Magnus, DA, Zouine M, Mazel, D. The adaptive genetic arsenal of pathogenic *Vibrio* species: the role of integrons. In: Thomson FL et al., editors. The biology of vibrios. Washington, DC: ASM Press; 2006. pp. 95–111.
11. Hall RM, Collis CM. Mobile gene-cassettes and integrons: capture and spread of gene by site-specific recombination. Mol Microbiol. 1995;15:593–600.
12. Nunes-Duby SE, Known HJ., Tirumalai RS, Ellenberger T, Landy A. Similarities and differences among 105 members of the *Int* family of site-specific recombinases. Nucleic Acids Res. 1998;26:391–406.
13. Rechia GD, Hall RM. Gene cassettes: a new class of mobile element. Microbiology. 1995;141:3015–27.
14. Stokes HW, O'Gorman DB, Recchia GD, Parsekhian M, Hall RM. Structure and function of 59-base element recombination sites associated with mobile gene cassettes. Mol Microbiol. 1997;26:731–45.
15. Collis CN, Kim MJ, Stokes HW, Hall RM. Binding of purified integron DNA integrase I*ntl*1 to integron – and cassette-associated recommendation sites. Mol Microbiol. 1998;29:477–90.

[1] Literature Survey Completed July 2008

16. Hall RM, Stokes, HW. Integrons or super integrons? Microbiology. 2004;150:3–4.
17. Boucher Y, Labbale M, Koenig JE, Stokes HN. Integrons: mobilizable platforms that promote genetic diversity in bacteria. Trend Microbiol. 2007;15:301–9.
18. Mazel D. Integrons: agents of bacterial evolution. Nat Rev Microbiol. 2006;4:608–20.
19. Leibert CA, Hall RM, Summers O. Tn21, flagship of the floating genome. Microbiol Mol Biol Rev. 1999;63:507–22.
20. Martinez-Freijo P, Fluit A, Schmitz F, Grek V, Verhoef J, Jones, M. Class 1 integrons in gram-negative isolates from European hospitals and association with decreased susceptibility to multiple antibiotic compounds. J Antimicrob Chemother. 1998;42:689–96.
21. Essen-Zandbergen A, Smith H, Veldman K, Mevius D. Occurrence and characteristics of class 1, 2 and 3 integrons in *Escherichia coli*, *Salmonella* and *Campylobacter* spp. in the Netherlands. J Antimicrob Chemother. 2007;59:746–50.
22. Stokes H., Nesbø C, Holley M, Bahl M, Gillings M, Boucher Y. Class 1 integrons predating the association with Tn402-like transposition genes are present in a sediment microbial community. J Bacteriol. 2006;188:5722–30.
23. Nesvera J, Hochmannova J, Patek M. An integron of class 1 is present on the plasmid CG4 from gram-positive bacterium *C. glutamicum*. FEMS Microbiol Lett. 1998;169: 391–5.
24. Nandi S, Maurer JJ, Hafacre C, Summers AO. Gram-positive bacteria are a major reservoir of class 1 antibiotic resistance integrons in poultry litter. Proc Natl Acad Sci USA. 2004;101:7118–22.
25. Gillings M, Boucher Y, Labbate M, Holmes A, Krishnan S, Holley M, Stokes HW. The evolution of class 1 integrons and the rise of antibiotic resistance. J Bacteriol. 2008;190: 5095–100.
26. Hall RM, Stokes, HW. Integrons: novel DNA elements which capture genes by site-specific recombination. Genetica. 1993;90:115–32.
27. Radstorm P, Skold O, Swedberg G, Flensburg J, Roy, PH, Sudstorm L. Transposon Tn5090 of plasmid R751, which carries an integron, is related to Tn7, Mu and the retroelements. J Bacteriol. 1994;176:3257–68.
28. Biskri L, Mazel D. Erythromycin esterase gene (*ere*A) is located in a functional gene cassette in an unusual class 2 integron. Antimicrob Agents Chemother. 2003;47:3326–31.
29. Ramirez MS, Quiroga C, Centron D. Novel rearrangement of a class 2 integron in two non-epidemiologically related isolates of *A. baumannii.* Antimicrob Agents Chemother. 2005;49:5179–81.
30. Gonzalez G, Sossa K, Bello H, Dominguez M, Mella S, Zemelman R. Presence of integrons in isolates of different biotypes of *A. baumannii* from Chilean hospitals. FEMS Microbiol Lett. 1998;161:125–8.
31. Orman BE, Pinero SA, Arduino S, Galas M, Melano R, Caffer MI, Sordelli DO, Centron D. Evolution of multiresistance in non-typhoid *Salmonella* serovars from 1984 to 1998 in Argentina. Antimicrob Agents Chemother. 2002;46:3963–70.
32. Ramirez MS, Vargas, LJ, Cagnoni V, Tokumota M, Centron D. Class 2 integron with a novel cassette array in a *B. cenocepacia* isolate. Antimicrob Agents Chemother. 2005;49: 4418–20.
33. Gassama Sow A, Diallo MH, Gatet M, Denis F, Aïdara-Kane A, Ploy MC. Description of an unusual class 2 integron in *Shigella sonnei* isolates in Senegal (sub-Saharan Africa). J Antimicrob Chemother. 2008;62:843–4.
34. Barlow RS, Gobius KS. Diverse Class 2 integrons in bacteria from beef cattle sources. J Antimicrob Chemother. 2006;58:1133–8.
35. Ahmed AM, Kawaguchi F, Shimamoto T. Class 2 integrons in *V. cholerae*. J Med Microbiol. 2006;55:643–4.
36. Arakawa Y, Murakami M, Suzuki K, Ito H, Wacharotayankun R, Oshuka S, Kato N, Ohta M. A novel integron like element carrying the metallo-beta-lactamase gene *blaIMP*. Antimicrob Agents Chemother. 1995;39:1612–5.

37. Correia M, Boavida F, Grosso F, Salgado MJ, Lito LM, Cristino JM, Mendo S, Duarte A. Molecular characterization of a new class 3 integron in *Klebsiella pneumoniae*. Antimicrob Agents Chemother. 2003;47:2838–43.
38. Xu H, Davies J, Miao V. Molecular characterization of class 3 integrons from *Delftia* spp. J Bacteriol. 2007;189:6276–83.
39. Collis CM, Kim MJ, Partridge SR, Stokes HW, Hall, RM. Characterization of the class 3 integron and the site-specific recombination system it determines. J Bacteriol. 202;184: 3017–26.
40. Hochhut B, Lotfi Y, Mazel D, Faruque SM, Woodgate R, Waldor MK. Molecular analysis of antibiotic resistance gene clusters in *Vibrio cholerae* O139 and O1 SXT constins. Antimicrob Agents Chemother. 2001;45:2991–3000.
41. Waldor MK, Tschape H, Mekalanos JJ. A new type of conjugative transposon encodes resistance to sulfamethoxazole, trimethoprim and streptomycin in *Vibrio cholerae* O139. J Bacteriol. 1996;178:4157–65.
42. Hochhut B, Waldor MK. Site-specific integration of the conjugal *Vibrio cholerae* element into *prfC*. Mol Microbiol. 1999;32:99–110.
43. Burrus V, Pavlovic B, Decaris B, Guedan G. Conjugative transposons: the tip of the iceberg. Mol Microbiol. 2002;46:601–10.
44. Beaber JW, Burrus V, Hochhut B, Waldor, MK. Comparison of SXT and R391, two conjugative integrating elements: definition of a genetic backbone for the mobilization of resistant determinants cell. Mol Life Sci. 2002;59:2065–70.
45. Barker A, Clark, CA Manning, PA. Identification of VCR, a repeated sequence associated with a locus encoding a hemagglutinin of *Vibrio cholerae* O1. J Bacteriol. 1994;176: 5450–58.
46. Clark CA, Purins L, Kaewrakon P, Manning PA. VCR repetitive sequence elements in the *Vibrio cholerae* chromosome constitute a mega-integron. Mol Microbiol. 1997;26: 1137–8.
47. Mazel D, Dychinco B, Webb VA, Davies J. A distinctive class of integron in the *Vibrio cholerae* genome. Science. 1998;280:605–8.
48. Heidelberg JF, Eisen JA, Nelson WC, Clayton RA, Gwinn ML, Dodson RJ, Haft DH, Hickey EK, Peterson JD, Umayam L, Gill SR, Nelson KE, Read TD, Tettelin H, Richardson D, Ermolaeva MD, Vamathevan J, Bass S, Qin H, Dragoi I, Sellers P, McDonald L, Utterback T, Fleishmann RD, Nierman WC, White O, Salzberg SL, Smith HO, Colwell RR, Mekalanos JJ, Venter JC, Fraser CM. DNA sequence of both chromosomes of the cholera pathogen *Vibrio cholerae*. Nature. 2006;406:477–83.
49. Rowe-Magnus DA, Guerout AM, Mazel D. Super-integrons. Res Microbiol. 1999;150: 641–51.
50. Rowe-Magnus DA, Guerout AM, Ploncard P, Dychinco B, Davies J, Mazel D. The evolutionary history of chromosomal super-integron provides an ancestry for multi-resistant integrons. Proc Natl Acad Sci USA. 2001;98:652–57.
51. Melano R, Petroni A, Garutti A, Saka HA, Mange L, Pasteran F, Rapoport M, Rossi A, Galas M. New Carbenicillin-hydrolyzing beta-lactamase (CARB-7) from *Vibrio cholerae* non-O1, non-O139 strains encoded by the VCR region of the *V. cholerae* genome. Antimicrob Agents Chemother. 2002;46:2162–8.
52. Petroni A, Corso A, Melano R, Cacace ML, Bru M, Rossi A, Galas M. Plasmidic extended-spectrum β-Lactamases in *Vibrio cholerae* O1 El Tor isolates in Argentina Antimicrob Agents Chemother. 2002;46:1462–8.
53. Fonseca EL, Freitas FS, Viera VV, Vicente, ACP. New *qnr* gene cassettes associated with superintegron repeats in *Vibrio cholerae* O1. Emerg Infect Dis. 2008;14:1129–31.
54. Dalsgaard A, Forslund A, Tam NV, Vinh DX, Cam PD. Cholera in Vietnam: changes in genotypes and emergence of class 1 integrons containing aminoglycoside resistance gene cassettes in *Vibrio cholerae* O1 strains isolated from 1979 to 1996. J Clin Microb. 1999;37: 734–41.

55. Ploy M-C, Denis F, Courvalin P, Lambert T. Molecular characterization of integrons in *Acinetobacter baumannii*: description of a hybrid class 2 integron. Antimicrob Agents Chemother. 2000;44:2684–8.
56. Dalsgaard A, Serichantalergs O, Forslund A, Pitarangsi C, Echverria P. Phenotypic and molecular characterization of *Vibrio cholerae* O1 isolated in Samutsakorn, Thailand before, during and after the emergence of *V. cholerae* O139. Epidemiol Infect. 1998;121:259–68.
57. Dalsgaard A, Forslund A, Serichantalergs O, Sandvang D. Distribution and content of class 1 integrons in different *Vibrio cholerae* O-serotype strains isolated in Thailand. Antimicrob Agents Chemother. 2000;44:1315–21.
58. Ehara M, Nguyen BM, Nguyen DT, Toma C, Higa N, Iwanaga M. Drug susceptibility and its genetic basis in epidemic *Vibrio cholerae* O1 in Vietnam. Epidemiol Infect. 2004;132: 595–600.
59. Iwanaga M, Toma C, Miyazato T, Insisengmay S, Nakasone N, Ehara M. Antibiotic resistance conferred by a class 1 integron and SXT constin in *Vibrio cholerae* O1 strains isolated in Laos. Antimicrob Agents Chemother. 2004;48:2364–9.
60. Bhattacharya MK, Dutta D, Bhattacharya SK, Deb A, Mukopadhyaya AK, Nair GB, Shimada T, Takeda Y, Chawdhury A, Mahalanabis D. Association of a disease approximating cholera caused by *Vibrio cholerae* of serogroups other than O1 and O139. Epidemiol Infect. 1998;120:1–5.
61. Thungapathra M, Amita, Sinha KK, Roy Chaudhuri S, Garg P, Ramamurthy T, Nair GB, Ghosh A. Occurrence of antibiotic resistance gene cassettes *aac(6′)-Ib*, *dfr*A5, *dfr*A12, and *ere*A2 in class 1 integrons in non-O1, non-O139 *Vibrio cholerae* strains in India. Antimicrob Agents Chemother. 2002;46:2948–55.
62. Dalsgaard A, Forslund A, Petersen A, Brown DJ, Dias F, Monterio S, Mϕlbak K, Aaby P, Rodriques A, Sandström A. Class 1 integron-borne, multiple antibiotic resistance encoded by a 150 Kb conjugative plasmid in epidemic *Vibrio cholerae* O1 strains isolated in Guinea-Bissau. J Clin Microbiol. 2000;38:3774–9.
63. Srinivasan VB. Studies on the drug resistance of *Vibrio cholerae*. Ph.D. Thesis, Jawaharlal Nehru University, New Delhi; 2006.
64. Sharma C, Ghosh A, Dalsgaard A, Forslund A, Ghosh RK, Bhattacharya SK, Nair GB. Molecular evidence that a distinct *Vibrio cholerae* O1 biotype El Tor strain in Calcutta may have spread to the African Continent. J Clin Microbiol. 1998;36:843–4.
65. Amita, Chowdhury SR, Thungapathra M, Ramamurthy T, Nair GB, Ghosh A. Class 1 integrons and SXT elements in El Tor strains isolated before and after 1992 *Vibrio cholerae* O139 outbreak, Calcutta, India. Emerg Infect Dis. 2003;9:500–2.
66. Shi L, Fujihara K, Sato T, Ito H, Garg P, Chakrabarty R, Ramamurthy T, Nair GB, Takeda Y, Yamasaki S. Distribution and characterization of integrons in various serogroups of *Vibrio cholerae* strains isolated from diarrheal patients between 1992 and 2000 in Kolkata, India. J Med Microbiol. 2006;55:575–83.
67. Dalsgaard A, Forslund A, Sandvang D, Arntzen L, Keddy K. *Vibrio cholerae* O1 outbreak isolates in Mozambique and South Africa in 1998 are multi-drug resistant, contain the SXT element and the *aad*A1 gene located on class 1 integrons. J Antimicrob Chemother. 2001;48:827–38.
68. Ceccarelli D, Bani S, Cappuccinelli P, Colombo MM. Prevalence of *aad*A1, *dfr*A15 class 1 integron cassettes and SXT Circulation in *Vibrio cholerae* O1 isolates from Africa. J Antimicrob Chemother. 2006;58:1095–7.
69. Taviani E, Ceccarelli D, Lazaro N, Bani S, Cappuccinelli R, Colwell RR, Colombo MM. Environmental *Vibrio* spp., isolated in Mozambique, contain a polymorphic group of integrative conjugative elements and class 1 integrons. FEMS Microbiol Ecol. 2008;63: 45–54.
70. Ammari H, Ruimi R, Skurnik D, Andremont A, Mazel D, Guerout A, Assaous F, Rahal K. Antibiotic resistance conferred by a class 1 integron and SXT element in different *Vibrio cholerae* O-serotype isolated in Algeria. Abstract No. 1733-20. ESCMID, 17th European

congress of clinical microbiology and infectious diseases, Munich, Germany, 31 Mar 2004, Apr 2007.

71. Ciccarelli D, Salvia AM, Sami J, Cappuccinelli P, Colombo MM. New cluster of plasmid-located class 1 integrons in *Vibrio cholerae* O1 and *dfr*A15 cassette containing integron in *Vibrio paratraemolyticus* isolated in Angola. Antimicrob Agents Chemother. 2006;50: 2493–9.
72. Mwansa JCL, Mwaba J, Lukwesa C, Bhuiyan NA, Ansaruzamman M, Ramamurthy T, Alam M, Nair GB. Multiply antibiotic-resistant *Vibrio cholerae* O1 biotype El Tor strains emerge during cholera outbreaks in Zambia. Epidemiol Infect. 2007;135:847–53.
73. World Health Organization. Cholera, Albania. Weekly Epidemiol Rec. 1994;69:280.
74. World Health Organization. Cholera, Italy. Weekly Epidemiol Rec. 1994;69:325.
75. World Health Organization. Cholera in 1993. Weekly Epidemiol Rec. 1995;70:201–8.
76. Greco D, Luzzi I, Sallabanda A, Dibra A, Kakarricy E, Shapo L. Cholera in Mediterranean: outbreak in Albania. Eurosurveillance. 1995;1:1–2.
77. Falbo V, Carattoli A, Tosini F, Pezzella C, Dionisi AM, Luzzi I. Antibiotic resistance conferred by a conjugative plasmid and a class 1 integron in *Vibrio cholerae* O1 El Tor strains isolated in Albania and Italy. Antimicrob Agents Chemother. 1999;43:693–6.
78. Sanchez, JL, Taylor, DN. Cholera. Lancet. 1997;349:1825–30.
79. Ries AA, Vugia DJ, Beingolia L, Palacios AM, Vasquez E, Wells JG, Baca NG, Swerdlow DL, Pollack M, Bean NH, Seminario L, Tauxe RV. Cholera in Piura, Peru: a modern urban epidemic. J Infect Dis. 1992;166:1429–33.
80. Campos LC, Zahner V, Avelar KES, Alves RM, Pereira DSG, Brazil JMV, Freitas FS, Salles CA, Karaolis DKR. Genetic diversity and antibiotic resistance of clinical and environmental *Vibrio cholerae* suggests that many serogroups are reservoirs of resistance. Epidemiol Infect. 2004;132:985–92.
81. Hall RM, Stokes HW. The structure of a partial duplication in the integron of plasmid DGO100. Plasmid. 1990;23:76–9.
82. Stokes HW, Tomaras C, Parasons Y, Hall RM. The partial 3′-conserved segment duplications in the integron In6 from pSa and In7 from pDGO100 have a common origin. Plasmid. 1993;30:39–50.
83. Toleman MA, Bennett PM, Walsh TR. ISCR elements: novel gene capturing systems of the 21st century. Microb Mol Biol Rev. 2006;70:296–316.
84. Verdet C, Arlet G, Barnaud G, Lagrange PH, Phillipon A. A novel integron in *Salmonella enterica* serovar Enteritidis, carrying the $bla_{\mathrm{DHA\text{-}1}}$ gene and its regulator gene *amp*R, originated from *Morganella morganii*. Antimicrob Agents Chemother. 2000;44: 222–5.
85. Arduino SM, Roy PH, Jacoby GA, Orman, BE, Pinoria SA, Centron D. *bla*CTX-M-2 is located in an unusual class 1 integron (In35) which includes *orf*513. Antimicrob Agents Chemother. 2002;46:2303–6.
86. Sabate M, Navarro E, Miro' E, Campoy S, Mirelis B, Barbé J, Prats G. Novel complex sut1-type integron in *Escherichia coli* carrying *bla* (CTX-M-9). Antimicrob Agents Chemother. 2002;46:2656–61.
87. Petroni A, Melano, RG, Saka, HA, Garutti A, Mange L, Pasteran, F, Rapoport, M., Miranda M, Faccone D, Rossi A, Hoffman PS, Galas MF. CARB-9, a carbenicillinase encoded in the VCR region of *Vibrio cholerae* belongs to a family of cassette-encoded β-lactamases. Antimicrob Agents Chemother. 2004;48:4042–6.
88. Bistué AJC, Martin FA, Petroni A, Faccone D, Galas M, Tolmasky, ME, Zorreguieta A. *Vibrio cholerae* InVII7, a class 1 integron harboring *aac(6′)-Ib* and *bla*CTX-M-2, is linked to transposition genes. Antimicrob Agents Chemother. 2006;50:1903–7.
89. Ahmed AM, Nakagawa T, Arakawa E, Ramamurthy T, Shinoda S, Shimamoto T. New aminoglycoside acetyltransferase gene, *aac*(3)-Id, in a class 1 integron from a multiresistant strain of *V. fluvialis* isolated from an infant aged 6 months. J Antimicrob Chemother. 2004;53: 947–51.

90. Srinivasan VB, Virk RK, Kaundal A, Chakrabarty R, Datta B, Ramamurthy T, Mukhopadhyay, AK, Ghosh A. Mechanism of drug resistance in clonally related clinical isolates of *Vibrio .fluvialis* isolated in Kolkata, India. Antimicrob Agents Chemother. 2006;50:2428–32.
91. Huq MI, Alam, AKMJ, Brenner DJ, Morris JJ. Isolation of *Vibrio*-like group EF-6, from patients with diarrhea. J Clin Microbiol. 1980;11:621–4.
92. Hlady WG, Klontz KC. The epidemiology of *Vibrio* infections in Florida, 1981–1993. J Infect Dis. 1996;5:1176–83.
93. Lesmana M, Subekti DS, Tjaniadi P, Simanjuntak CH, Punjabi NH, Campbell JR, Oyofo BA. Spectrum of *Vibrio* species associated with acute diarrhea in North Jakarta, Indonesia. Diagn Microbiol Infect Dis. 2002;2:91–7.
94. Doublet B, Weill FX, Fabre L, Chaslas-Dancla E, Cloeckaert A. Variant *Salmonella* genomic island/antibiotic resistance gene cluster containing a novel 3′-*N*-aminoglycoside acetyltransferase gene cassette, *aac*(3)-Id in *S. enterica* serovar Newport. Antimicrob Agents Chemother. 2004;10:3806–12.
95. Levy SB, Marshall B. Antimicrobial resistance worldwide: causes, challenges and response. Nat Med. 2004;10:5122–9.

Chapter 18
Aquatic Realm and Cholera

Anwar Huq, Chris J. Grim, and Rita R. Colwell

Abstract Cholera is an ancient disease that can be severe and life threatening. It occurs predominantly in areas of the world where populations lack safe drinking water. Epidemics of cholera are linked with malnutrition, poor sanitation, and conditions resulting from natural disasters such as severe flooding. According to a report published by WHO in 2000 [1], cholera remains a major public health problem and is becoming increasingly important since the number of countries in which cholera is endemic continues to increase. Unfortunately, outbreaks of the disease continue into the twenty-first century with ominous portent in the wake of global climate change [1]. Yet cholera is a preventable disease if people have access to safe drinking water and are properly educated how to protect themselves from the risk of infection with vibrios. Cholera also is an easily treatable disease. Oral rehydration therapy, a solution containing glucose and appropriate salts, has proven to be effective for treatment of most cholera victims [2]. Nevertheless, each year, tens of thousands of people are victims of the disease, bringing this "curse of humankind" to modern civilization. Present understanding of cholera is based on studies conducted over the past three decades and significant new information has been gained concerning environmental factors associated with this disease, especially how to detect the bacterium and where it lives in the natural environment, outside the human gut, and what triggers the annual outbreaks that occur with remarkable regularity. Environmental research on *Vibrio cholerae* and cholera has provided insights for prediction and prevention of the disease it causes, while the race for effective vaccines against cholera continues.

A. Huq (✉)
Maryland Pathogen Research Institute, University of Maryland, College Park, MD 20742, USA
e-mail: huq@umd.edu

T. Ramamurthy, S.K. Bhattacharya (eds.), *Epidemiological and Molecular Aspects on Cholera*, Infectious Disease, DOI 10.1007/978-1-60327-265-0_18,
© Springer Science+Business Media, LLC 2011

18.1 History of Cholera

18.1.1 Old Beliefs and Myths About Cholera

The evidence of cholera or cholera-like disease can be traced historically to 500–400 BC, in ancient Indian literature, based on the described symptoms. Cholera was often depicted as a "ghost" in older literature, reflecting the belief that "miasma" or "bad air" caused the disease [3]. Approximately 500 years ago, the name "cholera" was first applied to the disease, originating from the Greek word "chole" (bile) and "rein" (to flow), meaning the "flow of bile" [4]. Later it became evident that the disease is enteric in its symptoms, i.e., a disease of the intestine, as expressed by Robert Koch, who identified the causative agent of cholera in 1854. At that time, Koch also speculated that multiplication of the cholera bacterium took place in river water. He lacked data proving his assertion, but he was prescient in his observation. Prior to Koch's observations and identification, an Italian medical student described the cholera *Vibrio*. His work was relegated to obscurity, even though his stained slides of the cholera *Vibrio* can be seen in a museum display in Italy [5]. John Snow, the renowned British physician and epidemiologist, mapped cases of cholera in London more than 150 years ago and concluded that contaminated water was directly responsible for the disease.

People in villages in India and Bangladesh considered the disease as punishment from God. If a case of cholera was reported in a village, those villagers not afflicted with the disease fled from the village where cholera cases had occurred to another area far away, for protection from cholera. A common practice in villages of India and Bangladesh was to wash soiled clothing of cholera patients in a stream or river, but not in local ponds. This practice contributed to the spread of local epidemics in densely populated areas from village to village, since untreated river water served as a source of water for domestic purposes, including drinking. Interestingly, the linkage of cholera and religious beliefs was preached in other countries, as well as India and Bangladesh. In the United States in the mid-1800s, punishment by God for evil behavior was claimed when cholera outbreaks occurred [6].

18.1.2 Historical Background and Global Occurrence of Cholera

Although cholera is routinely reported in Asian and African countries, sporadic cases are reported in almost every country of the world [7]. Records on cholera, or cholera-like disease, in the Indian subcontinent trace back to the early sixteenth century. Beginning in the 1800s, reports of cholera gradually began to be documented in other parts of the world. Historically, cases of cholera in Europe, North America, and Australia were usually considered to have been imported from other countries. But, in reality, many of the cases were ultimately shown to be indigenous, with the majority of the cases associated with drinking untreated water or consumption of contaminated seafood. The last major epidemic of cholera in the United States was reported in 1911. In 1991, South America was hit with cholera outbreaks of extraordinary dimension, involving 21 countries and resulting in more than 800,000 cases [8]; thus, cholera can strike unexpectedly and severely, even after an absence of almost a century. With rare exception, cholera cases occur during the summer months, for example, in January through March in Australia and in August and September in the United States [9].

In Australia, *V. cholerae* serotype O1 was isolated from water samples collected from the Clarence and Georges Rivers in 1981, with some of the isolates proven to be toxigenic [9, 10]. Incidentally, in the same year, *V. cholerae* O1 was isolated from Chesapeake Bay, MD, but was ostensibly non-toxigenic [11]. In an extensive survey conducted in the marsh area of Kent, England, at about the same time, *V. cholerae* O1 was detected in water samples from a marsh area collected during the warmer months of summer [12, 13]. Thus, it became convincingly clear that *V. cholerae* occupies an ecological niche in the aquatic environment.

Similarity between pathogenic and non-pathogenic *V. cholerae* and between clinical and environmental isolates of *V. cholerae* was demonstrated by Colwell [14] and Citarella and Colwell [15], using DNA–DNA hybridization methods. Additionally, results from microcosm experiments demonstrated that the non-O1 serotype of *V. cholerae* can convert to O1 serotype [16]. Waldor and Mekalanos [17, 18] reported lateral gene transfer in *V. cholerae* serogroup O139, a major advancement in understanding the evolution of pathogenic bacteria since many virulence factors are encoded on mobile genetic elements. For example, structural genes for cholera toxin (CT) are encoded by a filamentous bacteriophage (designated CTXφ), which is related to coliphage M13. The CTXφ genome is chromosomally integrated or replicated as a plasmid.

In Bangladesh, the Ganges deltaic region reports cases of cholera, with two distinct peaks in its incidence almost every year. A smaller peak of cholera cases occurs in the late spring at the beginning of the monsoon season and the larger peak occurs in late fall [19]. However, the number of cases or the length and geographic range of the epidemic can vary significantly. Environmental factors have been correlated with peaks in incidence of cholera, but no apparent social or cultural correlation has been observed. Of the environmental parameters, season is strongly linked with occurrence of outbreaks (discussed later).

In 1991, a major epidemic of cholera occurred in Bangladesh, with 400,000 cases and 8,000 (2%) deaths [20]. Interestingly, during the same year, Latin America experienced a cholera outbreak with 800,000 cases and 4,000 (0.5%) deaths reported [8]. Also in that same year, Africa suffered 150,000 cases of cholera, with the largest number of deaths at 14,000 (9%) [21, 22]. According to a WHO report released in September 2007, there were approximately 132,000 cases of cholera in 2005 with 1.7% fatality [23].

Clearly, cholera is, at the present time, a global threat to public health and cholera is now recognized as a re-emerging disease, with a significant increase in the number of cases since 2005, notably affecting vulnerable populations living in conditions of poor sanitation. For those countries where at least the minimum hygiene standards are met, cholera has receded in importance, but it remains a threat in almost every developing country [24]. The number of cholera cases reported to WHO during 2006 rose dramatically, reaching the level of the late 1990s. A total of 236,896 cases were reported from 52 countries, including 6,311 (2.7%) deaths, an overall increase of 79% compared with the number of cases reported in 2005. It is interesting to note that many of the cases were reported as part of major outbreaks that occurred in countries where cases had not been reported in previous years. It is estimated that only a small proportion, perhaps <10%, of cholera cases worldwide are reported to the WHO [24]. Thus, the true burden of the disease may actually be in the millions and the number of deaths may also be grossly underestimated. This estimate is particularly disturbing if the probable effects of global warming are taken into account, e.g., the average sea surface temperature has increased by 4.4ºF over the last 30 years and is expected to increase another 12.5°F by the end of the century and sea surface height is expected to increase between 8 and 33 in. by year 2100 [25].

The most common risk factors associated with cholera epidemics include water source contamination, heavy rainfall and flooding, and population dislocation [26]. The number of cases of cholera has been shown to be related to sea surface temperature (SST) and sea surface height (SSH) [22, 27, 28]. According to available records, global outbreaks of cholera characteristically begin in coastal regions and those areas with poor sanitation and a lack of safe drinking water. Although records of the occurrence of cholera prior to the 1960s are not readily available, it has been established from results of a study published in 1992 that cholera does indeed begin in coastal areas and the number of cases gradually increases in time in the inland areas [29]. Such a situation was also observed in Latin America in 1991 when the outbreak of cholera occurred in Lima, Peru, with subsequent cases occurring further north along the coast [8].

Another interesting aspect of cholera is that, until 1992, all epidemics of cholera were caused by *V. cholerae* serogroup O1. However, during the later part of 1992, a new serogroup was isolated, first in Madras on the eastern coast of India [30] and then in the southern part of Bangladesh along the coast of the Bay of Bengal. This new serogroup of *V. cholerae* O139 caused a major epidemic [31]. Later, *V. cholerae* O139 was detected in countries neighboring India and Bangladesh and has since continued to persist in that geographic region of the world [30, 31].

18.2 Ecology of *Vibrio cholerae*

18.2.1 Environmental Factors Affecting the Organism

Vibrio cholerae and related vibrios comprise a significant component of the heterotrophic bacterial flora of the marine and estuarine aquatic environments [32]. Two environmental parameters, salinity and temperature, have been found to have a pronounced effect on the ecology of this group of organisms in the aquatic environment. Individual *Vibrio* species display a variable range of minimal sodium chloride concentrations required for optimal growth. These differences in salt tolerance/requirement influence the distribution of an individual species, with respect to its potential to inhabit environments of different salinities. For example, *V. cholerae* has a sodium ion requirement of 5–15 mM; however, this requirement can be significantly lower in highly eutrophic waters containing abundant organic nutrients (>1.0 mg/L) [33, 34]. The salinity optimum for *V. cholerae* survival and growth has been reported to be 1.8–2.5% [33–36]. These findings indicate that *V. cholerae* can inhabit both freshwater and estuarine environments [37]. The effect of temperature on the ecology of *V. cholerae* is more pronounced than salinity. *Vibrio cholerae* is more frequently isolated from natural waters and in laboratory microcosms prepared by inoculation with natural water samples when the water temperature is in the range of 15–20°C or higher [11, 33, 35].

18.2.2 Biological Factors Affecting the Organism

The aquatic environment exhibits a strong selective pressure on organisms including bacteria, since nutrients are usually either limited or abundant only within small intervals of time. Indeed, most environmental bacteria not attached to surfaces or present in biofilms are relegated to a feast-famine lifestyle. Because of this phenomenon, *V. cholerae* has evolved the capacity to utilize a large variety of nutrients and substrates. The genomic sequence of *V. cholerae* O1 El Tor N16961 revealed that a significant portion of the genes present in the *V. cholerae* genome function in transport and subsequent metabolism of those various substrates that are taken up by these transport systems [38]. In addition to limited nutrient availability, there are a multitude of environmental stresses that aquatic bacteria must endure in order to survive and flourish. Some of the biological factors affecting the ability of this organism to survive in the aquatic environment are discussed below.

18.2.2.1 Viable but Non-culturable State

Vibrio cholerae is able to survive or persist through periods of adverse environmental stress in the aquatic environment by entering into a "viable but non-culturable" (VBNC) state [39]. VBNC is a term that describes a condition in which (mainly) gram-negative bacteria exhibit detectable metabolic activity, such as uptake of radio-labeled substrate [39], but are not culturable by conventional laboratory culture methods. Initially described in *V. cholerae*, the VBNC phenomenon has been reported to occur in many other bacterial species, including *Vibrio vulnificus, Vibrio campbelli, Vibrio mimicus, Vibrio natriegens, Vibrio parahaemolyticus, Vibrio proteolyticus, Vibrio anguillarum, V. salmonicida, Salmonella enteritidis,* enterotoxigenic *Escherichia coli, Enterobacter aerogenes, Enterococcus faecalis, Klebsiella pneumoniae, Legionella pneumophila, Pseudomonas putida, Shigella sonnei, Shigella flexneri, Shigella dysenteriae*, the fish pathogen *Pasteurella piscicida*, and *Campylobacter jejuni* [40].

Cells of *V. cholerae* enter the VBNC state in response to nutrient deprivation or elevated salinity, especially at low temperature, through a definable physiological process that takes place in multiple steps, the first of which is a "rounding up" of cells and reduced metabolic activity [41]. This transition of *V. cholerae* from a curved rod to coccoid shape involves a reduction in cell volume and cell size and loss of its flagellum. Despite a significant drop in bio-volume, there is no evidence that reductive cell division (increase in cell numbers without an increase in cell biomass) occurs, as exhibited by cells entering the starvation state [42, 43]. Coccoid VBNC cells vary in the extent of reduction (15–300 times) based on the length of exposure to starvation and the intensity of starvation [44].

As evidenced by transmission electron microscopy of nonculturable cells of *C. jejuni,* ultrastructural changes accompanying the entry of bacteria into the VBNC state include retention of an intact but "asymmetric" formation of membranes, with a condensed cytosol [45]. Linder and Oliver [46] observed that VBNC cells of *V. vulnificus* contained a significantly reduced density of ribosomes and nucleic acid but maintained a normal cytoplasmic membrane. Their study showed that the non-culturable cells were rounded and reduced in size to about one-half that of culturable cells. These observations are summarized by Colwell as follows, "the bacterium maintains the integrity of the membrane, which is important for preserving its internal environment against the surroundings, keeping its genetic material (DNA) intact, and at the same time having its metabolic activity at the lowest rate but enough to promote the uptake of nutrients via appropriate transport systems when these substrates become available again in the environment" [47].

Asakura et al. [48] have shown that the distribution of induced or repressed genes in the VBNC state was similar to that of stationary phase cells using microarrays. The genes included those that code for transport and binding proteins, regulatory functions, cellular processes, energy metabolism, protein fate and synthesis, cell envelope, and other housekeeping functions. They found that expression of 100 genes encoding proteins involved in cellular processes is highly induced in the VBNC state (five-fold) as compared to stationary phase cells.

Still debated is the question whether the VBNC state is reversible, often termed "resuscitation." Resuscitation has been defined by Oliver [49] as a reversal of the metabolic and physiological processes that lead to non-culturability, resulting in the ability of the cells to be culturable on those media normally supporting growth of the cells. Since environmental stresses induce the viable but non-culturable state of bacteria, reversal or removal of the stress conditions should result in resuscitation of the bacteria. Wai et al. [50] reported resuscitation of viable but nonculturable *V. cholerae* by heating at 45°C for 1 min, followed by plating on nutrient agar. Culturable cells were recovered and the investigators suggested that ammonium chloride may have a role in growth and resuscitation of such cells. However, some investigators [51] attempting resuscitation of VBNC *V. vulnificus* by temperature upshift observed, instead, regrowth, i.e., a few cells retained the ability to take up substrate released by moribund cells, and then these cells began to grow rapidly. Ravel et al. [52] found that temperature change alone was insufficient for resuscitation of VBNC *V. cholerae.* Oliver [49] emphasized that, whether in the presence of exogenous nutrient or not, resuscitation of non-culturable cells appears to require a significant length of time. In the case of *V. vulnificus,* 2–4 days was required before culturable cells could be observed on agar plates, following temperature upshift.

In vivo resuscitation of VBNC has been demonstrated using tissue culture, animal models, and human volunteers. Even though VBNC cells of *V. cholerae* cannot be cultured by conventional plate techniques, they maintain their virulence. Viable but non-culturable cells of *V. cholerae* cause fluid accumulation by continued production of cholera toxin in ligated rabbit ileal loops and culturable cells can be recovered from the ileal loop fluid [53]. In another study, non-culturable cells of an attenuated strain of *V. cholerae* O1, prepared by incubation in a salt solution at 4°C, were fed to human volunteers. Two out of five volunteers produced clinical symptoms of cholera including diarrhea and passed culturable *V. cholerae* O1 cells in their stool after 2–5 days. At no time prior to, during, or after ingestion by the volunteers did aliquots of the VBNC cells incubated in nutrient medium yield culturable cells from the non-culturable population. The conclusion was that resuscitation of the cells occurred during passage in vivo [53]. Also noteworthy is the observation that elevated antibody titer against *V. cholerae* O1 was detected among divers after 30 days of diving in water where VBNC cells of *V. cholerae* O1 were readily observed by direct fluorescent antibody detection methods [54]. Thus, we propose that the VBNC state serves as a survival mechanism for *V. cholerae* O1 and O139, as well as non-O1, non-O139 *V. cholerae,* allowing the species to persist in the environment during inter-epidemic periods and in non-epidemic areas when environmental conditions are unfavorable.

18.2.2.2 Biofilm Formation

Historically, aquatic bacteria were believed to be uniformly unattached and free-swimming members of the planktonic community. Contrary to this premise, Costerton et al. [55] found that the predominant lifestyle for aquatic bacteria is attached to surfaces in a complex matrix, termed a biofilm. Several advantages have been hypothesized for bacteria in biofilms, such as access to nutrients adsorbed to surfaces or comprising the surface, chitin, for example, and protection from a variety of stresses including oxidants [56, 57], heavy metals [58–60], antibiotics [61–65], UV damage [66], and protozoan grazing [67]. *Vibrio cholerae* has been shown to be able to attach and form biofilms on a number of different biotic substrates, such as copepods [68] and other zooplanktons, water hyacinths [69], crustaceans [70, 71], insects [72, 73], plants, and phytoplanktons [74].

Biofilm formation takes place through a series of successive steps [75]. As a bacterial cell moves close to a surface, attachment is achieved through the use of pili and flagella. Soon a discrete monolayer of cells is formed on the surface, as more cells become attached. These cells migrate along the surface to form microcolonies, again by means of pili and flagella. Eventually, a fully mature biofilm, composed of an elaborate three-dimensional matrix of bacterial structures inundated with water channels, is formed [76]. Single cells often detach and are shed from both developing and mature biofilms into the environment.

Several genes and gene products of *V. cholerae* responsible for biofilm formation have been deduced. The single polar flagellum and the genes responsible for its structure, such as *flrC, motY, flgF* [75], *flaA, flgL, flgH, fliN, fliH, pomA*, and chemotaxis genes [77], have been shown to be essential in biofilm formation, presumably by allowing the bacterium to overcome repulsive forces experienced as the bacterium approaches a surface and as the bacterium migrates along the surface [75]. The mannose-sensitive hemagglutinin (MSHA), a type-IV pilus, aids in biofilm formation [78, 79], presumably by attaching the bacterium to the surface. However, MSHA is not absolutely required, since O1 classical biotype strains of *V. cholerae* do not produce MSHA [80] and attachment to chitinous surfaces can be achieved independently of MSHA [79]. Additionally, two other type IV pili promote adherence or colonization of surfaces, the chitin-regulated pilus, ChiRP [81], and toxin-coregulated pilus (TCP) [82]. ChiRP appears to contribute to initial attachment to a chitinous substrate, while TCP is important in bacterial interactions that lead to microcolony formation.

Biofilms have been proposed to be important in the dispersal and transmission of pathogenic bacteria [83]. VBNC cells of *V. cholerae* O1 present in biofilms in the environment have been suggested to play a role in the transmission of cholera [84]. The main mechanism proposed is the accumulation of high cell densities of *V. cholerae* O1 or O139 in a detached biofilm particle, which can essentially deliver the infective dose of the cholera bacterium to cause disease. In competition assays, aggregates of *V. cholerae*, showing biofilm-like structure, shed in the stools of cholera patients were hyperinfective compared to unattached, broth cultures [85, 86]. In these experiments, the number of *V. cholerae* cells is approximately equivalent, so the advantage is directly related to the morphological state of the bacteria. One explanation is that the aggregated clumps of cells in biofilm can pass through the stomach acid barrier with significantly less mortality [87]. However, Alam et al. [84] have shown that the biofilm in cholera stool is not as resistant as *V. cholerae* O1 or O139 cells in biofilms formed in the natural aquatic environment.

18.2.2.3 Rugosity and Colonial Opacity

The ability of *V. cholerae* to display multiple colony morphological types was first observed by Robert Koch. Typically, colony types of *V. cholerae* are described as opaque or translucent, and rugose or smooth. Most data relating to the colony variation phenomenon of *V. cholerae* are observational or do not hold true for every strain and serogroup, leading to considerable confusion and misunderstanding. What appear to be the most generally accepted data are as follows. Classical strains of serogroup O1 generally appear as opaque colonies on solid media [88]. Translucent colonies are usually more virulent than opaque colonies of the same strain [88, 89], but this is not always the case [90]. Freshly isolated strains of *V. cholerae* are predominantly translucent, whereas opaque colonies are seemingly produced in higher proportions in "aged" cultures, especially when grown on minimal media [91].

Finkelstein et al. [92] observed that opaque colonies possessed inclusion bodies containing poly-β-hydroxybutyrate (PHB), while translucent colonies did not. Additionally, the number of PHB-containing inclusion bodies was correlated with degree of opacity. For one strain of *V. cholerae* serogroup O139, opacity was correlated with extent of encapsulation, with highly opaque colonies possessing a thick capsule and less opaque colonies possessing a very thin capsule. Also, opaque colonies had a higher survival rate when transferred to saline microcosms, compared with translucent colonies.

In addition to displaying opaque and translucent colony types, *V. cholerae* is capable of displaying two very different phenotypic variants, smooth and rugose [93]. All serovars of *V. cholerae* have been shown to produce some rugose colonies, except classical strains of the O1 serogroup [94]. Morris et al. [94] observed that *V. cholerae* O1 El Tor rugose variants produce an exopolysaccharide (EPS) matrix that subsequently enhances cell aggregation. The production of the matrix-forming sugar has also been shown to result in biofilm formation. Wai et al. [95] characterized the EPS of a rugose variant of *V. cholerae* O1 El Tor TSI-4 and found that it was composed of *N*-acetyl-D-glucosamine, D-mannose, 6-deoxy-D-galactose, and D-galactose. Smooth variants do not produce a capsule. Unlike the two colony types, phenotypic variants of the same strain do not differ in their pathogenicity [96]. However, the two phenotypic variants differ in their response to certain oxidative and osmotic stresses [96] specifically; rugose phenotypes are resistant to chlorine [91, 94, 97], hydrogen peroxide, and high levels of extracellular salt [95]. Additionally, rugose variants are less motile than smooth variants [98]. Recently, Matz et al. [67] demonstrated that the rugose variant is also less susceptible to protozoan grazing, regardless of whether the bacteria were suspended in the water column or maintained in a biofilm. The grazing inhibition is controlled by quorum sensing (see below), facilitated through HapR.

The rugose phenotype is dependent upon several genes and gene products. The first identified operon *vps* (*Vibrio* polysaccharide synthesis) is divided into two clusters, *vpsA-K* and *vpsL-Q*, I and II, respectively, and all of these genes are found on chromosome I [91]. The expression of the operon is positively regulated by the two-component signal transduction pair VpsR [99] and VpsT [100] and negatively regulated by HapR [96]. The intergenic region between the two *vps* clusters contains six genes, *rbmA-F* (rugosity and biofilm modulators), which directly affect phenotypic variation as well as biofilm formation [101, 102]. A smaller operon, *vpvA-C*, has been recently found to be important in determining phenotypic variation, in particular a single nucleotide difference in *vpvC* [98].

18.2.2.4 Quorum Sensing

The traditional view of a bacterium was that it is an independent organism pursuing its own food sources, replicating, and carrying out an array of metabolic and physiological functions in an individualistic manner. In the early 1970s, Hastings and Greeberg [103] observed that cultures of the bioluminescent bacterium *Vibrio fischeri* are luminescent at approximately the late logarithmic phase of growth, but not so during the early phases of growth. Findings on *V. fischeri* luminescence behavior subsequently led to the realization that bacteria can perceive and respond to each other in a coordinated manner. Hastings and coworkers [103, 104] concluded that control of luminescence in *V. fischeri* and the related species of *Vibrio harveyi* was achieved by a cell density-dependent, gene regulation mechanism, termed quorum sensing.

Quorum sensing relies upon the interaction of a small, freely diffusible signal molecule, autoinducer (AI), usually an *N*-acyl homoserine lactone (AHL or HSL), with a sensor or a transcriptional activator (R protein) that couples gene expression with cell population density [105]. At low cell densities, AI passively diffuses out of cells and into the surrounding medium where the concentration is too low to stimulate expression. At high cell densities, the concentration of AI in the surrounding environment increases, resulting in the expression of the target genes, in the case of *V. fischeri*, triggering the expression of bioluminescent genes [106, 107]. Autoinducer concentrations as low as 10 nM have been shown to be adequate to initiate bioluminescence in *V. fischeri* [108].

de Kievit and Iglewski [105] reported 22 bacterial species that use quorum-sensing systems, homologous to the *luxI/R* regulated bioluminescence of *V. fischeri,* to regulate phenotypes that include conjugation, production of virulence and invasion factors, nodulation, swarming, and synthesis of antibiotics. For example, *Erwinia carotovora* and *Pseudomonas aeruginosa* use quorum sensing to synchronize the expression of a variety of extracellular virulence proteins. In *Agrobacterium tumefaciens*, quorum sensing determines nutrient availability and initiates transfer of Ti plasmids. The majority of bacteria that possess quorum-sensing regulation systems are found in association with plants or animals, suggesting that this type of regulation is important in the colonization of hosts by symbiotic or pathogenic bacteria.

The *lux* regulatory network of *V. harveyi* contains two separate signal/sensor systems for controlling luminescence, termed system 1 and system 2 [109], and it has been proposed that quorum sensing serves as a mechanism for both species-specific (system 1) and inter-specific (system 2) communication. LuxN and LuxQP are the sensor proteins for systems 1 and 2, respectively. They are two-component hybrid kinases, containing sensor kinase and response regulator domains. LuxN responds only to *N*-(3-hydroxybutanoyl)-L-homoserine lactone or AI-1 that is synthesized by LuxL and LuxM [110]. The LuxQP sensor-2 complex responds to a furanosyl borate diester, the system 2 autoinducer, the production of which is controlled by LuxS [111]. LuxO acts as a central shared integrator protein of the system that, when active, represses light production [112]. LuxU is a histidine phosphotransferase protein that is important in transducing the phosphorelay message from LuxN and LuxQP to LuxO. LuxR is a positive transcriptional factor of the system [113]. LuxT is a repressor of LuxO that binds to the *luxO* promoter [114]. LuxT mutants of *V. harveyi* showed increased LuxO expression, affecting the cell density-dependent induction of bioluminescence. LuxT shows homology with the AcrR/TetR family of transcriptional regulators.

Several other members of the genus *Vibrio* contain genes homologous to the *lux* regulatory genes of *V. harveyi*, indicating that these species possess autoinduction-controlled phenotypes. Bassler et al. [115] reported that *V. cholerae*, *V. parahaemolyticus, V. anguillarum, Vibrio alginolyticus,* and *V. natriegens* stimulate luminescence in a *V. harveyi luxS* reporter strain.

The genome of *V. cholerae* N16961 O1 El Tor contains several homologues of the *V. harveyi lux* regulatory network [39]. This strain possesses homologues to *luxO, P, Q, R* (*hapR* [116]), *S, U,* and *N* (*cqsS* [118], a *luxCDABEG* locus is not present in this strain), indicating that the bacterium possesses a functional system II quorum-sensing regulatory network. Although luminescent strains of the species exist [118], a *luxCDABEG* locus is not present in this strain. Furthermore, Miller et al. [117] identified the synthase for the system 1 signal, *cqsA*, and a yet undescribed system 3 input circuit in *V. cholerae*, although luminescent strains of the species exist [118]. The quorum-sensing circuit in *V. cholerae* O1 controls virulence, production of the hemagglutinin protease, motility, biofilm formation [119], and oxidative and nutritional stress response (RpoS) [120] through HapR. Surprisingly, the quorum-sensing system of *V. cholerae* represses the ToxR virulence gene regulon at high cell densities. One hypothesis is that during the initial stages of infection, virulence genes *tcp* and *ctx* are expressed so that successful colonization of the intestine results. In the aquatic environment, quorum-sensing regulation may function in a similar fashion to promote successful attachment to copepods and other chitinous zooplanktons. At higher cell densities, HapR production occurs, repressing TCP and CT but initiating production of Hap. The protease is proposed to aid in detachment, so the bacterium is shed and dispersed in large numbers back into the aquatic environment [121].

18.2.2.5 Chitinase and Chitin Utilization

A strategy *V. cholerae* has adopted to counteract carbon and nitrogen limitation in aquatic environments is to utilize chitin. Chitin is a long-chain biopolymer of repeating β(1,4)-linked *N*-acetylglucosamine units found in the cell walls of fungi and the exoskeleton of arthropods. It is the most abundant polysaccharide found in the aquatic environment. The chitin degradation pathway of *V. cholerae* includes production of the enzyme chitinase, which breaks the glycosidic bonds in chitin, an important first step in the metabolism of chitin. This enzyme is produced by the majority of *Vibrio* species [122], a strong indication that this is an important, but generalized, survival strategy for species of this genus.

Vibrio cholerae can often be found associated mainly with chitinous zooplankton, and occasionally with other aquatic flora and fauna [123]. Microscopic analysis of copepod-associated *V. cholerae* revealed that the bacterium preferentially accumulated in the oral region and on the egg sac of copepods [68]. It has been hypothesized that this is a symbiotic association with the *Vibrio* sp. acting as a mechanism for egg sac splitting and release of the copepod eggs. The steps involved in chitin utilization include chemotaxis, attachment, and degradation [125]. Meibom et al. [81] were able to identify three classes of chitin-regulated genes, depending upon the chitin or the chitin subunit substrate employed. They identified the sensor ChiS for the class which responds to chitin oligosaccharides. They also found that *V. cholerae* produces two extracellular chitinases, ChiA-1 and 2, a chitoporin, ChiP, and a chitin-regulated type-IV pilus, ChiRP [81].

Recently, Meibom et al. [125] have shown that cells of *V. cholerae* become competent in the presence of chitin. In their experiments, a recipient strain of *V. cholerae*, grown in the presence of chitin oligosaccharides or on a chitinous surface, was able to uptake and express a gene for antibiotic resistance present in the DNA of a donor *V. cholerae* strain. The transformation was found to be dependent upon ChiRP, the competence regulator $TfoX^{VC}$, HapR, and RpoS, indicating that natural competence in *V. cholerae* is controlled by chitin, quorum sensing, and stress response [125].

18.2.2.6 Intracellular Existence

The ability of free-living protozoans to harbor intracellular bacterial symbionts has been well studied and the list of protists which can harbor endosymbionts is extensive. Approximately 25% of *Acanthamoeba* isolates from clinical and environmental origins contain bacterial endosymbionts. One of the most studied cases of this symbiosis is that of *L. pneumophila* and *Acanthamoeba* [126–128], but a number of other human bacterial pathogens can be found in these associations [129]. Recently, *V. cholerae* has been added to this growing list of pathogenic bacteria that can survive and multiple inside higher order members of the aquatic environment. In separate experiments, *V. cholerae* serogroup O139 [130] and serogroup O1 classical and El Tor [131] were shown to intracellularly localize in vacuoles and the surrounding cytoplasm of *Acanthamoeba castellanii* trophozoites and within the walls of cysts. Previously, co-cultivation of *V. cholerae* with the freshwater amoeba *Acanthamoeba polyphaga* and *Naegleria gruberi* was found to enhance survival of the bacterium in laboratory microcosms, but the details of the interaction were unclear [132]. For both serogroups, internalization occurred in as little as 1 day postintroduction [132, 133]. After 2 weeks, the density of intracellular *V. cholerae* cells had increased by 4–5 logs and stabilized. This behavior is similar to that of the facultative intracellular bacterium *Francisella tularensis* [133] and is a new finding with important ecological implications. This interaction may play an important role in the ability of *V. cholerae* to survive in the aquatic environment, where it must evade predation by grazers and withstand adverse environmental conditions, such as low nutrient availability. Additionally, the ability of the bacterium to multiply within the amoebic host may facilitate the organism's re-entry into a human host by providing an infectious dose.

18.3 Detection of *Vibrio cholerae* from the Environments

Vibrio cholerae can be detected in marine, estuarine, and fresh water environments throughout the year, by a variety of methods [11, 22, 134–136]. The choice of method to employ is largely dependent on the type of sample and the desired outcome, i.e., whether it is detection or isolation or enumeration. Often combinations of different types or methods, including bacteriological, molecular, and immunological, are used for a specific type of sample. Here, we discuss the various types of detection and isolation protocols commonly used and widely accepted. For more detailed protocols for selected methods, please refer [137].

18.3.1 Conventional Bacteriological Culture Methods

Historically, detection of *V. cholerae* from natural waters has relied mostly on methods developed for clinical isolation of the organism, that is, enrichment in alkaline peptone water, pH 8.6, for 6–8 h and subsequent plating on selective media, such as thiosulfate citrate–bile salts–sucrose (TCBS) or taurocholate tellurite gelatin (TTG) agar [32, 138, 139]. The alkaline nature of the enrichment broth allows differential multiplication of *Vibrio* species. Presumptive isolates of *V. cholerae* appear as translucent, flat, yellow sucrose-fermenting colonies with elevated centers on TCBS and colorless colonies, often with a characteristic dark center after 2 days of growth, surrounded by a halo due to the hydrolysis of gelatin, on TTGA.

The identity of presumptive colonies is confirmed by specific biochemical tests designed to differentiate *V. cholerae* from other species of the genus and other members of the *Vibrionaceae* and *Enterobacteriaceae*. The most commonly used tests are oxidase, arginine dihydrolase, ornithine decarboxylase and lysine decarboxylase, growth in nutrient broth with 0, 6, and 8% NaCl, acid production from arabinose and mannitol, esculin hydrolysis, and the methyl red and Voges–Proskauer reactions. Choopun et al. [140] developed an abbreviated scheme for the isolation of *V. cholerae*, using only two biochemical tests arginine dihydrolase and esculin hydrolysis. Confirmed *V. cholerae* strains are subsequently serogrouped as O1 or O139 by slide agglutination. Strains of *V. cholerae* that do not agglutinate with specific O1 and O139 antisera are designated as non-O1, non-O139. Typically, strains of *V. cholerae* that are non-O1, non-O139 serogroups are easily isolated, while *V. cholerae* O1 or O139 are seldom isolated from environmental water samples. For isolation and detection of *V. cholerae* from the environment, concentrated water samples or plankton samples, or both, are recommended, since a combination of both types of samples has provided higher probability of *V. cholerae* detection [22, 134, 141].

18.3.2 Colony Blot Lift and Hybridization with DNA Probes

Often, results from biochemical tests are ambiguous and several of the tests require extended incubation times. An alternative, when isolation into culture is desired, is colony blot lift and hybridization. The colony lift procedure immobilizes DNA from bacterial colonies onto nitrocellulose or nylon filters to allow quick screening of a large number of colonies for genetic elements of interest by nucleic acid hybridization. The targets that have been used include 16S–23S rDNA intergenic spacer region [142], 16S rRNA [137, 143, 144], *stn* [145], and *ctxA* [146].

Despite intensive efforts, *V. cholerae* may not be detected in samples collected from the aquatic environment for several reasons: low concentration, inter-specific competition, and physiological state (VBNC, starved). The major drawback of the conventional culture method is that it is culture dependent. Successful culturing of an individual bacterial species from air, water, or soil samples can be challenging because that species may be easily overgrown (inter-species competition), it may be fastidious in its nutrient requirements, or it may have entered the VBNC (dormant) state. Additionally, the culture method is laborious and time consuming. To eliminate these drawbacks, direct methods of detection have been developed, primarily by borrowing and modifying methods from different fields, such as immunology (direct fluorescent antibody) and molecular biology (polymerase chain reaction, PCR). Several of such alternative methods are discussed below.

18.3.3 Immunological Methods

An immunofluorescent method for detection of *V. cholerae* serovar O1 in aquatic environmental samples was first introduced by Xu et al. [147]. Antiserum specific for the O1 somatic antigen, "A" factor of *V. cholerae* O1 lipopolysaccharide, that reacts with both biotypes, Ogawa and Inaba, produced in rabbits coupled with fluorescein isothiocyanate (FITC)-conjugated anti-rabbit globulin goat serum was used to detect cells of *V. cholerae* serogroup O1. When observed under an epifluorescent microscope, cells of *V. cholerae* O1 exhibit a bright green fluorescing periphery (the outer cell wall) with a darker interior. The protocol has been simplified and commercialized as a direct fluorescent antibody staining kit for *V. cholerae* O1, Cholera DFA® (New Horizon Diagnostics, Columbia, MD), by Hasan et al. [148]. Later, a kit for *V. cholerae* serogroup O139, Bengal DFA® (New Horizon Diagnostics, Columbia, MD), was developed [149]. These immunological methods have been found to be very useful for detecting organisms in samples which give negative results by culture [134, 150].

Using these immunological methods, the mystery concerning the inability to culture *V. cholerae* in environmental samples during inter-epidemic periods in Bangladesh was solved by the discovery of the VBNC, or essentially dormant, state of *V. cholerae* [134, 151]. In an effort to aid interpretation of DFA signals and further support the assertion of VBNC, Chowdhury et al. [152] coupled the direct viable count (DVC) method of Kogure et al. [153] with the direct fluorescent antibody (DFA) method. In this two-step protocol, samples are first incubated with yeast extract in the presence of nalidixic acid for a period of 6–16 h at ambient temperature and then assayed using the DFA method. Actively viable, substrate-responsive cells become enlarged and elongated [153]. This is a rapid method by which one can determine the presence of viable *V. cholerae* within 10 h if the DVC incubation is for at least 6 h. If overnight incubation with yeast extract and nalidixic acid is performed, then the time is extended to 20 h. *Vibrio cholerae* O1 and O139 DVC–DFA-positive samples can be confirmed by PCR [142, 154].

18.3.4 Direct Detection of **Vibrio cholerae** *by PCR*

Despite the ubiquitous nature of *V. cholerae*, isolation and detection by culture methods are difficult as stated above since the classical methods for isolating bacterial species rely on cultivating the organism. The PCR offers a molecular-based alternative to traditional culture and to the immunological methods discussed above. Three types of targets are used to detect *V. cholerae* in environmental samples: species-specific genes (16S rDNA, ITS, *ompW*, collagenase); serogroup-specific genes

Table 18.1 Commonly used PCR primers for *V. cholerae* and expected amplicon size

Target	Primer	Sequence (5-3′)	Amplicon	References
ITS	pVC-F2	TTAAGCSTTTTCRCTGAGAATG	295–310	[184]
	pVCM-R1	AGTCACTTAACCATACAACCCG		
ctxA	pCTA-94F	CGGGCAGATTCTAGACCTCCTG	563	[185]
	pCTA-614R	CGATGATCTTGGAGCATTCCCAC		
toxR	pToxR-101F	CCTTCGATCCCCTAAGCAATAC	778	[186]
	pToxR-837R	AGGGTTAGCAACCGATGCGTAAG		
tcpA	pTcpA-72F	CACGATAAGAAAACCGGTCAAGAG	452 621	[187]
	pTcpAET-477R	CGAAAGCACCTTCTTTCACGTTG		
	pTcpACL-647R	TTACCAAATGCAACGCCGAATG		
zot	pZot-225F	TCGCTTAACGATGGCGCGTTTT	946	[186]
	pZot-1129R	AACCCCGTTTCACTTCTACCCA		
ompU	pOmpU-80F	ACGCTGACGGAATCAACCAAAG	868	[186]
	pOmpU-906R	GCGGAAGTTTGGCTTGAAGTAG		
ctxA	VCT1	ACAGAGTGAGTACTTTGACC	308	[154]
	VCT2	ATACCATCCATATATTTGGGAG		
O1-*rfb*	O1F2-1	GTTTCACTGAACAGATGGG	192	[154]
	O1R2-2	GGTCATCTGTAAGTACAAC		
O139-*rfb*	O139F2	AGCCTCTTTATTACGGGTGG	449	[154]
	O139R2	GTCAAACCCGATCGTAAAGG		
ompW	ompW-F	CACCAAGAAGGTGACTTTATTGTG	588	[155]
	ompW-R	GAACTTATAACCACCCGCG		
ctxA	ctxA-F	CTCAGACGGGATTTGTTAGGCACG	302	[188]
	ctxA-R	TCTATCTCTGTAGCCCCTATTACG		
16S rDNA	16S-F	CAGCMGCCGCGGTAATWC	888	[189]
	16S-R	ACGGGCGGTGTGTRC		
collagenase	VC-F	CGGCGTGGCTGGATACATTG	389	[190]
	VC-R	GTCACACTTAAATAGTAGCGT		

(O1 and O139 *rfb*); and toxin and pathogenic factor genes (*ctx, tcpA*, etc.). We present the most commonly used PCR targets/primers in Table 18.1 and it is important to mention that several multiplex PCR methods have been developed for *V. cholerae* detection. Most notable are the multiplex PCRs targeting the *wbe* regions of O1 and O139 along with *ctxA* [154] and *ompW/ctxA* [155]. The polymerase chain reaction is also a useful alternative to labor-intensive biochemical tests traditionally used for presumptive colony confirmation. Additionally, PCR is often used to characterize a confirmed *V. cholerae* strain by targeting several genes associated with pathogenesis (*ctxA, toxR, tcpA, zot, ompU*, etc.).

18.4 Prediction and Prevention of Cholera

The oceans drive the earth's climate and influence global warming, with the potential to generate major natural disasters, e.g., hurricanes, tsunamis, and El Niño-related storms. The connection between oceans, climate, and public health has been a subject of scientific interest since the last century. An improved understanding of the role of the oceans in human health is possible if several key components are included. An established historical baseline of high-quality ocean and health observations (records of disease occurrence) and the application of technologies, such as satellite sensors, oceanographic instruments, along with molecular probes for detection of the pathogen, can provide a robust predictive model for waterborne diseases like cholera. Furthermore, improved conventional bacteriological methods, as well as more sophisticated oceanographic instruments that are currently under construction, including moored biosensors and biological oceanography instruments, will be invaluable tools for future modeling work. When these technologies are used to understand the pathogen and its ecology, relationships to the ocean can be discerned in greater detail and with more powerful prediction.

Hartley et al. [156] have proposed an epidemiological model for the dynamics of cholera that explicitly included the hyperinfectious stage of *V. cholerae* once the bacterium passes through the human intestine. We are in agreement with the model only as a cause of a localized epidemic, yielding an index case responsible for contaminating the surrounding waters of a village. Our studies have focused on developing a predictive model to determine the risk of cholera in a geographic region and, thereby, preventing the occurrence of that "index" case. It is known from volunteer studies that the clinical symptoms of cholera are dose dependent, with 10^4–10^6 cells of toxigenic *V. cholerae* required to produce disease in humans [157]. Thus, accurate detection and enumeration are key for assessing risk to the human population of a geographic region where *V. cholerae* is endemic in that aquatic environment [11, 134, 136]. In one study, it was determined that an individual planktonic crustacean, namely the copepod, can carry 10^3–10^4 bacteria and that attachment was a significant factor in survival, growth, and multiplication in natural aquatic systems [158].

18.4.1 Climatological Models for Prediction

Significant progress has been achieved in a very short time in developing climate models for cholera but only after scientists of different disciplines, e.g., microbiologists, oceanographers, atmospheric scientists, biostatisticians, and mathematic modelers, worked together to meet this challenge. Occurrence of *V. cholerae* in the environment is known to be directly related to environmental conditions that are dependent on climate. Among all the parameters studied, temperature has the strongest influence on *V. cholerae* and cholera epidemics. A strong correlation between cholera outbreaks and adjacent coastal conditions was detected, using data from two outbreaks of cholera that occurred in the early 1990s [8, 30]. Early warning of El Niño events offered the opportunity to test the relationship of climate on cholera epidemics. A close association was observed between sea surface temperature in the Bay of Bengal, measured by satellite remote sensing, and cases of cholera in Matlab, Bangladesh [27, 28]. In addition, sea surface height and plankton biomass have been shown to be additional significant factors in the link between appearance of cholera in the coastal region of the Bay of Bengal and cholera epidemics in Bangladesh. This was an important finding and was based on work conducted over the past 25 years that made it possible to link the seemingly sporadic nature of cholera epidemics to climate and climate events like El Niño [159]. Warm waters along the coast influence *V. cholerae* in the water, both free-swimming and associated with zooplankton [27, 160]. Phytoplankton blooms followed by zooplankton blooms have been shown to be related to the incidence of *V. cholerae* [135, 161, 162]. Following this initial discovery, investigators began to link environmental factors and the occurrence of cholera in different parts of the world.

The main objective of climate studies is to develop a powerful model to predict infectious disease, cholera in this case, so that prevention of infectious diseases that are environmentally driven will be possible. The epidemics of cholera in Bangladesh have been associated with the end of monsoon season, occurring annually and with El Niño (ENSO) on an inter-annual frequency. In a time series analysis for the period 1980–1998, an historical association was demonstrated between cases of cholera in Bangladesh and ENSO events [163]. Effect of environmental factors has been further studied and hypothetical models proposed for the occurrence of *V. cholerae* in the environment [136, 164], providing information to build predictive models. Seasonal factors, such as rainfall and hours of sunlight, contribute directly to the physical and chemical characteristics of water which have a direct influence on plankton populations.

Studies conducted in Bangladesh, Peru, and the Chesapeake Bay, MD, reveal temperature to be a key factor associated with increased counts of *V. cholerae* [27, 28, 136, 166–168)]. Based on an extensive study conducted in Bangladesh, direct correlation between water temperature and cases of cholera in the surrounding areas was established recently [22]. According to results of the study, a 5°C increase in water temperature yielded a 3.31-fold increase in risk of cholera when a lag period of 6 weeks was used. The 95% confidence interval for the relative risk was 2.28–4.59 times higher 6 weeks after the increase of 5°C occurred. It is interesting to note that observed number of cholera cases versus the cases predicted by the Poisson regression model and upper 95% prediction limit, when water temperature was used, had a 6 weeks lag, the *ctx* gene probe count and conductivity had no time lag, and rainfall showed an 8-week lag for the pond studied in Bakerganj, Bangladesh (Fig. 18.1). Reports of other investigators also support the conclusion that temperature has a major role in the survival and multiplication of *V. cholerae* [167]. Thus, the findings of this temperature-based model are a significant development toward achieving a predictive model for cholera. Another environmental factor rainfall had an 8-week lag time before outbreak of cholera was recorded. This was interesting because this period of time is necessary to complete the cycle, in sequence, from growth of phytoplankton followed by growth of zooplankton, thereby providing the optimal

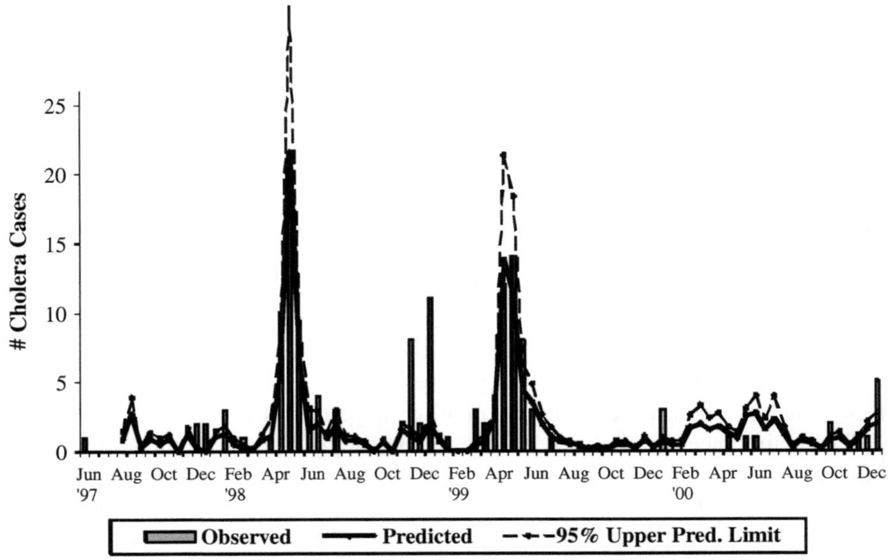

Fig. 18.1 Observed number of cholera cases vs. number of cases predicted by the Poisson regression model and upper 95% prediction limit, using water temperature (lag 6 weeks), *ctx*-gene probe count (lag 0 weeks), conductivity (lag 0 weeks), and rainfall (lag 8 weeks) for lake no. 2 in Bakerganj (Huq et al. [22]. Reproduced with permission from the publisher)

conditions necessary for the cholera bacterium to increase in the environment to the levels of an infectious dose in water routinely used for drinking as shown in many studies [168–170].

18.4.2 Simple Methods for Prevention and Intervention of Cholera

At the present time, there is no vaccine fully effective for protection from cholera. From human volunteer trials, $\sim 10^4$–10^6 *Vibrio cholerae* O1, if ingested, is likely to produce clinical cholera [157]. During the late spring and early fall of each year, phytoplankton blooms occur in Bangladesh, followed by significant zooplankton blooms, especially during September and October. Cholera outbreaks follow the zooplankton bloom [161]. The majority of the members of the zooplankton population consist of crustaceans, and these are known to support attachment of *Vibrio* [171, 172]. Furthermore, plankton demonstrates seasonal distribution, both in size of individuals and in species composition [173, 174].

Copepods, a dominant group of the zooplankton community in riverine and brackish water, have a remarkably characteristic seasonal distribution in size and species [175] and can carry a large number of *V. cholerae*, enough to cause cholera if $\sim$1–10 copepods are ingested via water [157, 176]. In addition, copepods feed on phytoplankton, with the result that *V. cholerae* attached to phytoplankton will be ingested by copepods, as demonstrated by Huq et al. [177], contributing to the *Vibrio* content of copepods. Copepods are large enough to be removed by simple filtration [177]. Filtering water at the time of collection for household use and just before drinking has been successful in removing copepods of the genus *Cyclops*, the planktonic host of the larvae of the guinea worm. The worm larvae are removed using a nylon net and this method has proven highly effective in preventing dracunculiasis, a life-threatening disease common in Africa [178]. Although boiling water prior to drinking will kill both *Cyclops* and guinea worm larvae, as well as pathogenic microorganisms, including *V. cholerae*, it is a time-consuming procedure and expensive as well. In a country like Bangladesh, where fuel wood is in very short supply, boiling the water is not followed. Furthermore, boiling water is not socially acceptable in most rural villages of Africa [178], a custom that also prevails in Bangladesh [179]. Moreover, during severe flooding, which occurs every year, there are some areas of Bangladesh that experience a reduction of conditions to those of mere survival, i.e., even the barest necessities become difficult to obtain and making fires to boil water is simply not possible.

In Bangladesh, a majority of the population in the villages still depends on untreated surface water for household consumption [179]. Surface water taken from ponds and rivers is preferred as a source of drinking water, for reasons of taste, convenience, or a local belief that "quality" water is "natural," i.e., not chemically treated [169, 179]. A family and neighborhood study of cholera transmission demonstrated that those who used *V. cholerae* O1 culture-positive water sources for cooking, bathing, or washing, but used culture-negative water for drinking, had the same rate of infection as those who used *V. cholerae* O1 culture-positive water for drinking [180]. Family-based clustering of cases is not generally observed (M. Yunus and R. Glass, unpublished data). However, there are ~14–25 "hotspot" villages in the International Center for Diarrhoeal Diseases Research Centre, Bangladesh (ICDDR,B) surveillance area, where the number of cases of cholera is higher than average. Once a case is reported in a family, the secondary case rate is approximately 10% among the remaining family members [181]. However, 29% of the family members have a risk of being colonized with *V. cholerae* O1 (M. Yunus, S. Islam, K.M.A. Aziz, personal communication). Once an index case is reported, it is most likely that further spread within the family takes place via food or other means, such as direct contact, which may not be preventable, even if water brought into the house is free of *V. cholerae* O1. Therefore, an intervention method was proposed at the index case level, which would also help prevent releasing large numbers of bacteria into the household environment when sanitary latrines are not available. The importance of safe water for all household purposes, i.e., cooking, bathing, washing and drinking, cannot be overemphasized. In addition, during monsoons, vast areas of Bangladesh are submerged by floods, resulting in a few shallow sources, where water is obtained by using hand-operated tube wells that are available to the villagers. Sanitary latrines are also flooded during monsoons, causing serious problems with hygiene and posing a direct threat of contamination and spread of enteric pathogenic bacteria, including *V. cholerae*. Based on previous studies and as described above, during late monsoons, phytoplankton blooms occur, followed by zooplankton blooms, to which vibrios are attached and multiply, amplifying the number of *V. cholerae* in the natural water of the environment [182, 166, 169, 183]. When consumption of surface water cannot be avoided, particularly during flooding or other natural disasters, which occur every year in Bangladesh, a simple method that is effective in reducing the number of *V. cholerae* is practical and useful. It is a common practice in villages in Bangladesh to use cloth, frequently a piece of an old sari cloth, to filter homemade drinks, usually containing sugar/molasses and served with rose water or lemon juice for flavoring, to guests. There is no ethical, cultural, or social barrier to using sari cloth to filter drinking water. Therefore, simple filtration is readily acceptable to villagers in their homes and used without reservation. This method was successfully introduced, as described below.

Based on the environmental data, a simple method of filtration that was effective in removing 99% of the cells of *V. cholerae* (notably those attached to plankton) was developed [177]. This method employed sari cloth folded 4–8 times to filter natural water for household purposes including drinking. This simple filtration method was deployed by demonstrating, teaching, and educating village women to place the folded sari over the mouth of a water collecting pot, filtering the water as it entered the pot. This sari filter method was successful when employed in a field trial carried out in Matlab, Bangladesh. The method proved effective in a 2-year study, reducing the number of cases of cholera by approximately 50% in the study population of 45,000 villagers in 65 villages in Matlab, Bangladesh. One of the questions of the study was whether the villagers would accept and sustain the practice of filtration. In a recent follow-up survey, it was determined that indeed the method is sustainable, most importantly, effective in reducing the number of cases of cholera in Matlab, Bangladesh [191].

18.5 Summary

It is clear from the numerous studies cited in this chapter that interdisciplinary studies made it possible to gain important understanding of the environmental habitat of *V. cholerae* and how the organism is transmitted to humans, causing disease. A significant step was the application of satellite remote sensing to enhance understanding of the ecology of this organism. This technology has allowed investigators to realize the potential for reliable prediction and, ultimately, prevention of the deadly disease cholera. Although cholera is the focus of this chapter, it provides an excellent example of how ecological understanding of this pathogenic microorganism whose natural habitat is the aquatic environment can provide new avenues of research for addressing those human pathogens present in the aquatic environment.

Acknowledgments The authors acknowledge support received from the following grants: National Institutes of Health grant #1RO1A139129-01, Thrasher Research Fund grant #028–219, and NGIA grant # 0000. CJG was supported by an Intelligence Community Postdoctoral Research Fellowship.

References

1. World Health Organization. Report on global surveillance of epidemic prone infectious diseases, communicable diseases and surveillance response. WHO; 2000. Report No. WHO/CDS/CSR/ISR/2000.
2. Barua D, Merson, MH. Prevention and control of cholera. In: Barua D, Greenough III, WB, editors. Cholera. New York, NY: Plenum; 1992. pp. 329–49.
3. Barua D, Greenough WB, III, eds. Cholera. New York: Plenum Medical Book Company; 1992.
4. Barua D. Laboratory diagnosis of cholera. In: Barua D, Burrows W, editors. Cholera: W.B. Saunders Company; 1974. pp. 85–126.
5. Pacini F. Observazioni microscopiche e deduzioni patologiche sul cholera asiatico. Tipografia Bencini. 1854;6.
6. Rosenberg CE. The cholera years. Chicago, IL: The University of Chicago Press; 1962.
7. WHO. Cholera, 2006. WHO Weekly Epidemiol Records. 2006;31:273–84.
8. Tauxe R, Seminario L, Tapia R, Libel M. The Latin American epidemic. In: Wachsmuth I, Blake P, Olsvik O, editors. *Vibrio cholerae* and cholera: molecular to global perspectives. Washington, DC: ASM; 1994. pp. 321–44.
9. Blake P. Endemic cholera in Australia and the United States. In: Wachsmuth I, Blake P, Olsvik O, editors. *Vibrio cholerae* and cholera: molecular to global perspectives. Washington, DC: American Society for Microbiology; 1994. pp. 309–20.
10. Anonymous. Detection of *Vibrio cholerae* O1 in the Georges River estuary-New South Wales. Commun Dis Intell Bull. 1981;81:2.
11. Colwell RR, Seidler RJ, Kaper J, Joseph SW, Garges S, Lockman H, Maneval D, Bradford H, Roberts N, Remmers E, Huq I, Huq A. Occurrence of *Vibrio cholerae* serotype O1 in Maryland and Louisiana estuaries. Appl Environ Micbrobiol. 1981;41:555–8.
12. West PA, Lee JV. Ecology of *Vibrio* spp. including *Vibrio cholerae* in natural waters of Kent, England. J Appl Bacteriol. 1982;52:435–48.
13. Lee J, Bashford D, Donovan T, Furniss A, West PA. The incidence and distribution of *Vibrio cholerae* in England. In: Colwell RR, editor. Vibrios in the environment. New York: Wiley; 1984. pp. 427–50.
14. Colwell RR. Polyphasic taxonomy of the genus *Vibrio*: numerical taxonomy of *Vibrio cholerae, Vibrio parahaemolyticus*, and related *Vibrio* species. J Bacteriol. 1970;104: 410–33.
15. Citarella R, Colwell RR. Polyphasic taxonomy of the genus *Vibrio*: polynucleotide sequence relationships among selected *Vibrio* species. J Bacteriol. 1970;104:434–42.
16. Colwell RR, Huq A, Chowdhury M, Brayton P, Xu B. Serogroup conversion of *Vibrio cholerae*. Can J Microbiol. 1995;41:946–50.
17. Waldor M, Mekalanos J. ToxR regulates virulence gene expression in non-O1 strains of *Vibrio cholerae* that cause epidemic cholera. Infect Immunol. 1994;62:72–8.
18. Waldor MK, Mekalanos JJ. Lysogenic conversion by a filamentous phage encoding cholera toxin. Science. 1996;272:1910–14.
19. Glass RI, Becker S, Huq MI, Stoll BJ, Khan MU, Merson MH, Lee JV, Black RE. Endemic cholera in rural Bangladesh, 1966–1980. Am J Epidemiol. 1982;116:959–70.
20. Siddique AK, Zaman K, Baqui AH, Akram K, Mutsuddy P, Eusof A, Haider K, Islam S, Sack RB. Cholera epidemics in Bangladesh: 1985–1991. J Diarrhoeal Dis Res. 1992;10:79–86.
21. Swerdlow DL, Mintz ED, Rodriguez M, Tejada E, Ocampo C, Espejo L, Greene KD, Saldana W, Seminario L, Tauxe RV, et al. Waterborne transmission of epidemic cholera in Trujillo, Peru: lessons for a continent at risk. Lancet. 1992;340:28–32.
22. Huq A, Sack RB, Nizam A, Longini IM, Nair GB, Ali A, Morris JG Jr, Khan MN, Siddique AK, Yunus M, Albert MJ, Sack DA, Colwell RR. Critical factors influencing the occurrence of *Vibrio cholerae* in the environment of Bangladesh. Appl Environ Microbiol. 2005;71:4645–54.

23. World Health Organization. Cholera, 2005. Weekly Epidemiol Record. 2006;31:297–308.
24. World Health Organization. Cholera factsheet No. 107. World Health Organization; 2007.
25. An introductory overview of climate change. (Accessed at http://www.dec.ny.gov/docs/remediation_hudson_pdf/hvcc0930ad.pdf.)
26. Griffith DC, Kelly-Hope LA, Miller MA. Review of reported cholera outbreaks worldwide, 1995–2005. Am J Trop Med Hyg. 2006;75:973–7.
27. Colwell RR. Global climate and infectious disease: the cholera paradigm. Science. 1996;274:2025–31.
28. Lobitz B, Beck L, Huq A, Wood B, Fuchs G, Faruque AS, Colwell RR. Climate and infectious disease: use of remote sensing for detection of *Vibrio cholerae* by indirect measurement. Proc Natl Acad Sci. 2000;97:1438–43.
29. Glass RI, Black RE. The epidemiology of cholera. In: Barua D, Greenough WB, editors. Cholera. New York: Plenum Medical Book Company; 1992. pp. 129–54.
30. Ramamurthy T, Garg S, Sharma R, Bhattacharya SK, Nair GB, Shimada T, Takeda T, Karasawa T, Kurazano H, Pal A, Takeda Y. Emergence of novel strains of *Vibrio cholerae* with epidemic potential in southern and eastern India. Lancet. 1993;341:703–4.
31. Cholera Working Group I, B. Large epidemic of cholera-like disease in Bangladesh caused by *Vibrio cholerae* O139 synonym Bengal. Lancet. 1993;342:387–90.
32. Colwell RR, Kaper J, Joseph SW. *Vibrio cholerae, Vibrio parahaemolyticus*, and other vibrios: occurrence and distribution in Chesapeake Bay. Science. 1977;198:394–6.
33. Singleton FL, Attwell R, Jangi S, Colwell RR. Effects of temperature and salinity on *Vibrio cholerae* growth. Appl Environ Microbiol. 1982;44:1047–58.
34. Singleton FL, Attwell RW, Jangi MS, Colwell RR. Influence of salinity and organic nutrient concentration on survival and growth of *Vibrio cholerae* in aquatic microcosms. Appl Environ Microbiol. 1982;43:1080–5.
35. Hood MA, Ness GE, Rodrick GE, Blake NJ. Distribution of *Vibrio cholerae* in two Florida estuaries. Microb Ecol. 1983;9:65–75.
36. Miller CJ, Drasar BS, Feachem RG. Response of toxigenic *Vibrio cholerae* O1 to physico-chemical stresses in aquatic environments. J Hyg (Lond). 1984;93:475–95.
37. Colwell RR, Spira WM. The ecology of *Vibrio cholerae*. In: Barua D, Greenough III, WB, editors. Cholera. New York, NY: Plenum; 1992. pp. 107–27.
38. Heidelberg JF, Eisen JA, Nelson WC Clayton RA, Gwinn ML, Dodson RJ, Haft DH, Hickey EK, Peterson JD, Umayam L, Gill SR, Nelson KE, Read TD, Tettelin H, Richardson D, Ermolaeva MD, Vamathevan J, Bass S, Qin H, Dragoi I, Sellers P, McDonald L, Utterback T, Fleishmann RD, Nierman WC, White O. DNA sequence of both chromosomes of the cholera pathogen *Vibrio cholerae*. Nature. 2000;406:477–83.
39. Roszak DB, Colwell RR. Survival strategies of bacteria in the natural environment. Microbiol Rev. 1987;51:365–79.
40. Huq A, Rivera I, Colwell RR. Epidemiological Significance of Viable but non-culturable microorganisms. In: Colwell RR, Grimes D, editors. Non-culturable microorganisms in the environment. Washington, DC: ASM Press; 2000. pp. 301–23.
41. Chaiyanan S, Grim C, Maugel T, Huq A, Colwell RR. Ultrastructure of coccoid viable but non-culturable *Vibrio cholerae*. Environ Microbiol. 2007;9:393–402.
42. Novitsky JA, Morita RY. Morphological characterization of small cells resulting from nutrient starvation of a psychrophilic marine *Vibrio*. Appl Environ Microbiol. 1976;32: 617–22.
43. Novitsky JA, Morita RY. Survival of a psychrophilic marine *Vibrio* under long-term nutrient starvation. Appl Environ Microbiol. 1977;33:635–41.
44. Colwell RR, Huq, A. Vibrios in the environment: viable but non-culturable *Vibrio cholerae*. In: Wachsmuth IK, Blake PA, Olsvik O, editors. *Vibrio cholerae* and cholera: molecular to global perspectives. Washington, DC: ASM Press; 1994.
45. Rollins DM, Colwell RR. Viable but non-culturable stage of *Campylobacter jejuni* and its role in survival in the natural aquatic environment. Appl Environ Microbiol. 1986;52:531–8.

46. Linder K, Oliver JD. Membrane fatty acid and virulence changes in the viable but nonculturable state of *Vibrio vulnificus*. Appl Environ Microbiol. 1989;55:2837–42.
47. Colwell RR. Bacterial death revisited. In: Colwell RR, Grimes D, editors. Nonculturable microorganisms in the environment. Washington, DC: ASM Press; 2000.
48. Asakura H, Ishiwa A, Arakawa E, Makino S, Okada Y, Yamamoto S, Igimi S. Gene expression profile of *Vibrio cholerae* in the cold stress-induced viable but non-culturable state. Environ Microbiol. 2006;9:869–79.
49. Oliver JD. Formation of viable but nonculturable cells. In: Kjelleberg S, editor. Starvation in bacteria. New York, NY: Plenum Press; 1993. pp. 239–72.
50. Wai SN, Moriya T, Kondo K, Misumi H, Amako K. Resuscitation of *Vibrio cholerae* O1 strain TSI-4 from a viable but nonculturable state by heat shock. FEMS Microbiol Lett. 1996;136:187–91.
51. Weichart D, Oliver JD, Kjelleberg S. Low temperature induced non-culturability and killing of *Vibrio vulnificus*. FEMS Microbiol Lett. 1992;79:205–10.
52. Ravel J, Knight IT, Monahan CE, Hill RT, Colwell RR. Temperature-induced recovery of *Vibrio cholerae* from the viable but nonculturable state: growth or resuscitation? Microbiology. 1995;141(Pt 2):377–83.
53. Colwell RR, Brayton PR, Grimes DJ, Roszak DR, Huq SA, Palmer LM. Viable, but nonculturable *Vibrio cholerae* and related pathogens in the environment: implications for release of genetically engineered microorganisms. Bio-Technology. 1985;3:817–20.
54. Huq A, Hasan JA, Losonsky G, Diomin V, Colwell RR. Colonization of professional divers by toxigenic *Vibrio cholerae* O1 and *V. cholerae* non-O1 at dive sites in the United States, Ukraine and Russia. FEMS Microbiol Lett. 1994;120:137–42.
55. Costerton JW, Cheng KJ, Geesey GG, Ladd TI, Nickel JC, Dasgupta M, Marrie TJ. Bacterial biofilms in nature and disease. Annu Rev Microbiol. 1987;41:435–64.
56. Stewart PS, Roe F, Rayner J, Elkins JG, Lewandowski Z, Ochsner UA, Hassett DJ. Effect of catalase on hydrogen peroxide penetration into *Pseudomonas aeruginosa* biofilms. Appl Environ Microbiol. 2000;66:836–8.
57. Elkins JG, Hassett DJ, Stewart PS, Schweizer HP, McDermott TR. Protective role of catalase in *Pseudomonas aeruginosa* biofilm resistance to hydrogen peroxide. Appl Environ Microbiol. 1999;65:4594–600.
58. Harrison JJ, Rabiei M, Turner RJ, Badry EA, Sproule KM, Ceri H. Metal resistance in *Candida* biofilms. FEMS Microbiol Ecol. 2006;55:479–91.
59. Teitzel GM, Parsek MR. Heavy metal resistance of biofilm and planktonic *Pseudomonas aeruginosa*. Appl Environ Microbiol. 2003;69:2313–20.
60. Schembri MA, Kjaergaard K, Klemm P. Global gene expression in *Escherichia coli* biofilms. Mol Microbiol. 2003;48:253–67.
61. Vorachit M, Lam K, Jayanetra P, Costerton JW. Resistance of *Pseudomonas pseudomallei* growing as a biofilm on silastic discs to ceftazidime and co-trimoxazole. Antimicrob Agents Chemother. 1993;37:2000–2.
62. Hoyle BD, Costerton JW. Bacterial resistance to antibiotics: the role of biofilms. Prog Drug Res. 1991;37:91–105.
63. Nichols WW, Evans MJ, Slack MP, Walmsley HL. The penetration of antibiotics into aggregates of mucoid and non-mucoid *Pseudomonas aeruginosa*. J Gen Microbiol. 1989;135:1291–303.
64. Nickel JC, Wright JB, Ruseska I, Marrie TJ, Whitfield C, Costerton JW. Antibiotic resistance of *Pseudomonas aeruginosa* colonizing a urinary catheter in vitro. Eur J Clin Microbiol. 1985;4:213–8.
65. Ketyi I. Resistance of *Escherichia coli* to some antibiotics and biocides in the intestinal biofilm of mice. Acta Microbiol Hung. 1991;38:33–41.
66. Espeland EM, Wetzel RG. Complexation, stabilization, and UV photolysis of extracellular and surface-bound glucosidase and alkaline phosphatase: implications for biofilm microbiota. Microb Ecol. 2001;42:572–85.

67. Matz C, McDougald D, Moreno AM, Yung PY, Yildiz FH, Kjelleberg S. Biofilm formation and phenotypic variation enhance predation-driven persistence of *Vibrio cholerae*. Proc Natl Acad Sci USA. 2005;102:16819–24.
68. Huq A, Small EB, West PA, Huq MI, Rahman R, Colwell RR. Ecological relationships between *Vibrio cholerae* and planktonic crustacean copepods. Appl Environ Microbiol. 1983;45:275–83.
69. Spira WM, Huq A, Ahmed QS, Saeed YA. Uptake of *Vibrio cholerae* biotype El Tor from contaminated water by water hyacinth (*Eichornia crassipes*). Appl Environ Microbiol. 1981;42:550–3.
70. Shukla BN, Singh DV, Sanyal SC. Attachment of non-culturable toxigenic *Vibrio cholerae* O1 and non-O1 and *Aeromonas* spp. to the aquatic arthropod *Gerris spinolae* and plants in the River Ganga, Varanasi. FEMS Immunol Med Microbiol. 1995;12:113–20.
71. Castro-Rosas J, Escartin EF. Adhesion and colonization of *Vibrio cholerae* O1 on shrimp and crab carapaces. J Food Prot. 2002;65:492–8.
72. Broza M, Halpern M. Pathogen reservoirs. Chironomid egg masses and *Vibrio cholerae*. Nature. 2001;412:40.
73. Halpern M, Broza YB, Mittler S, Arakawa E, Broza M. Chironomid egg masses as a natural reservoir of *Vibrio cholerae* non-O1 and non-O139 in freshwater habitats. Microb Ecol. 2004;47:341–9.
74. Islam MS, Drasar BS, Bradley DJ. Attachment of toxigenic *Vibrio cholerae* O1 to various freshwater plants and survival with a filamentous green alga, *Rhizoclonium fontanum*. J Trop Med Hyg. 1989;92:396–401.
75. Watnick PI, Kolter R. Steps in the development of a *Vibrio cholerae* El Tor biofilm. Mol Microbiol. 1999;34:586–95.
76. Costerton JW, Lewandowski Z, Caldwell DE, Korber DR, Lappin-Scott HM. Microbial biofilms. Annu Rev Microbiol. 1995;49:711–45.
77. Mueller RS, McDougald D, Cusumano D, Sodhi N, Kjelleberg S, Azam F, Bartlett DH. *Vibrio cholerae* strains possess multiple strategies for abiotic and biotic surface colonization. J Bacteriol. 2007;189:5348–60.
78. Chiavelli DA, Marsh JW, Taylor RK. The mannose-sensitive hemagglutinin of *Vibrio cholerae* promotes adherence to zooplankton. Appl Environ Microbiol. 2001;67:3220–5.
79. Watnick PI, Fullner KJ, Kolter R. A role for the mannose-sensitive hemagglutinin in biofilm formation by *Vibrio cholerae* El Tor. J Bacteriol. 1999;181:3606–9.
80. Finkelstein RA, Mukerjee S. Hemagglutination: a rapid method for differentiating *Vibrio cholerae* and El Tor vibrios. Proc Soc Exp Biol Med. 1963;112:355–9.
81. Meibom KL, Li XB, Nielsen AT, Wu CY, Roseman S, Schoolnik GK. The *Vibrio cholerae* chitin utilization program. Proc Natl Acad Sci USA. 2004;101:2524–9.
82. Reguera G, Kolter R. Virulence and the environment: a novel role for *Vibrio cholerae* toxin-coregulated pili in biofilm formation on chitin. J Bacteriol. 2005;187:3551–5.
83. Hall-Stoodley L, Stoodley P. Biofilm formation and dispersal and the transmission of human pathogens. Trends Microbiol. 2005;13:7–10.
84. Alam M, Sultana M, Nair GB, Siddique AK, Hasan NA, Sack RB, Sack DA, Ahmed KU, Sadique A, Watanabe H, Grim CJ, Huq A, Colwell RR. Viable but nonculturable *Vibrio cholerae* O1 in biofilms in the aquatic environment and their role in cholera transmission. Proc Natl Acad Sci. 2007;104:17801–6.
85. Merrell DS, Butler SM, Qadri F, Dolganov NA, Alam A, Cohen MB, Calderwood SB, Schoolnik GK, Camilli A. Host-induced epidemic spread of the cholera bacterium. Nature. 2002;417:642–5.
86. Faruque SM, Biswas K, Udden SM, Ahmad QS, Sack DA, Nair GB, Mekalanos JJ. Transmissibility of cholera: in vivo-formed biofilms and their relationship to infectivity and persistence in the environment. Proc Natl Acad Sci. 2006;103:6350–5.
87. Nalin DR, Daya V, Reid A, Levine MM, Cisneros L. Adsorption and growth of *Vibrio cholerae* on chitin. Infect Immunol. 1979;25:768–70.

88. Lankford CE. Factors of virulence of *Vibrio cholerae*. Ann NY Acad Sci. 1960;88:1203–12.
89. Holmes RK, Vasil ML, Finkelstein RA. Studies on toxinogenesis in *Vibrio cholerae*. III. Characterization of non-toxinogenic mutants in vitro and in experimental animals. J Clin Invest. 1975;55:551–60.
90. Finkelstein RA, Boesman-Finkelstein M, Chang Y, Hase CC. *Vibrio cholerae* hemagglutinin/protease, colonial variation, virulence, and detachment. Infect Immunol. 1992;60: 472–8.
91. Yildiz FH, Schoolnik GK. *Vibrio cholerae* O1 El Tor: identification of a gene cluster required for the rugose colony type, exopolysaccharide production, chlorine resistance, and biofilm formation. Proc Natl Acad Sci. 1999;96:4028–33.
92. Finkelstein RA, Boesman-Finkelstein M, Sengupta DK, Page WJ, Stanley CM, Phillips TE. Colonial opacity variations among the choleragenic vibrios. Microbiology. 1997;143(Pt 1):23–34.
93. White BP. The rugose variant of vibrios. J Pathol. 1938;46:1–6.
94. Morris JG Jr, Sztein MB, Rice EW, Nataro JP, Losonsky GA, Panigrahi P, Tacket CO, Johnson JA. *Vibrio cholerae* O1 can assume a chlorine-resistant rugose survival form that is virulent for humans. J Infect Dis. 1996;174:1364–8.
95. Wai SN, Mizunoe Y, Takade A, Kawabata SI, Yoshida SI. *Vibrio cholerae* O1 strain TSI-4 produces the exopolysaccharide materials that determine colony morphology, stress resistance, and biofilm formation. Appl Environ Microbiol. 1998;64:3648–55.
96. Yildiz FH, Liu XS, Heydorn A, Schoolnik GK. Molecular analysis of rugosity in a *Vibrio cholerae* O1 El Tor phase variant. Mol Microbiol. 2004;53:497–515.
97. Rice EW, Johnson CJ, Clark RM, Fox KR, Reasoner DJ, Dunnigan ME, Panigrahi P, Johnson JA, Morris JG Jr. Chlorine and survival of "rugose" *Vibrio cholerae*. Lancet. 1992;340(8821):740.
98. Beyhan S, Yildiz FH. Smooth to rugose phase variation in *Vibrio cholerae* can be mediated by a single nucleotide change that targets c-di-GMP signalling pathway. Mol Microbiol. 2007;63:995–1007.
99. Yildiz FH, Dolganov NA, Schoolnik GK. VpsR, a member of the response regulators of the two-component regulatory systems, is required for expression of *vps* biosynthesis genes and EPS(ETr)-associated phenotypes in *Vibrio cholerae* O1 El Tor. J Bacteriol. 2001;183: 1716–26.
100. Casper-Lindley C, Yildiz FH. VpsT is a transcriptional regulator required for expression of *vps* biosynthesis genes and the development of rugose colonial morphology in *Vibrio cholerae* O1 El Tor. J Bacteriol. 2004;186:1574–8.
101. Fong JC, Karplus K, Schoolnik GK, Yildiz FH. Identification and characterization of RbmA, a novel protein required for the development of rugose colony morphology and biofilm structure in *Vibrio cholerae*. J Bacteriol. 2006;188:1049–59.
102. Fong JC, Yildiz FH. The *rbmBCDEF* gene cluster modulates development of rugose colony morphology and biofilm formation in *Vibrio cholerae*. J Bacteriol. 2007;189: 2319–30.
103. Hastings JW, Greenberg E. Quorum sensing: the explanation of a curious phenomenon reveals a common characteristic of bacteria. J Bacteriol. 1999;181:2667–8.
104. Nealson K, Platt T, Hastings J. Cellular control of the synthesis and activity of the bacterial luminescence system. J Bacteriol. 1970;104:313–22.
105. de Kievit TR, Iglewski BH. Bacterial quorum sensing in pathogenic relationships. Infect Immunol. 2000;68:4839–49.
106. Eberhard A. Inhibition and activation of bacterial luciferase synthesis. J Bacteriol. 1972;109:1101–5.
107. Nealson KH. Autoinduction of bacterial luciferase. Occurrence, mechanism and significance. Arch Microbiol. 1977;112:73–9.
108. Kaplan HB, Greenberg EP. Diffusion of autoinducer is involved in regulation of the *Vibrio fischeri* luminescence system. J Bacteriol. 1985;163:1210–4.

109. Bassler BL, Wright M, Silverman MR. Multiple signalling systems controlling expression of luminescence in *Vibrio harveyi*: sequence and function of genes encoding a second sensory pathway. Mol Microbiol. 1994;13:273–86.
110. Bassler BL, Wright M, Showalter RE, Silverman MR. Intercellular signalling in *Vibrio harveyi*: sequence and function of genes regulating expression of luminescence. Mol Microbiol. 1993;9:773–86.
111. Chen X, Schauder S, Potier N, Van Dorsselaer A, Pelczer I, Bassler BL, Hughson FM. Structural identification of a bacterial quorum-sensing signal containing boron. Nature. 2002;415:545–9.
112. Freeman JA, Bassler BL. A genetic analysis of the function of LuxO, a two-component response regulator involved in quorum sensing in *Vibrio harveyi*. Mol Microbiol. 1999;31:665–7.
113. Swartzman E, Silverman M, Meighen EA. The *luxR* gene product of *Vibrio harveyi* is a transcriptional activator of the *lux* promoter. J Bacteriol. 1992;174:7490–3.
114. Lin YH, Miyamoto C, Meighen EA. Cloning and functional studies of a *luxO* regulator LuxT from *Vibrio harveyi*. Biochim Biophys Acta. 2000;1494:226–35.
115. Bassler BL, Greenberg EP, Stevens AM. Cross-species induction of luminescence in the quorum-sensing bacterium *Vibrio harveyi*. J Bacteriol. 1997;179:4043–5.
116. Jobling MG, Holmes RK. Characterization of *hapR*, a positive regulator of the *Vibrio cholerae* HA/protease gene *hap*, and its identification as a functional homologue of the *Vibrio harveyi luxR* gene. Mol Microbiol. 1997;26:1023–34.
117. Miller MB, Skorupski K, Lenz DH, Taylor RK, Bassler BL. Parallel quorum sensing systems converge to regulate virulence in *Vibrio cholerae*. Cell. 2002;110:303–14.
118. West PA, Brayton PR, Bryant TN, Colwell RR. Numerical taxonomy of vibrios isolated from aquatic environments. Int J Sys Bacteriol. 1986;36:531–43.
119. Zhu J, Miller MB, Vance RE, Dziejman M, Bassler BL, Mekalanos JJ. Quorum-sensing regulators control virulence gene expression in *Vibrio cholerae*. Proc Natl Acad Sci. 2002;99:3129–34.
120. Joelsson A, Kan B, Zhu J. Quorum sensing enhances the stress response in *Vibrio cholerae*. Appl Environ Microbiol. 2007;73:3742–6.
121. Camara M, Hardman A, Williams P, Milton D. Quorum sensing in *Vibrio cholerae*. Nat Genet. 2002;32:217–8.
122. Holt JG, Krieg NR, Sneath PHA, Staley JT, Williams ST. Bergey's manual of determinative bacteriology, 9th ed. Baltimore, MD: Williams & Wilkins; 1994.
123. Colwell RR, Huq A. Environmental reservoir of *Vibrio cholerae*. The causative agent of cholera. Ann NY Acad Sci. 1994;740:44–54.
124. Keyhani NO, Roseman S. Physiological aspects of chitin catabolism in marine bacteria. Biochim Biophys Acta. 1999;1473:108–22.
125. Meibom KL, Blokesch M, Dolganov NA, Wu CY, Schoolnik GK. Chitin induces natural competence in *Vibrio cholerae*. Science. 2005;310:1824–7.
126. Rowbotham TJ. Preliminary report on the pathogenicity of *Legionella pneumophila* for freshwater and soil amoebae. J Clin Pathol. 1980;33:1179–83.
127. Rowbotham TJ. Isolation of *Legionella pneumophila* from clinical specimens via amoebae, and the interaction of those and other isolates with amoebae. J Clin Pathol. 1983;36:978–86.
128. Rowbotham TJ. Current views on the relationships between amoebae, legionellae and man. Isr J Med Sci. 1986;22:678–89.
129. Marciano-Cabral F, Cabral G. *Acanthamoeba* spp. as agents of disease in humans. Clin Microbiol Rev. 2003;16:273–307.
130. Abd H, Weintraub A, Sandstrom G. Intracellular survival and replication of *Vibrio cholerae* O139 in aquatic free-living amoebae. Environ Microbiol. 2005;7:1003–8.
131. Abd H, Saeed A, Weintraub A, Nair GB, Sandstrom G. *Vibrio cholerae* O1 strains are facultative intracellular bacteria, able to survive and multiply symbiotically inside the aquatic free-living amoeba *Acanthamoeba castellanii*. FEMS Microbiol Ecol. 2007;60:33–9.

132. Thom S, Warhurst D, Drasar BS. Association of *Vibrio cholerae* with fresh water amoebae. J Med Microbiol. 1992;36:303–6.
133. Abd H, Johansson T, Golovliov I, Sandstrom G, Forsman M. Survival and growth of *Francisella tularensis* in *Acanthamoeba castellanii.* Appl Environ Microbiol. 2003;69: 600–6.
134. Huq A, Colwell RR, Rahman R, Ali A, Chowdhury MA, Parveen S, Sack DA, Russek-Cohen E. Detection of *Vibrio cholerae* O1 in the aquatic environment by fluorescent monoclonal antibody and culture method. Appl Environ Microbiol. 1990;56:2370–3.
135. Louis V, Choopun N, Rivera ING, et al. Detection of *Vibrio cholerae* in the aquatic environment by culture methods, bacterial plate count/hybridization, and direct fluorescence assay to evaluate its ecology and public health risk. In: ASM Annual Conference, Los Angeles, CA; 2000.
136. Louis VR, Russek-Cohen E, Choopun N, Rivera IN, Gangle B, Jiang SC, Rubin A, Patz JA, Huq A, Colwell RR. Predictability of *Vibrio cholerae* in Chesapeake Bay. Appl Environ Microbiol. 2003;69:2773–85.
137. Huq A, Grim C, Colwell RR, Nair GB. Detection, isolation, and identification of *Vibrio cholerae* from the environment. In: Coico R, Kowalik T, Quarles J, Stevenson B, Taylor R, editors. Current protocols in microbiology. New York: Wiley; 2006.
138. Farmer JJ III, Hickman-Brenner, FW, Kelly, MT. *Vibrio*. In: Balows A, Hausler WJ Jr., Shadomy HJ, editors. Manual of clinical microbiology. Washington, DC: ASM Press; 1970. pp. 282–301.
139. Monsur KA. A highly selective gelatin–taurocholate–tellurite medium for the isolation of *Vibrio cholerae*. Trans R Soc Trop Med Hyg. 1961;55:440–2.
140. Choopun N, Louis V, Huq A, Colwell RR. Simple procedure for rapid identification of *Vibrio cholerae* from the aquatic environment. Appl Environ Microbiol. 2002;68: 995–8.
141. Binsztein N, Costagliola MC, Pichel M, Jurquiza V, Ramírez FC, Akselman R, Vacchino M, Huq A, Colwell RR. Viable but nonculturable *Vibrio cholerae* O1 in the aquatic environment of Argentina. Appl Environ Microbiol. 2004;70:7481–6.
142. Jiang SC, Fu W. Seasonal abundance and distribution of *Vibrio cholerae* in coastal waters quantified by a 16S–23S intergenic spacer probe. Microb Ecol. 2001;42:540–8.
143. Heidelberg JF, Heidelberg KB, Colwell RR. Seasonality of Chesapeake Bay bacterioplankton species. Appl Environ Microbiol. 2002;68:5488–97.
144. Heidelberg JF, Heidelberg KB, Colwell RR. Bacteria of the gamma-subclass Proteobacteria associated with zooplankton in Chesapeake Bay. Appl Environ Microbiol. 2002;68: 5498–507.
145. Pal A, Ramamurthy T, Bhadra RK, Takeda T, Shimada T, Takeda Y, Nair GB, Pal SC, Chakrabarti S. Reassessment of the prevalence of heat-stable enterotoxin (NAG-ST) among environmental *Vibrio cholerae* non-O1 strains isolated from Calcutta, India, by using a NAG-ST DNA probe. Appl Environ Microbiol. 1992;58:2485–9.
146. Franco AA, Fix AD, Prada A, Paredes E, Palomino JC, Wright AC, Johnson JA, McCarter R, Guerra H, Morris JG Jr. Cholera in Lima, Peru, correlates with prior isolation of *Vibrio cholerae* from the environment. Am J Epidemiol. 1997;146:1067–75.
147. Xu HS, Roberts N, Singleton FL, Attwell RW, Grimes DJ, Colwell RR. An indirect fluorescent antibody staining procedure for detection of *Vibrio cholerae* serovar O1 cells in aquatic environmental samples. J Microb Methods. 1984;2:221–31.
148. Hasan J, Bernstein D, Huq A, Loomis L, Tamplin M, Colwell RR. Cholera DFA: an improved direct fluorescent monoclonal antibody staining kit for rapid detection and enumeration of *Vibrio cholerae* O1. FEMS Microbiol Lett. 1994;120:143–8.
149. Hasan JA, Huq A, Nair GB, Garg S, Mukhopadhyay AK, Loomis L, Bernstein D, Colwell RR. Development and testing of monoclonal antibody-based rapid immunodiagnostic test kits for direct detection of *Vibrio cholerae* O139 synonym Bengal. J Clin Microbiol. 1995;33:2935–9.

150. Brayton PR, Roszak D, Palmer LM, Huq SA, Grimes DJ, Colwell RR. Fluorescent antibody enumeration of *Vibrio cholerae* in the marine environment. IFREMER. 1986;3: 507–14.
151. Roszak D, Colwell RR. Survival strategies of bacteria in the natural environment. Microbiol Rev. 1987;51:365–79.
152. Chowdhury MAR, Hasan JAK, Huq A, Xu B, Montilla R. DVC–DFA: a simplified technique for detection of viable *Vibrio cholerae* O1 and O139. Can J Microbiol. 1994;42:87–93.
153. Kogure K, Simidu U, Taga N. A tentative direct microscopic method for counting living marine bacteria. Can J Microbiol. 1979;25:415–20.
154. Hoshino K, Yamasaki S, Mukhopadhyay AK, Chakraborty S, Basu A, Bhattacharya SK, Nair GB, Shimada T, Takeda Y. Development and evaluation of a multiplex PCR assay for rapid detection of toxigenic *Vibrio cholerae* O1 and O139. FEMS Immunol Med Microbiol. 1998;20:201–7.
155. Nandi B, Nandy RK, Mukhopadhyay S, Nair GB, Shimada T, Ghose AC. Rapid method for species-specific identification of *Vibrio cholerae* using primers targeted to the gene of outer membrane protein OmpW. J Clin Microbiol. 2000;38:4145–51.
156. Hartley DM, Morris JG Jr, Smith DL. Hyperinfectivity: a critical element in the ability of *Vibrio cholerae* to cause epidemics? PLoS Med. 2006;3(1):e7.
157. Cash R, Music S, Libonati J, Snyder M, Wenzel R, Hornick R. Response of man to infection with *Vibrio cholerae* I. Clinical, serologic, and bacteriologic responses to a known inoculum. J Infect Dis. 1974;129:45–52.
158. Huq A. The role of planktonic copepods in the survival and multiplication of *Vibrio cholerae* in the aquatic environment. College Park, MD: University of Maryland; 1984.
159. Watkins JD, Huq A. Relationship between oceans and human health. In: Koop CE, Pearson CE, Schwarz MR, editors. Critical issues in global health. San Francisco: Jossey-Bass, A Wiley Company; 2001. pp.356–63.
160. Mourino-Perez RR. Oceanography and the seventh cholera pandemic. Epidemiology. 1998;9:355–7.
161. Huq A, Small EB, West PA, Huq MI, Rahman R, Colwell RR. Ecological relationship between *Vibrio cholerae* and planktonic copepods. Appl Environ Microbiol. 1983;45: 275–83.
162. Tamplin M, Gauzens A, Huq A, Sack D, Colwell RR. Attachment of *Vibrio cholerae* serogroup O1 to zooplankton and phytoplankton of Bangladesh waters. Appl Environ Miobiol. 1990;56:1977–80.
163. Pascual M, Rodo X, Ellner SP, Colwell RR, Bouma MJ. Cholera dynamics and El Nino–Southern Oscillation. Science. 2000;289:1766–9.
164. Colwell RR, Huq A. Environmental reservoir of *Vibrio cholerae*. The causative agent of cholera. In: Wilson M, Levins R, Spielman A, editors. Disease in evolution: global changes and emergence of infectious diseases. New York: New York Academy of Sciences; 1994. pp. 44–54.
165. Franco AA, Fix AD, Prada A, Paredes E, Palomino JC, Wright AC, Johnson JA, McCarter R, Guerra H, Morris JG Jr. Cholera in Lima, Peru correlates with prior isolation of *Vibrio cholerae* from the environment. Am J Epidemiol. 1997;146:1067–75.
166. Huq A, West PA, Small EB, Huq MI, Colwell RR. Influence of water temperature, salinity and pH on survival and growth of toxigenic *Vibrio cholerae* serovar O1 associated with live copepods in laboratory microcosm. Appl Environ Microbiol. 1984;48:420–4.
167. Pascual M, Bouma M, Dobson A. Cholera and climate: revisiting the quantitative evidence. Microbes Infect. 2002;4:237–45.
168. Huq A, Chowdhury M, Felsenstein A, Colwell RR, Rahman R, Hossain K. Detection of *Vibrio cholerae* from aquatic environments in Bangladesh. In: Yasuno M, Whitton BA, editors. Biological monitoring of environment pollution. Tokyo: Tokai University; 1988. pp. 259–64.
169. Colwell R, Huq A. Vibrios in the environment: Viable but non-culturable *Vibrio cholerae*. In: Wachsmuth I, Blake P, Olsvik O, editors. *Vibrio cholerae* and cholera: molecular to global perspectives. Washington, DC: ASM Press; 1994. pp. 117–33.

170. Lipp EK, Huq A, Colwell RR. Effects of global climate on infectious disease: the cholera model. Clin Microbiol Rev. 2002;15:757–70.
171. Sochard MR, Wilson DF, Austin B, Colwell RR. Bacteria associated with the surface and gut of marine copepods. Appl Environ Microbiol. 1979;37:750–9.
172. Kaneko T, Colwell RR. Adsorption of *Vibrio parahaemolyticus* onto chitin and zooplanktonic copepods. Appl Microbiol. 1975;29:269–74.
173. Kiorboe T, Neilson TJ. Regulation of zooplankton biomass and production and production in a temperate coastal ecosystem. 1. Copepods. Limnol Oceanogr. 1994;39.
174. McCarthy SA, Khambaty FM. International dissemination of epidemic *Vibrio cholerae* by cargo ship ballast and other non-potable waters. Appl Environ Microbiol. 1994;60: 2597–601.
175. Sverdrup HV, Johnson MW, Fleming RH. The oceans. Englewood Cliffs, NJ: Prentice Hall; 1947.
176. Colwell RR, Tamplin ML, Brayton PR, et al. Environmental aspects of *Vibrio cholerae* in transmission of cholera. In: Sack RB, Zinnaka Y, editors. Advances in research on cholera and related diarrhoeas. Tokyo: K.T.K. Scientific Publishers; 1990. pp. 327–43.
177. Huq A, Xu B, Chowdhury M, Islam M, Montilla R, Colwell RR. A simple filtration method to remove plankton-associated *Vibrio cholerae* in raw water supplies in developing countries. Appl Environ Microbiol. 1996;62:2508–12.
178. WASH. Orientation to Guinea worm disease: a guide for use in pre-service and in-service training. Sponsored by US-AID; 1991.
179. Briscoe J. The role of water supply in improving health in poor countries, with special reference to Bangladesh. Am J Clin Nutr. 1978;31:2100–13.
180. Hughes JM, Boyce JM, Levine RJ, Khan M, Aziz KM, Huq MI, Curlin GT. Epidemiology of El Tor cholera in rural Bangladesh: importance of surface water in transmission. WHO Bull. 1982;60:395–404.
181. Glass RI, Holmgren J, Khan MR, Hossain KMB, Huq MI, Greenough WB. A randomized control trial of the toxin-blocking effects of B submitted in family members of patients with cholera. J Infect Dis. 1984;149:495–500.
182. Cabrini M, Fonda Umani S, Honsell G. Mucilaginous aggregates in the Gulf of Trieste (Northern Adriatic Sea) analysis of the phytoplanktonic communities in the period June–August 1989.Sci Total Environ Suppl. 1992:557–68.
183. Islam MS, Draser BS, Bradley DJ. Attachment of toxigenic *Vibrio cholerae* O1 to various freshwater plants and survival with filamentous green algae *Rhizoclonium fontanum*. J Trop Med Hyg. 1989;92:396–401.
184. Chun J, Huq A, Colwell RR. Analysis of 16S–23S rRNA intergenic spacer regions of *Vibrio cholerae* and *Vibrio mimicus*. Appl Environ Microbiol. 1999;65:2202–8.
185. Fields PI, Popovic T, Wachsmuth K, Olsvik O.. Use of polymerase chain reaction for detection of toxigenic *Vibrio cholerae* strains from the Latin American cholera epidemic. J Clin Microbiol. 1992;30:2118–21.
186. Rivera ING, Chun J, Huq A, Sack RB, Colwell RR. Genotypes associated with virulence in environmental isolates of *Vibrio cholerae*. Appl Environ Microbiol. 2001;67:2421–9.
187. Keasler S, Hall R. Detecting and biotyping *Vibrio cholerae* O1 with multiplex polymerase chain reaction. Lancet. 1993;341:1661.
188. Shirai H, Nishibuchi M, Ramamurthy T, Bhattacharya SK, Pal SC, Takeda Y. Polymerase chain reaction for detection of cholera enterotoxin operon of *Vibrio cholerae*. J Clin Microbiol. 1991;29:2517–21.
189. Amann R, Ludweg W, Schleifer K. Phylogenetic identification and *in situ* detection of individual microbial cells without cultivation. Microbiol Rev. 1995;59:143–69.
190. Di Pinto A, Ciccarese G, Tantillo G, Catalano D, Forte VT. A collagenase-targeted multiplex PCR assay for identification of *Vibrio alginolyticus, Vibrio cholerae,* and *Vibrio parahaemolyticus*. J Food Prot. 2005;68:150–3.
191. Huq A, Yunus M, et al. Simple Sari Cloth Filtration of Water is Sustainable and Continues to Protect Villagers from Cholera in Matlab, Bangladesh. mBio. 2010;1(1):1–5.

Chapter 19
Management of Cholera

P. Dutta, D. Sur, and S.K. Bhattacharya

Abstract Cholera is caused by *Vibrio cholerae* O1 and O139. Sporadic cholera cases usually occur throughout the year in endemic areas. Epidemics of cholera are also observed. This disease is transmitted mostly by faecal-oral route through contaminated water. Clinically, cholera is characterized by uncontrolled purging of watery stool leading to life-threatening dehydration, hypovolaemic shock, acidosis, and if left untreated, it can also lead to death. Asymptomatic carriers may also play a role in the spread of infection. Management of cholera without dehydration is aimed to prevent dehydration by giving more fluid than usual in the form of home-available fluids or oral rehydration salt (ORS) solution. Patient with mild dehydration can be managed with ORS solution for correction of initial dehydration and severely dehydrated patient should be treated with intravenous Ringer's lactate solution for correction of initial dehydration. After correction of initial dehydration, patient should receive "ORS" as maintenance therapy till diarrhoea stops. Antibiotics are adjunct to fluid therapy, which is recommended for the treatment of severely dehydrated cholera cases only. Oral tetracycline, azithromycin, single-dose doxycycline, and quinolone groups of drugs are usually recommended for the treatment of cholera. However, a good percentage of *V. cholerae* strains became resistant to these drugs, hence erythromycin is recommended. Breastfeeding should not be stopped during management of cholera. Food should be reintroduced after correction of initial dehydration.

19.1 Background

Cholera is the dreaded ancient disease characterized by dehydrating acute watery diarrhea and vomiting caused by *Vibrio cholerae* [1]. The history of cholera has four distinct phases. Initially (prior to 1817) the disease was confined to the eastern part of the world. This was followed by six pandemics (1817–1923) of which

S.K. Bhattacharya (✉)
Indian Council of Medical Research, New Delhi 110029, India
e-mail: sujitkbhattacharya@yahoo.com

T. Ramamurthy, S.K. Bhattacharya (eds.), *Epidemiological and Molecular Aspects on Cholera*, Infectious Disease, DOI 10.1007/978-1-60327-265-0_19,
© Springer Science+Business Media, LLC 2011

five had originated from India. In the third phase, cholera again retreated to Asia (1923–1960). Finally in 1961, cholera El Tor biotype which was isolated at El Tor quarantine station in Egypt and localized to Sulawesi (Celebes) in Indonesia for decades spread worldwide causing the seventh pandemic which is still continuing.

The causative organism of cholera was predominantly *V. cholerae* O1 El Tor; in 1992, a cholera epidemic in Chennai, India, revealed a new organism designated as *V. cholerae* O139 Bengal [2–5]. Cholera is characterized by uncontrolled purging of watery stools leading to life-threatening dehydration, hypovolemic shock, acidosis, and if left untreated, it can also lead to death. *Vibrio cholerae* is classified into two major serogroups, cholera toxin (CT)-producing serogroups, which include *V. cholerae* O1 [6, 7] and O139 [8, 9], and non-CT-producing serogroups that are collectively known as *V. cholerae* non-O1, non-O139 and include serogroups O2 to O206. *Vibrio cholerae* O1 is further subdivided into two biotypes (classical and El Tor) and two serotypes (Ogawa and Inaba). The CT-producing strains (O1 and O139) are capable of causing epidemics of cholera.

Cholera may usually be confirmed by isolating these organisms from stool or rectal swab samples. However, typical cases of cholera do not require confirmation of diagnosis by laboratory examination. Any patient develops severe dehydration from acute watery diarrhea in an area where cholera is endemic may be diagnosed as cholera [10]. Patients with mild diarrhea and very young children may also suffer from cholera but that illness needs to be differentiated from other causes of diarrhea by laboratory investigations. Laboratory confirmation of cholera is not needed for case management but it is essential for the public health administrators to control the epidemic of cholera. Cholera may occur as sporadic, epidemic and pandemic forms.

During 2005 a total of 131,943 confirmed cholera cases have been notified to World Health Organization (WHO) including 2,272 deaths from 52 countries worldwide [11]. However, actual number of cholera cases is expected to be much higher; this discrepancy is due to underreporting and other limitations of surveillance systems in most of the developing countries. Many of these cases are labeled as acute watery diarrhea in spite of having WHO-recommended clinical case definition.

Man is the only known reservoir of cholera [12]. Cholera is transmitted from man to man by fecal–oral route through environment. Like that of other infectious diseases, cholera also has characteristic seasonal pattern which varies from place to place. In Kolkata, India, though sporadic cholera cases usually occur throughout the year, it increases in number in hot summer and early winter months. However, in Bangladesh, it is common in rainy season [13].

Two types of epidemics are usually observed: (i) explosive form (common source epidemic where large number of cases appears in a short period of time) and (ii) protracted form (small number of cases per day or week over several weeks). Contamination of groundwater sources and damaged pipelines of drinking water often result in explosive outbreak of cholera [14]. Stagnant water in low lying areas sometimes becomes contaminated with indiscriminate defecation or overflowing of sewerage system which may also cause explosive epidemic [15, 16]. In endemic areas, protracted pattern of epidemic is more common. Spread of this infection usually occurs from asymptomatic individuals or mild cases through contaminated food or drinks. Person to person transmission is not so common. Cholera is common in low socioeconomic groups due to poor hygienic status and poor sanitation. Adult men are affected more at the time of epidemic due to greater mobility but in endemic areas, children are affected more than adults as the population develop natural immunity [17].

Cholera can range from asymptomatic infection to mild diarrhea or severe diarrhea with life-threatening dehydration. It is now well recognized that *Vibrios* actually produce many more asymptomatic and mild cases than severe cases [17]. Roughly 10% cases become severe and require hospitalization which represents tip of the iceberg. Diarrhea and vomiting are the main presentations of cholera [17]. In initial phase, diarrhea is usually mild in nature, which consists of fecal matter and watery stool. Majority of these cases remain at this stage which are indistinguishable from diarrhea caused by other enteropathogens and also remain untreated [17]. They can be diagnosed only by laboratory examination. They do not have clinical importance but they are important epidemiologically as they maintain bacterial load to the community. However, some patients have abrupt onset of watery diarrhea and vomiting which become severe with the passage of painless and effortless rice-watery stool (resembling discarded white water after washing rice) without any fecal matter.

After ingestion of *V. cholerae*, the organisms multiply in the small intestine [18, 19] and produce a potent enterotoxin (cholera toxin). With the influence of cholera toxin in the small intestine, there is increased chloride and bicarbonate secretion in the crypts which is also associated with an increase in passive transfer of water across the mucous surface in the gut lumen [20, 21]. On the other hand, cholera toxin inhibits the absorption of sodium chloride and water by villous cells. Ultimate effect is accumulation of huge amount of fluid and electrolytes in the lumen of small intestine which is beyond the absorption capacity of the colon [21]. Accumulated fluid and electrolytes pass out as profuse watery diarrhea in cholera. Cholera patient loses a significant amount of electrolytes (sodium, potassium, chloride, bicarbonate) through liquid stool. This loss of fluid and electrolytes from body leads to dehydration which is the main cause of death due to cholera. Cholera stool is isotonic to plasma with low protein content. Cholera stool contains electrolytes with following concentrations (mmol/l): sodium 90–140, potassium 20–35, chloride 90–100 and bicarbonate 30–45. Sodium and chloride concentrations of cholera stool are roughly proportionate to that of plasma; however, concentration of potassium is twofold and bicarbonate is fourfold higher than plasma [22–26].

19.2 Clinical Presentations

Before development of clinical signs and symptoms of dehydration, patient is usually alert with moist lips, mouth, tongue, normal eyes and skin elasticity which is termed as 'no dehydration'. At this stage patient may lose body fluid less than 5% of body weight. Clinical signs and symptoms of dehydration develop when loss of

Table 19.1 Assessment of patient for dehydration

Look at	Condition	Well, alert	Restless, irritable	Lethargic or unconscious; floppy
	Eyes	Normal	Sunken	Very sunken and dry
	Tears	Present	Absent	Absent
	Mouth and tongue	Moist	Dry	Very dry
	Thirst	Drinks normally, not thirsty	Thirsty, drinks eagerly	Drinks poorly or not able to drink
Feel	Skin pinch	Goes back quickly	Goes back slowly	Goes back very slowly
Decide		The patient has *no signs of dehydration*	If the patient has two or more signs, there is *some dehydration*	If the patient has two or more signs, there is *severe dehydration*

Adapted from WHO Manual for Treatment of Diarrhoea WHO/CDD/95.3

body fluid is more than 5% of body weight. Dehydration can be classified clinically as 'some' or 'severe' according to the loss of body fluid which can be assessed by detecting the clinical signs and symptoms following the guidelines of WHO [10] as depicted in Table 19.1. Patient with 'some' dehydration usually presents with thirst, irritability, sunken eyes, dry mouth, lips and tongue and sign of loss of skin elasticity (pinched abdominal skin retracts back slowly). Loss of body fluid is equivalent to 5–9% of body weight.

Patient suffering from 'severe' dehydration becomes lethargic, floppy, or comatose. Patient also has very sunken eye, very dry mouth, lips, and tongue. Sign of loss of skin elasticity is very much prominent (pinched abdominal skin retracts back very slowly; it takes more than 3 s to be normal). Pulse volume and systolic blood pressure may be low and sometimes undetectable. Peripheral body temperature in arms and legs may be cold. Cyanosis may occur in the finger tips, lips, and tongue. Some patients complain cramps in the hands and feet. Wrinkling of the skin of the fingers (washerwomen's hands) and hoarseness of voice may be present. Respiratory rate may be rapid; sometimes respiration may be deep with raising chest wall and Kussmaul type of breathing due to base-deficit acidosis. Severely dehydrated patient loses fluid equivalent to 10% or more of body weight.

Laboratory finding shows marked saline depletion, hemoconcentration, and increased plasma-specific gravity [27, 28]. Acidosis with low blood pH and low bicarbonate content is detected. Serum potassium, sodium, and chloride levels are within normal limit. Some children develop additional symptoms like altered sensorium, seizures, and fever. Occurrence of seizures is due to marked hypoglycemia caused by prolonged starvation before initiation of treatment.

19.3 Management

Management of cholera involves immediate replacement and maintenance of fluid and electrolyte losses due to diarrhea and vomiting. This can be achieved by administration of adequate and proper fluid in the form of either oral rehydration therapy or intravenous therapy depending upon the degree of dehydration. Effective management with adequate fluid therapy reduces death rate to less than 1% in all age groups [10, 29].

19.3.1 Management of Patients with 'No' Dehydration

Management of cholera without dehydration is aimed to prevent dehydration by giving more fluid than normal in the form of home-available fluids like salt–sugar solution, rice water, weak tea, etc., preferably from beginning of onset of illness [10, 29]. Oral rehydration salt (ORS) solution dissolved in appropriate amount of drinking water can also be used if available at that time. Any of these fluids should be given approximately 50–100 ml and 100–200 ml per loose stool for the children below 2 years and between 2 and 10 years of age, respectively. The amount should be given by mouth slowly, one to two teaspoonfuls every 1–2 min. If patient has repeated vomiting, administration of fluid should be stopped for 10–15 min and should be reintroduced later. Older children and adults should drink fluid as much as they want.

19.3.2 Management of Patient with 'Some' Dehydration

Patient with 'some' dehydration can be managed with ORS solution recommended by WHO in two phases. Firstly, existing loss of body fluid and electrolytes should be replaced by administration of ORS as quick as possible in adults and older children. In smaller children, fluid deficit should be replaced in a dose of 75 ml/kg of body weight within 4–6 h. Fluid can be administered by cup and spoon. Use of feeding bottle should be discouraged primarily because feeding bottles are often contaminated and secondly bottle feeding of ORS may detract the programme of breastfeeding promotion in some countries. After that period, child should be reassessed for hydration status. If the patient is still in 'some' dehydration, same amount of fluid should be continued for another cycle of 4–6 h and it should be continued till the patient becomes fully hydrated. When the hydration status of the patient becomes normal, administration of ORS should be continued matching the stool loss as maintenance therapy till the diarrhea stops. Management of patient with 'some' dehydration is preferred at hospital set-up but patient should not be kept in hospital throughout the period of illness. Patient can be discharged after correction of initial dehydration by providing few packets of ORS and advising to return back to the hospital if diarrhea worsens or if signs of dehydration reappear. Management of 'some' dehydration can also be done at the community level by direct supervision of health workers who are experienced to detect further deterioration of the condition by watching for the signs and symptoms of dehydration and other complications. If the patient develops signs and symptoms of severe dehydration or stoppage of urine for more than 6 h or becomes unconscious, it is advisable to refer the patient to nearby health center where facilities of intravenous fluid administration are available.

19.3.3 Oral Rehydration Salt (ORS) Solution

Discovery of glucose-enhanced absorption of sodium and water in the gut during cholera leads to development of the oral rehydration salt (ORS) solution. This discovery is considered as one of the most important advancements in medicine in the last century. ORS having sodium concentration of 90 mmol/l and glucose 111 mmol/l (total osmolarity 311 mmol/l) has been used for last four decades as cheap and effective therapy for prevention and treatment of dehydration caused by acute watery diarrhea including cholera [30–34]. However, this fluid has some demerits, as it fails to reduce duration, frequency, and volume of diarrhea. Furthermore, sometimes it may produce hypernatremia in very young and severely malnourished children. Recently several clinical trials have shown that ORS solution having low osmolarity (total osmolarity 245 mmol/l by reducing sodium and glucose concentrations) is more effective than ORS of osmolarity of 311 mmol/l for the management of acute watery diarrhea of all etiologies including cholera [35–38]. This fluid therapy has the capacity to reduce stool output, duration, and frequency of diarrhea in children suffering from cholera. It also reduces vomiting and need of unscheduled intravenous fluid therapy. This solution is also effective for the management of adult cholera; however, sometimes it may produce transient asymptomatic hyponatremia. WHO and UNICEF jointly recommended that the single ORS solution containing 75 mmol/l of sodium and 75 mmol/l of glucose with a total osmolarity of 245 (reduced ORS or hypo-osmolar ORS) should be used as single ORS solution throughout the world for the management of all diarrhea cases including cholera and in all age groups [39]. This new formula (Table 19.2) was launched by WHO and its partners during United Nation General Assembly, special session of children on May 2002.

Table 19.2 Composition of oral rehydration salts (ORS)

Composition by weight			Composition osmolarity		
Ingredients	Old ORS (g)	Hypo ORS (g)		Old ORS (mmol/l)	Hypo ORS (mmol/l)
Sodium chloride	3.5	2.6	Sodium	90	75
Potassium chloride	1.5	1.5	Potassium	20	20
Trisodium citrate dihydrate	2.9	2.9	Chloride	80	65
Glucose, anhydrous	20	13.5	Glucose	111	75
Total	27.9	20.5	Total mosmol	311	245

19.3.4 Management of Patient with Severe Dehydration

Severely dehydrated patient should be treated with intravenous fluid for correction of initial dehydration. Ringer's lactate solution (mEq/l; sodium 130, chloride 109, potassium 4, lactate 28. osmolality 271) is the preferred fluid as it contains adequate amount of sodium, chloride, some potassium and citrate (which is converted to bicarbonate in liver and corrects base-deficit acidosis), and most commonly available commercially produced solution [40, 41]. If Ringer's lactate solution is not available, normal saline (sodium chloride solution 8.9 g/l) may be used [42] but it contains neither bicarbonate nor potassium for which simultaneous administration of ORS by mouth to replace potassium and bicarbonate losses is advisable. There was a controversy of the use of large amount of potassium during rehydration phase as the patient does not pass urine for a considerable period. Presently, it is documented that absence of urine production is not contraindicated for administration of potassium in the initial rehydration therapy; rather, without administration of potassium, patient may develop severe hypokalemia. Patients should be assessed for hydration status very frequently (every half an hour). Rate of infusion of intravenous fluid is depicted in Table 19.3.

Accesses to the veins in hands are sometimes difficult due to vascular collapse in severely dehydrated patients. Anticubital or external jugular veins may alternatively be selected for infusion. In small children, scalp vein needle may be used in scalp. If venipuncture fails, it is better to have surgical exposure of vein (cut down) by a trained doctor or nursing staff.

After correction of initial dehydration by intravenous fluid, patient should receive 'ORS' as maintenance therapy till the diarrhea stops. Use of ORS in maintenance phase should be sufficient to replace ongoing losses from diarrhea, vomiting, and normal insensible fluid losses. Insensible losses should be calculated as being 100 ml/kg of first 10 kg of body weight, 50 ml/kg of body weight for next 10 kg of body weight, and 10 ml/kg of body weight over 20 kg. Frequent monitoring of stool loss and fluid intake is essential which may prevent reappearance of severe dehydration. Monitoring of ongoing fluid losses can be easily done by collecting the stool into a calibrated bucket. Use of 'cholera cot' is the most suitable method.

Table 19.3 Guidelines for intravenous treatment of children and adults with severe dehydration

Age	First give 30 ml/kg in	Then give 70 ml/kg in
Infants (under 12 months)	1 h	5 h
Older	30 min	2½ h

- Reassess the patient every 1–2 h. If hydration is not improving, give i.v. drip more rapidly
- After 6 h (infants) or 3 h (older patient) evaluate for dehydration status, then choose the treatment plan

Adopted from WHO Manual for Physician and other Senior Health Workers (WHO/CDD/SER/95.3)

'Cholera cot' is a simple plastic-covered cot with a central hole through which stool drains into the bucket. Administration of intravenous fluid during maintenance phase should be avoided unless the patient has severe purging, protracted vomiting, or reappearance of the signs and symptoms of severe dehydration. If intake of ORS is not sufficient to match the stool loss, intravenous fluid should be restarted.

19.3.5 Management with Antibiotic

Antibiotic, as an adjunct to fluid therapy, is recommended for the treatment of severely dehydrated cholera cases only which helps in the reduction of stool volume, shortening of hospital stay, and duration of excretion of *V. cholerae*. Oral tetracycline is the drug of choice for the treatment of cholera [43]. It can be used in adults (500 mg) and also in pediatric cases (12.5 mg/kg) four times a day for 3 days. Tetracycline is thought to be contraindicated for pediatric population because it may cause staining of permanent teeth; however, it does not harm much as it is used in short period of time. Single dose of doxycycline (300 mg) is also useful for adults [44]. Furazolidone, at the dose of 1.25 mg/kg in four divided doses, is recommended as an alternative to tetracycline for pediatric patients. However, a good percentage of *V. cholerae* O1 and O139 isolated from different parts of the world showed resistance to this drug. In such situation, erythromycin is recommended at the dose 12.5 mg/kg in four divided doses for 3 days [2]. Higher quinolones like norfloxacin (400 mg twice daily), ciprofloxacin (500 mg twice daily), ofloxacin (200 mg twice daily) for 3 days are also useful for adult cases [10, 45].

Selection of antibiotic for management of cholera becomes difficult as many of the *V. cholerae* strains in different parts of the world become resistant to commonly used antibiotics [46–48]. Tetracycline-resistant strains are also detected frequently from India and many other developing countries [49–53]. Recently, a good percentage of *V. cholerae* strains isolated from India become resistant to quinolone group of drugs [54, 55]. Selection of antibiotic for cholera should be according to the susceptibility pattern of local strains of *V. cholerae*.

19.3.6 Management of Vomiting

Vomiting is a commonly associated symptom in cholera [56]. Vomiting can be safely managed by proper correction of dehydration by administration of adequate amount of 'ORS' or intravenous fluid. Furthermore, use of hypo-osmolar ORS reduces the chance of vomiting. If patient suffers from repeated vomiting (more than four times per hour), even after proper correction of dehydration, intravenous fluid should be continued as maintenance therapy. Routine use of anti-emetic is discouraged as it may cause sedation which interferes administration of ORS. It may also give false impression on dehydration status. However, it can be used for severe and protracted vomiting even after continuation of intravenous fluid. Domperidone is the safest anti-emetic. It can be used at the dose of 0.1–0.3 mg/kg/dose once only and it should not be used repeatedly [10].

19.4 Feeding

Breastfeeding should not be stopped. Mothers should be encouraged to continue uninterrupted breastfeeding throughout the treatment period. However, other food should be temporarily withdrawn during correction of initial dehydration in both 'some' and 'severely' dehydrated children. Food should be reintroduced after correction of initial dehydration. Children should continue formula or other milk feed if they are used to have it before diarrhea but it should not be diluted [10].

Children older than 6 months and who are already habituated to take cereal- and vegetable-based foods should be continued to take the same. Food should be easily digestible, culturally acceptable and readily available and contain high energy and micronutrients. To make the food energy dense, 5–10 ml vegetable should be added to these foods during serving [10]. Meat, fish, or egg should not be withheld if habituated to have these items. Potassium-rich fruits like bananas, green coconut water, and fresh fruit juice are beneficial. One extra energy-rich meal is recommended after cessation of diarrhea to catch up growth in children.

19.5 Other Drugs or Agents

Some of the antidiarrheal drugs are commonly used even for the treatment of cholera but they are not useful; rather sometimes they harm the patients [10]. Adsorbents like kaolin and pectin are commonly believed to have the property of binding and inactivating bacterial toxins but they do not have the proven efficacy. It is also claimed that anti-motility drugs like loperamide, opiate, and opium derivatives have the capacity to reduce frequency and volume of stool by decreasing intestinal motility. However, clinical trials fail to document efficacy. Rather, sometimes they produce paralytic ileus in children and may cause prolonged illness by delaying elimination of pathogens. Steroids, cardiac stimulants, and vasoactive drugs like adrenaline and nicotinamide are also used for the treatment of shock in severely dehydrated cholera patients but these drugs have no role as shock is due to loss of fluid (hypovolemia). Prompt replacement of fluid and electrolytes is ideal treatment of this shock. Purgatives are also used sometimes but they should never be used as they make the disease worse. Recently zinc and probiotics have shown promising results for the treatment of diarrhea in children; however, they are not evaluated for the management of cholera. Some studies showed that racecadotril, an antisecretory agent, is effective in acute secretary diarrhea in children but that drug needs to be evaluated in cholera.

19.6 Complications

Severely dehydrated cholera patients sometimes develop acute renal failure with tubular necrosis if they suffer from hypovolemic shock for prolonged period or do not receive adequate amount of intravenous fluid for correction of initial dehydration [57, 58]. However, the clinical condition is reversible with judicious fluid management. Presently, acute renal failure is the rarest complication as most of the patients receive considerable amount of fluid therapy. Clinically significant hypokalemia is the problem of children suffering from cholera. Hypokalemic children may present with cardiac and gastrointestinal consequences like atonic bowel, cardiac arrhythmia, hypotension, and ultimately to cardiac arrest if adequate amount of potassium is not replaced. These complications are also very rare nowadays as most of the children receive considerable amount of potassium in their replacement fluid. Septicemia is not common in cholera infection [59] but some children develop septicemia and that may be due to the use of contaminated intravenous fluid or infusion sets. Several studies on early twentieth century showed that death rate among cholera-infected pregnant women was much higher than that among non-pregnant women [60]. Presently, judicious use of fluid therapy reduces the mortality rate.

References

1. Barua D. History of cholera. In: Barua D, Greenough WB, III, editors. Cholera. New York: Plenum Medical Book Company; 1992. pp. 1–36.
2. Ramamurthy T, Garg S, Sharma R, Bhattacharya SK, Nair GB, Shimada T, Takeda T, Karasawa T, Kurazano H, Pal A, Takeda Y. Emergence of novel strain of *Vibrio cholerae* with epidemic potential in southern and eastern India. Lancet. 1993;341:703–4.
3. Bhattacharya MK, Bhattacharya SK, Garg S, Saha PK, Nair GB, Deb BC. Outbreaks of *Vibrio cholerae* non-O1 in India and Bangladesh. Lancet. 1993;341:1346–7.
4. Bhattacharya SK, Bhattacharya MK, Nair GB, Deb BC, Ramamurthy T, et al. Clinical profile of acute diarrhea cases infected with the epidemic strains of *Vibrio cholerae* O139: designation of diseases as cholera. J Infect. 1993;27:11–5.
5. Albert MJ, Siddique AK, Islam MS, Faruque ASG, Ansaruzzaman M, Faraque SM, Sack RB. Large outbreak of clinical cholera due to *Vibrio cholerae* non-O1 in Bangladesh. Lancet. 1993;341:704.
6. Baumann P, Furniss AL, Lee JV. Genus 1, *Vibrio* P. In: Krieg NR, Holt JG, editors. Bergey's manual of systematic bacteriology, vol. 1. Baltimore: Williams and Wilkins; 1984. pp. 518–38.
7. Sakazaki R, Donovan TJ. Serology and epidemiology of *Vibrio cholerae* and *Vibrio mimicus*. In: Bergan T, editor. Methods in microbial, vol. 16. London: Academia Press; 1985. pp. 271–89.
8. Morris GJ Jr. and Cholera Laboratory Task Force *Vibrio cholerae* O139 Bengal. In: Wachsmuth IK, Blake PA, Olsvik O, editors. *Vibrio cholerae* and cholera molecular to global perspective. Washington, DC: ASM Press; 1994. pp. 95–102.
9. Nair GB, Albert MJ, Shimada T, Takeda Y. *Vibrio cholerae* O139 Bengal the serogroup causing cholera. Rev Med Microbiol. 1996;7:43–51.
10. World Health Organization: The treatment of diarrhoea: a manual for physicians and other senior health workers. Geneva: WHO. WHO/CDR/95.3; 1995.
11. World Health Organization. Weekly Epidemiol Rec. 2006;81:297–308.
12. Barrett, TJ, Blake PA. Epidemiological usefulness of changes in hemolytic activity of *Vibrio cholerae* biotype El Tor during the seventh pandemic. J Clin Microbiol. 1981;13:126–9.
13. Principles and practice of cholera control. World Health Organization public health paper, No. 40, Geneva; 1970.
14. Sirkar BK, Deb BC, Sengupta PG, Mondal S, De SP, Sen D, Saha MR, Ghosh S, Sikdar SN, Pal SC. Longitudinal study of diarrhea among children in Calcutta communities. Indian J Med Res. 1984;80:546–50.

15. Deb BC, De SP, Pal SC. Study of an extensive outbreak of cholera in Cooch Behar during 1974. Indian J Med Res. 1979;70:691–6.
16. Sur D, Sengupta PG, Mondal SK, Dutta P, Gupta DN, Ghosh S, Pandey A, Ramamurty T, Bhattacharya SK. A localized outbreak of *Vibrio cholerae* O139 in Kolkata, West Bengal. Indian J Med Res. 2002;115:149–52.
17. World Health Organization. Guidelines for cholera control, revised ed. Report WHO/CDD/SER/80.4 REV. 4. Geneva: World Health Organization; 1992.
18. Owen RL, Pierce NF, Apple RT, Cray WC Jr. M cell transport of *Vibrio cholerae* from intestinal lumen into Peyer's patches: a mechanism for antigen sampling and for microbial transepithelial migration. J Infect Dis. 1986;153:1108–18.
19. Gangarosa EJ, Beisel WR, Benyajati C, Sprinz H, Piyaratn P. The nature of the gastrointestinal lesion in Asiatic cholera and its relation to pathogenesis. A biopsy study. Am J Trop Med Hyg. 1960;9:125–35.
20. Field M, Fromm D, Al-Awqati Q, Greenough III WB. Effects of cholera enterotoxin on ion transport across isolated ileal mucosa. J Clin Invest. 1972;51:796–804.
21. Field M, Rao MC, Chan EB. Intestinal electrolyte transport and diarrheal diseases. N Engl J Med. 1989;321:800–6, 879–83.
22. Griffith LSC, Fresh JW, Watten RH, Villaroman MP. Electrolyte replacement in paediatric cholera. Lancet. 1967;I:1197–9.
23. Mahalanabis D, Wallace CK, Kallen RJ, Mondal A, Pierce NF. Water and electrolyte losses due to cholera in infants and small children: a recovery balance study. Pediatrics. 1970;45:374–85.
24. Molla AM, Rahaman M, Sarker SA, Sack DA, Molla A. 1981. Stool electrolyte content and purging rates in diarrhea caused by rotavirus, enterotoxigenic *E. coli*, and *V. cholerae* in children. J Pediatr. 1981;98:825–38.
25. Phillips RA. Water and electrolyte losses in cholera. Fed Proc. 1964;23:610–3.
26. Watten RH, Morgan FM, Songkhla YN, Vankiati B, Phillips RA. Water and electrolyte studies in cholera. J Clin Invest. 1959;38:1879–89.
27. Love AHG, Phillips RA. Measurement of dehydration in cholera. J Infect Dis. 1969; 119:39–42.
28. Dutta P, Dutta D, De S, De SP, Sen D, Pal SC. Clinical gradation of severity in hospitalised cholera cases due to ElTor biotype. Indian J Med Res. 1978;68:896–900.
29. Deb BC, Bhattacharya SK, Pal SC. Epidemiology of cholera its treatment and control. Cur Sci. 1990;59:702–7.
30. Mahalanabis D. Oral rehydration therapy among Bangladeshi refugees. John Hopkins Med J. 1973;132:197–205.
31. Nalin DR, Cash RA, Islam R, Molla M, Phillips RA. Oral maintenance therapy for cholera in adults. Lancet. 1968;ii:370–3.
32. World Health Organization. Diarrhoeal Diseases Control Programme – Report of the first meeting of the Scientific Working Group on drug development and management of acute diarrhoea, Geneva, 15–18 September 1980. WHO/CDD/DPM/81.2.
33. Dutta P, Bhattacharya SK, Dutta D, Mitra U, Bhattacharya MK, Rasaily R, Sen D, Saha MR, Mukherjee A, Pal SC. Oral dehydration solution containing 90 mmol sodium is safe and useful in treating diarrhoea in severely malnourished children. J Diarr Dis Res. 1991;9:118–22.
34. Dutta P, Dutta D, Bhattacharya SK, Bhattacharya MK, Sinha AK, Monal BC, Pal SC. Comparative efficacy of three different oral dehydration solutions for the treatment of dehydrating diarrhoea in children. Indian J Med Res. 1988;87:229–33.
35. Faruque ASG, Mahalanabis D, Hamadini J, Zetter Strom R. Reduced osmolarity oral rehydration salt in cholera. Scand J Infect Dis. 1996;28:87–90.
36. Dutta P, Mitra U, Manna B, Niyogi SK, Roy K, Mondal C, Bhattacharya SK. Double blind, randomized controlled clinical trial of hypo-osmolar oral rehydration salts solution in dehydrating acute diarrhoea in severely malnourished (marasmic) children. Arch Dis Child. 2001;84:237–40.

37. Rautanen T, El-Ridhi S, Vesikari T. Clinical experience with a hypotonic oral rehydration solution in acute diarrhoea. Acta Pediatr. 1993;82:52–4.
38. International Study Group on Reduced Osmolarity ORS Solutions. Multicentre evaluation of reduced – osmolarity oral rehydration salt solution. Lancet. 1995;345:282–5.
39. Reduced osmolarity oral rehydration salts (ORS) formulation – report from a meeting of experts jointly organized by UNICEF and WHO. New York (USA) WHO/CAH/01.22.
40. Mahalanabis D, Brayton JB, Mondal A, Pierce NF. The use of Ringer's lactate in the treatment of children with cholera and acute non cholera diarrhea. Bull WHO. 1972;46:311–9.
41. Choudhuri RN, Srivastava D, Mondal A, Sircar BK, Ganguly R, Samanta B, De SP, De S. A study of pediatric cholera rehydrated with ringer lactate. Indian J Med Res. 1971;59: 544–51.
42. Watten RH, Morgan FM, Songkhla YN, Vanikiati B, Phillips RA. Water and electrolyte studies in cholera. J Clin Invest. 1959;38:1879–89.
43. Pierce NF, Banwell JG, Mitra RC, Caranasos GJ, Keimaqutz RI, Thomas J, Monda A. Controlled comparison of tetracycline and furazolidone in cholera. Br Med J. 1968;3:277–80.
44. De S, Chowdhuri A, Dutta P, Dutta D, De SP, Pal SC. Doxycycline in the treatment of cholera. Bull WHO. 1976;54:177–9.
45. Bhattacharya SK, Bhattacharya MK, Dutta P, Dutta D, De SP, Sikder SN, Moitra A, Dutta A, Pal SC. Double blind, randomized, controlled clinical trial of norfloxacin for cholera. Antimicrob Agents Chemother. 1990;34:939–40.
46. Jesudasan MV, John TJ. Transferable trimethoprim resistance of *Vibrio cholerae* O1 encountered in southern India. Trans R Soc Trop Med Hyg. 1990;84;136–7.
47. Rammurthy T, Pal A, Bhattacharya MK, Bhattacharya SK, Chowdhury AS, Takeda Y, Takeda T, Pal SC, Nair GB. Serovar, biotype, phage type, toxigenicity and antibiotic susceptibility pattern of *Vibrio cholerae* isolated during two consecutive cholera session (1989–1990) in Calcutta. Indian J Med Res. 1992;95:125–9.
48. Tabtieng R, Wattanasri S, Echeverria P, Seriwatana J, Bodhidatta L, Chatkaeomorakot A, Rowe B. An epidemic of *Vibrio cholerae* El Tor Inaba resistant to several antibiotics with a conjugative group C plasmid coding for type II dihydrofolate reductase in Thailand. Am J Trop Med Hyg. 1989;41:680–6.
49. Thungapathra M, Amita, Sinha KK, Chaudhuri SR, Garg P, Ramamurthy T, Nair GB, Ghose A. Occurrence of antibiotic resistance gene cassettes *aac* (6′)-Ib, *dfrA5*, *dfrA12*, and *ereA2* in class I integrons in non-O1, non-O139 *Vibrio cholerae* strains in India. Antimicrob Agents Chemother. 2002;46:2948–55.
50. Faruque AS, Alam K, Malek MA, Khan MG, Ahmed S, Saha D, Khan WA, Nair GB, Salam MA, Luby SP, Sack DA. Emergence of multidrug-resistant strain of *Vibrio cholerae* O1 in Bangladesh and reversal of their susceptibility to tetracycline after two years. J Health Popul Nutr. 2007;25:241–3.
51. Roychowdhury A, Pan A, Dutta D, Mukhopadhyay AK, Ramamurthy T, Nandy RK, Bhattacharya SK, Bhattacharya MK. Emergence of tetracycline-resistant *Vibrio cholerae* O1 serotype Inaba, in Kolkata, India. Jpn J Infect Dis. 2008;61:128–9.
52. Mwansa JC, Mwaba J, Lukwesa C, Bhuiyan NA, Ansaruzzaman M, Ramamurthy T, Alam M, Balakrish Nair G. Multiply antibiotic-resistant *Vibrio cholerae* O1 biotype El Tor strains emerge during cholera outbreaks in Zambia. Epidemiol Infect. 2007;135:847–53.
53. Garg P, Chakraborty S, Basu I, Datta S, Rajendran K, Bhattacharya T, Yamasaki S, Bhattacharya SK, Takeda Y, Nair GB, Ramamurthy T. Expanding multiple antibiotic resistance among clinical strains of *Vibrio cholerae* isolated from 1992 to 1997 in Calcutta, India. Epidemiol Infect. 2000;124:393–9.
54. Okuda J, Ramamurthy T, Yamasaki S. The potent antibacterial activity of sitafloxacin against fluoroquinolone-resistant clinical isolates of *Vibrio cholerae* O1. Microbiol Immunol. 2007;51:467–9.
55. Garg P, Sinha S, Chakraborty R, Bhattacharya SK, Nair GB, Ramamurthy T, Takeda Y. Emergence of fluoroquinolone-resistant strains of *Vibrio cholerae* O1 biotype El Tor

among hospitalized patients with cholera in Calcutta, India. Antimicrob Agents Chemother. 2001;45:1605–6.
56. Nalin DR, Levine MM, Hornick RB, Berquist EJ, Hoover D, Holley HP, Watetman D, VanBlerk J, Matheny S, Sotman S , Rennels M. The problem of emesis during oral glucose–electrolyte therapy given from the onset of severe cholera. Trans R Soc Trop Med Hyg. 1978;73:10–4.
57. Benyajati C, Keoplug M, Beisel WR, Gangrosa EJ, Sprinz H, Sitprija V. Acute renal failure in Asiatic cholera: clinicopathologic correlations with acute tubular necrosis and hypokalemic nephropathy. Ann Intern Med. 1960;52:960–75.
58. De SN, Sengupta KP, Chandra NN. Renal changes including total cortical necrosis in cholera. Arch Pathol. 1954;37:505–15.
59. Jamil B, Ahmed A, Sturm AW. *Vibrio cholerae* O1 septicaemia. Lancet. 1992;340:910–1.
60. Hirschhorn N, Chowdhury AK, Lindenbaum J. Cholera in pregnant women. Lancet. 1969;1 (7608):1230–2.

Subject Index

Note: The letters 'f' and 't' following locators refer to figures and tables respectively.

T. Ramamurthy, S.K. Bhattacharya (eds.), *Epidemiological and Molecular Aspects on Cholera*, Infectious Disease, DOI 10.1007/978-1-60327-265-0,
© Springer Science+Business Media, LLC 2011

M

S

MIX
Papier aus verantwortungsvollen Quellen
Paper from responsible sources
FSC® C105338

If you have any concerns about our products,
you can contact us on
ProductSafety@springernature.com

In case Publisher is established outside the EU,
the EU authorized representative is:
Springer Nature Customer Service Center GmbH
Europaplatz 3, 69115 Heidelberg, Germany

Printed by Libri Plureos GmbH
in Hamburg, Germany